DISCRETE MATHEMATICS
WITH APPLICATIONS

DISCRETE MATHEMATICS
WITH APPLICATIONS

WILLIAM BARNIER
and
JEAN B. CHAN
SONOMA STATE UNIVERSITY

WEST PUBLISHING COMPANY
St. Paul   New York   Los Angeles   San Francisco

COPYRIGHT © 1989 By WEST PUBLISHING COMPANY
50 W. Kellogg Boulevard
P.O. Box 64526
St. Paul, MN 55164-1003

All rights reserved

Printed in the United States of America

96   95   94   93   92   91   90   89   8   7   6   5   4   3   2   1   0

LIBRARY OF CONGRESS CATALOGING-IN-PUBLICATION DATA

Barnier, William J.
   Discrete mathematics with applications/William J. Barnier, Jean B. Chan.
     p. cm.
   Includes index.
   ISBN 0-314-45966-9
    1. Mathematics–1961–   2. Electronic data processing—Mathematics.
I. Chan, Jean B.   II. Title.
QA39.2.B369   1989
511—dc19   88-26150
                     CIP

TO CATHY AND MY TEACHERS,
ESPECIALLY CARROLL ANDERSON, NEWTON SMITH,
AND ROBERT BROWN

W. B.

TO
WING SHU AND CHI SUN CHAN,
FREDERICK A. VALENTINE,
PETE, DAVE, AND MARTHE STANEK

J. B. C.

CONTENTS

Preface   xi

CHAPTER 1
INTRODUCTION TO SETS, ALGORITHMS, AND LOGIC   1

- **1.1** Introduction to Sets   2
- **1.2** Elementary Set Theory   8
- **1.3** Algorithms   17
- **1.4** Reading Proofs   25
- **1.5** Introduction to Logic   30
- Key Terms   36
- Review Exercises   36
- References   38

CHAPTER 2
DIRECTED GRAPHS AND RELATIONS   39

- Application   39
- **2.1** Directed Graphs   41
- **2.2** Relations   48
- **2.3** Transitive Closure and the Connectivity Relation   58
- **2.4** Matrix Representation of Digraphs and Relations   65
- **2.5** Reachability and Warshall's Algorithm   78
- **2.6** Application: Deadlock Detection   89
- Key Terms   90
- Review Exercises   90
- References   91

CHAPTER 3
RELATIONS AND FUNCTIONS   93

- Application   93
- **3.1** Equivalence Relations and Partial Orderings   94
- **3.2** Extremal Elements in a Partially Ordered Set   101
- **3.3** Functions   107

  3.4 Special Functions 116
  3.5 Application: Time-Complexity Functions 125
    Key Terms 128
    Review Exercises 128
    References 130

CHAPTER 4
COMBINATORICS AND FINITE PROBABILITY 131

    Application 131
  4.1 Basic Counting Techniques 132
  4.2 Permutations 140
  4.3 Combinations 149
  4.4 Introduction to Finite Probability 162
  4.5 Application: Stack Permutations and Catalan Numbers 174
    Key Terms 177
    Review Exercises 177
    References 179

CHAPTER 5
LOGIC AND PROOF 180

    Application 180
  5.1 Propositional Logic 182
  5.2 Logical Equivalence and Tautologies 187
  5.3 Proof Techniques 194
  5.4 Introduction to Mathematical Induction 204
  5.5 Predicates and Quantifiers 211
  5.6 Application: Artificial Intelligence—Automated Theorem Proving 223
    Key Terms 227
    Review Exercises 227
    References 228

CHAPTER 6
INTEGERS AND BINARY NUMBERS 230

    Application 230
  6.1 Factorization in the Integers 231
  6.2 The Euclidean Algorithm 237
  6.3 Modular Arithmetic and Binary Numbers 243

6.4 Application: Cryptography and Factoring 250
Key Terms 254
Review Exercises 254
References 255

CHAPTER 7
BOOLEAN ALGEBRA 256

Application 256
7.1 Boolean Algebra 257
7.2 Boolean Functions and Boolean Expressions 264
7.3 Logic Networks and Karnaugh Maps 275
7.4 The Structure of Boolean Algebra (Optional) 289
7.5 Application: Seven-Segment Display 296
Key Terms 299
Review Exercises 299
References 301

CHAPTER 8
GRAPHS AND TREES 302

Application 302
8.1 Graphs 303
8.2 Paths, Circuits, and Cycles 313
8.3 Trees 324
8.4 Spanning Trees 337
8.5 Application: Heapsort 343
Key Terms 348
Review Exercises 348
References 351

CHAPTER 9
RECURRENCE RELATIONS 352

Application 352
9.1 Recursion and Recurrence Relations 353
9.2 Recurrence Relations and the Characteristic Equation Method 364
9.3 Recurrence Relations and Generating Functions 373
9.4 Application: Divide and Conquer Algorithms 382
Key Terms 385
Review Exercises 385
References 386

CHAPTER 10
AUTOMATA THEORY AND FORMAL LANGUAGES 387

        Application 387
- **10.1** Finite State Recognizers and Regular Expressions 388
- **10.2** Phrase Structure Grammars 399
- **10.3** Regular Grammars and Regular Languages 407
- **10.4** Context-Free Grammars and Parse Trees 413
- **10.5** Application: Push-Down Automata and Stacks 426

        Key Terms 431
        Review Exercises 432
        References 433

SOLUTIONS TO SELECTED PRACTICE PROBLEMS A-1
SOLUTIONS FOR THE EXERCISES A-15
INDEX I-1

PREFACE

Discrete mathematics is the study of the logical and algebraic relationships between discrete objects. Its roots lie deep in set theory, probability, combinatorics, matrix algebra, graph theory, formal languages, and automata theory. It is therefore positioned at what the mathematician Norbert Wiener calls the "cracks," or boundaries, between intellectual specialties, where he found the most intriguing and productive areas for investigation.

Applications of discrete mathematics are extensive. Although computer science has motivated recent interest in this subject, discrete mathematics is also important in other fields including economics, engineering, and operations research. Furthermore, discrete mathematics is itself an interesting and fertile area of study.

We aim to provide an introduction to discrete mathematics that serves the needs of many communities. Students of mathematics, computer science, and engineering require an introduction to the topics of discrete mathematics as background for their chosen field of study. Indeed, the introduction of computers and software as a common artifact of modern technological life has made acquisition of knowledge in the topics of discrete mathematics absolutely essential for every educated individual. Our text offers not only an introduction to discrete mathematics, but also an array of powerful tools with important applications that will support undergraduate students of computer science and mathematics in an academic program and sustain their professional activities thereafter. Discrete mathematics has also been recommended as a course of study for secondary school teachers, and business and management majors.

This book is intended for a one-semester, one-quarter, two-quarter, or one-year introductory course in discrete mathematics. Some mathematical experience and sophistication is required of students taking such a course. We believe the student should have a strong precalculus background. We recommend, but do not require, that students have taken one semester of calculus. A few examples and exercises in this book refer to concepts from first-semester calculus, but they can be omitted without loss of continuity.

Chapter 1 includes preliminary material; it may be covered as rapidly as the students' background permits. Directed graphs appear in Chapter 2 for many reasons: They are intuitive and relatively easy to comprehend, they provide a natural introduction to binary relations, they have important applications, and they provide an opportunity to study several important algorithms early in the book.

Chapter 3 provides all the necessary background in relations and functions for any mathematics or computer science student. Chapter 4 is an introduction to the most useful methods of combinatorics and probability. Bit strings are used to explain the connection between sets and characteristic functions in Chapter 3; they are a conceptual tool for several counting techniques discussed in Chapter 4. Propositional logic, methods of proof including mathematical induction, and predicate logic are discussed in Chapter 5. Methods of proof are treated as natural consequences of the laws of logic. They are used to prove many theorems concerning concepts discussed in Chapters 2–4.

Chapter 6 is an introduction to the integers and number theory. Boolean algebra and its applications are thoroughly discussed in Chapter 7. Chapter 8 treats the theory and applications of graphs and trees with an emphasis on algorithms. Chapter 9 introduces recurrence relations with many examples and applications. Two solutions for linear second-order homogeneous recurrence relations are given. Finite state machines and their role as language recognizers and grammars for generating formal languages are discussed in Chapter 10.

Suggested Course Organization

The figure below shows the relationships among chapters. We recommend that Chapters 1–5 be covered in order. These five chapters form the core of the book. Within those chapters, Sections 2.6, 3.5, 4.4, 4.5, 5.5, and 5.6 can be omitted without loss of continuity. Because Chapters 6–10 are independent, material can be selected from them with great flexibility. For example, if time permits only partial coverage of any of Chapters 7–10, we recommend Sections 7.1–7.3, 8.1–8.2, 9.1–9.2, or 10.1.

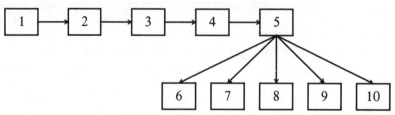

CHAPTER PREREQUISITES

One semester with an emphasis on combinatorics: Chapters 1–5, 9, either 7 or 8, and (time permitting) from two to four sections of the remaining material

One semester with an emphasis on graph theory: Chapters 1–5, 8, either 7 or 9, and (time permitting) from two to four sections of the remaining material

One semester with an emphasis on algebra: Chapters 1–5, 6, 7, and (time permitting) from two to four sections of the remaining material

One quarter with no special emphasis: Chapters 1–5

Two quarters with no special emphasis: Chapters 1–5, 7, 8, 9, and (time permitting) from two to four sections of the remaining material

Two semesters:
  First semester: Chapters 1–5 with no omitted sections and extra time spent on exercises and applications
  Second semester: Chapters 6–10 with no omitted sections and extra time spent on exercises and applications

Features of the Text

There are more than 1150 exercises and over 175 Practice Problems in this text. In addition to exercise sets for each section, all chapters provide review exercises for further practice. Within each section, Practice Problems are designed to encourage active student participation in understanding the material. Solutions to Selected Practice Problems are given at the back of the book. In our experience, those students who diligently work along with the text are most likely to succeed. We are firm believers in the old adage "Mathematics is not a spectator sport."

Each chapter (except Chapter 1) begins and ends with an application of concepts to highlight important areas of application and to provide a focus for the concepts introduced in the chapter. These applications can be covered in class or assigned as reading.

Programming Notes throughout the book amplify our examples. We use Pascal language constructs to show how programming and mathematics are related, especially how mathematics contributes to computer science, and, in some cases, how programming sheds light on mathematical concepts. The notes can be omitted without loss of continuity, but we have found that many students will read them whether assigned or not. In any case, class time need not be spent on the Programming Notes. Exercises corresponding to the Programming Notes are clearly labeled as Programming Exercises.

Remarks on Pedagogy

We believe that some repetition is necessary and beneficial. Many concepts are discussed first at a basic level, and later, further applications or proofs of these concepts are discussed in more depth. For example, functions are first discussed in Chapter 3 and then applied in Chapter 4; proofs regarding functions are given in Chapter 5. Of course, functions and many other basic concepts introduced in Chapters 1–5 are used in later chapters. We believe that mathematics is best learned with a spiral approach to concepts along with a gradual increase in the demand for rigor placed on a student.

In particular, our approach to proofs is gradual. The method of reading proofs is discussed in Section 1.4. We expect that students can read and understand proofs in Chapters 2–4 and work an occasional proof exercise at the end of some sections. In Chapter 5, after sets, directed graphs, relations, functions, and combinatorics have been introduced, we thoroughly discuss proof techniques. At this point students have seen many proofs and, more importantly, have seen many statements in need of proof. We have found that students are more motivated and better prepared to learn how to create and write a correct proof with the background gained in Chapters 1–4.

Acknowledgments

Without the generous help and patient support of many colleagues, this project would have been more difficult and the results would have been of substantially lower quality. Each of the following has read the manuscript, criticized it, and made useful suggestions: David Butcher, Richard Gordon, George Ledin, and Michael Lyle, Computer Science Department, Sonoma State University; Donald Duncan, Clement Falbo, Norman Feldman, Robert Johnson, Thomas Nelson, Tom Volk, Mathematics Department, Sonoma State University; John Barnier, English Department, Grossmont College; Richard Hamming, Naval Postgraduate School, Monterey, California; James Jantosciak, Mathematics Department, Brooklyn College.

We especially appreciate the skilled and diligent clerical help of Sally Cochran and Connie Eagle.

The following reviewers offered many useful suggestions and insights:

Carol G. Crawford, U.S. Naval Academy

David Cusick, Clemson University

Herbert B. Enderton, University of California, Los Angeles

Ross A. Gagliano, Georgia State University

James H. Gaunt, Florida State University

Ronald Gould, Emory University

Richard Hitt, University of South Alabama

Moana Karsteter, Florida State University

Edith H. Luchins, Rensselaer Polytechnic Institute

John K. Luedeman, Clemson University

Elmo Moore, Humboldt State University

Don Small, Colby College

Lynne Walling, St. Olaf College

We are also grateful to our editor, Richard Jones, at West Publishing Company for his consistent professionalism, insightful suggestions, and unflagging encouragement.

DISCRETE MATHEMATICS
WITH APPLICATIONS

CHAPTER 1

INTRODUCTION TO SETS, ALGORITHMS, AND LOGIC

The notions of set theory are found everywhere in mathematics. We start with a review of fundamental and useful concepts of sets, subsets, and their operations. In Chapter 3 we use our knowledge of sets to discuss relations and functions. Sets provide a basis for our study of combinatorial problems and probability in Chapter 4. Furthermore, set theory provides an important and interesting model of Boolean algebra, discussed in Chapter 7. We study proofs for set theory theorems in Chapter 5.

The ideas of proofs and algorithms lie at the confluence of mathematics and computer science. Section 1.3 introduces the important notion of algorithms. After Chapter 1, the reader will be expected to read and construct simple algorithms. In Section 1.4 we discuss how to read and understand proofs. Section 1.5 introduces basic concepts of logic, which will be useful for understanding elementary proofs. In Chapter 5, we discuss proof techniques in more detail and give readers an opportunity to create proofs.

A thorough understanding of the introductory material in this chapter is prerequisite for the remainder of this book and, indeed, for any study of mathematics and computer science. We may as well begin at the beginning.

1.1 INTRODUCTION TO SETS

The concept of sets is attributed to the mathematician Georg Cantor (1845–1918). Cantor's first paper on the theory of sets in 1874 was very controversial, because it differed radically from the mathematics of that time. Today the idea of sets is widely accepted and used in every branch of mathematics.

Intuitively, a **set** is a collection of objects. The objects are called **elements** or **members** of the set. For example, all the students currently enrolled in our discrete mathematics course form a set. Each student is an element or a member of this set.

There are several ways we can describe a set. One way is to list the elements, separated by commas, inside two braces as shown in the following example.

EXAMPLE 1.1.1 $\{a, b, c, d, e\}$ is the set containing the letters a, b, c, d, and e.

$\{0, 1, 2, \ldots, 9\}$ is the set of all decimal digits.

$\{0, 1\}$ is the set of binary digits.

$\{a, b, c, \ldots, x, y, z\}$ is the set containing the letters of the English alphabet. The three dots between c and x are called an **ellipsis**. We use an ellipsis for brevity, instead of listing elements, when it is clear from the context what elements are included in the set.

Ellipses are necessary for listing infinite sets. For example, the set of all positive even integers is $\{2, 4, 6, \ldots\}$ and the set of all positive odd integers is $\{1, 3, 5, \ldots\}$. ∎

We usually use capital letters for sets and lowercase letters for elements of sets. When the sets A and B are identical, we say A and B are equal and write $A = B$. In other words, $A = B$ whenever the sets A and B contain exactly the same elements. We write $A \neq B$ when A and B do not contain the same elements.

EXAMPLE 1.1.2   Let $A = \{1, 2, 3, 4, 5, 6\}$ and $B = \{x : x \text{ is a positive integer less than } 7\}$. Then $A = B$ since A and B contain exactly the same elements. ∎

EXAMPLE 1.1.3   Let $S = \{x : x \in \mathbb{R} \text{ and } x^2 = 100\}$ and $T = \{10\}$. Then $S \neq T$. ∎

The order in which the elements in a set are listed is immaterial. For example, the sets $\{a, b, c\}$ and $\{c, a, b\}$ are equal. Each element only needs to be listed once. Thus, $\{a, b, c\}$, $\{c, a, b\}$, and $\{a, b, c, b\}$ denote the same set. However, we will not write $\{a, b, c, b\}$ to designate $\{a, b, c\}$.

Consider a set A. We write $x \in A$ to denote that x is an element (or a member) of A. When x is not an element of A, we write $x \notin A$. For brevity, we also say "x is in A" for $x \in A$, and "x is not in A" for $x \notin A$.

Some sets are used frequently, so special letters are reserved for them. For example,

$\mathbb{N} = \{0, 1, 2, 3, \ldots\}$, the **natural numbers** or **nonnegative integers**

$\mathbb{Z} = \{\ldots, -3, -2, -1, 0, 1, 2, 3, \ldots\}$, the **integers**

$\mathbb{R} =$ the **real numbers**

$\mathbb{R}^+ =$ the **positive real numbers**

Another way of describing a set is to use the **set-builder notation**. In other words, we describe the elements of a set by stating the property or properties of the elements. For example,

$C = \{x : x \in \mathbb{N} \text{ and } x < 12\}$

represents the set $\{0, 1, 2, 3, 4, 5, 6, 7, 8, 9, 10, 11\}$. The above notation $\{x : x \in \mathbb{N} \text{ and } x < 12\}$ is read "the set of all x such that x is in $\mathbb{N}$ and x is less than 12."

EXAMPLE 1.1.4 The set $E = \{x : x \in \mathbb{R} \text{ and } 0 < x < 5\}$ describes the set of all real numbers that are greater than 0 and less than 5. This set has infinitely many elements; it is not possible to list all of its elements. ∎

If a set S has a finite number of elements, **Card(S)** denotes the number of elements in S, or the **cardinality** of S. Cardinality is a mathematical synonym for "number of elements." Card(S) is read "cardinality of S." In Example 1.1.2, Card(A) = 6. The cardinality of finite sets will be discussed more thoroughly in Chapter 4.

PRACTICE PROBLEM 1 Let $F = \{x : x \in \mathbb{N} \text{ and } x < 20\}$ and $G = \{x : x \in \mathbb{Z} \text{ and } -3 < x < 6\}$.

a. List the elements in F and the elements in G, respectively.
b. Find Card(F) and Card(G).

A set may contain no elements. The set $E = \{x : x \in \mathbb{R} \text{ and } x^2 = -1\}$ is such a set. Since the square of every real number is nonnegative, E has no elements. A set with no elements is called an **empty set** or a **null set**, denoted by the symbol $\emptyset$ or $\{\ \}$.

We note that $\{0\}$ is a set containing the digit 0 and is not the same as the empty set $\emptyset$. The set $\{0\}$ contains exactly one element; such a set is called a **singleton set**. For example, $\{a\}$, $\{*\}$, and $\{9\}$ are all singleton sets. Note that the singleton set $\{a\}$ is not the same as the element a; that is, $\{a\} \neq a$. As the mathematician Paul Halmos remarks in his classic text *Naive Set Theory*: A box containing a hat is not the same as the hat alone.

If every element of a set A is also an element of a set B, then we say A is a **subset** of B and write $A \subseteq B$. If $A \subseteq B$ and $A \neq B$, then A is a **proper subset** of B. When A is a proper subset of B, we will sometimes write $A \subset B$. If there is an element of A that does not belong to B, then A is not a subset of B and we write $A \nsubseteq B$.

Consider any set B. Since $\emptyset$ contains no element, every element of $\emptyset$ is an element of B. Hence, $\emptyset$ is a subset of B. In other words, $\emptyset \subseteq B$ for any set B. Since every element of B is an element of B, we have $B \subseteq B$. Hence, $B \subseteq B$ for any set B.

EXAMPLE 1.1.5 Let $A = \{1, 2, 3, 4, 5, 6, 7\}$, $B = \{2, 4, 6\}$, and $C = \{a, b, 2, d, 8\}$. Then $B \subset A$, $B \neq C$, and $A \nsubseteq B$. ∎

PRACTICE PROBLEM 2
a. Let $S = \{x : x \in \mathbb{R} \text{ and } x^2 - 5x + 6 = 0\}$ and $T = \{2, 3\}$. Show that $S = T$.
b. Let $A = \{\text{BASIC, FORTRAN, Pascal, C, Ada}\}$ and $B = \{\text{Pascal, Ada}\}$. Show that $\emptyset \subseteq A$ and $B \subseteq A$.

EXAMPLE 1.1.6 Let $A = \{a, b\}$. Find all the subsets of A.

Solution The subsets of A are: $\emptyset$, $\{a\}$, $\{b\}$, A. We note again that $\emptyset$ and A are always subsets of A. ∎

Set Operations

We will discuss intersection, union, and the complement of sets. The results of these operations are sets.

Let $D = \{0, 1, 2, 3, 4, 5, 6, 7, 8, 9\}$ and $E = \{-3, -2, -1, 0, 1, 2, 3\}$. The elements 0, 1, 2, and 3 but no others belong to both D and E. We say $\{0, 1, 2, 3\}$ is the intersection of D and E and write $D \cap E = \{0, 1, 2, 3\}$.

DEFINITION

> The **intersection** of two sets A and B, denoted by $A \cap B$, is the set of all elements that belong to both A and B. Hence,
>
> $A \cap B = \{x : x \in A \text{ and } x \in B\}$

Note that $A \cap B = B \cap A$, and $A \cap \emptyset = \emptyset$.

EXAMPLE 1.1.7 Let $C = \{a, b, c, d, e, f\}$, $D = \{x, y, d, f\}$, $E = \{a, d, t\}$, and $F = \{s, t\}$. We have $C \cap D = \{d, f\}$, $C \cap E = \{a, d\}$, $E \cap F = \{t\}$, $D \cap F = \emptyset$, and $D \cap \emptyset = \emptyset$. ∎

DEFINITION

> Two sets A and B are **disjoint** if $A \cap B = \emptyset$. So, two sets are disjoint if they have no elements in common.

In Example 1.1.7, the sets D and F are disjoint, and the sets E and F are not disjoint.

DEFINITION

> The **union** of two sets A and B, denoted by $A \cup B$, is the set of all elements that belong to A or to B or to both. Hence,
>
> $A \cup B = \{x : x \in A \text{ or } x \in B\}$

1.1 INTRODUCTION TO SETS

We note that $A \cup B = B \cup A$, and $A \cup \emptyset = A$. Also note that any element in both A and B is also in A or in B. Therefore, $A \cap B \subseteq A \cup B$.

EXAMPLE 1.1.8 Let set $A = \{1, 2, 3, 4, 5\}$ and set $B = \{-3, -4, 3, 4\}$. Then $A \cup B = \{-3, -4, 1, 2, 3, 4, 5\}$ and $A \cap B = \{3, 4\}$. Note that $A \cap B \subseteq A \cup B$. ∎

PRACTICE PROBLEM 3 Let $A = \{a, b, c, d, e, f\}$ and $B = \{1, 2, d, f\}$. Find $A \cap \emptyset$, $B \cup \emptyset$, $A \cap B$, and $A \cup B$.

Both the union and the intersection can be generalized to more than two sets.

DEFINITION

The **union of n sets** $A_1, A_2, \ldots, A_n$, written $A_1 \cup A_2 \cup \cdots \cup A_n$, is defined by:

$x \in A_1 \cup A_2 \cup \cdots \cup A_n$ if $x \in A_j$ for at least one j, $1 \leq j \leq n$

The **intersection of n sets** $A_1, A_2, \ldots, A_n$, written $A_1 \cap A_2 \cap \cdots \cap A_n$, is defined by:

$x \in A_1 \cap A_2 \cap \cdots \cap A_n$ if $x \in A_j$ for every j, $1 \leq j \leq n$

PRACTICE PROBLEM 4 Let $A = \{a, b, c, 3, 5\}$, $B = \{a, 3, 4\}$, and $C = \{a, b, 3, 5, 6\}$.

a. Find $A \cup B \cup C$.   **b.** Find $A \cap B \cap C$.

In any discussion of sets $A, B, \ldots$, the elements in $A, B, \ldots$ are assumed to belong to a fixed set, called a **universal set**, describing the context for discussion. The universal set is frequently called the universe of discourse and contains all the elements under discussion. We will always denote the universal set by the capital letter U.

DEFINITION

Let A be a subset of a universal set U. The **complement** of A, denoted by A', is the set of all elements that are in U but not in A.

EXAMPLE 1.1.9 Let $A = \{2, 4, 6, 8\}$ and $U = \{0, 1, 2, 3, 4, 5, 6, 7, 8\}$. Then $A' = \{0, 1, 3, 5, 7\}$. Note that it is important how the universal set is specified. If, for example, we had chosen U to be the set of all digits, then A' would include the element 9 also. ∎

Given two sets A and B, the **difference set**, denoted by $A - B$, is the set of all elements that belong to A but not B. That is,

$A - B = \{x : x \in A \text{ and } x \notin B\}$

In general, $A - B$ and $B - A$ are not the same. Also note $U - B = B'$. The difference set $A - B$ is defined whether or not B is a subset of A. See Example 1.1.10 following.

EXAMPLE 1.1.10 Let $A = \{2, 4, 6, 8\}$ and $B = \{0, 1, 2\}$. Then $A - B = \{4, 6, 8\}$ and $B - A = \{0, 1\}$. Note that $A - B$ is defined, although B is not a subset of A. ∎

PRACTICE PROBLEM 5 Let $U = \{-5, -4, \ldots, 4, 5\}$, $A = \{-1, 0, 2, 4\}$, and $B = \{1, 2, 3, 4\}$.

  a. Find A', B', $A \cup B$, and $A \cap B$.
  b. Find $A - B$ and $B - A$. Observe that $A - B \neq B - A$.
  c. Show that $A - B \subseteq A \cup B$.

■ **PROGRAMMING NOTES**

In the programming language Pascal, sets are implemented as data structures. In Pascal notation, COLORS = [red, blue, gray, black, white] denotes the set of hues listed, and ODDSET = [1, 3, 5, 7, 9, 11] denotes the set of odd integers greater than 0 and less than 12. The brackets, [and], are used in place of braces when writing set notation in Pascal. The empty set $\emptyset$ is denoted by [ ] in Pascal.

PRACTICE PROBLEM 6 Refer to the sets COLORS and ODDSET in the programming notes above.

  a. Find Card(COLORS).
  b. Find all the subsets of ODDSET that contain exactly two elements.

Set Operations in Pascal In Pascal, the union of two sets A and B is written as $A + B$, and the intersection of A and B is written as $A * B$. The difference set of A and B is written as $A - B$. In Pascal notation, if $A = [1, 3, 5, 7, 9]$, $B = [1, 2, 3, 4, 5]$, and $C = [2, 4, 6, 8]$, then

$A + B = [1, 2, 3, 4, 5, 7, 9]$

$A * B = [1, 3, 5]$

$A - B = [7, 9]$

$A - C = A$

$A + C = [1, 2, 3, 4, 5, 6, 7, 8, 9]$

$A * C = [\ ]$

PRACTICE PROBLEM 7 Let $A = [1, 3, 5, 7, 9]$, $B = [1, 2, 3, 4, 5]$, and $C = [2, 4, 6, 8]$. Find $B * C$, $B - A$, $B - C$, and $B + C$. Express answers in Pascal notation.

EXERCISE SET 1.1

1. Let $A = \{a, b, c, d, 2, 4, 6\}$. Determine whether each of the following is true or false.
  a. $b \in A$ **b.** $7 \in A$ **c.** $\emptyset \subseteq A$
  d. $\{2, 4, 6, 8\} \subseteq A$ **e.** $\{d\} \in A$ **f.** $\{a, b, c, d\} \subseteq A$

2. Let $S = \{x : x \in \mathbb{N} \text{ and } x \leq 10\}$. Determine whether each of the following is true or false.
  a. $5 \in S$ **b.** $13 \in S$ **c.** $S \subseteq \mathbb{N}$ **d.** $S \subseteq \mathbb{Z}$ **e.** $8 \notin S$ **f.** $4.5 \in S$

3. Let $A = \{x : x \in \mathbb{R} \text{ and } x^2 = 25\}$. List the elements of A.

1.1 INTRODUCTION TO SETS

4. Let $B = \{x : x \in \mathbb{R} \text{ and } x^2 - 6x + 8 = 0\}$. List the elements of B.

5. Form the set consisting of the letters in each of the following words.
   a. NOON   b. MADAM   c. MATHEMATICS   d. RACECAR

6. Write the following sets in set-builder notation.
   a. $A = \{2, 3, 4, 5, 6, 7\}$ b. $B = \{-3, 3\}$ c. $C = \{0, 1, 2, 3, 4, 5, 6, 7, 8\}$

7. Find the cardinality of each of the following sets.
   a. $A = \{x : x \in \mathbb{Z} \text{ and } x^2 = -9\}$ b. $B = \{x : x \in \mathbb{R} \text{ and } x^2 = 7\}$
   c. $C = \{x : x \in \mathbb{N} \text{ and } -3 < x < -6\}$ d. $D = \{x : x \in \mathbb{N} \text{ and } -4 < x < 8\}$

8. Find the cardinality of each of the following sets.
   a. $A = \{x : x \in \mathbb{N} \text{ and } 3 < x < 15\}$ b. $B = \{x : x \in \mathbb{R} \text{ and } x^3 = -27\}$
   c. $C = \{x : x \in \mathbb{Z} \text{ and } -5 < x < 4\}$ d. $D = \{x : x \in \mathbb{N} \text{ and } 0 \leq x \leq 5\}$

9. Refer to the sets A, B, C, and D in Exercise 7. Determine whether each of the following is true or false.
   a. $A \subseteq D$ b. $D \subseteq C$ c. $A \neq D$ d. $D \subseteq B$ e. $-2 \in C$

10. Refer to the sets A, B, C, and D in Exercise 8. Determine whether each of the following is true or false.
    a. $D = \{1, 2, 3, 4\}$ b. $B \subseteq C$ c. $D \subseteq A$ d. $3 \in B$ e. $B = \emptyset$

11. For each of the following pairs of sets, determine whether the two sets are equal.
    a. $\{1, 2, 3, 4\}$, $\{1, 3, 4, 2\}$ b. $\{x : x \in \mathbb{R} \text{ and } x^2 - 3x - 18 = 0\}$, $\{3, -6\}$
    c. $\{x : x \in \mathbb{R} \text{ and } |x| = 5\}$, $\{5\}$ d. $\{3, 4, 5, 6\}$, $\{x : x \in \mathbb{N} \text{ and } 2 < x < 7\}$

12. For each of the following pairs of sets, determine whether the two sets are equal.
    a. $\{1/x : x \in \mathbb{N} \text{ and } 2 < x < 7\}$, $\{4, 5, 6, 7\}$
    b. $\{x : x \in \mathbb{R} \text{ and } |x + 2| = -1\}$, $\{-3\}$
    c. $\{x : x \in \mathbb{Z} \text{ and } -5 < x < 1\}$, $\{-4, -3, -2, -1, 0\}$
    d. $\{2, 9\}$, $\{x : x \in \mathbb{N} \text{ and } x^2 - 11x + 18 = 0\}$

13. Find all subsets of the set $\{a\}$.

14. List all proper subsets of the set $\{a, b, c\}$.

15. Let $S = \{x : x \in \mathbb{Z} \text{ and } x^2 \leq 100\}$. Determine whether each of the following is true or false.
    a. $\{0, 1, 2, 3\} \subseteq S$ b. $100 \in S$ c. $12 \notin S$
    d. $\mathbb{Z} \subseteq S$ e. $\{x : x \in \mathbb{Z} \text{ and } x < 10\} \subseteq S$ f. $\text{Card}(S) = 21$

16. List the elements of each set.
    a. $\{x : x \in \mathbb{Z} \text{ and } 2 < |x| < 9\}$ b. $\{3x : x \in \mathbb{N} \text{ and } x^2 = 36\}$
    c. $\{2x - 5 : x \in \mathbb{Z} \text{ and } -2 \leq x \leq 3\}$

17. Let $U = \{0, 1, 2, 3, 4, 5, 6, 7, 8, 9, 10\}$, $A = \{2, 6, 10\}$, and $B = \{6, 9, 10\}$.
    a. Find $A \cup B$.   b. Find $A \cap B$.   c. Find $A \cup B'$.
    d. Find $B \cap A'$.   e. Are A and B disjoint?

18. Let $U = \{a, b, c, d, e, f, g\}$, $C = \{a, b, c, d\}$, and $D = \{d, e, f\}$.
    a. Find $C \cup D$.   b. Find $C \cap D$.   c. Find $D \cup C'$.
    d. Find $D' \cap C$.   e. Are C and D disjoint?

19. Let $A = \{x : x \in \mathbb{R} \text{ and } x^2 - 3x + 2 = 0\}$ and $B = \{x : x \in \mathbb{R} \text{ and } x^2 - 2x + 1 = 0\}$. Find:
    a. $A \cap B$ b. $A \cup B$ c. $A - B$ d. $B - A$

20. Let $C = \{x : x \in \mathbb{R} \text{ and } x^2 + 1 = 0\}$ and $D = \{x : x \in \mathbb{R} \text{ and } x^2 - x - 6 = 0\}$. Find:
    a. $C \cup D$ b. $C \cap D$ c. $C - D$ d. $D - C$

21. Let $A = \{$Peter, Joan, Eva, Jean, Mark, Steve, Martha, Bruce, David, Anastasia$\}$ be the set of top computer programmers of the year 1987. Let $B = \{$David, Peter, Anna, Daisy, Patty, Martha$\}$ be the set of top computer programmers of the year 1988.
    a. Find the set of people who were top computer programmers for two years in a row (that is, in 1987 and 1988). What is the set operation used to find this answer?
    b. Find the set of people who were top computer programmers in 1987 or 1988 or both years. What is the set operation used to find this answer?

22. Let U be the set of all college students in the United States. Let A, B, and C be subsets of U defined as follows:

    A is the set of students who are under 20.
    B is the set of students who are 20 or over.
    C is the set of students who own a personal computer.

    Determine whether each of the following is true or false.
    a. $A \cap B = \emptyset$ b. $B \cap C = \emptyset$ c. $A \in U$ d. $A = B'$
    e. $A \cap C = \emptyset$ f. $A \cup B = U$ g. $C \subseteq A \cup B$

23. Find conditions on sets A and B as subsets of $\mathbb{R}$ to make each of the following statements true.
    a. $A \cap B = B$ b. $A \cup B = B$ c. $A \cap A = A$ d. $A - B = \emptyset$

24. Find conditions on sets A and B as subsets of some universal set U to make each of the following statements true.
    a. $A \subseteq A \cap B$ b. $B - A = B$ c. $A - B = B - A$

25. Consider the following subsets of $\mathbb{Z}$:

    $I = \{n : |n| \leq 5\}$
    $J = \{n : n = 2k + 1 \text{ for } k \in \mathbb{Z}\}$
    $K = \{n : n > 5\}$

    Describe each of the following sets in terms of I, J, and K and set operations:
    a. $\{n : n \leq 5 \text{ and } n \text{ is odd}\}$ b. The set of all even integers
    c. $\{-5, -3, -1, 1, 3, 5\}$ d. $\{-5, -4, -3, -2, -1, 0, 1, 2, 3, 4, 5, 6, 7, \ldots\}$

26. Repeat Exercise 25 for each of the following:
    a. The set of all even integers greater than or equal to 6
    b. The empty set
    c. $\mathbb{Z}$

1.2
ELEMENTARY SET THEORY

Venn Diagrams

A helpful device to picture relationships among sets is to represent the universal set by a rectangle, and to represent proper subsets in the universal set by circular or other regions inside the rectangle. (See Figure 1.2.1.) The figures are called **Venn diagrams.** The logician John Venn (1834–1923) used diagrams to illustrate logic in his 1881 work, *Symbolic Logic*. He was an ordained priest but left the clergy in 1883 to spend the rest of his life studying logic and teaching mathematics.

1.2 ELEMENTARY SET THEORY

EXAMPLE 1.2.1 Figure 1.2.1 shows Venn diagrams for the sets A and B, the intersection of A and B, the union of A and B, the difference of A and B, and the complement of the set A, respectively. The shaded area in each diagram shows the location of elements in each set specified.

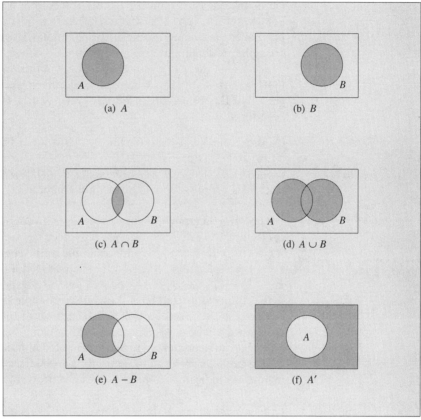

FIGURE 1.2.1

EXAMPLE 1.2.2 Figure 1.2.2 shows $A \subseteq B$ and $A \cap B = \emptyset$.

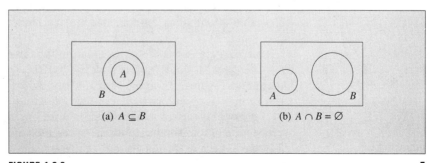

FIGURE 1.2.2

Conjectures and Counterexamples

Drawing Venn diagrams is a good method for conceiving conjectures of your own or testing a conjecture made by someone else. A **conjecture** is an assertion that may or may not be true. If a conjecture about some mathematical fact is proved true, then it becomes a theorem. A conjecture is false if it is not always true. To show that a conjecture is false usually requires an example called a **counterexample**. In Chapter 5 we will discuss counterexamples in detail and give many proofs of theorems.

Because of the way a minus sign works in arithmetic, we might conjecture that $B - (B - A) = A$. The Venn diagram in the following example suggests that this conjecture is false. In other words, $B - (B - A) = A$ does not hold for all sets.

EXAMPLE 1.2.3 Consider the Venn diagram for $B - (B - A)$ in Figure 1.2.3. Based on the Venn diagram in Figure 1.2.3, it is more reasonable to conjecture that $B - (B - A) = A \cap B$. The conjecture $B - (B - A) = A \cap B$ can be proved true. This proof will be deferred until Chapter 5. ∎

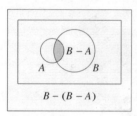

FIGURE 1.2.3

The Venn diagram in Figure 1.2.3 suggests a counterexample to the original conjecture $B - (B - A) = A$. We only need to choose sets A and B so that $A \neq A \cap B$. Here is one counterexample to the conjecture $B - (B - A) = A$: Let $A = \{1, 3\}$ and $B = \{1, 2\}$. Then $B - A = \{2\}$ and $B - (B - A) = B - \{2\} = \{1\}$. So $B - (B - A) \neq A$ for this example.

It is important to note that a counterexample to a conjecture about sets is a *specific* list of sets so that, when each set in the list is substituted in the conjecture, a false statement results. Although a Venn diagram can be useful for suggesting counterexamples, it cannot stand alone as a counterexample.

In general, each Venn diagram for two intersecting sets has four disjoint regions, as indicated in the next practice problem.

PRACTICE PROBLEM 1 Draw a Venn diagram for two intersecting sets A and B showing four disjoint regions. Write each set represented by a disjoint region in terms of A, B, ′, and ∩.

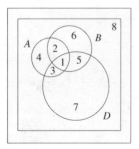

FIGURE 1.2.4

Let us look at Venn diagrams for three sets A, B, and D. Figure 1.2.4 shows that a Venn diagram for three sets has eight disjoint regions indicated by 1, 2, ..., 8.

Table 1.2.1 describes the Venn diagram for three sets. In this table, "yes" is denoted by "1" and "no" is denoted by "0." Two facts about the Venn diagram in Figure 1.2.4 and Table 1.2.1 should be noted. Because the eight disjoint regions form a partition of the points in the rectangle (the elements of U), every possible combination of the three sets A, B, and D can be written as the union of some of the eight sets represented by the eight disjoint regions. Also, the pattern of 0's and 1's in the table will surface again when we discuss truth tables in Chapter 5.

1.2 ELEMENTARY SET THEORY

TABLE 1.2.1

REGION	SET REPRESENTED	AN ELEMENT OF THE REGION IS IN		
		A	B	D
1	$A \cap B \cap D$	1	1	1
2	$A \cap B \cap D'$	1	1	0
3	$A \cap B' \cap D$	1	0	1
4	$A \cap B' \cap D'$	1	0	0
5	$A' \cap B \cap D$	0	1	1
6	$A' \cap B \cap D'$	0	1	0
7	$A' \cap B' \cap D$	0	0	1
8	$A' \cap B' \cap D'$	0	0	0

PRACTICE PROBLEM 2 Draw the Venn diagrams for $D \cup (A \cap B)$ and $D \cap (A \cup B)$.

    a. Make a conjecture based on the diagrams.
    b. Write both $D \cup (A \cap B)$ and $D \cap (A \cup B)$ as the union of some of the eight sets represented by the disjoint regions $A \cap B \cap D, A \cap B \cap D', \ldots, A' \cap B' \cap D'$ in Figure 1.2.4.

PRACTICE PROBLEM 3 Let $A = \{a, b, c, d\}$, $B = \{a, d, e, f\}$, and $C = \{d, f, g, k\}$.

    a. Find $A \cup B$ and $B \cup C$.
    b. Find $(A \cup B) \cup C$ and $A \cup (B \cup C)$ and verify that these two sets are equal.
    c. Find $A \cap B$ and $B \cap C$.
    d. Verify that the sets $(A \cap B) \cap C$ and $A \cap (B \cap C)$ are equal.

Cartesian Products

It is frequently important to consider elements in a given order. For two elements x and y, we can write the **ordered pair** (x, y), where x is the **first coordinate** and y is the **second coordinate** of the ordered pair. Two ordered pairs are identical when their first coordinates are the same and their second coordinates are the same. In other words, $(x, y) = (z, w)$ if $x = z$ and $y = w$.

The **product of two sets** A and B, written $A \times B$, is defined to be the set of ordered pairs $\{(a, b): a \in A \text{ and } b \in B\}$. In other words, $A \times B$ consists of all ordered pairs with first coordinate from A and second coordinate from B. The set $A \times B$ is also called the **Cartesian product** of A and B. The product $A \times A$ is frequently written A^2.

In general, $A \times B \neq B \times A$, as illustrated in Example 1.2.4, following. If one or both of the sets A and B are empty, then $A \times B = \emptyset$. Also, if $A \times B = \emptyset$, then at least one of A and B is empty.

CHAPTER 1 INTRODUCTION TO SETS, ALGORITHMS, AND LOGIC

EXAMPLE 1.2.4 Let $A = \{a, b, c, d\}$ and $B = \{1, 2, 8\}$. Then $A \times B = \{(a, 1), (a, 2), (a, 8), (b, 1), (b, 2), (b, 8), (c, 1), (c, 2), (c, 8), (d, 1), (d, 2), (d, 8)\}$. There are 12 elements in $A \times B$, whereas there are 4 elements in A and 3 elements in B. Note Card$(A \times B) = 12 = 4 \cdot 3$. Now $B \times A = \{(1, a), (1, b), (1, c), (1, d), (2, a), (2, b), (2, c), (2, d), (8, a), (8, b), (8, c), (8, d)\}$, and Card$(B \times A) = 12 = 3 \cdot 4$. We note that $A \times B \neq B \times A$, whereas Card$(A \times B) =$ Card$(B \times A)$. ■

We observe that, in Example 1.2.4, Card$(A \times B) =$ Card$(A) \cdot$ Card(B). We will justify this fact in Chapter 4.

PRACTICE PROBLEM 4 Let $A = \{a, b, c, d, e\}$ and $B = \{3, 4\}$.

  a. List all the elements in $A \times B$.
  b. List all the elements in $B \times A$.
  c. Observe that $A \times B \neq B \times A$.
  d. Verify that Card$(A \times B) =$ Card$(B \times A)$.

Power Set

The elements of a product set are ordered pairs of elements from other sets. It will also be useful to study a kind of set with elements that are themselves sets of elements. Let S be a set. The **power set** of S, $P(S)$, is the set of all subsets of S.

EXAMPLE 1.2.5 Let $S = \{a, b, c\}$. The subsets of S are: $\emptyset$, $\{a\}$, $\{b\}$, $\{c\}$, $\{a, b\}$, $\{a, c\}$, $\{b, c\}$, $\{a, b, c\}$. There are eight subsets of S. Thus, $P(S) = \{\emptyset, \{a\}, \{b\}, \{c\}, \{a, b\}, \{a, c\}, \{b, c\}, \{a, b, c\}\}$. Note that Card$(P(S)) = 8$, and each member of $P(S)$ is a subset of S. Also note that Card$(S) = 3$ and Card$(P(S)) = 8 = 2^3$. ■

In general, if S is a set and $n =$ Card(S), then Card$(P(S)) = 2^n$. We will prove this fact in Chapter 4.

PRACTICE PROBLEM 5 Let $A = \{1, 2, 3, 4\}$. Find:

  a. $P(\emptyset)$ and Card$(P(\emptyset))$ **b.** $P(A)$ and Card$(P(A))$
  c. Card$(A \times \emptyset)$ **d.** Card$(A \times A)$

PRACTICE PROBLEM 6 Fill in the missing numbers in the following table.

S	Card(S)	Card$(P(S))$	Card$(S \times S)$
$\emptyset$	0	1	
$\{a\}$	1		
$\{a, b\}$	2		
$\{a, b, c\}$	3		

1.2 ELEMENTARY SET THEORY

Some of the theorems on cardinality will be studied in Section 4.1 when we discuss basic counting techniques.

Boolean Laws for Set Theory

The Boolean laws for set theory, named after George Boole (1815–1864), are listed here for reference. A similar list of ten rules will appear again as the axioms for Boolean algebra and the laws of propositional logic when we discuss Boolean algebra in Chapter 7 and propositional logic in Chapter 5. These ten laws provide a unifying framework for several apparently different structures.

In the Boolean laws described below, U denotes the universal set, $\emptyset$ is the empty set, and A, B, and C are arbitrary subsets of U.

BOOLEAN LAWS FOR SETS

UNION	INTERSECTION
1. a. $A \cup B = B \cup A$	**b.** $A \cap B = B \cap A$
2. a. $(A \cup B) \cup C = A \cup (B \cup C)$	**b.** $(A \cap B) \cap C = A \cap (B \cap C)$
3. a. $A \cup (B \cap C) = (A \cup B) \cap (A \cup C)$	**b.** $A \cap (B \cup C) = (A \cap B) \cup (A \cap C)$
4. a. $A \cup \emptyset = A$	**b.** $A \cap U = A$
5. a. $A \cup A' = U$	**b.** $A \cap A' = \emptyset$

The ten laws above are listed in parallel columns to exhibit their duality. These laws will be proved in Chapter 5. Law 1a is the commutative law for union, 2a is the associative law for union, 3a is the distributive law for union over intersection, 4a is the identity law for union, and 5a is the complement law for union. Laws 1b through 5b for intersection have similar names.

■ **PROGRAMMING NOTES**

The Use of Sets in Pascal The programming language Pascal is notable for its implementation of sets as a data structure. Let us summarize and compare set notations in mathematics and in Pascal.

MATHEMATICS	PASCAL
$\emptyset$	[ ]
$A \cup B$	$A + B$
$A \cap B$	$A * B$
$A - B$	$A - B$
$x \in A$	x in A
$A \subseteq B$	$A <= B$
$A = B$	$A = B$
$A = \{a, b, c\}$	$A = $ ['a', 'b', 'c']

Characters in Pascal are enclosed in single quotation marks when written in a program. The data type **char** is used to represent single characters. The character set differs from computer to computer but always includes the letters of the alphabet, the digits, and symbols such as +, !, #, =, etc.

Sets can be very useful in certain programming situations. Suppose we want to read a line of text and print the distinct letters the line contains. Following are two procedures written in Pascal that will accomplish this task. We will assume the universal set includes the set of all upper- and lowercase letters. It is declared as follows:

Type
   Symbols = Set of 'A' .. 'z';

The following procedure takes the letters in a line of text and places them in a set called Lineletters. The set Lineletters is initialized as the empty set by the assignment statement "Lineletters := [ ]," where := is the Pascal assignment operator. "Lineletters := [ ]" means to assign the empty set [ ] to the set Lineletters. Letters from the line of text are added to the set Lineletters by the set union operation indicated by the assignment statement "Lineletters := Lineletters + [Ch]." Since a set lists an element exactly once, the set Lineletters becomes a list of the distinct letters from the line of text.

```
procedure Getletters (var Lineletters: Symbols);
    var
        Ch : char;
        Letters : Symbols;
    begin
        Lineletters := [ ];
        Letters := ['A' .. 'Z'] + ['a' .. 'z'];
        while not eoln do
            begin
                read(Ch);
                if Ch in Letters then
                    Lineletters := Lineletters + [Ch]
            end
    end;
```

The reserved word EOLN stands for "end of line," and it is a Boolean variable that indicates whether an end-of-line marker has been read. We are assuming the line of text has an end-of-line marker.

The following procedure will print the letters from the line of text in alphabetical order.

```
procedure Printletters (Lineletters: Symbols);
    var
        Ch : char;
    begin
        for Ch := 'A' to 'z' do
            if Ch in Lineletters then
                writeln(Ch:2)
    end;
```

1.2 ELEMENTARY SET THEORY

EXERCISE SET 1.2

1. Let $U = \mathbb{Z}$, $A = \{-1, 7\}$, $B = \{x : x \leq 4\}$, and $D = \{1, 2, 4, 7, 9\}$. Find:
   a. B' b. $A \cup B$ c. $A \cup D$ d. $D - B$ e. $A \cap D$ f. $A \cap B$

2. Refer to the sets in Exercise 1. Find:
   a. $A \cap (B \cup D)$ b. $A \times \emptyset$ c. $(A \cap B) \cup D$
   d. $A \times D$ e. $P(A)$ f. $B' \cap (A \cup D)$

3. Let $U = \mathbb{N}$, $A = \{1, 3, 5\}$, $B = \{x : x^2 - 6x + 5 = 0\}$, and $D = \{x : x > 5\}$. Find:
   a. D' b. $D - A$ c. $P(A)$ d. $A \cup B$ e. $A - D$ f. $A \times A$

4. Refer to the sets in Exercise 3. Find:
   a. $A \cap D'$ b. Card(B) c. $B \times B$
   d. $A \cap B$ e. $A \times B$ f. $D' \cap (A \cup B)$

5. Let $A = \{1, \{b\}, \{1, b\}\}$.
   a. Place an $\in$ or $\subseteq$ in the blank to make each of the following statements true.

   $1 \underline{\quad} A$; $\{1\} \underline{\quad} A$; $\{b\} \underline{\quad} A$

   b. List the eight elements of $P(A)$.

6. Use the set A in Exercise 5.
   a. Place an $\in$ or $\subseteq$ in the blank to make each of the following statements true.

   $\{\{b\}\} \underline{\quad} A$; $\{1, b\} \underline{\quad} A$; $\{\{b\}, 1\} \underline{\quad} A$

   b. List the nine elements of $A \times A$.

7. Rewrite each of the following sets as simply as possible.
   a. $\emptyset \cup \{\emptyset\}$ b. $\{\emptyset\} - \emptyset$ c. $P(\{\emptyset\}) - \{\emptyset\}$

8. Rewrite each of the following sets as simply as possible.
   a. $\emptyset \cap \{\emptyset\}$ b. $\{\{\emptyset\}, \emptyset\} - \{\emptyset\}$ c. $\emptyset \times \{\emptyset\}$

9. Let $A = \{a\}$. Find:
   a. $P(A)$ b. $P(\{A\})$

10. Find:
    a. $P(\{\emptyset\})$ b. $P(P(\{\emptyset\}))$

11. Decide whether each of the following conjectures is true or false. If the statement is false, give a counterexample. Assume that the sets A and B are finite.
    a. Card($A \times B$) = Card(A) · Card(B)   b. Card($A - B$) = Card(A) - Card(B)

12. Decide whether each of the following conjectures is true or false. If the statement is false, give a counterexample. Assume that the sets A and B are finite.
    a. Card($A \cup B$) = Card(A) + Card(B)   b. Card($P(A)$) = $2^{\text{Card}(A)}$

13. Draw Venn diagrams for the following sets.
    a. $A \cap (B \cup C)$ b. $(A \cap B) \cup C$

14. Draw Venn diagrams for the following sets.
    a. $A \cup (B \cap C)$ b. $(A \cup B) \cap C$

15. Draw Venn diagrams for the following sets.
    a. $(A \cap B)'$ b. $A' \cup B'$

16. Draw Venn diagrams for the following sets.
    a. $(A \cup B)'$ b. $A' \cap B'$

17. Refer to Exercises 13 and 14. Decide whether each of the following conjectures is false (in other words, not always true). If the statement is false, give a counterexample suggested by a Venn diagram.
    a. $A \cap (B \cup C) = (A \cap B) \cup C$ b. $A \cup (B \cap C) = (A \cup B) \cap C$

18. Refer to Exercises 15 and 16. Decide whether each of the following conjectures is false. If the statement is false, give a counterexample suggested by a Venn diagram.
    a. $(A \cap B)' = A' \cup B'$ b. $(A \cup B)' = A' \cap B'$

19. Decide whether the following conjecture is false. If the statement is false, give a counterexample suggested by a Venn diagram.

    $$(A \cup B) \cap (C \cup D) = (A \cap C) \cup (B \cap D)$$

20. Let $A = \{a, b, c\}$ and $B = \{c, d\}$.
    a. Find $A \times B$.
    b. Find $B \times A$.
    c. Verify Card$(A \times B)$ = Card$(B \times A)$.
    d. Find $A - B$ and $B - A$ and verify $A - B \neq B - A$.

21. Let $C = \{1, 2, 3, 4\}$ and $D = \{8\}$.
    a. Find $C \times D$.
    b. Find $D \times C$.
    c. Verify Card$(C \times D)$ = Card$(D \times C)$.
    d. Find $C - D$ and $D - C$ and verify $C - D \neq D - C$.

22. Solve for s and t in each of the following.
    a. $(6s, t) = (12, 4)$
    b. $(8, s + 4) = (t, 10)$
    c. $(s + 2t, 4t) = (6, 12)$
    d. $(t - s, t + s) = (4, 8)$

23. Solve for u and v in each of the following.
    a. $(u - 7, v + 3) = (10, -4)$
    b. $(3, u + 8v) = (2u - v, 6)$
    c. $(9, 3) = (u + v, v)$
    d. $(u + v, 3v) = (8, 9)$

24. Let A and B be subsets of a universal set U. The **symmetric difference** of A and B, denoted by $A \Delta B$, is defined to be the set

    $\{x : x \in A \text{ or } x \in B \text{ but not both}\}$

    Find $A \Delta B$ for each of the following pairs of subsets of $\mathbb{Z}$.
    a. $A = \{x : x \in \mathbb{Z} \text{ and } x < -10\}$
       $B = \{x : x \in \mathbb{Z} \text{ and } |x| = |9|\}$
    b. $A = \{x : x \in \mathbb{Z} \text{ and } x^2 - 2x - 15 = 0\}$
       $B = \{x : x \in \mathbb{Z} \text{ and } |x| < 6\}$
    c. A is the set of all integers.
       B is the set of all even integers.

25. Repeat Exercise 24 for each of the following pairs of subsets of $\mathbb{R}$.
    a. $A = \{x : x \in \mathbb{R} \text{ and } |x| < 1\}$
       $B = \{x : x \in \mathbb{R} \text{ and } 0 < x < 2\}$
    b. $A = \{x : x \in \mathbb{R} \text{ and } x^2 = 2\}$
       $B = \{x : x \in \mathbb{R} \text{ and } |x| > 1\}$

26. See Exercise 24 for the definition of $A \Delta B$. Use Venn diagrams to illustrate each of the following.
    a. $A \Delta B = (A \cup B) - (A \cap B)$
    b. $A \Delta B = (A - B) \cup (B - A)$

■ PROGRAMMING EXERCISES

27. Use Pascal set notation to write each of the following as a single set.
    a. [i, j, k, l, m] + [j, m, n, p]
    b. [x, y, z, w] * [x, w, t, v]
    c. [6, 9, 21, 30, 32] − [1, 2, 3, 32, 49]

28. Use Pascal set notation to write each of the following as a single set.
   a. [FORTRAN, Ada, C, BASIC] + [Pascal, COBOL]
   b. [0, 1, 2, 5] ∗ [2, 3, 4, 5, 6, 7]
   c. [a, 1, b, 2, c, 3, d, 4] − [a, b, c, d]

1.3 ALGORITHMS

A finite set of precise instructions to solve a specific problem is called an **algorithm**. Although algorithms have been a part of mathematics since ancient times, the field of computer science has stimulated an intensive study of algorithms more recently. Any finite set of instructions precise enough to be executed by a computer may be considered an algorithm. The term *algorithm* is synonymous with the term *effective procedure*.

Fundamentally, an algorithm is directed to some purpose; an algorithm is used to solve some problem. The solution may depend on information given to the algorithm and results obtained from the operations of the algorithm.

Here are three problems with algorithmic solutions. We will use these problems to illuminate the nature of algorithms. Each has a given set of information called "input," and a set of derived information for the solution called "output."

PROBLEM 1.3.1 Given a nonnegative integer n and a real number a, find a^n. The solution to this problem requires computation. The computation can be tedious for large n and for a real number a given as a decimal with many significant figures.

PROBLEM 1.3.2 Find the largest number in a list of n integers. Although obvious in principle, the solution to this problem will require nontrivial mathematical techniques.

PROBLEM 1.3.3 Decide whether a given positive integer is a prime. The solution to this problem is a simple "true" or "false." This is an example of a **decision problem**.

Problem 1.3.1 specifies the input. In Problem 1.3.2 the input is assumed to be a finite list of integers. For Problem 1.3.3 the input is a positive integer. In general, inputs and outputs for a specific problem will be from specific sets or of specific types.

The output for Problem 1.3.1 will be a single real number. The solution to Problem 1.3.2 will yield an integer as output. For Problem 1.3.3 the output is either true or false.

Any algorithm designed to solve one of the three preceding problems must accommodate all acceptable inputs for that problem. In particular, the algorithm must accept a list of numbers with any finite length and return output of an appropriate type.

We will design algorithms to solve each of the preceding problems. First, we outline a solution for each problem. Each solution is then refined until it is a precisely written algorithm. Our goal is to have an algorithm that can easily be translated into a sequence of instructions for execution by a computer.

We will assume, for Problem 1.3.1, that multiplication (·) of two real numbers is available to us but taking the nth power of a real number is not.

Trial Solution 1a Given a nonnegative integer n and a nonzero real number a, this algorithm computes a^n.

If $n = 0$, set $a^n = 1$.
If $n \neq 0$, let a^n be a multiplied by itself n times. ∎

For designing and writing this and other algorithms, we will use a language called **pseudocode**, which is somewhere between ordinary English and a formal computer programming language. The steps of algorithms are named for easy reference at the initial stage. We will also usually give the output a meaningful name. In this case, we let a^n be "Power."

The assignment operator ":=" will also provide a useful shorthand. For example, "set $b = 1$" can be written simply as "$b := 1$." The assignment statement "$b := 1$" means "assign to the variable b the value 1." It is helpful to contrast "$b := 1$" with "$b = 1$." The expression "$b = 1$" is an expression that is either true or false depending on the value of b. On the other hand, "$b := 1$" has no truth value; it is an action statement.

With these conventions and some documentation for clarity, we have Trial Solution 1b.

Trial Solution 1b Given a nonnegative integer n and a nonzero real number a, this algorithm computes Power $= a^n$.

Begin
　Basis Step: If n = 0, Power := 1.
　Next Power Step: Otherwise, Power := a multiplied by itself n times.
End. ∎

We are assuming that steps are executed in order and that after each step we go to the following step unless otherwise specified. For example, if $n = 3$ above, the Basis Step will be executed with no action and then the Next Power Step will be executed.

The Basis Step in Trial Solution 1b is clear and precise. The phrase "a multiplied by itself n times" has meaning in ordinary English but is entirely inadequate for an algorithm. The Next Power Step needs refinement. Since $a^0 = 1$, it will be helpful to involve the integer 1 in the multiplications. Note that, for $n = 1$, we have Power $= 1 \cdot a$; for $n = 2$, we have Power $= 1 \cdot a \cdot a$; and, for an arbitrary positive n, the computation of Power requires n multiplications. Furthermore, note that, if, say, Power $= a^3$ has already been computed, then $a^4 =$ Power $\cdot a$. With this in mind, we arrive at Trial Solution 1c.

1.3 ALGORITHMS

Trial Solution 1c Given a nonnegative integer n and a nonzero real number a, this algorithm computes Power $= a^n$.

>   Begin
>   Basis Step: Power := 1.
>   Next Power Step: Repeat n times the assignment Power := Power · a.
>   Output Step: Return Power.
>   End.

For greater precision we will use a variable called "Count" to keep track of the repeated multiplications in the Next Power Step. Finally, we have a solution worthy of the name *algorithm*.

ALGORITHM 1 Given a nonnegative integer n and a nonzero real number a, this algorithm computes Power $= a^n$.

>   Begin
>   Power := 1.   (*Basis*)
>   Count := 1.   (*Initialize*)
>   While Count ≤ n do   (*Next Power and Iterate*)
>   Power := Power · a and
>   Count := Count + 1.
>   Return Power.   (*Output*)
>   End.

Note that "While Count ≤ n do" is a loop structure in which the assignments "Power := Power · a" and "Count := Count + 1" are executed repeatedly as long as Count ≤ n. There will be more on loop structures at the end of this section when we introduce control structures.

At this point we list the important features of an algorithm for easy reference. In addition to being a finite set of instructions to solve a specific problem, an algorithm has the following characteristics:

1. An algorithm must terminate after a finite number of steps.
2. Each instruction is clear and unambiguous.
3. Each instruction is deterministic. That is, the result of every step depends only on the input and the results of the previous steps. No chance occurrences are possible.
4. An algorithm accepts zero or more inputs from some specified set of objects.
5. An algorithm produces one or more outputs.
6. An algorithm is expected to be effective. In other words, all the algorithm's operations are sufficiently simple that they can, in principle, be performed exactly and in a finite amount of time by a person using pencil and paper.

We can write an alternate form of Algorithm 1 by using another type of loop structure. Note that, in the loop structure below, the assignment "Power := Power · a" is repeated n times as Count increases from 1 to n.

CHAPTER 1 INTRODUCTION TO SETS, ALGORITHMS, AND LOGIC

ALGORITHM 1 Given a nonnegative integer n and a real number a, this algorithm computes
(Alternate Version) Power $= a^n$.

    Begin
        Power $:= 1$.   (*Initialize*)
        If $n = 0$ then
            Return Power   (*Basis*)
        Else
            For Count $= 1$ to n do
                Power $:=$ Power $\cdot$ a.   (*Next Power*)
        Return Power.   (*Output*)
    End.

Next we consider Problem 1.3.2. Here is a first attempt at a solution.

Trial Solution 2 Given a list of n integers, this algorithm finds the largest integer in the list and returns it as Max.

    Begin
        Set Max equal to the first number in the list.   (*Initialize*)
        For each successive number in the list, set Max equal to
        that number if Max is smaller than the number;
        otherwise do nothing.   (*New Max*)
        Return Max.   (*Output*)
    End.

One way to make this solution more precise is to assume the list of integers is given as a sequence $A(1), A(2), \ldots, A(n)$. Now we write our solution as Algorithm 2.

ALGORITHM 2 Given a list of n integers, $A(1), A(2), \ldots, A(n)$, this algorithm finds the largest integer in the list and returns it as Max.

    Begin
        Max $:= A(1)$.   (*Initialize*)
        For $j = 2$ to n do
            If Max $< A(j)$ then   (*New Maximum*)
                Max $:= A(j)$.
        Return Max.   (*Output*)
    End.

Before solving Problem 1.3.3, we define the term *prime*. A **prime** or **prime number** is any positive integer with exactly two divisors, 1 and itself. In particular, the positive integer 1 is not a prime. Examples of primes include 2, 3, 5, 7, 11, 13, 17, ... Note that 2 is the only even prime.

We will solve Problem 1.3.3 by counting the number of divisors for each positive integer. If the number of divisors is two, our algorithm will output "true." Otherwise it will output "false."

ALGORITHM 3 Given a positive integer n, this algorithm decides whether n is a prime.
```
Begin
    Count := 1.                 (*Initialize*)
    For k = 2 to n do
        If k divides n then
            Count := Count + 1.   (*New Count*)
    If Count = 2, then
        return true
    else                        (*Output*)
        return false.
End.
```

Algorithm 3 is a clear and precise algorithm as long as "k divides n" is well-defined for positive integers k and n. We will assume for now that there is an algorithm that takes two positive integers a and b as input and outputs true when a divides b and outputs false otherwise. In fact, we leave it as an exercise to design such an algorithm (see Exercise 12 at the end of this section).

In designing Algorithms 1, 2, and 3, we have used a top-down approach. That is, we started with an outline of the solution and added detail and precision to the structure as we progressed toward the final solution, the algorithm. This concept of moving from the top level of the solution's design toward more detailed levels of design is called the **top-down design** methodology.

Another aspect of design methodology used in the design of Algorithm 3 is that of modularization. We embedded within Algorithm 3 a subprocess needed for deciding whether one integer divides another. Our design simply assumes that the subprocess or module can be defined in detail later.

Given a specific algorithm, it is important to know:

1. Is the algorithm correct? That is, does it yield correct output for valid input? Put another way, does it do what it was designed to do?
2. Is the algorithm efficient? What is the cost of running the algorithm in terms of both storage space (computer memory) and time (execution time)?

Question 1 addresses the issue of algorithm correctness. We discuss algorithm correctness when we consider loop invariants in Chapter 5. Question 2 refers to the complexity of algorithms. As indicated, there are two aspects of complexity. It is beyond the scope of this text to consider space complexity. However, we discuss time complexity of algorithms in Section 3.5, and again when we discuss the big-oh notation in Chapter 5.

Historical Notes on the Word *Algorithm*

A Persian textbook author named Mohammed ibn Mūsā al-Khowārizmi (Mohammed, the son of Moses, native of Khwarezm) wrote a book on Hindu numerals in A.D. 825. In 1857, a Latin translation of this book was found, which began "Spoken has Algoritmi, ..." Here the author's name al-Khowārizmi had become Algoritmi, from which the mathematical word *algorithm* was derived.

Most mathematics students probably first encounter the word *algorithm* when they study the method for finding the greatest common divisor of two integers. This method, which appeared in Book 7 of Euclid's *Elements*, is called the Euclidean algorithm in modern mathematics textbooks. Currently, the word *algorithm* is omnipresent in many computer science and mathematics books.

Control Structures

We used four types of **control structures** in Algorithms 1–3. These control structures are: "If . . . then . . . ," "If . . . then . . . else . . . ," "For . . . do . . . ," and "While . . . do . . . ." Control structures deserve a short discussion.

An example of the "If . . . then . . ." structure is "If k divides n, then Count := Count + 1." The form for an "If . . . then . . ." control structure is

If Boolean expression then
   Statement Block.

The phrase "k divides n" is a Boolean expression. A **Boolean expression** is any valid expression that is either true or false at the time it is evaluated. When it is true, Statement Block is executed; when it is false, the algorithm passes to the next step. A **statement block** may contain any legitimate pseudocode including assignment statements and control structures.

A closely related control structure is the "If . . . then . . . else . . ." structure. Its form is

If Boolean expression then
   Statement Block 1
else
   Statement Block 2.

Statement Block 1 will be executed when the Boolean expression is true, and Statement Block 2 will be executed when the Boolean expression is false.

The "For . . . do . . ." structure repeatedly executes a statement block as the variable increases from some initial value to a final value. This control statement has the form

For Index = Initial Value to Final Value do
   Statement Block.

The "While . . . do . . ." control structure was used in Algorithm 1. It has the form

While Boolean expression do
   Statement Block.

1.3 ALGORITHMS

This control structure repeatedly executes Statement Block as long as the Boolean expression is true. Frequently the statement block includes a method of changing the truth value of a Boolean expression under certain conditions. Refer to Algorithm 1 and note that "Count := Count + 1" will ultimately change the truth value of the Boolean expression "Count ≤ n."

Let us rewrite Algorithm 3 using a "While...do..." control structure.

ALGORITHM 3
(Alternate Version)

Given a positive integer *n*, this algorithm decides whether *n* is a prime.

```
Begin
    Count := 1.                    (*Initialize*)
    For k = 2 to n do
        While Count < 3 do
            If k divides n then
                Count := Count + 1.    (*New Count*)
    If Count = 2, then
        return true
    else                           (*Output*)
        return false.
End.
```

Note that the use of "While Count < 3" may result in fewer executions of the statement block "If k divides n then Count := Count + 1" than would the statement "For k = 2 to n do" alone.

Developing an algorithm to perform a specific task can be a trial-and-error process. After an initial attempt and review, we may modify the existing algorithm or possibly reject it and start anew. We aim to improve the algorithm until we are satisfied it is correct and ready to be expressed in a suitable programming language. An algorithm written in a computer programming language is called a **program** or **computer program**. We should keep in mind that modifications of algorithms and programs continue as long as the programs are in use.

Most of the algorithms in this book are expressed in pseudocode or in Pascal. The pseudocode used here is very much like Pascal. Nevertheless, even students not familiar with Pascal can read the pseudocode and the Pascal we use for writing algorithms.

We will not, as a rule, write every algorithm in pseudocode as formal as that used in Algorithms 1, 2, and 3.

EXERCISE SET 1.3

1. In a certain state the income tax is 3% of all income in excess of $10,000. No tax is paid on the first $10,000 of income. Design an algorithm that takes a variable Income as input and outputs the total tax. Write the algorithm in pseudocode.

2. In a given state the sales tax is 6% of all nonfood purchases. No tax is paid on the purchase of food. Design an algorithm that takes two variables FoodCost and TotalCost as input and outputs the total tax. Write the algorithm in pseudocode.

For Exercises 3–6 assume A is a sequence $A(1), A(2), \ldots, A(n)$ of n real numbers, where n is a given positive integer.

3. Design an algorithm to compute and output the sum of the numbers in the sequence A. Write the algorithm in pseudocode.

4. Design an algorithm to compute and output the sum of the squares of the numbers in the sequence A. Write the algorithm in pseudocode.

5. Design an algorithm to compute and output the average of the numbers in the sequence A. Write the algorithm in pseudocode.

6. Design an algorithm to compute and output the square root of the sum of the numbers in the sequence A if the sum is nonnegative; otherwise, the algorithm should return the message "Square root not real." You may use the function SQRT(x), which returns the square root of any nonnegative real number x. Write the algorithm in pseudocode.

In Exercises 7–10 write an algorithm in pseudocode to compute and output the quantity specified.

7. The sum of the first n positive odd integers

8. The sum of the first n positive even integers

9. The sum of the squares of the first n positive integers

10. The product of the first n positive odd integers

11. Design an algorithm that takes as input one positive integer a and outputs true when a is even and outputs false otherwise. Assume that the only arithmetic operations at your disposal are addition and subtraction. Write the algorithm in pseudocode.

12. Design an algorithm that takes as input two positive integers a and b and outputs true when a divides b and outputs false otherwise. For positive integers a and b we say that a **divides** b if there is a positive integer m so that $b = ma$. Assume that the only arithmetic operations at your disposal are addition and subtraction. Write the algorithm in pseudocode.

In Exercises 13–17 algorithms are given in pseudocode. Describe what each algorithm does.

13. Let x and y be two real numbers.
    Begin
       If $x < y$ then
          $M := x$
       else
          $M := y$.
       Output M.
    End.

14. Begin
       $S := 0$.
       For $k = 1$ to n do
          $S := S + 1/k$.
       Output S.
    End.

15. Let n be a positive integer and let $A(1), A(2), \ldots, A(n)$ be n real numbers.
    Begin
       $P := 1$.
       $k := 0$.
       While $k < n$ do
          $k := k + 1$ and $P := P \cdot A(k)$.
       Output P.
    End.

16. Let *n* be a positive integer.
   Begin
     S := n.
     j := 1.
       While j < n do
         S := S + n and j := j + 1.
     Output S.
   End.

17. Let *n* be a positive integer.
   Begin
     S := 0.
     For k = 1 to n do
       S := 2 · S + 1.
     Output S.
   End.

In Exercises 18–26 determine the value of each given variable after the given control statement is executed.

18. J := 2 and A := 2.
   If (J = 5 or A = 2) then
     J := J + 1
   else
     J := 5.

19. J := 2 and A := 3.
   If (not (J = 5) and A = 3) then
     J := 7
   else
     A := 7.

20. J := 2 and A := 3.
   If not (J = 2 and A = 3) then
     A := 5
   else
     A := 2.

21. J := 2 and A := 3.
   For k = 1 to 5 do
     J := J + 1.

22. J := 2 and A := 5.
   For k = 1 to 7 do
     A := A + 1.

23. J := 2 and A := 5.
   While J < 7 do
     A := A + 1 and J := J + 1.

24. J := 2 and A := 5.
   While A ≠ 7 do
     A := A + 1 and J := J + 1.

25. J := 1.
   For k = 2 to 8 do
     While J < 3 do
       If k divides 8 then
         J := J + 1.

26. J := 1.
   For k = 2 to 7 do
     While J < 3 do
       If k divides 7 then
         J := J + 1.

1.4 READING PROOFS

In all mathematical proofs there will be a collection of statements called the **hypotheses** and a statement called the **conclusion**, which must be proved to follow logically from the hypotheses. Although a proof can take many forms, every proof is a finite sequence of statements that are either hypotheses, previously proved statements, or statements that follow logically from previous statements. The final statement is the conclusion. In this section, we give an overview of writing and reading proofs.

    The first step to understanding a proof is to know what is to be proved. Secondly, we must know what the proof is "all about." For example, is it a proof about sets or about functions? A proof about sets will require facts and definitions of set theory. A proof involving functions will require definitions such as for "bijective function" and facts such as "Every bijective function has an inverse."

Questions to answer while reading a proof:

What is the goal of the proof?
What are the hypotheses?
What definitions are necessary?
What previously proved facts or laws of logic are used in the proof?

The goal of a proof includes whatever is needed to prove the conclusion using laws of logic in the context of the given facts. For example, if the conclusion of a proof is "The square of every even integer is also even," the goal will be to prove an integer n^2 is even; the statement "n is even" will be one of the statements assumed true. We will need to refer to concepts that have been defined earlier, and also to statements that have been proved earlier.

To illustrate these ideas, consider the following examples. The first is from set theory. The steps in this proof are numbered for easy reference.

EXAMPLE 1.4.1 Let A and B be arbitrary sets. Prove $A \cap B \subseteq A$.

Proof
1. Let $x \in A \cap B$.
2. It follows that $x \in A$ and $x \in B$.
3. Hence, $x \in A$.

The goal is to prove $A \cap B \subseteq A$. The proof takes the form shown because of the definition of $\subseteq$ (subset). From that definition, we know $A \cap B \subseteq A$ whenever, for every element x, if $x \in A \cap B$, then $x \in A$. So we must prove, for any element x, if $x \in A \cap B$, then $x \in A$. This means Step 1 should be: Assume $x \in A \cap B$. Step 2 follows directly from the definition of set intersection. Step 3 follows from a law of logic: If both p and q are assumed true, then p must be true.

Another way to look at this proof is first to consider what must be proved. We need to prove $A \cap B \subseteq A$. After we consider the definition of $\subseteq$, it is clear we need to prove that, for any element x, $x \in A \cap B$ implies that $x \in A$. From this, we see that we need to assume $x \in A \cap B$ and prove $x \in A$. What we have done is a goal assessment. That is, we have assessed and refined the goal until we have a more concrete notion of what really must be proved. ∎

The proof in Example 1.4.1 uses the "pick-a-point" method for proving that one set is a subset of another set. The **pick-a-point method** for proving $A \subseteq B$ is as follows: Take an arbitrary element u in the universal set U and assume $u \in A$. Then use any known true statements, including properties of A and B, to prove $u \in B$. This method of proof is based on the definition: $A \subseteq B$ if $u \in A$ implies that $u \in B$ for all $u \in U$. The pick-a-point method will be discussed again in Chapter 5.

Refining the goal of a proof sometimes leads to working backwards from the conclusion toward the assumptions, as illustrated in the next example.

EXAMPLE 1.4.2 Let x and y be nonnegative numbers. Prove $\sqrt{xy} \leq (x+y)/2$. Let us see how refining the goal can help construct a proof.

The goal is to prove $\sqrt{xy} \leq (x+y)/2$. So, we need to prove $2\sqrt{xy} \leq (x+y)$. But, since all quantities on both sides of the inequality are nonnegative, an equivalent statement is $(2\sqrt{xy})^2 \leq (x+y)^2$. Hence, we need to prove $4xy \leq x^2 + 2xy + y^2$. After subtracting $4xy$ from both sides, we obtain the equivalent statement $0 \leq x^2 - 2xy + y^2 = (x-y)^2$. And $0 \leq (x-y)^2$ is true, since the square of any real number is nonnegative.

By working backwards from $\sqrt{xy} \leq (x+y)/2$, we are led to the true statement $0 \leq (x-y)^2$.

This scratch work enables us to construct the following proof.

Proof Since the square of every real number is nonnegative, we have $0 \leq (x-y)^2$. Hence, $0 \leq x^2 - 2xy + y^2$. Adding $4xy$ to both sides yields $4xy \leq x^2 + 2xy + y^2 = (x+y)^2$. Taking the square root of both sides, we obtain $2\sqrt{xy} \leq (x+y)$. After dividing both sides by 2, we obtain $\sqrt{xy} \leq (x+y)/2$. ∎

A comparison of the scratch work with the finished proof in Example 1.4.2 makes it apparent that the scratch work contains information that makes the proof easier to understand. The proof is correct and logically complete but lacks information that motivates each step. For example, the first step of the proof seems to come out of the blue. It is clear why that first step is taken only after the proof is completely read and the total picture absorbed.

The main goal of a proof writer is to convince others that the statement being proved follows logically from certain specific assumptions. A secondary goal is to write an elegant and concise proof. In some ways, these goals conflict. What is clear, concise, and elegant to some readers will be a terse muddle to others. The less experienced you are at reading proofs, the more important it is to do your own scratch work in order to understand the proof.

Here is another example. We will give the proof in its final form first and then outline an analysis of it.

EXAMPLE 1.4.3 Let a, b, and c be the lengths of the sides and the hypotenuse, respectively, of a right triangle as shown in Figure 1.4.1 (page 28). Prove: If the area of the triangle is equal to $c^2/4$, then the triangle is isosceles.

Proof Since the area is $c^2/4$, we have $c^2 = 2ab$. By the Pythagorean Theorem, $a^2 + b^2 = 2ab$. Hence, $(a-b)^2 = 0$ and $a - b = 0$. Therefore, $a = b$, and the triangle is isosceles.

Discussion   This proof is very terse, almost "bare bones." (We promise to give proofs in subsequent chapters of this text with more flesh on them.) However, the proof is correct. With a little close reading, we can understand it. Consider each of the questions that should be asked when reading a proof.

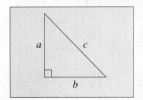

FIGURE 1.4.1

What is the goal of the proof?

The goal is to prove that a triangle is isosceles by assuming that the area of a right triangle is $c^2/4$. A refinement of that goal, using Figure 1.4.1, tells us we need to prove $a = b$.

What is the hypothesis?

We are given that the area of a right triangle is equal to $c^2/4$, where c is the length of the hypotenuse of the triangle.

What definitions are necessary?

The definition of isosceles triangle was used in the preceding goal assessment.

What previously proved facts are used in the proof?

The area of a triangle with legs of length a and b is $ab/2$. The Pythagorean Theorem says that $a^2 + b^2 = c^2$ for the right triangle in Figure 1.4.1.

Now let us fill in some of the missing details for this proof.

Annotated Proof   Since the area is $c^2/4$, we have $c^2 = 2ab$.

We can assume the area is $c^2/4$. But the formula for the area of such a triangle gives the area as $ab/2$. Hence, $ab/2 = c^2/4$. Multiplying both sides by 4 yields $c^2 = 2ab$.

By the Pythagorean Theorem, $a^2 + b^2 = 2ab$.

The Pythagorean Theorem says that $a^2 + b^2 = c^2$ for a right triangle with hypotenuse c. Using $c^2 = 2ab$ from the step above gives us $a^2 + b^2 = 2ab$.

Hence, $(a - b)^2 = 0$.

Subtracting $2ab$ from both sides of $a^2 + b^2 = 2ab$ yields $a^2 - 2ab + b^2 = 0$. So, $(a - b)^2 = a^2 - 2ab + b^2 = 0$.

Therefore, the triangle is isosceles.

Since $(a - b)^2 = 0$, it follows that $a = b$.   ∎

This more complete version of the original proof helps make the proof easier to understand. However, it still may not be clear why or how we might think of the steps in the first place. What went into the creation of this proof? For example, why was the Pythagoren Theorem first considered important for this proof? Perhaps because c^2 appears in the hypothesis. Perhaps because it is a right triangle with sides a and b and hypotenuse c. Of course, we may have followed a few bad leads before finding this proof.

If we were creating this proof, we might have been brought to the goal of proving that the angles adjacent to the hypotenuse are equal. However, we would soon find it difficult to use the given information to arrive at any conclusion regarding angles. If we persevered, we would go back and revise our goal so that we could use the hypothesis effectively.

The creation of a proof sometimes requires that we work backward from the original goal, by means of a goal assessment, to a new goal that is closer to the given facts. Also, we may need to rewrite the given facts in the hypotheses, being careful not to change the hypotheses, so that they are closer to the new goal. This process is sometimes called the **backwards/forwards method**.

EXERCISE SET 1.4

In each of Exercises 1–12 a proof will be given or you will be asked to supply a proof. If a proof is given, outline the proof and fill in any details that will make the proof easier to understand. Hints are given for the exercises in which you must supply the proof. Answer each of the following questions in every exercise:

a. What is the goal of the proof?
b. What is the hypothesis?
c. What definitions are necessary?
d. What axioms, previously proved facts, or laws of logic are used in the proof?

You may need to look up relevant definitions or theorems in this text or in a calculus book. Exercises 11 and 12 require knowledge of first-semester calculus.

1. Let A and B be arbitrary sets. Prove $A \subseteq A \cup B$.
   Proof Let $x \in A$. Then $x \in A$ or $x \in B$. Hence, $x \in A \cup B$.

2. Let A and B be arbitrary sets. Prove $B - (B - A) \subseteq B$.
   Proof Let $x \in B - (B - A)$. Then $x \in B$ and $x \notin B - A$. Hence, $x \in B$.

3. Let x and y be nonnegative numbers. Prove: If $\sqrt{xy} = (x + y)/2$, then $x = y$.
   Proof Assume $\sqrt{xy} = (x + y)/2$. Hence, $2\sqrt{xy} = x + y$. So, $(2\sqrt{xy})^2 = (x + y)^2$. This yields $4xy = x^2 + 2xy + y^2$. After subtracting $4xy$ from both sides, we obtain $0 = x^2 - 2xy + y^2 = (x - y)^2$. Therefore, $x = y$.

4. Let x and y be nonnegative numbers. Prove: If $x = y$, then $\sqrt{xy} = (x + y)/2$. [*Hint:* Reverse the steps of the proof in Exercise 3.]

5. Let a, b, and c be the lengths of the sides and the hypotenuse, respectively, of a right triangle. (See Figure 1.4.1.) Prove: If the triangle is isosceles, then the area of the triangle is equal to $c^2/4$. [*Hint:* The steps of this proof will be the reverse of the steps in the proof of Example 1.4.3.]

6. Let a, b, and c be the lengths of the sides of an isosceles triangle, with $a = b$. Assume that the area of the triangle is $c^2/4$. Prove that the triangle is a right triangle with hypotenuse of length c. (The Pythagorean Theorem states that, if a right triangle has legs of length a and b and hypotenuse of length c, then $a^2 + b^2 = c^2$. Conversely, it is also true that a triangle with sides a, b, and c such that $a^2 + b^2 = c^2$ is a right triangle.)

Proof Draw a triangle with sides of length a and b and a third side of length c at the base with $a = b$. Since the area is $c^2/4$, we have $c^2/4 = (c/4)\sqrt{4a^2 - c^2}$. Hence, $c = \sqrt{4a^2 - c^2}$ and $c = 2a^2$. Since $a = b$, we have $c^2 = a^2 + b^2$. Therefore, the triangle is a right triangle.

7. Let x and y be integers. Prove: If x and y are odd, then $x + y$ is even.
  Proof Assume x and y are odd integers. Then $x = 2j + 1$ and $y = 2k + 1$ for some integers j and k. Hence, $x + y = 2j + 2k + 2 = 2(j + k + 1)$. Therefore, $x + y$ is an even integer.

8. Let x and y be integers. Prove: If x and y are odd, then $x \cdot y$ is odd. [*Hint:* The proof is similar to that for Exercise 7.]

9. Let x be a positive real number. Prove $x + 1/x \geq 2$. [*Hint:* Do a goal assessment and work backwards. Start by multiplying both sides of the inequality by x.]

10. Let f be any real-valued function. Prove: If f is an increasing function, then f is a one-to-one function.
  Proof Assume $x \neq z$. Then either $x < z$ or $z < x$. Hence, $f(x) < f(z)$ or $f(z) < f(x)$, since f is an increasing function. In either case, $f(x) \neq f(z)$. Therefore, f is a one-to-one function.

11. Prove that $\sin x \leq x$ for all nonnegative real numbers x.
  Proof Consider the function $g(x) = x - \sin x$. The derivative $g'(x) = 1 - \cos x$, and so $g'(x) \geq 0$. Hence, g is a nondecreasing function. But $g(0) = 0$. Therefore, $g(x) \geq 0$ for all $x \geq 0$ and the result follows.

12. Prove that $\cos x = x$ for some nonnegative real number x.
  Proof Consider the function $f(x) = x - \cos x$. It follows that $f(0) = -1 < 0$ and $f(\pi/2) = \pi/2 > 0$. Therefore, by the Intermediate Value Theorem, there is a number x, with $0 < x < \pi/2$, such that $f(x) = 0$. The result follows.

13. Prove: If a^2 is an even integer, then a is an even integer. [*Hint:* Assume a^2 is an even integer and a is an odd integer. Now use Exercise 8 to derive a contradiction.]

14. Prove that $\sqrt{2}$ is not a quotient of two integers. [*Hint:* Suppose $\sqrt{2} = a/b$ and a and b are not both even. Use Exercise 13 to derive a contradiction.]

1.5
INTRODUCTION TO LOGIC

Everybody needs the ability to read and understand valid arguments and to recognize invalid arguments. A basic knowledge of logic is indispensable for analyzing and constructing proofs. Symbolic logic can be described as the analytical study of the art of reasoning. The two most important branches of symbolic logic are propositional logic and predicate logic. In this section we give a brief introduction to the propositional logic having immediate application in Chapters 2–4. We will delay the detailed study of symbolic logic until Chapter 5.

Propositional logic is sometimes called the statement calculus or propositional calculus. A **statement** is a declarative sentence that is either true or false but not both. In other words, a statement can be assigned the truth

1.5 INTRODUCTION TO LOGIC

value true (T) or the truth value false (F) but not both. Examples of sentences that are statements include:

"All horses are green."
"$2 - 1 = 7$."
"There is one even prime number."

The sentence "Is this concept important?" is not a statement because it is not a declarative sentence. The equation "$x + 1 = 7$" is not a statement because its truth value depends on the numeric value of x.

We leave it as an exercise to verify that the sentence

"This sentence is false."

is not a statement because asserting it is true (or false) leads to a logical paradox. (See Exercise 5, Exercise Set 1.5.)

We will use p, q, r, and sometimes other letters to symbolize statements. Possible combinations of truth values for one or two statements are shown in Tables 1.5.1 and 1.5.2. For example, when p symbolizes a statement, p can take on either of two truth values, T or F. Two statements p and q have four possible truth value combinations.

More complicated statements are constructed from simpler ones using statement connectives. The three basic **statement connectives** are the **AND**, **OR**, and **NOT** connectives, abbreviated $\wedge$, $\vee$, and $\neg$, respectively. They are defined by the truth values in Tables 1.5.3–1.5.5.

TABLE 1.5.1

p
T
F

TABLE 1.5.2

p	q
T	T
T	F
F	T
F	F

TABLE 1.5.3

p	q	AND $p \wedge q$
T	T	T
T	F	F
F	T	F
F	F	F

TABLE 1.5.4

p	q	OR $p \vee q$
T	T	T
T	F	T
F	T	T
F	F	F

TABLE 1.5.5

p	NOT $\neg p$
T	F
F	T

For simplicity, we say "p is true" for "p has truth value T" and "p is false" for "p has truth value F." From the preceding tables, we see that $p \wedge q$ is true exactly when both p and q are true; $p \vee q$ is false exactly when both p and q are false; and $\neg p$ is true when p is false, whereas $\neg p$ is false when p is true.

A comparison of these basic logic connectives with basic concepts of set theory is helpful. Set intersection and union are defined in terms of the AND and OR connectives, since

$$A \cap B = \{x : x \in A \wedge x \in B\} \quad \text{and} \quad A \cup B = \{x : x \in A \vee x \in B\}$$

Of course, $x \notin A$ means $\neg(x \in A)$, and $A \nsubseteq B$ means $\neg(A \subseteq B)$.

The Implication

Many theorems are stated in the form "If p, then q." A theorem of this type is asserting that, if the statement p is true, then the statement q must also be true. We also write "p implies q" for "If p, then q."

The statement connective **IMPLIES**, abbreviated $\rightarrow$, is defined by the truth values in Table 1.5.6.

The statement $p \rightarrow q$ is called an **implication** and is read "If p, then q." The statement p is called the **hypothesis**, whereas q is called the **conclusion**. An implication is sometimes called a **conditional**.

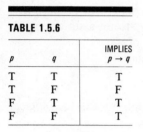

TABLE 1.5.6

p	q	IMPLIES $p \rightarrow q$
T	T	T
T	F	F
F	T	T
F	F	T

To understand the meaning of *implication*, let us consider an example. Suppose you buy a car with a guarantee that says: "If the car breaks down, then the mechanic will fix it." Let p be "the car breaks down" and q be "the mechanic will fix it." So the guarantee says $p \rightarrow q$. To say the guarantee is not valid is to say $p \rightarrow q$ is false. Under what conditions can we conclude that the guarantee is not valid? Only when the car breaks down (p is true) and the mechanic will not fix it (q is false). Note that, by the second line of the truth table in Table 1.5.6, $p \rightarrow q$ is false when p is true and q is false.

Note that the guarantee is valid when the car breaks down (p is true) and the mechanic does fix it (q is true). That is shown in the first line of the truth table. Lines 3 and 4 of the truth table define $p \rightarrow q$ to be true when p is false, no matter what truth value q has. This is perfectly reasonable. The guarantee is always valid when the car does not break down (p is false). In this case it does not matter whether the mechanic will fix the car.

Lines 3 and 4 of the truth table in Table 1.5.6 are called the **vacuously true cases**. In other words, the vacuously true cases occur when the hypothesis is false. To illustrate this idea further, consider the statement "If I play with my new tennis racket, then I'll win the match." The person making this statement cannot be called a liar, no matter how the match turns out, as long as that person plays with the old racket.

Comparing the implication with a concept of set theory is also useful. Recall that a set A is a subset of B whenever every element of A is also an element of B. This can be concisely written using an implication:

$A \subseteq B$ whenever, for every element x, $x \in A \rightarrow x \in B$.

A statement that is true because it satisfies a vacuously true case is said to be a **vacuously true statement**. For example, the statement $\emptyset \subseteq A$ is true for every set A precisely because it is vacuously true.

PRACTICE PROBLEM 1 Explain why the statement $\emptyset \subseteq A$ is vacuously true.

In order to get a better sense of the meaning of each of these statement connectives, especially the implication, we consider some examples.

EXAMPLE 1.5.1 Let $A = \{a, b, c\}$, $B = \{b, c\}$, and $C = \{a, e, f\}$. Then the statements in Table 1.5.7 have the truth values given in the table.

■ 1.5 INTRODUCTION TO LOGIC

TABLE 1.5.7

STATEMENT	TRUTH VALUE
$b \in A \land b \in B$	True
$c \in B \lor c \in C$	True
$a \in A \land a \in B$	False
$f \notin B \land f \in C$	True
$a \in A \to a \in C$	True
$a \in A \to a \in B$	False
$a \in B \to a \in C$	(Vacuously) True
If $e \in B$, then $e \in A$.	(Vacuously) True

PRACTICE PROBLEM 2 Refer to Example 1.5.1. Supply the truth value for each of the following statements.

a. If $e \in C$, then $e \in A$.   **b.** $f \in A \cap C \to f \in C$.   **c.** $a \in A \to a \notin C$.

The implication $q \to p$ is the **converse** of $p \to q$. An example will serve to illustrate the distinction between an implication and its converse. Let q be "it is cloudy" and p be "it is raining." Then $p \to q$ is "If it is raining, then it is cloudy." The converse is "If it is cloudy, then it is raining." Here is another example from algebra. The implication "If a real number is negative, then its square is positive" is true. However, its converse "If the square of a real number is positive, then the real number is negative" is false.

Proving an Implication

The basic strategy for proving an implication $p \to q$ always includes the assumption that the hypothesis p is true. This is so because $p \to q$ is only false when p is true and q is false; it is true otherwise.

In Example 1.4.3 we proved that if the area of a right triangle is equal to the square of the hypotenuse divided by 4, then the triangle is isosceles. The first step in the proof is to assume the area is equal to $c^2/4$, where c is the hypotenuse of the right triangle.

In order to apply these logical principles in a specific mathematical context, we assume the standard rules of algebra for the real numbers and the following five rules for "less than" ($<$) in the real numbers. We can write $x < y$ and $y > x$ interchangeably.

RULES FOR "LESS THAN" IN THE REAL NUMBERS

A1. If $x < y$ and $y < z$, then $x < z$.
A2. Exactly one of the following is true: $x < y$, $y < x$, or $x = y$.
A3. If $x < y$, then $x + z < y + z$.
A4. If $x < y$ and $z > 0$, then $xz < yz$.
A5. If n is any positive integer, then $n > 0$.

EXAMPLE 1.5.2 Prove: If $a < 3$, then $a + 2 < 5$.

Proof Assume $a < 3$. By rule A3, $a + 2 < 3 + 2$. But $3 + 2 = 5$. Hence, $a + 2 < 5$. Therefore, $a < 3$ implies $a + 2 < 5$. ∎

As we know, the implication $p \rightarrow q$ is read "If p, then q." Two alternative ways of reading $p \rightarrow q$ are "q if p" and "p only if q."

The Biconditional

A theorem of the form "p if and only if q" is asserting both that "p implies q" and "q implies p." For "p if and only if q," we sometimes say "If q, then p, and conversely."

The statement connective **IF AND ONLY IF**, abbreviated $\leftrightarrow$, is defined by the truth values in Table 1.5.8. The statement "p if and only if q" is also called the **biconditional**.

TABLE 1.5.8

p	q	IF AND ONLY IF $p \leftrightarrow q$
T	T	T
T	F	F
F	T	F
F	F	T

EXAMPLE 1.5.3 Let $A = \{a, b, c\}$, $B = \{b, c\}$, and $C = \{a, c, b\}$. Then the statements in Table 1.5.9 have the truth values as given in the table.

TABLE 1.5.9

STATEMENT	TRUTH VALUE
$b \in A \leftrightarrow b \in B$	True
$f \in B \leftrightarrow f \in C$	True
$a \in A \leftrightarrow a \in B$	False

∎

As we will see in Chapter 5, to prove a biconditional we must prove two implications. In other words, to prove a statement $p \leftrightarrow q$, we must prove $p \rightarrow q$ and $q \rightarrow p$. In proofs we will frequently label the proof of "p implies q" with "($\rightarrow$)" and the proof of "q implies p" with "($\leftarrow$)." Since $p \rightarrow q$ and $q \rightarrow p$ are converses of each other, we must prove an implication and its converse to prove a biconditional.

EXAMPLE 1.5.4 Prove $a < 3$ if and only if $a + 2 < 5$.

Proof ($\rightarrow$) Prove $a < 3$ implies $a + 2 < 5$. This proof is given in Example 1.5.2.
($\leftarrow$) Conversely, we prove $a + 2 < 5$ implies $a < 3$. Assume $a + 2 < 5$. Then $(a + 2) + (-2) < 5 + (-2)$ by Rule A3. Since $5 + (-2) = 5 - 2 = 3$ and $(a + 2) + (-2) = a + (2 - 2) = a$, we obtain $a < 3$. ∎

PRACTICE PROBLEM 3 Prove: If $a < b$ and $u < 0$, then $a + u < b$.

EXAMPLE 1.5.5 Prove $x < y$ if and only if $x + z < y + z$.

1.5 INTRODUCTION TO LOGIC

Proof ($\rightarrow$) This is Rule A3.

($\leftarrow$) Assume $x + z < y + z$. Then $(x + z) + (-z) < (y + z) + (-z)$ by Rule A3. Hence $x < y$. ∎

EXERCISE SET 1.5

Determine whether each of the sentences in Exercises 1–4 is a statement. If it is a statement, determine the truth value of the statement.

1. **a.** $3 = 1 + 2$ **b.** If $3 = 2 + 2$, then $1 = 0$.
2. **a.** If $x = 3$, then $x + 2 = 5$. **b.** If $x = 3$, then $x + 2 = 5$, for all real numbers x.
3. **a.** $(3 = 1 + 2) \wedge (3 < 7)$ **b.** If $3 < 2 + 2$, then $1 = 0$.
4. **a.** $(x = 3) \vee (x + 2 = 5)$ **b.** $(x = 3) \vee (x + 2 = 5)$, for all real numbers x
5. Verify that the sentence "This sentence is false" is not a statement because asserting it is true (or false) leads to a logical paradox.
6. Explain how the pick-a-point method discussed in Section 1.4 is justified by the method of proving an implication discussed in this section.
7. Let $A = \{x, y, z, w\}$, $B = \{z, w\}$, $C = \{x, y, t, a\}$, and $D = \{a, b, c\}$. Fill in the truth values of the statements in Table 1.5.10.

TABLE 1.5.10

STATEMENT	TRUTH VALUE
$x \in A \wedge t \in B$	
$w \in B \vee k \in C$	
$y \in A \wedge z \in B$	
$f \notin B \wedge a \in D$	
$a \in D \rightarrow g \in C$	
$x \in A \rightarrow a \in B$	
$z \in B \rightarrow a \in D$	
If $h \in D$, then $e \in A$.	
$x \in B \leftrightarrow x \in A$	

8. Let $A = \{a, b, c\}$, $B = \{b, c\}$, and $C = \{a, e, f\}$. Fill in the truth values of the statements in Table 1.5.11.

TABLE 1.5.11

STATEMENT	TRUTH VALUE
$c \in B \vee j \in C$	
$b \in A \wedge a \in B$	
$f \notin B \wedge f \in C$	
$a \in A \leftrightarrow a \in C$	
$r \in A \rightarrow b \in B$	
$b \in B \rightarrow f \in C$	
If $e \in B$, then $e \in A$.	

Decide whether each implication given in Exercises 9–12 is true. Is its converse true?

9. If a real number x is negative, then x^3 is negative.
10. If a real number x is positive, then $x < x^2$.
11. If a real number $x > 1$, then $x < x^2$.
12. If a real number x is positive, then $\sqrt{x} < x$.

Give a proof in Exercises 13–20.

13. $-1 < 0$
14. If $x < y$ and $u < v$, then $x + u < y + v$.
15. $v < 0$ if and only if $0 < -v$
16. If $x < y$ and $v < 0$, then $yv < xv$.
17. If $x > 1$, then $x^2 > 1$.
18. If $x < -1$, then $x^2 > 1$.
19. If $0 < x$ and $x < y$, then $x^2 < y^2$.
20. If $x < 0$ and $y < 0$ and $x < y$, then $y^2 < x^2$.

KEY TERMS

Set
Elements, or members, of a set
Ellipsis
Natural numbers, or nonnegative integers
Integers
Real numbers
Positive real numbers
Set-builder notation
Cardinality of S, Card(S)
Equality of two sets
Empty set, or null set
Singleton set
Subset
Proper subset
Intersection ∩
Disjoint sets
Union ∪
Universal set
Complement of a set
Difference set
Venn diagram
Conjecture
Counterexample
Ordered pair
First coordinate
Second coordinate
Product of two sets
Cartesian product
Power set
Char (character)
Algorithm
Decision problem
Pseudocode
Prime, or prime number
Top-down design
Control structures
Boolean expressions
Statement block
Computer program
Hypothesis
Conclusion
Questions to answer while reading a proof
Pick-a-point method
Backwards/forwards method
Statement
Statement connectives AND, OR, NOT, IMPLIES
Implication
Conditional
Vacuously true case
Vacuously true statement
Converse
IF AND ONLY IF
Biconditional

REVIEW EXERCISES

1. Let $A = \{a, b, c\}$, $B = \{b, c\}$, and $C = \{a, e, f\}$. Determine which of the following statements are true.
   a. $c \in A$ b. $\{c\} \in A$ c. $\{c\} \subseteq A$ d. $B \subseteq A$

REVIEW EXERCISES

  e. $A \subseteq B$ f. $\emptyset \subseteq C$ g. $\emptyset \in B$ h. $A \cap C \subseteq B$

2. Let $A = \{1, 2, 3, 4\}$, $B = \{1, 4\}$, and $C = \{4, 7, 9, 10\}$. Determine which of the following statements are true.
  a. $4 \in B$ b. $\{4\} \in B$ c. $\{4\} \subseteq B$ d. $B \subseteq A$
  e. $A \subseteq B$ f. $\emptyset \subseteq B$ g. $\emptyset \in A$ h. $A \cap B \subseteq C$

3. Let $A = \{a, b, c, d, e\}$, $U = \{a, b, c, d, e, 1, 2, 3\}$. Find:
  a. A' b. $A - U$ c. $U - A'$ d. $A \cap U$ e. U'

4. Let $C = \{1, 2, 3, 4, 5, 6\}$ and $U = \{1, 2, 3, 4, 5, 6, 7, 8, 9\}$. Find:
  a. C' b. U' c. $U - C'$ d. $C' \cap U$ e. $C \cup C'$

5. Let $A = \{a\}$, $B = \{b, c\}$, $C = \{d, e, f\}$, and $D = \{g, h, i, j\}$. Find:
  a. $P(\emptyset)$, $P(A)$, and $P(B)$ b. $P(C)$ and $\text{Card}(P(C))$
  c. $P(D)$ and $\text{Card}(P(D))$ d. $B \times C$ and $\text{Card}(B \times C)$
  e. $A \times P(B)$ and $\text{Card}(A \times P(B))$

6. Which of the following statements are true for all sets A, B, and C?
  a. $A \in P(A)$ b. $A \cap B \cap C \subseteq B$ c. $\emptyset \subseteq B \cap C$

7. Which of the following statements are true for all sets E, F, and G?
  a. $E \cap F \subseteq E \cup G$ b. $P(F) = F$ c. $\emptyset \in P(E)$

8. Which of the following statements is true for all sets A, B, and C? If the statement is false, give a counterexample.
  a. If $A = B$ and $B = C$, then $A = C$.   b. If $A \neq B$ and $B \neq C$, then $A \neq C$.

9. Which of the following statements is true for all sets A, B, and C?
  a. If $A \subset B$ and $B \subseteq C$, then $A \subset C$.   b. If $A \in B$ and $B \subseteq C$, then $A \subseteq C$.

10. Design an algorithm to compute and output the distance between two points in the xy-plane. You may input the points as ordered pairs of real numbers. Write the algorithm in pseudocode.

11. Design an algorithm to compute and output the square root of the product of the real numbers in a sequence A if the product is nonnegative; otherwise the algorithm should return the message "Square root not real." You may use the function SQRT(x), which returns the square root of any nonnegative real number x. Write the algorithm in pseudocode.

12. Write an algorithm in pseudocode to compute and output the alternating sum of the reciprocals of the first n positive integers,
$$1 - \frac{1}{2} + \frac{1}{3} - \cdots + (-1)^{n+1}\left(\frac{1}{n}\right)$$

13. Use the pick-a-point method to prove $A \triangle B = (A \cup B) - (A \cap B)$. (See Exercise 24 in Section 1.2 for the definition of $A \triangle B$.)

14. Refer to Practice Problem 2 in Section 1.2. Use the pick-a-point method to prove $D \cap (A \cup B) \subseteq D \cup (A \cap B)$.

15. Prove: If $x > 1$, then $x < x^2$.

16. Prove: If $x < 0$, then $x < x^2$.

17. Prove $(x - y)^2 = 0$ if and only if $x = y$. [*Hint:* You may assume that, if $a^2 = 0$, then $a = 0$.]

18. Prove $(x + y)^2 - (x - y)^2 = 0$ if and only if $xy = 0$.

REFERENCES

Eves, H. *An Introduction to the History of Mathematics*, 5th ed. New York: Saunders, 1983.

Halmos, P. R. *Naive Set Theory*. New York: Springer-Verlag, 1974.

Knuth, D. E. *The Art of Computer Programming: Volume 1, Fundamental Algorithms*, 2nd ed. Reading, MA: Addison-Wesley, 1973.

Lipschutz. S. *Theory and Problems of Set Theory and Related Topics*. New York: Schaum, 1964.

Nance, D. W. *Pascal: Understanding Programming and Problem Solving*. St. Paul, MN: West, 1986.

Solow, D. *How to Read and Do Proofs*. New York: Wiley, 1982.

Stoll, R. R. *Sets, Logic, and Axiomatic Theories*, 2nd ed. New York: Freeman, 1974.

CHAPTER 2

DIRECTED GRAPHS AND RELATIONS

Directed graphs arise naturally in studying maps of transportation routes. Applications of directed graphs occur in many fields, including computer science, electrical engineering, economics, and operations research. In computer science, directed graphs are used in areas such as computer design, systems analysis, data structures, and program optimization.

In addition, we explore the concept of binary relations and their connection with directed graphs. The important issue of reachability—"Can we get there from here?"—will also be discussed.

APPLICATION

Developing efficient telecommunications networks has become an important part of computer science research. The figure shows the well-known

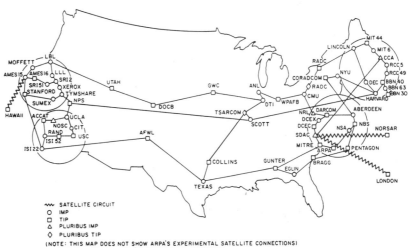

ARPANET GEOGRAPHIC MAP, SEPTEMBER 1979

Advanced Research Project Agency Network (ARPANET), a network developed to interconnect computers at widely separated ARPA-sponsored research projects. ARPANET itself provides an environment for research in network design, resource sharing, and satellite communications. Each ARPANET node represents a computer facility; each line segment represents a transmission link.

The complete standstill of all or part of a communications network is called a deadlock. Detecting potential deadlocks in a telecommunications network is important for network management. Preferably, potential deadlocks are detected in the network design stage in order to avoid pitfalls. How do we detect possible deadlocks? What method can identify nodes in a network where deadlocks could occur? We discuss these questions in Section 2.6.

2.1
DIRECTED GRAPHS

As a first example of a directed graph, consider a diagram of an oil refinery showing pipelines between various facilities v_1, v_2, v_3, v_4, and v_5 of the refinery. See Figure 2.1.1. Vertices represent facilities, and edges represent pipelines. The arrows on the edges indicate the direction of the oil flow.

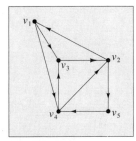

FIGURE 2.1.1

The idea of a diagram of pipelines between locations generalizes to the concept of a directed graph. In a directed graph, each dot represents a vertex, and a directed line segment between two vertices represents an edge.

DEFINITION

A **directed graph** $G = (V, E)$ consists of a nonempty finite set V of **vertices** (**nodes**) and a finite set E of **edges** (**arcs**) where each edge is an ordered pair of vertices. In an edge (v, w), v is called the **initial vertex** and w the **terminal vertex**; we say the edge (v, w) is "from v to w." Order is important in the description of an edge; for $v \neq w$, the edges (v, w) and (w, v) are different. For simplicity, a directed graph is usually called a **digraph**.

EXAMPLE 2.1.1 Let $V = \{a, b, c\}$ and $E = \{(a, b), (a, c), (b, c), (c, b)\}$. Figure 2.1.2 shows a picture of the digraph $G = (V, E)$. For example, the directed line segment from a to b represents the edge (a, b), and the directed curve from b to c represents the edge (b, c). ∎

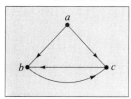

FIGURE 2.1.2

In the execution of computer programs, the sequential flow of control and of data transformation is important. A convenient device for analyzing sequential execution is a directed graph, where the edge direction indicates the order of execution.

EXAMPLE 2.1.2 Figure 2.1.3 (page 42) is a flowchart for a computer program to add 100 numbers, specified as $A(1), A(2), \ldots, A(100)$, and to calculate their average. This flowchart is an example of a digraph. The control, assignment, and decision statements are shown as vertices. The edges indicate the order and direction of execution.

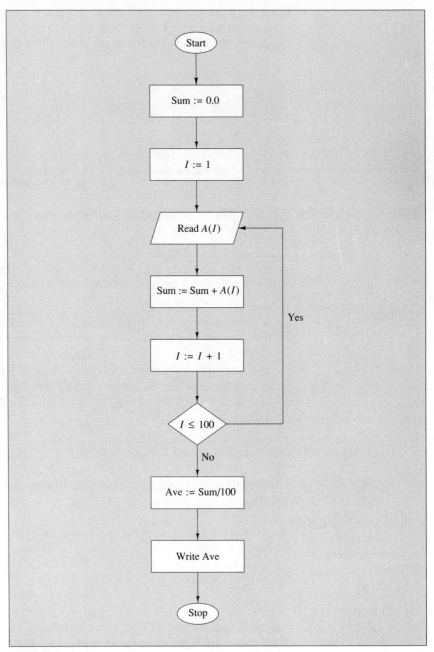

FIGURE 2.1.3

2.1 DIRECTED GRAPHS

A **path** from v_0 to v_n in a digraph is a finite sequence of edges $((v_0, v_1), (v_1, v_2), \ldots, (v_{i-1}, v_i), \ldots, (v_{n-1}, v_n))$. The vertex v_0 is called the **initial vertex** of the path and the vertex v_n the **terminal vertex** of the path. We also say the path is "from v_0 to v_n." The path $((v_0, v_1), (v_1, v_2), \ldots, (v_{n-1}, v_n))$ is called a **closed path at v_0** if $v_0 = v_n$. A path with n edges is said to have **length** n. We will assume that $n > 0$ and will not be concerned with any path of length 0.

A path with distinct edges is called a **simple path**. A closed simple path is called a **circuit**. If the initial vertices of all the edges in a circuit are distinct, the circuit is called a **cycle**. When it is important to identify the initial vertex v of a circuit or a cycle, we designate the circuit (or cycle) as the circuit at v (or cycle at v). See Figure 2.1.4 for a circuit at v that is not a cycle at v.

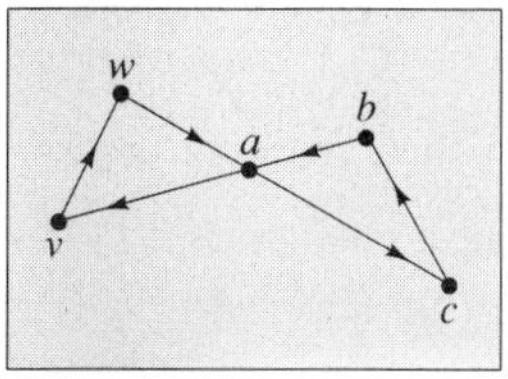

An edge of the form (v, v) is called a **self-loop**. A self-loop is a path of length 1 and is therefore both a circuit and a cycle. A self-loop should not be confused with the loop statement in a computer program. The concepts of path, closed path, circuit, cycle, and self-loop are illustrated by Example 2.1.3.

FIGURE 2.1.4

When there is no cause for confusion, we often express a path $((v_0, v_1), (v_1, v_2), \ldots, (v_{n-1}, v_n))$ in the form $(v_0, v_1, v_2, \ldots, v_{n-1}, v_n)$ by listing in order the initial vertices of all edges and the terminal vertex of the last edge of the path.

EXAMPLE 2.1.3 Let $V = \{v_1, v_2, v_3, v_4, v_5\}$ be a set of vertices, and let

$$E = \{(v_1, v_2), (v_2, v_2), (v_2, v_3), (v_3, v_1), (v_3, v_2), (v_3, v_5), (v_4, v_3), (v_4, v_5), \\ (v_1, v_4), (v_5, v_1), (v_5, v_5)\}$$

be a set of edges. The digraph (V, E) is shown in Figure 2.1.5.

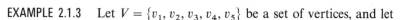

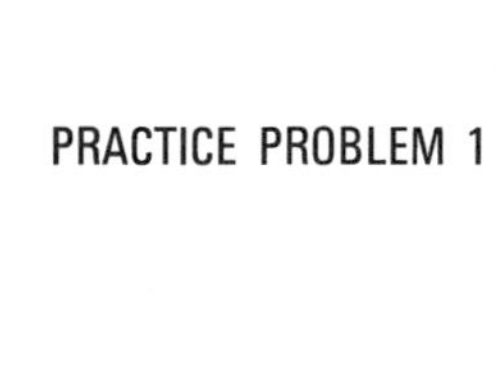

a. (v_2, v_2) and (v_5, v_5) are self-loops.
b. $((v_2, v_3), (v_3, v_1), (v_1, v_2))$ is a cycle at v_2 of length 3.
c. $((v_3, v_2), (v_2, v_2), (v_2, v_3))$ is a circuit of length 3; it is not a cycle because the initial vertices v_3, v_2, v_2 are not all distinct.
d. $((v_4, v_3), (v_3, v_5), (v_5, v_1), (v_1, v_4))$ is a cycle at v_4 of length 4.
e. $((v_1, v_2), (v_2, v_2), (v_2, v_3), (v_3, v_2), (v_2, v_3), (v_3, v_1))$ is a closed path but it is neither a circuit nor a cycle at v_1. ∎

FIGURE 2.1.5

PRACTICE PROBLEM 1 Refer to Figure 2.1.5.

a. Find a cycle of length 4 at v_3.
b. Find a circuit of length 5 at v_4.

PRACTICE PROBLEM 2 Construct a digraph containing a circuit that is not a cycle and that contains no self-loops.

In a complicated digraph, finding a path between two given vertices is an intricate problem. This task is known as the **reachability problem**, and it seeks to answer the crucial question "Can we get there from here?" In Section 2.5, we will investigate the reachability problem in more detail. A systematic method of solving the reachability problem also gives us a method for detecting closed paths, since the existence of a closed path from vertex v back to vertex v is an affirmative answer to the question "Can we get back to v after departing from v?"

A vertex w is **reachable** from a vertex v in a digraph G if $v = w$ or if there is some path in G from v to w. Thus, every vertex is reachable from itself. In the digraph of Figure 2.1.5, v_5 is reachable from v_1 by the path $((v_1, v_2), (v_2, v_3), (v_3, v_5))$ of length 3 and by the path $((v_1, v_4), (v_4, v_5))$ of length 2. Is v_5 reachable from v_1 by any other paths?

PRACTICE PROBLEM 3 In the digraph of Figure 2.1.5, find at least two paths in G from v_2 to v_4.

Project Networks

Managing a highly complex industrial or scientific project for efficient operation is an extremely difficult task. Often, the best approach is to divide the project into smaller activities that are more easily controlled, and then order these activities into a suitable network. Roughly speaking, a **project network** is a digraph for describing activities and their relationships. An activity starting at a vertex v cannot begin until all activities ending at vertex v are completed. The "flow" through this network is measured in consumption of limited resources, such as time or money. The project network is an extremely useful tool in systems analysis and project management. We will describe a very simple example of a project network.

EXAMPLE 2.1.4 **The Manufacturing Problem** A manufacturing operation is devoted to producing a widget. The completed widget has five parts, $S_1, S_2, S_3, S_4,$ and S_5. These may be fabricated independently, subject to the following constraints:

S_2 may not be started until S_1 is completed.
S_3 may be started at any time but requires both S_1 and S_2 for completion.
S_4 may be started at any time but requires S_3 for completion.
S_5 may be started only after S_1 through S_4 are completed.

A widget is produced when all parts are completed.

Activities: S_i indicates the production of part S_i.
Dummy activities: A indicates the finish of S_3 and the start of C.
  B indicates the finish of S_2 and the start of C.
  C indicates the finish of A and B and the start of S_5.
  D indicates the finish of S_4 and the start of S_5.

The digraph in Figure 2.1.6 is a project network illustrating how the various activities are related. Activities are represented by directed line segments. The beginning and the ending of an activity are represented by vertices.

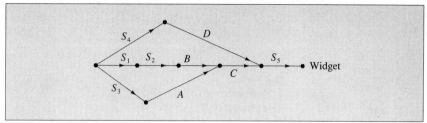

FIGURE 2.1.6

PROGRAMMING NOTES

A Flowchart Describing a Loop Statement in a Computer Program Digraphs as flowcharts are useful for understanding the syntax of a loop statement in a computer program. The digraph in Figure 2.1.7 illustrates the structure of a "for" loop. Assume that I is an integer and that the initial value for I is less than or equal to the final value. We note that each statement in the flowchart represents a vertex and each directed line segment in the flowchart represents an edge in the digraph.

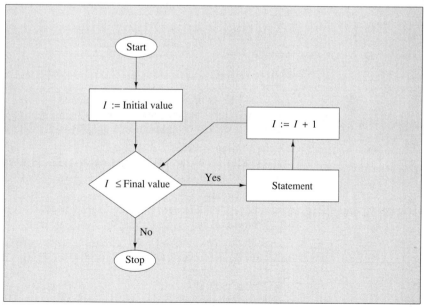

FIGURE 2.1.7

A Program Fragment Part of a Pascal program implementing the flowchart in Example 2.1.2 is shown at the top of the next page. We use indentations in writing the program fragment to show the important structure of the program.

```
Sum := 0.0;
for I := 1 to 100 do
   begin
      read(A[I]);
      Sum := Sum + A[I]
   end
Ave := Sum/100.0;
writeln(Ave)
```

EXERCISE SET 2.1

1. Write the set of vertices V and the set of edges E for the digraph in Figure 2.1.8.

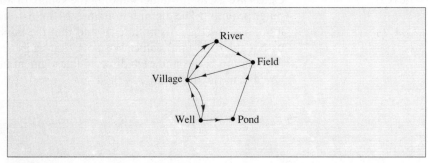

FIGURE 2.1.8

2. Refer to Figure 2.1.8.
   a. Find the set of vertices that are reachable from the vertex Village.
   b. Find all possible paths of length less than 6 from Village to Field.
   c. Find a cycle of length 4 from Village back to Village.

3. Refer to Figure 2.1.8.
   a. Find the set of vertices that are reachable from Well.
   b. Find all possible paths of length less than 5 from Well to Field.

4. Refer to Figure 2.1.8.
   a. Find a circuit of length 6 from Well back to Well such that this circuit is not a cycle.
   b. Find a cycle of length 4 from Well back to Well.

5. Consider a digraph $G = (V, E)$, where $V = \{a, b, c, d\}$ and $E = \{(a, a), (a, b), (b, d), (c, a), (c, c), (d, a), (d, c), (d, d)\}$.
   a. Draw the digraph.
   b. Find a path of length 3 from a to c with all distinct vertices.
   c. Find the set of all vertices reachable from a.

6. Refer to Exercise 5.
   a. Find the set of all vertices reachable from b.
   b. Find the set of all vertices reachable from c.
   c. Is there a circuit that is not a cycle in this digraph? If so, specify it.

7. Consider a digraph $G = (V, E)$, where $V = \{a, b, c, d\}$ and $E = \{(b, a), (a, b), (b, c), (c, a)\}$.

2.1 DIRECTED GRAPHS

    a. Draw the digraph.
    b. Find the set of all vertices reachable from a.
    c. Is there a circuit that is not a cycle in this digraph? If so, specify it.

8. Consider the digraph G in Figure 2.1.9.
    a. List the set V of all vertices of G.
    b. List the set E of all edges of G.
    c. List all paths of length 3 starting from vertex v_6.

9. Refer to Figure 2.1.9.
    a. Find a closed path of length 4 starting at vertex v_2.
    b. Find a cycle of length 5 starting at vertex v_1.
    c. Find a circuit that is not a cycle at vertex v_4.

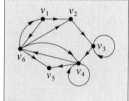

FIGURE 2.1.9

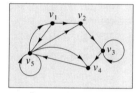

FIGURE 2.1.10

10. Consider the digraph in Figure 2.1.10.
    a. List all paths of length 3 starting from v_2.
    b. Find a closed path of length 4 that is not a circuit at v_5.
    c. Find a cycle of length 4 at v_5.
    d. Find a circuit that is not a cycle starting at v_1.

11. Consider the digraph in Figure 2.1.11.
    a. Find a closed path at v_1.
    b. List all the vertices that are not reachable from any other vertices.
    c. Find a cycle of length 4 at v_2.

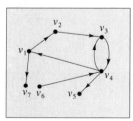

FIGURE 2.1.11

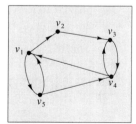

FIGURE 2.1.12

12. Consider the digraph in Figure 2.1.12.
    a. Find a closed path of length 9 at v_5.
    b. List all the vertices that are not reachable from any other vertices.
    c. Find circuits of lengths 4, 5, and 6 at v_3.
    d. Which of the circuits found in part c are also cycles?

13. True or false? If there is a path from v to w in a digraph G, then there is a simple path from v to w in G. Supply a convincing argument with an answer of true or an example with an answer of false.

14. True or false? If there is a circuit from v to v in a digraph G, then there is a cycle from v to v in G. Supply a convincing argument with an answer of true or an example with an answer of false.

15. Mother, father, and child want to prepare toast with butter and jelly for breakfast. The father will get the bread from the bread box and toast the bread. The child will get the jelly from the pantry and spoon the jelly onto the buttered toast. The mother will get the butter from the refrigerator and spread butter on the hot toast. Construct a project network to illustrate how various activities in the breakfast project are related. Describe and label all activities involved.

16. Construct a project network to describe the following manufacturing operation. The operation has six suboperations S_1, S_2, S_3, S_4, S_5, and S_6; it is finished when S_6 is completed. The suboperations may be carried out independently, subject to the following constraints:

    S_2 and S_3 may be started after S_1 is completed.

    S_4 may be started at any time after S_1 is completed, and it requires S_3 for completion.

    S_5 may be started after S_2 is completed, and it requires S_4 for completion.

    S_6 may be started only after S_5 is completed.

2.2
RELATIONS

Digraphs are useful tools for analyzing relations among objects. We will discuss the correspondence between digraphs and binary relations after we introduce the concept of binary relation.

Data such as names, addresses, Social Security numbers, occupations, and number of dependents can be effectively described by a relational model. The relational model is based on a mathematical concept called *relation*, which has proved to be very useful in the theory of data base management.

Before defining binary relation, we give two examples.

EXAMPLE 2.2.1 Mr. Smith is the father of Tommy and Amy, Mrs. Jones is the mother of Julie, and Mr. Moore is the grandfather of James. To express the special relationships of these people, we construct a set of ordered pairs

$R = \{$(Mr. Smith, Tommy), (Mr. Smith, Amy), (Mrs. Jones, Julie),

(Mr. Moore, James)$\}$

In the set R, the second member in each ordered pair is a descendant of the first member. The set R is a binary relation from the set

$A = \{$Mr. Smith, Mrs. Jones, Mr. Moore$\}$

to the set

$B = \{$Tommy, Amy, Julie, James$\}$

Note that R is a subset of the product set $A \times B$.

EXAMPLE 2.2.2 Let $A = \{4, 6\}$ and $B = \{2, 8, 12\}$. Then

$$A \times B = \{(4, 2), (4, 8), (4, 12), (6, 2), (6, 8), (6, 12)\}$$

The set $R = \{(4, 8), (6, 12)\}$ is a binary relation from A to B. Note that $R \subseteq A \times B$. We observe that an ordered pair $(a, b) \in A \times B$ is a member of R whenever $b = 2a$. Hence, the set of ordered pairs R is a way to specify the relationship $b = 2a$. ∎

The relations in Examples 2.2.1 and 2.2.2 are binary relations, which are relations between two sets of objects.

DEFINITION

> Given nonempty sets A and B, a **binary relation** (or **relation**) R from A to B is any subset of $A \times B$. That is, $R \subseteq A \times B$. When $A = B$, we say R is a binary relation on A. In other words, a binary relation on A is a set of ordered pairs of elements from A. If $(a, b) \in R$, we can write $a\, R\, b$, and conversely.

EXAMPLE 2.2.3 Let $A = \{a, b, c, d\}$ and $B = \{x, y, t\}$.

  a. The set $R = \{(a, x), (b, t), (d, y), (c, x)\}$ is a binary relation from A to B.
  b. The set $A \times B$ is also a binary relation from A to B.
  c. The set $S = \{(a, b), (a, c), (c, d), (d, b), (d, d)\}$ is a binary relation on A.
  d. The set $T = \{(x, y), (y, t), (t, x)\}$ is a binary relation on B. ∎

EXAMPLE 2.2.4 Let $A = B = \mathbb{R}$. Thus, $A \times B$ is the Cartesian plane $\mathbb{R} \times \mathbb{R}$. The following subsets R, S, and T of $\mathbb{R} \times \mathbb{R}$ are binary relations on $\mathbb{R}$:

$$R = \{(x, y) : x = y\}$$
$$S = \{(x, y) : |x| + |y| \leq 1\}$$
$$T = \{(x, y) : y = 1.5\}$$

The relations R, S, and T are shown as subsets of the xy-plane in Figure 2.2.1.

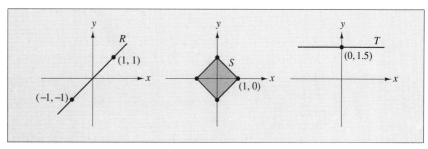

FIGURE 2.2.1 ∎

In order to discuss relations among more than two sets, we introduce the notions of *n*-tuples, the product of *n* sets, and the *n*-ary relation. These are generalizations of ordered pairs, the product of two sets, and the binary relation, respectively. The concept of *n*-ary relation is particularly useful in the relational data base model, which we will discuss shortly.

An **n-tuple** $(a_1, a_2, \ldots, a_n)$ is an ordered list of *n* elements, which are called the **coordinates** of the *n*-tuple. We write a 1-tuple (a_1) more simply as a_1. A 2-tuple is an ordered pair. As with ordered pairs, *n*-tuples are equal whenever all their corresponding coordinates are identical. For example, $(x, y, z) = (2, -5, 7)$ if and only if $x = 2$, $y = -5$, and $z = 7$.

The **product** or **Cartesian product** of *n* sets $A_1, A_2, \ldots, A_n$ (not necessarily distinct), written $A_1 \times A_2 \times \cdots \times A_n$, is the set of all *n*-tuples with the *j*th coordinate an element of A_j for each j, $1 \leq j \leq n$.

Given *n* nonempty sets $A_1, A_2, \ldots, A_n$ (not necessarily distinct), an **n-ary relation** R on the *n* sets $A_1, A_2, \ldots, A_n$ is a subset of the product $A_1 \times A_2 \times \cdots \times A_n$. Hence, each element of an *n*-ary relation R is an *n*-tuple $(a_1, a_2, \ldots, a_n)$, where $a_1 \in A_1, a_2 \in A_2, \ldots, a_n \in A_n$. The integer *n* is called the **degree** of the relation. The relation R is called a **unary relation** when $n = 1$, a **binary relation** when $n = 2$, and a **ternary relation** when $n = 3$. An *n*-ary relation on $A_1, A_2, \ldots, A_n$ is sometimes called a **table**. See Example 2.2.7 for a discussion of tables.

If R is a subset of $A^n = A \times A \times \cdots \times A$ (*n* times), then we call R an *n*-ary relation on A.

EXAMPLE 2.2.5 Let $A = \{a, b, c\}$, $B = \{m, n\}$, and $C = \{x, y, t, w\}$.

a. The set $R = \{(a, m, x), (b, m, y), (a, m, t), (c, m, w)\}$ is a ternary relation on the sets A, B, and C.
b. The set $T = \{(a, a, a, a), (b, a, a, c), (a, c, b, a)\}$ is a 4-ary relation on A.
c. The set $S = \{x, y, w\}$ is a unary relation on C. ■

EXAMPLE 2.2.6 Let $A_1 = A_2 = A_3 = \mathbb{R}$. The subset $R \subseteq \mathbb{R} \times \mathbb{R} \times \mathbb{R}$, where $R = \{(x, y, z): |x| \leq 1, |y| \leq 1, z = 0\}$, is a ternary relation on $\mathbb{R}$. The relation R is pictured in Figure 2.2.2. R is a square on the *xy*-plane in 3-dimensional space (or 3-space).

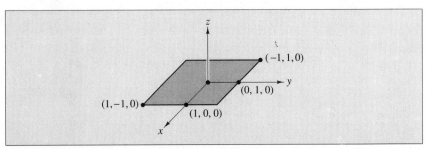

FIGURE 2.2.2 ■

■ 2.2 RELATIONS

PRACTICE PROBLEM 1   Let $R \subseteq \mathbb{N} \times \mathbb{N} \times \mathbb{N}$ be a ternary relation, where $R = \{(x, y, z): x^2 + y^2 = z^2\}$. Which of the triples (1, 1, 2), (3, 4, 5), (4, 5, 5), (0, 5, 5), (8, 6, 10) belong to R?

A Relational Model for Data Bases

As an application of relations, we give a brief introduction to a relational model for data bases. Computer systems are capable of handling large amounts of data. Investment account transactions for clients in a brokerage firm, inventory records of a chain of department stores, time schedules of course offerings at a university, or personnel data in a large company are a few examples.

These data must be organized in a way that can be accessed, updated, deleted, and searched effectively. One general way to represent data is by an n-ary relation on $A_1 \times A_2 \times \cdots \times A_n$, which can be visualized as a table.

EXAMPLE 2.2.7   Let

$A_1 = \text{Course} = \{\text{C.S. 101, C.S. 102, C.S. 408, C.S. 514}\}$

$A_2 = \text{Credit} = \{1, 2, 3, 4, 5, 6, 7, 8, 9, 10\}$

$A_3 = \text{Time} = \{8\text{:}00 \text{ A.M.}, 9\text{:}00 \text{ A.M.}, 10\text{:}00 \text{ A.M.}, 11\text{:}00 \text{ A.M.}, 1\text{:}00 \text{ P.M.},$
$\qquad\qquad\qquad\qquad 2\text{:}00 \text{ P.M.}, 3\text{:}00 \text{ P.M.}, 4\text{:}00 \text{ P.M.}, 5\text{:}00 \text{ P.M.}\}$

$A_4 = \text{Location} = \{\text{Albert Hall, Darwin Hall, Eckhart Hall}\}$

Table 2.2.1 shows a 4-ary relation called Schedule on A_1, A_2, A_3, and A_4.

TABLE 2.2.1 SCHEDULE

COURSE	CREDIT	TIME	LOCATION
C.S. 101	4	9:00 A.M.	Albert Hall
C.S. 102	3	11:00 A.M.	Darwin Hall
C.S. 408	3	3:00 P.M.	Darwin Hall
C.S. 514	3	5:00 P.M.	Eckhart Hall

We may express our relation Schedule as a subset of $A_1 \times A_2 \times A_3 \times A_4$. Each row in our table is a member of the relation. For example, the ordered 4-tuple (C.S. 408, 3, 3:00 P.M., Darwin Hall) is a member of Schedule. In this example, it is more convenient to display Schedule as a table.

If we require only information about the course number, time, and location, then we can take a subtable of Schedule as shown in Table 2.2.2. This

TABLE 2.2.2 $P_{134}(\text{Schedule})$

COURSE	TIME	LOCATION
C.S. 101	9:00 A.M.	Albert Hall
C.S. 102	11:00 A.M.	Darwin Hall
C.S. 408	3:00 P.M.	Darwin Hall
C.S. 514	5:00 P.M.	Eckhart Hall

subtable is called a projection of Schedule and uses only the first, third, and fourth columns from our table Schedule. This projection of Schedule is denoted P_{134}(Schedule). ∎

PRACTICE PROBLEM 2 Refer to Example 2.2.7. Find the projection P_{12}(Schedule).

In general, if R is an n-ary relation on $A_1, A_2, \ldots, A_n$, then a **projection** of R is a k-ary relation, $k \leq n$, obtained from R by keeping a specified and fixed subset of k components from each n-tuple in R. If $i_1, i_2, \ldots, i_k$ are the desired k indices among $1, 2, \ldots, n$, then the projection of R is denoted by $P_{i_1 i_2 \cdots i_k}(R)$. In P_{134}(Schedule) of Example 2.2.7, we have $i_1 = 1$, $i_2 = 3$, $i_3 = 4$, and $k = 3$.

Digraphs and Binary Relations

There is an important connection between a digraph and a binary relation. A binary relation R on a finite set A can be represented by a digraph, and conversely. The elements in A are the vertices in the digraph; and the set of ordered pairs R are the edges in the digraph. There is a directed line segment (or an edge) from vertex a to vertex b if $(a, b) \in R$. Conversely, if (V, E) is a digraph, then E is a set of ordered pairs of elements from V.

We state this connection between binary relations and digraphs as a theorem.

THEOREM 2.2.8 If R is a binary relation on a finite set A, then (A, R) is a digraph. Conversely, if (V, E) is a digraph, then E is a binary relation on V.

We illustrate Theorem 2.2.8 by the next example.

EXAMPLE 2.2.9 Let $A = \{2, 3, 4, 9, 12\}$, and let R be a binary relation on A defined by $R = \{(x, y) : x \in A \text{ and } y \in A \text{ and } y \text{ is a multiple of } x\}$. Therefore, $R = \{(2, 2), (2, 4), (2, 12), (3, 3), (3, 9), (3, 12), (4, 4), (4, 12), (9, 9), (12, 12)\}$. This binary relation R is represented by the digraph in Figure 2.2.3. There are five vertices and ten directed line segments representing the ten ordered pairs in R.

Conversely, if the digraph in Figure 2.2.3 is given, we see that $A = \{2, 3, 4, 9, 12\}$ and that the binary relation R consists of ordered pairs corresponding to the set of edges in the digraph. ∎

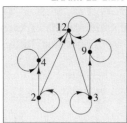

FIGURE 2.2.3

Properties of Binary Relations

A binary relation R on a set A may possess certain interesting properties. The four most important properties for a binary relation are the reflexive, symmetric, antisymmetric, and transitive properties.

2.2 RELATIONS

DEFINITION

Let R be a binary relation on a set A. The defining condition of each property is listed below:

PROPERTY OF R	DEFINING CONDITION
R is **reflexive**.	For all $x \in A$, $(x, x) \in R$.
R is **symmetric**.	For all $x, y \in A$, if $(x, y) \in R$, then $(y, x) \in R$.
R is **antisymmetric**.	For all $x, y \in A$, if $x \neq y$, then $(x, y) \notin R$ or $(y, x) \notin R$.
R is **transitive**.	For all $x, y, z \in A$, if $(x, y) \in R$ and $(y, z) \in R$, then $(x, z) \in R$.

Note that x, y, and z need not be distinct in the definitions above. Also, R is reflexive whenever $(x, x) \in R$ for *every* $x \in A$. For example, if $R = \{(a, a), (b, b)\}$, then R on $\{a, b\}$ is reflexive, but R on $\{a, b, c\}$ is not reflexive. The defining condition for R being antisymmetric may also be written in the following way. The relation R is antisymmetric if:

For all $x, y \in A$, $[(x, y) \in R$ and $(y, x) \in R]$ implies $x = y$.

It is important to note that each of the four properties is true only when the defining condition is true for *all* elements of the set A. Hence, a given property is false if the defining condition fails to hold for even one element of A.

Here are two examples to illustrate the four properties.

EXAMPLE 2.2.10 Let $R = \{(a, a), (c, c), (a, b), (b, a), (a, c)\}$ be a relation on the set $A = \{a, b, c\}$.

a. R is not reflexive since $(b, b) \notin R$.
b. R is not symmetric since $(a, c) \in R$ but $(c, a) \notin R$.
c. R is not antisymmetric since $(a, b) \in R$ and $(b, a) \in R$ but $a \neq b$.
d. R is not transitive since $(b, a) \in R$ and $(a, c) \in R$ but $(b, c) \notin R$. ∎

EXAMPLE 2.2.11 Let $R = \{(a, a), (b, b), (c, c), (b, a)\}$ be a relation on the set $A = \{a, b, c\}$.

a. R is reflexive because (a, a), (b, b), and (c, c) are in R.
b. R is not symmetric since $(b, a) \in R$ but $(a, b) \notin R$.
c. R is antisymmetric because (b, a) is the only ordered pair in R with unequal coordinates and $(a, b) \notin R$.
d. R is transitive because $[(b, a) \in R$ and $(a, a) \in R]$ implies $(b, a) \in R$ is true and so is every similar statement. ∎

Note that "antisymmetric" does not mean "not symmetric." A relation can be both symmetric and antisymmetric. It can be symmetric without being antisymmetric, and vice versa. Also, a relation can be neither symmetric nor antisymmetric. Since this concept may be confusing, a few concrete examples are helpful.

The following example illustrates some combinations of properties a given binary relation can possess.

EXAMPLE 2.2.12 Let $A = \{a, b, c\}$. Then:

a. $R = \{(a, a), (b, b), (c, c)\}$ is both symmetric and antisymmetric.
b. $S = \{(a, a), (a, c), (b, b), (c, a)\}$ is symmetric but is not antisymmetric. It is not antisymmetric because $(a, c) \in S$ and $(c, a) \in S$ but $a \neq c$.
c. $T = \{(a, a), (b, a), (b, b), (c, a), (c, b), (c, c)\}$ is antisymmetric, but it is not symmetric because $(c, a) \in R$ but $(a, c) \notin R$.
d. $U = \{(a, a), (a, c), (b, a), (b, b), (b, c), (c, b)\}$ is neither symmetric nor antisymmetric.
e. Both of the relations R and T are reflexive, whereas neither of the relations S and U is reflexive. ∎

PRACTICE PROBLEM 3 Let $A = \{1, 2, 3, 4\}$. Find binary relations R, S, and T on A so that:

a. R is reflexive but is neither antisymmetric nor symmetric.
b. S is antisymmetric but is neither reflexive nor symmetric.
c. T is both symmetric and antisymmetric but is not reflexive.

In what follows, we often use the word *relation* for *binary relation*.

Digraph Representation of the Four Properties

"R is reflexive" means that every vertex in the digraph for R has an edge from the vertex to itself. See Figure 2.2.4(a).

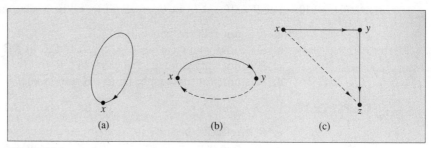

FIGURE 2.2.4

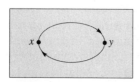

FIGURE 2.2.5

"R is symmetric" means that, for every edge (x, y) in the digraph, there must also be an edge (y, x) in the digraph as shown in Figure 2.2.4(b).

"R is transitive" means that, for every pair of edges (x, y) and (y, z) in the digraph, (x, z) must also be an edge in the digraph. See Figure 2.2.4(c).

"R is antisymmetric" means *no* distinct vertices x and y can have both the edge (x, y) and the edge (y, x) in the digraph. That is, Figure 2.2.5 is not part of the digraph of an antisymmetric relation.

■ 2.2 RELATIONS

The digraph pictured in Figure 2.2.6 represents a relation on $A = \{1, 2, 3, 4\}$ that has none of the four properties defined earlier (page 53).

PRACTICE PROBLEM 4 Use the digraph in Figure 2.2.6 to explain why the relation $R = \{(1, 1), (2, 2), (1, 2), (2, 3), (3, 2), (4, 1), (4, 2)\}$ on $A = \{1, 2, 3, 4\}$ is not reflexive, not symmetric, not antisymmetric, and not transitive.

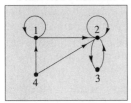

FIGURE 2.2.6

A relation R on a set A may lack one or more of the properties described in the definition on page 53. In what follows, we will be especially interested in augmenting a given relation R so that it becomes transitive. If R is not transitive, then we can try to add just enough ordered pairs to R so that the enlarged relation becomes transitive. This enlarged relation is called the *transitive closure of R*. A definition of the transitive closure of R is given below; other closures such as symmetric closure and reflexive closure can be similarly defined.

DEFINITION

> Let R be a binary relation on a set A. A **transitive closure** of R is a binary relation T such that
>
> **a.** R is a subset of T;
> **b.** T is transitive; and,
> **c.** If R is a subset of S and S is transitive, then T is a subset of S.

Note that there is at most one transitive closure of a relation R. For if both T and W are transitive closures of R, then T and W are transitive, $R \subseteq T$, and $R \subseteq W$. Hence, by part c of the definition, $T \subseteq W$. Similarly, $W \subseteq T$. Therefore, $W = T$.

It follows from the definition that the transitive closure of R is the *smallest* transitive relation containing R. For example, let $A = \{a, b, c\}$, $R = \{(a, b), (b, c)\}$, and $T = \{(a, b), (b, c), (a, c)\}$. We see that R is not transitive, but T is transitive. It turns out that T is the transitive closure of R. We will give more details and a geometric interpretation of transitive closure in the next section.

Does a transitive closure of a given relation always exist? Can we find it? The answer to both questions is yes for a relation R on a finite set A. This result will follow from our work in Sections 2.3–2.5.

PRACTICE PROBLEM 5 Refer to Example 2.2.12b. What edges need to be added to the digraph (A, S) to obtain the transitive closure of the relation S?

Here are two examples of relations that are reflexive and symmetric but not transitive.

EXAMPLE 2.2.13 **THE ACQUAINTANCE RELATION** Let A be a group of people. Let $R \subseteq A \times A$ such that $R = \{(a, b) : a \text{ is an acquaintance of } b\}$. R is called the **acquaintance**

relation. R is reflexive since $(a, a) \in R$, for each $a \in A$ knows himself or herself. R is symmetric for, if $a\ R\ b$ then $b\ R\ a$. Note that R is not necessarily transitive.

The host at a party where the acquaintance relation is not transitive has a problem. For example, among a, b, and c such that a knows b and b knows c, perhaps a does not know c. The host can introduce certain pairs of people until enough new acquaintances have been added to reach the transitive closure of the acquaintance relation. But how do we decide which pairs to introduce?

A systematic method for finding the transitive closure of any relation on a finite set will solve this problem. We will discuss the transitive closure and an algorithm for finding it in the remaining sections of this chapter. But, for now, finding the transitive closure of an arbitrary relation is a nontrivial problem. ■

EXAMPLE 2.2.14 Let $A = \{a, b, c, d\}$ and $R = \{(a, a), (a, b), (a, c), (b, a), (b, b), (b, c), (b, d), (c, a), (c, b), (c, c), (c, d), (d, b), (d, c), (d, d)\}$. R is reflexive since R contains (a, a), (b, b), (c, c), and (d, d). It is easily checked that R is also symmetric. Note that R is not transitive. ■

PRACTICE PROBLEM 6 Refer to Example 2.2.14.

a. Draw the digraph for R.
b. Verify that R is symmetric.
c. Find two ordered pairs $(x, y) \in R$ and $(y, z) \in R$ but $(x, z) \notin R$. In other words, verify that R is not transitive.

PRACTICE PROBLEM 7 Refer to the definition of transitive closure.

a. Give a definition of reflexive closure.
b. For a given relation R on $A = \{a_1, a_2, \ldots, a_n\}$, what is the reflexive closure of R?

EXERCISE SET 2.2

1. Let $A = \{1, 2, 3\}$ and $B = \{4, 9, 21, 25\}$. Define the relation R from A to B by $a\ R\ b$ if a divides b. (We say a **divides** b if there is an integer m such that $b = am$. In other words, a is a factor of b or b is a multiple of a.) List all the members of R. [*Hint:* (2, 4) belongs to R since 2 divides 4.]

2. Use sets A and B in Exercise 1. Define the relation S from A to B by $a\ S\ b$ if a does not divide b. List all the members of S.

3. Let $A = \{1, 2, 3, 4\}$. Define the relation R on A by $x\ R\ y$ if $x \leq y$.
   a. List all the members of R.
   b. Determine if R is reflexive, symmetric, antisymmetric, transitive.

4. Let $\mathbb{R}$ be the set of all real numbers, and let S be the relation on $\mathbb{R}$ defined by $S = \{(x, y): y = -2x\}$. Graph the set S in the xy-plane.

2.2 RELATIONS

5. Let $A = \{2, 3\}$ and $B = \{6, 9, 10, 13\}$.
   a. Let R be a relation from A to B defined by $a\,R\,b$ if $b = 3a$. List all the members of R.
   b. Let S be a relation from A to B defined by $a\,S\,b$ if $a < b - 4$. List all the members of S.

6. Use sets A and B in Exercise 5. Let T be a relation from A to B defined by $a\,T\,b$ if $b - a$ is an even integer. List all the members of T.

7. Let $A = \{2, 3\}$ and $B = \{6, 9, 10, 12\}$.
   a. Let T be a relation from A to B defined by $a\,T\,b$ if $a + b$ is an even integer. List all the members of T.
   b. Let S be a relation from A to B defined by $a\,S\,b$ if b is a multiple of a. List all the members of S.

8. Let S be a relation on the set $\mathbb{R}$ of all real numbers such that $S = \{(x, y) : y < x\}$.
   a. Graph the set S in the xy-plane.
   b. Determine if S is reflexive, symmetric, antisymmetric, transitive.

9. Let $A = \{1, 2, 5, 10\}$. Define R on A by $x\,R\,y$ if y is a multiple of x.
   a. List all the members of R.
   b. Draw the digraph representing R.
   c. Determine if R is reflexive, symmetric, antisymmetric, transitive.

10. Let $A = \{1, 2, 3, 4, 5\}$. Define R on A by $x\,R\,y$ if $x - y$ is a multiple of 2.
   a. List all the members of R.
   b. Determine if R is reflexive, symmetric, antisymmetric, transitive.

11. Let $A = \{a, b, c, d\}$ and $R = \{(a, b), (b, c), (c, c), (a, a)\}$.
   a. Find the transitive closure of R.   **b.** Find the reflexive closure of R.

12. Let $A = \{a, b, c, d\}$ and $R = \{(a, b), (b, a), (c, b), (b, c), (d, d)\}$.
   a. Find the transitive closure of R.   **b.** Find the reflexive closure of R.

13. Let

$A_1 = \{\text{Cathy, Connie, Sally, Barbara, John}\}$

$A_2 = \{x : x \in \mathbb{Z} \text{ and } 80 \leq x \leq 150\}$

$A_3 = \{\text{Swing, Tango, Cha-cha, Madison, Fox-trot}\}$

$A_4 = \{\text{Black, Blonde, Brown, Red}\}$

A 4-ary relation called Personal is a subset of $A_1 \times A_2 \times A_3 \times A_4$, where Personal is presented as Table 2.2.3.

TABLE 2.2.3   PERSONAL

NAME	WEIGHT	FAVORITE DANCE	HAIR COLOR
Cathy	110	Tango	Brown
Connie	100	Swing	Red
Sally	115	Madison	Blonde
Barbara	98	Cha-cha	Black
John	135	Fox-trot	Black

    a. Express the 4-ary relation as a set of 4-tuples.
    b. Find the projection P_{134}(Personal).
    c. Find the projection P_{13}(Personal).

14. Refer to Exercise 13. Suppose Cathy and Connie are first-year students majoring in computer science, Sally and John are seniors majoring in chemistry, and Barbara is a junior majoring in mathematics. Let $A_5 = $ {First-year student, Sophomore, Junior, Senior, Graduate} and $A_6 = $ {Mathematics, Computer science, Chemistry, Physics}.
    a. Express the 3-ary relation called Student as a table showing the students' name in the first column, their class level in the second column, and their major in the third column.
    b. Express the 3-ary relation Student as a set of 3-tuples, which is a subset of $A_1 \times A_5 \times A_6$.
    c. Find the projection P_{16}(Student).

15. Recall the definition of the acquaintance relation given in Example 2.2.13. Let $A = $ {Cathy, Bill, John, Laura}. Suppose Cathy and Bill are acquaintances, John and Laura are acquaintances, and Bill and John are acquaintances. Find the transitive closure of this acquaintance relation R.

16. Find the reflexive closure of the acquaintance relation R of Exercise 15.

In Exercises 17–20 give examples different from those presented in this section.

17. Let $A = \{1, 2, 3\}$. Give an example of a relation R on A such that R is symmetric but is not antisymmetric.

18. Let $B = \{a, b, c\}$. Give an example of a relation S on B such that S is neither symmetric nor antisymmetric.

19. Let $C = \{a, b, c, d\}$. Give an example of a relation T on C such that T is antisymmetric and symmetric.

20. Let $D = \{1, 2, 3, 4\}$. Give an example of a relation W on D such that W is antisymmetric but not symmetric.

21. Let $\mathbb{Z}$ be the set of all integers. Let $R = \{(a, b): a + b \text{ is divisible by 3}\}$.
    a. Show that R is not reflexive.   **b.** Show that R is not transitive.

22. Refer to Exercise 21. Graph the relation R in the xy-plane as a subset of $\mathbb{Z} \times \mathbb{Z}$. You need to graph only the points (a, b) in R where $-6 \leq a \leq 6$ and $-6 \leq b \leq 6$.

2.3
TRANSITIVE CLOSURE AND THE CONNECTIVITY RELATION

Some relations are transitive, whereas others are not. We defined the transitive closure for a relation R in the previous section. In what follows, we will prove that every relation R on a finite set has a transitive closure. The transitive closure concept is essential in the discussion of reachability in Section 2.5.

Let R be a binary relation on a set A. The **connectivity relation** for R is the set $R^+ = \{(a, b): a, b \in A \text{ and there is some path in } R \text{ from } a \text{ to } b\}$.

2.3 TRANSITIVE CLOSURE AND THE CONNECTIVITY RELATION

From a geometric point of view, the connectivity relation R^+ specifies which vertices are connected by some paths in R. We will introduce a systematic method for finding R^+ later in this section. First, let us consider an example.

EXAMPLE 2.3.1 Let $A = \{1, 2, 3, 4\}$ and $R = \{(1, 2), (2, 3), (3, 1), (4, 4)\}$. Let us find R^+. First, R is a subset of R^+ since each ordered pair in R is a path of length 1 in R. Next, since $(1, 2)$ and $(2, 3)$ are in R, we have $(1, 3)$ in R^+; since $(2, 3)$ and $(3, 1)$ are in R, we have $(2, 1)$ in R^+; since $(3, 1)$ and $(1, 2)$ are in R, we have $(3, 2)$ in R^+. In other words, $(1, 3)$, $(2, 1)$, and $(3, 2)$ are in R^+ by paths in R of length 2.

In addition, $(1, 1)$ is in R^+ because $(1, 2)$, $(2, 3)$, and $(3, 1)$ are in R; $(2, 2)$ is in R^+ because $(2, 3)$, $(3, 1)$, and $(1, 2)$ are in R; $(3, 3)$ is in R^+ because $(3, 1)$, $(1, 2)$, and $(2, 3)$ are in R. The ordered pairs $(1, 1)$, $(2, 2)$, and $(3, 3)$ are in R^+ by paths in R of length 3. Further searching produces no new ordered pairs in R^+. Therefore, $R^+ = \{(1, 2), (2, 3), (3, 1), (1, 3), (2, 1), (3, 2), (1, 1), (2, 2), (3, 3), (4, 4)\}$. ∎

PRACTICE PROBLEM 1 Refer to Example 2.3.1. Draw the digraphs for R and R^+.

Proving R^+ Is the Transitive Closure of R

After proving the next two theorems, we can conclude that the connectivity relation R^+ is the transitive closure of R for any relation R.

THEOREM 2.3.2 Let R be a relation on a set A, with $a, b \in A$. If R is transitive and if there is a path in R from a to b, then $a\,R\,b$.

Proof Assume there is a path in R from a to b. Then there is a sequence $x_0, x_1, x_2, \ldots, x_m$ of points in A, with $x_0 = a$ and $x_m = b$, such that $a\,R\,x_1$, $x_1\,R\,x_2, \ldots,$ and $x_{m-1}\,R\,b$. We are also assuming R is transitive. Since $a\,R\,x_1$ and $x_1\,R\,x_2$, we have $a\,R\,x_2$; and, since $a\,R\,x_2$ and $x_2\,R\,x_3$, we have $a\,R\,x_3$; and so on. Finally, since $a\,R\,x_{m-1}$ and $x_{m-1}\,R\,b$, we have $a\,R\,b$. ∎

Another proof of Theorem 2.3.2 can be constructed after we study mathematical induction in Chapter 5.

In Section 2.2 we defined the transitive closure T of a relation R to be the smallest transitive relation containing R. Let us find a transitive relation S containing $R = \{(1, 2), (2, 4), (2, 3), (3, 4)\}$. With a little thought, we see that S must include all pairs in R along with $(1, 3)$ and $(1, 4)$. We may also include other pairs in S, such as $(1, 1)$ and $(2, 2)$. Note that $S = R \cup \{(1, 3), (1, 4), (1, 1), (2, 2)\}$ is a transitive relation containing R. However, $T = R \cup \{(1, 3), (1, 4)\}$ is the *smallest* transitive relation containing R.

PRACTICE PROBLEM 2 Let $R = \{(1, 2), (2, 4), (2, 3), (3, 4)\}$. Find R^+ and verify that $R^+ = T$, where $T = R \cup \{(1, 3), (1, 4)\}$.

We will prove in the next theorem that the connectivity relation R^+ is the smallest transitive relation containing R. Therefore, R^+ is the transitive closure of R.

THEOREM 2.3.3 The connectivity relation R^+ is equal to the transitive closure of R.

Proof We prove that R^+ satisfies the three conditions of the definition of transitive closure.

a. R is a subset of R^+ because every ordered pair (x, y) in R is a path of length 1 in R.
b. R^+ is a relation on A since R^+ is a subset of $A \times A$ by definition. R^+ is transitive because, if there is a path from x to y in R and if there is a path from y to z in R, then there is a path from x to z in R.
c. Assume $R \subseteq S$ and S is transitive, and let $(x, y) \in R^+$. By definition of R^+, there exists a path in R from x to y. This path in R from x to y is also a path in S from x to y because R is a subset of S. Since S is transitive, (x, y) is in S by Theorem 2.3.2. Therefore, R^+ is a subset of S. ∎

We will refer to R^+ as either the connectivity relation or the transitive closure of R.

The Composition of Relations

The connectivity relation R^+ can be determined by means of composition of the relation R with itself. First we define the composition of two binary relations.

Let A, B, and C be sets. Let $R \subseteq A \times B$ be a relation from A to B. Let $S \subseteq B \times C$ be a relation from B to C. The **composition** of R and S, denoted $R \circ S$, is the relation from A to C defined to be

$$R \circ S = \{(x, y) : (x, b) \in R \text{ and } (b, y) \in S \text{ for some } b \in B\}$$

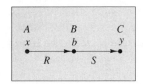

FIGURE 2.3.1

We can picture an element $(x, y) \in R \circ S$ as in Figure 2.3.1. A pair (x, y) is an element of $R \circ S$ whenever there is an element $b \in B$ such that $(x, b) \in R$ and $(b, y) \in S$.

An important special case of the composition of two relations occurs when both R and S are relations on a set A. Then $(x, y) \in R \circ S$ if there is an element $a \in A$ such that $(x, a) \in R$ and $(a, y) \in S$.

There is another useful graphical representation of a relation R that will be helpful for finding the composition of two relations on a set A. We illustrate this representation with the relation $R = \{(1, 2), (2, 2), (2, 3)\}$ on $A = \{1, 2, 3\}$. The elements of A are placed in two identical columns. The ordered pairs in R are represented by directed line segments from the elements of A in the first column to the elements of A in the second column. See Figure 2.3.2.

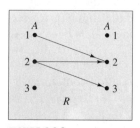

FIGURE 2.3.2

EXAMPLE 2.3.4 Let $A = \{1, 2, 3\}$, $R = \{(1, 2), (2, 2), (2, 3)\}$, and $S = \{(2, 1), (3, 3)\}$. We shall find $R \circ S$ by using the graphical representations corresponding to R and S.

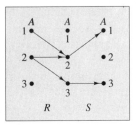

FIGURE 2.3.3

To find $R \circ S$, we look for paths of length 2 from the first column of elements of A to the third column of elements of A in Figure 2.3.3. We see that $(1, 1) \in R \circ S$ since $(1, 2) \in R$ and $(2, 1) \in S$; $(2, 1) \in R \circ S$ because $(2, 2) \in R$ and $(2, 1) \in S$; $(2, 3) \in R \circ S$ because $(2, 3) \in R$ and $(3, 3) \in S$. Hence, $R \circ S = \{(1, 1), (2, 3), (2, 1)\}$.

Similarly, we use the graphical representations of S and R in Figure 2.3.4 to find $S \circ R = \{(2, 2)\}$. Note that $R \circ S \neq S \circ R$. In general, $R \circ S$ and $S \circ R$ are not equal.

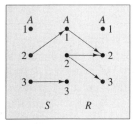

FIGURE 2.3.4

When $R = S$, we have

$$R \circ R = \{(x, y) : (x, a) \in R \text{ and } (a, y) \in R \text{ for some } a \in A\}$$

We define $R^2 = R \circ R$ and $R^3 = R \circ R^2$, where

$$R \circ R^2 = \{(x, y) : (x, a) \in R \text{ and } (a, y) \in R^2 \text{ for some } a \in A\}$$

We note that $R \circ R^2 = R^2 \circ R$, as illustrated by the following practice problem.

PRACTICE PROBLEM 3 Let $A = \{a, b, c\}$ and $R = \{(a, a), (a, b), (b, a), (c, c)\}$. Find R^2, $R \circ R^2$, and $R^2 \circ R$. Observe that $R \circ R^2 = R^2 \circ R$.

EXAMPLE 2.3.5 Let $A = \{1, 2, 3, 4\}$ and $R = \{(1, 2), (1, 3), (2, 2), (3, 4)\}$. We find $R \circ R$ by the graphical representation of R repeated twice, as shown in Figure 2.3.5.

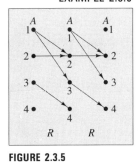

FIGURE 2.3.5

To find $R \circ R$, we look for paths of length 2 from the first column of elements of A to the third column of elements of A. We see that $(1, 2) \in R \circ R$ since $(1, 2) \in R$ and $(2, 2) \in R$; $(1, 4) \in R \circ R$ because $(1, 3) \in R$ and $(3, 4) \in R$; $(2, 2) \in R \circ R$ since $(2, 2) \in R$ and $(2, 2) \in R$. No other paths of length 2 go from the first column to the third column. Hence, $R \circ R = \{(1, 2), (1, 4), (2, 2)\}$.

PRACTICE PROBLEM 4 Let $A = \{1, 2, 3, 4\}$, $R = \{(1, 3), (2, 1), (3, 4), (4, 3)\}$, and $S = \{(1, 2), (3, 3)\}$. Use graphical representations to find $R \circ R$, $R \circ S$, and $S \circ R$.

In general, we define

$$R^1 = R, \quad R^2 = R \circ R^1, \quad R^3 = R \circ R^2, \quad \ldots, \quad R^m = R \circ R^{m-1},$$
$$\text{for any integer } m > 1$$

Note that $x R^m y$ whenever there are $x_0, x_1, x_2, \ldots, x_m$ in A such that $x_0 R x_1, x_1 R x_2, \ldots, x_{m-1} R x_m$, where $x = x_0$ and $y = x_m$. In other words, $x R^m y$ whenever there is a path in R of length m from x to y.

A Systematic Method for Finding R^+

Suppose R is a relation on a set $A = \{a, b, c, d, e\}$. If there exists some path from b to d, is there a path from b to d of length 5 or less? If so, we can find the elements R^+ by finding the elements of $R, R^2, R^3, R^4,$ and R^5. Theorem 2.3.7 gives an affirmative answer.

Before considering the proof of Theorem 2.3.7, we give an example to show that the existence of a path from x to y in a relation R on a set with n elements cannot guarantee a path of length less than n from x to y.

EXAMPLE 2.3.6 Consider $A = \{a, b, c, d, e\}$ and let $R = \{(a, b), (b, c), (c, d), (d, e), (e, a)\}$. It is easy to see that the shortest path from a to a is of length 5. ∎

By considering the digraph corresponding to R in Example 2.3.6, we see that every path in R is of length 1, 2, 3, 4, or 5. In other words, the elements of R^+ are the elements of $R, R^2, R^3, R^4,$ or R^5. From Theorems 2.3.7 and 2.3.8 following, we can conclude that, if R is a relation on a set with n elements, then the elements of R^+ are the elements of $R, R^2, \ldots,$ or R^n. This result makes it possible to find R^+ by finding $R, R^2, \ldots, R^n$.

THEOREM 2.3.7 Let R be a relation on A. If there is a path from x to y in R, then there is a path $((x, x_1), (x_1, x_2), \ldots, (x_j, y))$ from x to y such that the elements $x, x_1, x_2, \ldots, x_j$ are distinct.

The proof of Theorem 2.3.7 is left as an exercise. (See Exercise 21, Exercise Set 2.3.)

The pigeonhole principle is needed for the proof of Theorem 2.3.8.

PIGEONHOLE PRINCIPLE

Suppose m pigeons occupy n pigeonholes. If no pigeonhole contains more than one pigeon, then $m \leq n$.

For example, if in a group of people no two people have birthdays in the same month, then the group contains 12 or fewer people.

The mathematical statement of the pigeonhole principle involves functions; we will discuss it in Chapter 3. This principle is sometimes called the Dirichlet drawer principle because it was widely used by the 19th century mathematician Peter Gustav Lejeune Dirichlet (1805–1859).

THEOREM 2.3.8 Let A be a set with n elements with x and y in A, and let R be a relation on A. Then $x R^+ y$ if and only if there is some path in R from x to y of length less than or equal to n.

Proof ($\rightarrow$) Assume $x R^+ y$. Then there exists some path in R from x to y. By Theorem 2.3.7, there exist distinct elements $x, x_1, x_2, \ldots, x_{k-1}$ such that

2.3 TRANSITIVE CLOSURE AND THE CONNECTIVITY RELATION

$x R x_1, x_1 R x_2, \ldots, x_{k-1} R y$. This means $x R^k y$. By the pigeonhole principle, $k \leq n$. Therefore, there is some path in R from x to y of length less than or equal to n.

($\leftarrow$) Conversely, if there is some path in R from x to y of length less than or equal to n, then it follows from the definition of R^+ that $x R^+ y$. ∎

COROLLARY 2.3.9 If A is a set with n elements and R is a relation on A, then $R^+ = R^1 \cup R^2 \cup \cdots \cup R^n$.

Proof From Theorem 2.3.8, $(x, y) \in R^+$ if and only if $(x, y) \in R^1 \cup R^2 \cup \cdots \cup R^n$. ∎

Corollary 2.3.9 provides a method of computing the transitive closure R^+ of a relation R.

EXAMPLE 2.3.10 Let $A = \{1, 2, 3\}$ and let $R = \{(1, 2), (2, 1), (2, 3)\}$. We will find R^+. By Corollary 2.3.9, $R^+ = R^1 \cup R^2 \cup R^3$. Note $R^1 = R$. We will find R^2 and R^3 by using the graphical representation of R repeated three times, as shown in Figure 2.3.6.

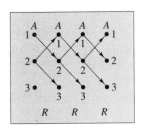

FIGURE 2.3.6

To find $R \circ R$, we look for paths of length 2 from the first column of elements of A to the third column of elements of A. We see that $(1, 1) \in R \circ R$ since $(1, 2)$ and $(2, 1)$ are in R; $(2, 2) \in R \circ R$ since $(2, 1)$ and $(1, 2)$ are in R; and $(1, 3) \in R \circ R$ since $(1, 2)$ and $(2, 3)$ are in R. Hence, $R^2 = R \circ R = \{(1, 1), (2, 2), (1, 3)\}$.

To find $R \circ R \circ R$, we look for paths of length 3 from the first column to the fourth column. We see that $R^3 = R \circ R \circ R = \{(2, 3), (1, 2), (2, 1)\}$. Hence, $R^+ = R^1 \cup R^2 \cup R^3 = \{(1, 2), (2, 1), (2, 3), (1, 1), (1, 3), (2, 2)\}$. Note that $(x, y) \in R^+$ whenever there is some path in R from x to y. The path is of length 1, 2, or 3. Furthermore, we observe that R^+ is the transitive closure of R. ∎

Although Corollary 2.3.9 gives us a method for finding the transitive closure of a relation R on a finite set A, this method can be tedious when the number of elements in A is large. In Section 2.4, we will use matrices and Corollary 2.3.9 as an alternate method of finding R^+.

EXERCISE SET 2.3

1. Let $A = \{1, 2, 3\}$, $B = \{a, b, c\}$, and $C = \{6, 8\}$. Let $R = \{(1, a), (2, c), (3, b)\}$ be a relation from A to B, and let $S = \{(a, 6), (a, 8), (b, 8), (c, 8)\}$ be a relation from B to C. Find the composition relation $R \circ S$ from A to C.

2. Let $A = \{x, y, t, w\}$, $B = \{1, 2, 3\}$, and $C = \{a, b\}$. Let $R = \{(x, 1), (x, 2), (t, 3), (w, 2)\}$ be a relation from A to B, and let $S = \{(1, b), (2, a), (3, b)\}$ be a relation from B to C. Find the composition relation $R \circ S$ from A to C.

3. Let $R = \{(1, a), (1, b), (a, 1), (b, 5), (5, b)\}$ be a relation on $\{1, a, b, 5\}$. Find the composition relation $R^2 = R \circ R$ on A.

4. Let $R = \{(x, y), (y, t), (t, x), (w, w)\}$ be a relation on $\{x, y, t, w\}$. Find the composition relation $R^3 = R \circ R^2$.

5. Let $R = \{(1, 2), (2, 3), (2, 4)\}$ be a relation on $\{1, 2, 3, 4\}$.
   a. Show that R is not transitive.   b. Find the transitive closure of R.

6. Let $R = \{(a, b), (b, c), (c, d), (b, a)\}$ be a relation on $\{a, b, c, d\}$. Find R^+.

7. Let $R = \{(a, b), (b, c), (c, a)\}$ be a relation on $\{a, b, c\}$. Show that $R^+ = A \times A$. [Note that $(a, a) \in R^3$ but is neither in R^1 nor in R^2. This example shows why we may need to find R^n when A has n elements in order to find R^+.]

8. Let $R = \{(1, 2), (2, 3), (3, 4)\}$ be a relation on $\{1, 2, 3, 4\}$. Find R^+ by using the graphical representation of R.

9. Let $R = \{(a, b), (b, a), (a, c), (c, d), (d, a)\}$.
   a. Is $(a, c) \in R^3$?   b. Is $(a, a) \in R^2$?
   c. Is $(c, c) \in R^5$?   d. For what values of $k \leq 5$ is $(a, d) \in R^k$?

10. Let $R = \{(a, b), (a, d), (b, a), (b, c), (c, d), (d, b)\}$.
    a. Is $(a, c) \in R^3$?   b. Is $(b, d) \in R^2$?
    c. Is $(c, b) \in R^5$?   d. For what values of $k \leq 6$ is $(b, b) \in R^k$?

11. Find a relation R on $\{1, 2, 3\}$ so that R^2 is not a subset of R and $R^3 = \emptyset$.

12. Find a transitive relation R on $\{1, 2, 3\}$ so that R is not a subset of R^2.

In Exercises 13 and 14 analyze the designated proof and answer the following questions about each proof.

a. What are the hypotheses of the proof?
b. What is the conclusion of the proof?
c. What definitions and facts are necessary?

13. Part b of the proof of Theorem 2.3.3
14. Part c of the proof of Theorem 2.3.3

Exercises 15 and 16 require the pigeonhole principle.

15. How many people must be chosen to be certain that at least two of them will have:
    a. A birthday on the same day of the year?
    b. The same last digit in their telephone numbers?

16. Consider a digraph on a set of n vertices. Suppose $x \neq y$. Find the maximum number of edges in a path from x to y with distinct vertices.

17. For a relation R, R is transitive if and only if $R \circ R \subseteq R$. However, it is not true in general that, if R is transitive, then $R \subseteq R \circ R$. Find an example of a set A and a transitive relation R on A such that R is not a subset of $R \circ R$.

18. Suppose R is a symmetric and transitive relation on $\{1, 2, \ldots, n\}$. Assume $(j, 2) \in R$ for $j = 1, 2, \ldots, n$. Prove R is reflexive.

19. Suppose R is a symmetric and transitive relation on $\{1, 2, \ldots, n\}$. Assume that, for each $j = 1, 2, \ldots, n$, there is an element $x \in \{1, 2, \ldots, n\}$ such that $(j, x) \in R$. Prove R is reflexive.

20. True or false? If R is a relation on a finite set and $(v, v) \in R^+$, then there is a cycle at v. Supply a convincing argument with an answer of true or an example with an answer of false.

21. Prove Theorem 2.3.7.

2.4
MATRIX REPRESENTATION OF DIGRAPHS AND RELATIONS

We explored the connection between digraphs and relations in the last section. Every digraph can be represented as a relation. Conversely, a relation R on a finite set A can be represented as a digraph (A, R). In this section, we will use matrices to study relations.

EXAMPLE 2.4.1 Let $A = \{3, 6, 9\}$, and let R be the relation on A defined by $m \, R \, n$ whenever m divides n. Hence, $R = \{(3, 6), (3, 9), (3, 3), (6, 6), (9, 9)\}$.

This relation R can be represented as a digraph $G = (V, E)$, where $V = \{3, 6, 9\}$ and $E = R$. Figure 2.4.1 indicates the three vertices and the five edges of G.

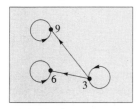

FIGURE 2.4.1

We will discuss how to represent a relation with a matrix shortly. As an introduction to this important idea, we show how the relation R can be represented by a matrix. We assume the elements of A are given in order. Let $a_1 = 3$, $a_2 = 6$, and $a_3 = 9$. Then R is represented by the matrix

$$\begin{array}{c} \\ a_1 \\ a_2 \\ a_3 \end{array} \begin{array}{ccc} a_1 & a_2 & a_3 \end{array} \\ \begin{bmatrix} 1 & 1 & 1 \\ 0 & 1 & 0 \\ 0 & 0 & 1 \end{bmatrix}$$

For example, the entry in row 1, column 2 is 1 because $(a_1, a_2) \in R$; the entry in row 2, column 3 is 0 because $(a_2, a_3) \notin R$. ∎

Any digraph $G = (V, E)$ can be represented by a matrix. Let us begin with a brief discussion of matrices. In general, an $n \times m$ **matrix** A is a rectangular array with n rows and m columns of numbers:

$$A = \begin{bmatrix} a_{11} & a_{12} & a_{13} & \cdots & a_{1m} \\ a_{21} & a_{22} & a_{23} & \cdots & a_{2m} \\ a_{31} & a_{32} & a_{33} & \cdots & a_{3m} \\ \vdots & \vdots & \vdots & \vdots & \vdots \\ a_{n1} & a_{n2} & a_{n3} & \cdots & a_{nm} \end{bmatrix}$$

As a useful shorthand, we write $A = [a_{ij}]$, where i ranges from 1 to n and j ranges from 1 to m. The first subscript i is called the **row index** and the second subscript j the **column index** of the matrix A. For example, the entry

a_{23} is located at row 2 and column 3. We also say a_{23} is the (2, 3) entry of A, or write $A[2, 3] = a_{23}$. In general, we write $A[i, j]$ or a_{ij} for the entry located at row i and column j.

When $n = m$, we have an $n \times n$ matrix A. An $n \times n$ matrix is called a **square matrix of order n**. The **main diagonal** of an $n \times n$ matrix consists of the entries $a_{11}, a_{22}, a_{33}, \ldots, a_{nn}$.

EXAMPLE 2.4.2
$$M = \begin{bmatrix} 5 & 4 & 7 & 8 \\ 1 & 0 & 4 & -6 \\ 9 & 10 & -1 & 16 \end{bmatrix} \text{ is a } 3 \times 4 \text{ matrix}$$

$$N = \begin{bmatrix} 1 & 2 & 1 \\ 0 & 4 & 6 \\ 7 & 9 & -3 \end{bmatrix} \text{ is a } 3 \times 3 \text{ matrix}$$

The main diagonal of the square 3×3 matrix N consists of the entries 1, 4, and -3. ∎

Adjacency Matrices

In what follows, we will be concerned only with matrices that have entries 0 or 1. These matrices are very useful for analyzing digraphs. Every digraph $G = (V, E)$ has a representation by a matrix as described below.

Let $V = \{v_1, v_2, \ldots, v_n\}$. The $n \times n$ matrix $A = [a_{ij}]$, where

$$a_{ij} = 1 \quad \text{if } (v_i, v_j) \text{ is an edge in } E$$

and

$$a_{ij} = 0 \quad \text{if } (v_i, v_j) \text{ is not an edge in } E$$

is called the **adjacency matrix** for the digraph $G = (V, E)$.

Note that the number of nonzero entries in the adjacency matrix is equal to the number of edges in the digraph. The adjacency matrix depends on the order of vertices $v_1, v_2, \ldots, v_n$ in V. We assume the order of $v_1, v_2, \ldots, v_n$ is fixed unless otherwise stated.

EXAMPLE 2.4.3 Let $G = (V, E)$ be a digraph, where $V = \{v_1, v_2, v_3, v_4\}$ and $E = \{(v_1, v_2), (v_2, v_4), (v_1, v_3), (v_1, v_4), (v_4, v_3)\}$, as shown in Figure 2.4.2. The adjacency matrix for G is:

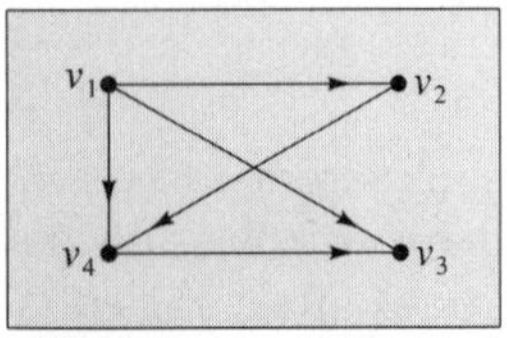

$$M = \begin{array}{c} \\ v_1 \\ v_2 \\ v_3 \\ v_4 \end{array} \begin{array}{c} \begin{matrix} v_1 & v_2 & v_3 & v_4 \end{matrix} \\ \begin{bmatrix} 0 & 1 & 1 & 1 \\ 0 & 0 & 0 & 1 \\ 0 & 0 & 0 & 0 \\ 0 & 0 & 1 & 0 \end{bmatrix} \end{array}$$

FIGURE 2.4.2

2.4 MATRIX REPRESENTATION OF DIGRAPHS AND RELATIONS

The four vertices are placed vertically on the left and horizontally on the top of the matrix M for easy reference. They may be omitted if the order of the vertices is clearly understood. Note that the (4, 3) entry of M is 1 because (v_4, v_3) is in E, and the (3, 4) entry of M is 0 because (v_3, v_4) is not in E. ∎

PRACTICE PROBLEM 1 Find the adjacency matrix for the digraph in Example 2.1.3 on page 43.

Since every relation corresponds to a digraph, any relation R on a finite set can be represented by the adjacency matrix of the corresponding digraph. We denote the adjacency matrix for R by M_R. It is important to note that $M_R = M_T$ if and only if R and T are identical relations on a set A.

EXAMPLE 2.4.4 Let $A = \{a, b, c, d\}$, and let the relation on A be $R = \{(a, a), (b, b), (c, c), (d, d), (b, c), (c, b), (b, d), (d, b)\}$. The adjacency matrix representing R is

$$M_R = \begin{array}{c} \\ a \\ b \\ c \\ d \end{array} \begin{array}{c} \begin{array}{cccc} a & b & c & d \end{array} \\ \left[\begin{array}{cccc} 1 & 0 & 0 & 0 \\ 0 & 1 & 1 & 1 \\ 0 & 1 & 1 & 0 \\ 0 & 1 & 0 & 1 \end{array} \right] \end{array}$$

∎

The fact that a relation is reflexive and symmetric is clearly displayed in its adjacency matrix. A relation is reflexive if every entry on the main diagonal is 1. A relation is symmetric if the adjacency matrix entries are symmetric with respect to the main diagonal. Geometrically speaking, a relation is symmetric if folding its adjacency matrix along the main diagonal matches entries off the main diagonal with each other. In other words, a "0" will match a "0" and a "1" will match a "1."

Note that R in Example 2.4.4 is reflexive since every entry on the main diagonal of the matrix M_R is 1. In addition, R is symmetric, since the matrix entries are symmetric with respect to the main diagonal.

The relation R in Example 2.4.4 is not antisymmetric. For, when the matrix M_R is folded over along the main diagonal, the "1" at the (b, c) position is matched with the "1" at the (c, b) position. That is, b ≠ c but both (b, c) and (c, b) are in R. In order that R be antisymmetric, every "1" off the main diagonal must be matched by a "0."

Whether or not R is transitive can also be detected from M_R. For example, R is not transitive if the (c, b) entry is 1 and the (b, d) entry is 1 but the (c, d) entry is 0. However, verifying that R has the transitive property by inspecting M_R is usually tedious.

PRACTICE PROBLEM 2 Let $A = \{a, b, c\}$ and $R = \{(a, a), (a, b), (a, c), (b, b), (c, b), (c, c)\}$. Find the adjacency matrix M_R that represents R. By studying M_R, determine if R is reflexive, symmetric, or antisymmetric.

We have seen that a digraph or a relation on a finite set can be represented by an adjacency matrix. Conversely, given a square matrix with entries of 0's and 1's, we can construct a digraph or relation that has M as its adjacency matrix.

In an adjacency matrix M for $G = (V, E)$, each entry a_{ij} of M is 1 or 0, according to whether there is an edge (a path of length 1) from vertex v_i to vertex v_j or not. In short, M gives all the information about paths of length 1.

In order to gain information on paths of all lengths, we will define addition and multiplication operations on adjacency matrices. First, we define **Boolean addition** ($+$) and **Boolean multiplication** ($\cdot$) for the elements 0 and 1 as follows:

a. BOOLEAN ADDITION

$0 + 0 = 0$
$0 + 1 = 1$
$1 + 0 = 1$
$1 + 1 = 1$

b. BOOLEAN MULTIPLICATION

$0 \cdot 0 = 0$
$0 \cdot 1 = 0$
$1 \cdot 0 = 0$
$1 \cdot 1 = 1$

We note three important properties of Boolean addition and multiplication on the elements of $\{0, 1\}$, where $x, y, z \in \{0, 1\}$:

BOOLEAN ADDITION AND MULTIPLICATION ON $\{0, 1\}$		
$x + y = y + x$	$x \cdot y = y \cdot x$	Commutative laws
$(x + y) + z = x + (y + z)$	$(x \cdot y) \cdot z = x \cdot (y \cdot z)$	Associative laws
$x + 0 = x$	$x \cdot 1 = x$	Identity laws

It is routine but tedious to verify these properties by checking all possible cases. For example, when $x = 0$ and $y = 1$, we have $x + y = 0 + 1 = 1 = 1 + 0 = y + x$.

The commutative laws say that we can add or multiply two elements in either order and obtain the same result. The associative laws allow us to add and multiply any number of elements without ambiguity. For example, we can perform the operation $x + y + z$ by adding x to y first or by adding y to z first. In other words, $(x + y) + z = x + (y + z)$.

Boolean Addition of Matrices

Let $N = [n_{ij}]$ and $T = [t_{ij}]$ be two $n \times m$ adjacency matrices. The **Boolean sum** of N and T is the $n \times m$ matrix $N \oplus T = [a_{ij}]$, where

$$a_{ij} = n_{ij} + t_{ij} \quad \text{for } 1 \leq i \leq n \text{ and } 1 \leq j \leq m$$

The addition ($+$) in the formula for a_{ij} is the Boolean addition for 0 and 1. Note that N and T must be of the same size for $N \oplus T$ to be defined.

EXAMPLE 2.4.5 Let

$$N = \begin{bmatrix} 1 & 0 & 1 \\ 0 & 0 & 1 \\ 1 & 1 & 0 \end{bmatrix} \quad \text{and} \quad T = \begin{bmatrix} 0 & 0 & 1 \\ 1 & 0 & 1 \\ 0 & 1 & 0 \end{bmatrix}$$

Then

$$A = N \oplus T = \begin{bmatrix} 1+0 & 0+0 & 1+1 \\ 0+1 & 0+0 & 1+1 \\ 1+0 & 1+1 & 0+0 \end{bmatrix} = \begin{bmatrix} 1 & 0 & 1 \\ 1 & 0 & 1 \\ 1 & 1 & 0 \end{bmatrix}$$

using the Boolean addition for 0 and 1. ∎

PRACTICE PROBLEM 3 Let

$$P = \begin{bmatrix} 1 & 0 \\ 0 & 0 \end{bmatrix} \quad \text{and} \quad Q = \begin{bmatrix} 1 & 1 \\ 1 & 0 \end{bmatrix}$$

Find $P \oplus Q$.

Boolean Multiplication of Matrices

Let $N = [b_{ij}]$ be an $n \times m$ matrix, and let $T = [t_{ij}]$ be an $m \times t$ matrix. The **Boolean product** of N and T is defined to be the $n \times t$ matrix $N \otimes T = [a_{ij}]$, where

$$a_{ij} = (b_{i1} \cdot t_{1j}) + (b_{i2} \cdot t_{2j}) + \cdots + (b_{im} \cdot t_{mj}), \quad \text{for } 1 \le i \le n \text{ and } 1 \le j \le t$$

The addition (+) and multiplication (·) in the formula for a_{ij} are the Boolean addition and Boolean multiplication for 0 and 1. Note that $M = N \otimes T$ is an $n \times t$ matrix when N is an $n \times m$ matrix and T is an $m \times t$ matrix. The column number of N must equal the row number of T for $N \otimes T$ to be defined.

EXAMPLE 2.4.6 Consider the adjacency matrices

$$P = \begin{bmatrix} 1 & 0 \\ 0 & 0 \end{bmatrix} \quad \text{and} \quad Q = \begin{bmatrix} 1 & 1 \\ 1 & 0 \end{bmatrix}$$

The Boolean product of P and Q is

$$M = P \otimes Q = \begin{bmatrix} (1 \cdot 1) + (0 \cdot 1) & (1 \cdot 1) + (0 \cdot 0) \\ (0 \cdot 1) + (0 \cdot 1) & (0 \cdot 1) + (0 \cdot 0) \end{bmatrix} = \begin{bmatrix} 1 & 1 \\ 0 & 0 \end{bmatrix}$$

For instance, the a_{12} entry is obtained by operating the first row of P on the second column of Q:

$$a_{12} = p_{11} \cdot q_{12} + p_{12} \cdot q_{22} = (1 \cdot 1) + (0 \cdot 0) = 1 + 0 = 1$$

This operation is easily remembered by imagining that you move across row 1 while you move down column 2. As you do so, multiply the corresponding entries and then add the resulting products:

$$\begin{bmatrix} 1 & 0 \\ 0 & 0 \end{bmatrix} \otimes \begin{bmatrix} 1 & 1 \\ 1 & 0 \end{bmatrix} = \begin{bmatrix} 1 & (1 \cdot 1) + (0 \cdot 0) \\ 0 & 0 \end{bmatrix}$$

∎

PRACTICE PROBLEM 4 Let

$$P = \begin{bmatrix} 1 & 0 & 1 \\ 1 & 1 & 0 \\ 0 & 1 & 1 \end{bmatrix} \quad \text{and} \quad Q = \begin{bmatrix} 1 & 0 \\ 1 & 1 \\ 1 & 0 \end{bmatrix}$$

Find $P \otimes Q$.

EXAMPLE 2.4.7 Let

$$N = \begin{bmatrix} 1 & 1 & 0 & 1 \\ 0 & 1 & 1 & 1 \\ 0 & 0 & 1 & 0 \end{bmatrix} \quad \text{and} \quad T = \begin{bmatrix} 1 & 0 \\ 1 & 1 \end{bmatrix}$$

Since N is a 3×4 matrix and T is a 2×2 matrix, neither the Boolean addition nor the Boolean multiplication of these two matrices is defined. ∎

When M is a square matrix, $M \otimes M$ is always defined. We will usually denote $M \otimes M$ as M^2 and call it "M squared." Similarly, $M^3 = M \otimes M^2$. In general, we define $M^n = M \otimes M^{n-1}$ for $n > 1$.

The $n \times m$ **zero matrix** is a matrix O with all entries zero.

The $n \times n$ **identity matrix** is a square matrix I in which every entry on the main diagonal is 1 and each entry off the main diagonal is 0.

PRACTICE PROBLEM 5 Let

$$M = \begin{bmatrix} 0 & 1 & 1 \\ 0 & 0 & 1 \\ 1 & 1 & 0 \end{bmatrix}, \quad N = \begin{bmatrix} 0 & 1 & 0 \\ 1 & 0 & 1 \\ 0 & 1 & 0 \end{bmatrix},$$

$$O = \begin{bmatrix} 0 & 0 & 0 \\ 0 & 0 & 0 \\ 0 & 0 & 0 \end{bmatrix}, \quad \text{and} \quad I = \begin{bmatrix} 1 & 0 & 0 \\ 0 & 1 & 0 \\ 0 & 0 & 1 \end{bmatrix}$$

a. Find $M \otimes O$ and $N \oplus O$.
b. Find $M \otimes I$ and $I \otimes N$.
c. Find $M \oplus N$ and $N \oplus M$.
d. Find $M \otimes N$ and $N \otimes M$.
e. Find M^2 and N^2.
f. Let $A = \{1, 2, 3\}$. Find relations R, S, T, W on A so that $M, N, O,$ and I are adjacency matrices for $R, S, T,$ and W, respectively.

2.4 MATRIX REPRESENTATION OF DIGRAPHS AND RELATIONS

As in the case of a relation on a finite set, any relation R from a finite set to another finite set can be represented by an adjacency matrix.

If $A = \{a_1, \ldots, a_r\}$ and $B = \{b_1, \ldots, b_s\}$, the **adjacency matrix** for a relation R from A to B is an $r \times s$ matrix $M_R = [m_{ij}]$, where

$m_{ij} = 1$ if (a_i, b_j) is in R, and
$m_{ij} = 0$ if (a_i, b_j) is not in R

If R and T are relations from A to B, then $M_R = M_T$ whenever $R = T$.

Matrix Multiplication and Composition of Relations

The product of two adjacency matrices provides an easy method to calculate the composition of two relations on finite sets. Let A, B, and C be sets. Suppose R and S are relations such that $R \subseteq A \times B$ and $S \subseteq B \times C$. Recall that the composition relation $R \circ S = \{(a, c) : (a, b) \in R$ and $(b, c) \in S$ for some $b \in B\}$ is a relation from A to C.

If the sets A, B, and C are finite, we can represent R and S by adjacency matrices M_R and M_S, respectively. It turns out that the Boolean product matrix $M_R \otimes M_S$ equals the adjacency matrix of the composition relation $R \circ S$. In other words, $M_R \otimes M_S = M_{R \circ S}$, a fact to be established in Theorem 2.4.9. We first illustrate this by the next example.

EXAMPLE 2.4.8 Let $A = \{a_1, a_2, a_3\}$, $B = \{b_1, b_2\}$, and $C = \{c_1, c_2, c_3, c_4\}$. Let $R = \{(a_1, b_1), (a_2, b_1), (a_3, b_2)\} \subseteq A \times B$ and $S = \{(b_1, c_1), (b_2, c_4)\} \subseteq B \times C$.

The adjacency matrices are

$$M_R = \begin{bmatrix} 1 & 0 \\ 1 & 0 \\ 0 & 1 \end{bmatrix} \quad \text{and} \quad M_S = \begin{bmatrix} 1 & 0 & 0 & 0 \\ 0 & 0 & 0 & 1 \end{bmatrix}$$

We find the Boolean product of M_R and M_S to be

$$M_R \otimes M_S = \begin{array}{c} \\ a_1 \\ a_2 \\ a_3 \end{array} \begin{array}{c} \begin{array}{cccc} c_1 & c_2 & c_3 & c_4 \end{array} \\ \begin{bmatrix} 1 & 0 & 0 & 0 \\ 1 & 0 & 0 & 0 \\ 0 & 0 & 0 & 1 \end{bmatrix} \end{array}$$

Next, we see from Figure 2.4.3(a) on page 72 that the composition relation is $R \circ S = \{(a_1, c_1), (a_2, c_1), (a_3, c_4)\}$, as shown in Figure 2.4.3(b).

The adjacency matrix of $R \circ S$ is

$$M_{R \circ S} = \begin{array}{c} \\ a_1 \\ a_2 \\ a_3 \end{array} \begin{array}{c} \begin{array}{cccc} c_1 & c_2 & c_3 & c_4 \end{array} \\ \begin{bmatrix} 1 & 0 & 0 & 0 \\ 1 & 0 & 0 & 0 \\ 0 & 0 & 0 & 1 \end{bmatrix} \end{array}$$

We see that $M_{R \circ S} = M_R \otimes M_S$.

CHAPTER 2 DIRECTED GRAPHS AND RELATIONS

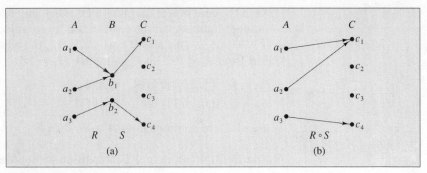

FIGURE 2.4.3

THEOREM 2.4.9 Let $A = \{a_1, \ldots, a_n\}$, $B = \{b_1, \ldots, b_r\}$, and $C = \{c_1, \ldots, c_p\}$. Let R and S be relations with $R \subseteq A \times B$ and $S \subseteq B \times C$. Then

$$M_{R \circ S} = M_R \otimes M_S$$

Proof Let $M_R = [r_{ij}]$, $M_S = [s_{ij}]$, and $M_{R \circ S} = [t_{ij}]$. For fixed i and j, $t_{ij} = 1$ if and only if $(a_i, b_k) \in R$ and $(b_k, c_j) \in S$ for some $b_k \in B$. In other words, $t_{ij} = 1$ if and only if $r_{ik} = 1$ and $s_{kj} = 1$ for some k. Thus, $t_{ij} = 1$ if and only if at least one of the following r equations holds:

$$r_{i1} \cdot s_{1j} = 1, \quad r_{i2} \cdot s_{2j} = 1, \quad \ldots, \quad r_{ir} \cdot s_{rj} = 1$$

Hence, $t_{ij} = (r_{i1} \cdot s_{1j}) + (r_{i2} \cdot s_{2j}) + (r_{i3} \cdot s_{3j}) + \cdots + (r_{ir} \cdot s_{rj})$, for $1 \leq i \leq n$ and $1 \leq j \leq p$, since the sum is equal to 1 if and only if at least one of the terms is equal to 1. Thus, t_{ij} is the (i, j) entry of the Boolean product $M_R \otimes M_S$. Since t_{ij} is also the (i, j) entry of $M_{R \circ S}$, we have $M_{R \circ S} = M_R \otimes M_S$. ∎

PRACTICE PROBLEM 6 Refer to Practice Problem 5.

a. Use Theorem 2.4.9 and Practice Problem 5 to find $R \circ S$.
b. Find M^2 and use it to find R^2.

It follows from Theorem 2.4.9 that $M_{R \circ R} = M_R \otimes M_R$, $M_{R^2} = (M_R)^2$. In general, we have $M_{R^n} = (M_R)^n$ for $n \geq 1$, as stated in Corollary 2.4.11. The next example illustrates how to obtain R^2 from $(M_R)^2$.

EXAMPLE 2.4.10 Let $V = \{v_1, v_2, v_3, v_4\}$ and $R = \{(v_1, v_2), (v_1, v_4), (v_2, v_1), (v_2, v_2), (v_3, v_3), (v_4, v_1), (v_4, v_2), (v_4, v_3)\}$. Find R^2 from $(M_R)^2$.

Solution The adjacency matrix is

$$M_R = \begin{bmatrix} 0 & 1 & 0 & 1 \\ 1 & 1 & 0 & 0 \\ 0 & 0 & 1 & 0 \\ 1 & 1 & 1 & 0 \end{bmatrix}$$

The information about paths of length 1 is displayed in M_R. The digraph is shown in Figure 2.4.4.

The product of M_R and M_R is

$$(M_R)^2 = \begin{bmatrix} 0 & 1 & 0 & 1 \\ 1 & 1 & 0 & 0 \\ 0 & 0 & 1 & 0 \\ 1 & 1 & 1 & 0 \end{bmatrix} \otimes \begin{bmatrix} 0 & 1 & 0 & 1 \\ 1 & 1 & 0 & 0 \\ 0 & 0 & 1 & 0 \\ 1 & 1 & 1 & 0 \end{bmatrix} = \begin{matrix} v_1 \\ v_2 \\ v_3 \\ v_4 \end{matrix} \begin{bmatrix} v_1 & v_2 & v_3 & v_4 \\ 1 & 1 & 1 & 0 \\ 1 & 1 & 0 & 1 \\ 0 & 0 & 1 & 0 \\ 1 & 1 & 1 & 1 \end{bmatrix}$$

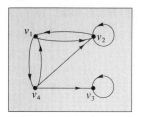

FIGURE 2.4.4

The (2, 4) entry in $(M_R)^2$ is 1 because the (2, 1) and (1, 4) entries in M_R are both 1. In other words, there is a path of length 2 from v_2 to v_4 because there is an edge from v_2 to v_1 and an edge from v_1 to v_4. This illustrates why $(M_R)^2$ designates paths of length 2 in R.

By Theorem 2.4.9, $(M_R)^2 = M_{R^2}$, and that implies that $(M_R)^2$ designates paths of length 2. From the matrix $(M_R)^2$, we obtain

$$R^2 = \{(v_1, v_1), (v_1, v_2), (v_1, v_3), (v_2, v_1), (v_2, v_2), (v_2, v_4), (v_3, v_3), (v_4, v_1), \\ (v_4, v_2), (v_4, v_3), (v_4, v_4)\}$$ ■

COROLLARY 2.4.11 If R is a relation on a finite set, then

$$M_{R^n} = (M_R)^n \qquad \text{for all } n \geq 1$$

A proof of this corollary will be given after we study mathematical induction in Chapter 5.

It follows from Corollary 2.4.11 that

$$R^n = \{(v_i, v_j) : (i, j) \text{ entry of } (M_R)^n \text{ is } 1\}$$

Let R and S be relations on a finite set A. Then, $R \cup S$ is a relation on A. It is not difficult to verify that the adjacency matrix representing $R \cup S$ is equal to $M_R \oplus M_S$. We illustrate $M_{R \cup S} = M_R \oplus M_S$ by the next example.

EXAMPLE 2.4.12 Let $A = \{a, b, c\}$, and let $R = \{(a, b), (b, b), (c, c)\}$ and $S = \{(a, a), (b, c), (c, a), (c, c)\}$ be relations on A. Thus, $R \cup S = \{(a, b), (b, b), (c, c), (a, a), (b, c), (c, a)\}$. From R, S, and $R \cup S$, we obtain

$$M_R = \begin{bmatrix} 0 & 1 & 0 \\ 0 & 1 & 0 \\ 0 & 0 & 1 \end{bmatrix}, \qquad M_S = \begin{bmatrix} 1 & 0 & 0 \\ 0 & 0 & 1 \\ 1 & 0 & 1 \end{bmatrix}, \quad \text{and} \quad M_{R \cup S} = \begin{bmatrix} 1 & 1 & 0 \\ 0 & 1 & 1 \\ 1 & 0 & 1 \end{bmatrix}$$

But

$$M_R \oplus M_S = \begin{bmatrix} 0 & 1 & 0 \\ 0 & 1 & 0 \\ 0 & 0 & 1 \end{bmatrix} \oplus \begin{bmatrix} 1 & 0 & 0 \\ 0 & 0 & 1 \\ 1 & 0 & 1 \end{bmatrix}$$

$$= \begin{bmatrix} 1 & 1 & 0 \\ 0 & 1 & 1 \\ 1 & 0 & 1 \end{bmatrix}$$

Hence, $M_{R \cup S} = M_R \oplus M_S$. ■

In general, if $R_1, R_2, \ldots, R_n$ are relations on a finite set A, then

$$M_{R_1 \cup R_2 \cup \cdots \cup R_n} = M_{R_1} \oplus M_{R_2} \oplus \cdots \oplus M_{R_n}$$

COROLLARY 2.4.13 If R is a relation on a set with n elements, then

$$M_{R^+} = M_R \oplus (M_R)^2 \oplus \cdots \oplus (M_R)^n$$

Proof From Corollary 2.3.9, $R^+ = R^1 \cup R^2 \cup \cdots \cup R^n$. Hence,

$$M_{R^+} = M_{R^1 \cup R^2 \cup \cdots \cup R^n} = M_{R^1} \oplus M_{R^2} \oplus \cdots \oplus M_{R^n}$$

By Corollary 2.4.11, $M_{R^+} = M_R \oplus (M_R)^2 \oplus \cdots \oplus (M_R)^n$. ∎

The matrix M_{R^+} is called the **connectivity matrix** for R; it specifies all vertices that can be connected by paths in R. Corollary 2.4.13 provides an effective method of computing R^+, as follows:

1. Find the adjacency matrix M_R.
2. Compute $(M_R)^2, (M_R)^3, \ldots, (M_R)^n$.
3. Compute $M_R \oplus (M_R)^2 \oplus (M_R)^3 \oplus \cdots \oplus (M_R)^n$.

In the next section, we study a more efficient method of computing M_{R^+}.

■ **PROGRAMMING NOTES**

A Pascal Procedure for Computing the Square of a Given Adjacency Matrix In order to use Pascal to manipulate adjacency matrices, the matrices are declared as two-dimensional Boolean arrays. So an entry of 0 in a matrix is assigned the value "false," and an entry of 1 is assigned the value "true." The Pascal implementation of the Boolean addition and multiplication for 0 and 1 is OR and AND, respectively.

First let us write an algorithm for computing the square of an $n \times n$ matrix M where $MSQ = M^2 = M \otimes M$. We will use pseudocode and the mathematical notation for Boolean addition and multiplication.

ALGORITHM 2.4.14 **MATRIXSQUARE** This algorithm computes the square of an $n \times n$ matrix M and returns $MSQ = M^2 = M \otimes M$.

for i = 1 to n do
  for j = 1 to n do
    MSQ[i, j] := 0
    for k = 1 to n do
      MSQ[i, j] := MSQ[i, j] + (M[i, k] · M[k, j])
              (*Where + and · are Boolean addition and
              Boolean multiplication, respectively*)
Return MSQ.

This algorithm is easily translated into a Pascal procedure. It is assumed that MSQ and M are $n \times n$ Boolean arrays.

```
procedure MatrixSquare;
    var
        i, j, k : integer;
    begin
        for i := 1 to n do
            for j := 1 to n do
                begin
                    MSQ[i, j] := false;
                    for k := 1 to n do
                        MSQ[i, j] := MSQ[i, j] or (M[i, k] and M[k, j])
                end
    end;
```

PRACTICE PROBLEM 7 Using paper and pencil, apply Algorithm 2.4.14 (Matrixsquare) to the matrix
$$M = \begin{bmatrix} 1 & 1 \\ 1 & 0 \end{bmatrix}$$

A Pascal Procedure to Determine Whether a Given Relation on a Finite Set Is Symmetric
Suppose we are given a relation on a set with n elements. We want a procedure, in Pascal, that will tell us whether the relation is symmetric. Let ADJ be the adjacency matrix for the given relation. Suppose maxnodes is a previously declared variable of type integer with an assigned value of n; and ADJ is a previously declared maxnodes by maxnodes array of type Boolean in a Pascal program. So we are assuming that $ADJ[i, j]$ is true if and only if the pair (i, j) is in the given relation. The following procedure will determine whether the relation is symmetric.

To see this we need to recall that the relation will be symmetric whenever "If $ADJ[i, j]$, then $ADJ[j, i]$" is true for all i, j. The Pascal equivalent to the statement "If $ADJ[i, j]$, then $ADJ[j, i]$" is $ADJ[i, j]) <= ADJ[j, i]$. Therefore, the Boolean variable SCheck will remain assigned as true only as long as the statement $ADJ[i, j] <= ADJ[j, i]$ is true.

```
procedure Symmcheck;
    var
        i, j : integer;
        SCheck : Boolean;
    begin
        SCheck := true;
        for j := 1 to maxnodes do
            for i := 1 to maxnodes do
                SCheck := SCheck and (ADJ[i, j] <= ADJ[j, i]);
        if SCheck then
            writeln('The relation is symmetric.')
        else
            writeln('The relation is not symmetric.')
    end;
```

EXERCISE SET 2.4

1. Let $A = \{a, b, c, d\}$ and $R = \{(a, a), (a, b), (b, d), (c, a), (c, c), (d, c), (d, d)\}$.
   a. Write the adjacency matrix M_R. b. Find $M_R \oplus M_R$.

2. Refer to Exercise 1.
   a. Find the Boolean product $M_R \otimes M_R$.
   b. Find $R \circ R$ using the graphical representation of R.
   c. Verify that the relation obtained from the matrix $M_R \otimes M_R$ is $R \circ R$.

3. Let
$$P = \begin{bmatrix} 0 & 1 & 1 \\ 0 & 1 & 1 \\ 1 & 1 & 0 \end{bmatrix} \quad \text{and} \quad Q = \begin{bmatrix} 1 & 1 & 0 \\ 1 & 0 & 1 \\ 0 & 1 & 0 \end{bmatrix}$$

   a. Find $P \oplus Q$, $P \otimes Q$, $Q \otimes P$, P^2, and Q^2.
   b. Let $A = \{1, 2, 3\}$. Find relations R and S on A such that $P = M_R$ and $Q = M_S$.
   c. Use Theorem 2.4.9 and parts a and b above to find $R \circ S$ and $S \circ R$.
   d. Use P^2 to find R^2.

4. Let
$$P = \begin{bmatrix} 0 & 1 & 0 \\ 0 & 1 & 0 \\ 1 & 0 & 1 \end{bmatrix} \quad \text{and} \quad Q = \begin{bmatrix} 1 & 0 & 0 \\ 1 & 1 & 1 \\ 0 & 1 & 0 \end{bmatrix}$$

   a. Find $P \otimes Q$ and P^2.
   b. Find relations R and S on $A = \{1, 2, 3\}$ such that $M_R = P$ and $M_S = Q$.
   c. Use Theorem 2.4.9 and parts a and b above to find $R \circ S$.
   d. Use P^2 to find R^2.

5. Let $A = \{a, b, c, d\}$ and $R = \{(a, b), (b, a), (b, c), (c, d)\}$.
   a. Find the adjacency matrices M_R, $(M_R)^2$, $(M_R)^3$, and $(M_R)^4$.
   b. Find M_{R^+}.
   c. Find R^+ from M_{R^+}.

6. Let $A = \{1, 2, 3, 4\}$ and $R = \{(1, 1), (1, 2), (1, 3), (2, 1), (2, 2), (3, 1), (3, 3)\}$.
   a. Write M_R.
   b. Determine whether R is reflexive by studying M_R.
   c. Determine whether R is symmetric by studying M_R.
   d. Find M_{R^+}.
   e. Find R^+ from M_{R^+}.

7. Let $A = \{1, 2, 3, 4, 5\}$ and $R = \{(1, 1), (1, 2), (1, 3), (1, 4), (1, 5), (2, 2), (2, 3), (2, 4), (2, 5), (3, 3), (3, 4), (3, 5), (4, 4), (4, 5), (5, 5)\}$.
   a. Write M_R.
   b. Determine whether R is symmetric by studying M_R.
   c. Determine whether R is antisymmetric by studying M_R.
   d. Find M_{R^+}.
   e. Find R^+ from M_{R^+}.

8. Let $A = \{1, 2, 3, 4, 5\}$ and $R = \{(1, 1), (1, 3), (1, 5), (2, 2), (2, 4), (3, 1), (3, 3), (3, 5), (4, 2), (4, 4), (5, 1), (5, 3), (5, 5)\}$.

a. Write M_R.
b. Determine whether R is symmetric by studying M_R.
c. Determine whether R is antisymmetric by studying M_R.
d. Find M_{R^+}.
e. Find R^+ from M_{R^+}.

9. Let $A = \{1, 2, 3, 4\}$, $R = \{(1, 2), (1, 3), (3, 4)\}$, and $S = \{(2, 1), (2, 2), (2, 4), (3, 4)\}$.
   a. Find M_R and M_S.
   b. Find $R \cup S$ and $M_{R \cup S}$.
   c. Verify that $M_{R \cup S} = M_R \oplus M_S$.

10. Let $A = \{1, 2, 3\}$, $B = \{2, 3\}$, $C = \{1, 3\}$, and $D = \{a, b, c\}$. Let R, S, and T be relations with $R \subseteq A \times B$, $S \subseteq B \times C$, and $T \subseteq C \times D$, where $R = \{(1, 2), (2, 2), (3, 3)\}$, $S = \{(2, 1), (2, 3), (3, 3)\}$, and $T = \{(1, b), (3, a)\}$.
    a. Find $R \circ S$ and $M_{R \circ S}$.
    b. Find M_R and M_S.
    c. Verify that $M_{R \circ S} = M_R \otimes M_S$.
    d. Verify that $M_{(R \circ S) \circ T} = M_{R \circ (S \circ T)}$. [Therefore, it follows that $(R \circ S) \circ T = R \circ (S \circ T)$.]

11. Let $A = \{a, b, c, d, e\}$ and $R = \{(a, a), (a, b), (a, d), (b, c), (c, e), (d, b)\}$.
    a. Find M_R.
    b. Find R^2 and R^3.
    c. Find M_{R^2} and M_{R^3}.
    d. Find $(M_R)^2$ and $(M_R)^3$.
    e. Use parts c and d to verify that $M_{R^2} = (M_R)^2$ and $M_{R^3} = (M_R)^3$.

12. Let $A = \{a, b, c, d\}$. Find an adjacency matrix for each of the following:
    a. A relation on A that is reflexive and antisymmetric but not symmetric
    b. A relation on A that is reflexive and symmetric but not antisymmetric

13. Let $A = \{a, b, c, d\}$. Find an adjacency matrix for each of the following:
    a. A relation on A that is antisymmetric but is neither reflexive nor symmetric
    b. A relation on A that is reflexive and symmetric but not transitive

14. Verify that $M_{R \cup S} = M_R \oplus M_S$ for arbitrary relations R and S on $\{1, 2, 3\}$. [Hint: Consider the cases $(i, j) \in R \cup S$ and $(i, j) \notin R \cup S$ for an arbitrary pair (i, j).]

15. Let $A = \{1, 2, 3, 4, 5\}$ and $R = \{(1, 1), (1, 2), (1, 3), (1, 4), (1, 5), (2, 2), (2, 3), (2, 4), (2, 5), (3, 3), (3, 4), (3, 5), (4, 4), (4, 5), (5, 5)\}$. Find the $(2, 5)$ entry in $(M_R)^3$ by considering paths in R.

16. Let $A = \{1, 2, 3, 4, 5\}$ and $R = \{(1, 1), (1, 3), (1, 5), (2, 2), (2, 4), (3, 1), (3, 3), (3, 5), (4, 2), (4, 4), (5, 1), (5, 3), (5, 5)\}$. Find the $(3, 4)$ entry in $(M_R)^5$ by considering paths in R.

17. Verify that Boolean addition of matrices satisfies the following properties in the special case of 2×3 matrices.
    a. $N \oplus T = T \oplus N$ Commutative law
    b. $(N \oplus T) \oplus P = N \oplus (T \oplus P)$ Associative law
    c. $M \oplus 0 = M = 0 \oplus M$ Additive identity law
    [Hint: Use the properties given for Boolean addition of 0 and 1.]

18. Verify that Boolean multiplication of $n \times n$ matrices satisfies the following properties for the special case of $n = 2$.
    a. $(N \otimes T) \otimes M = N \otimes (T \otimes M)$ Associative law
    b. $M \otimes I = M = I \otimes M$, where I is the $n \times n$ identity matrix   Identity law

19. Show that there is no commutative law for matrix multiplication by finding 2×2 matrices M and T such that $M \otimes T \neq T \otimes M$.

PROGRAMMING EXERCISES

20. Given a relation R and its adjacency matrix $M_R = M$, write a procedure in Pascal that will:
   a. Determine whether R is reflexive.
   b. Determine whether R is antisymmetric.
   c. Determine whether R is transitive.

21. Write a procedure in Pascal that takes as input two $n \times n$ Boolean matrices M and N and computes the Boolean product $M \otimes N$.

2.5
REACHABILITY AND WARSHALL'S ALGORITHM

In any network involving the flow of information or some other commodity, whether one can go from one place to another is a significant issue. This is the question of reachability.

The connectivity relation R^+ for a relation R on a finite set A specifies which vertices are connected to other vertices by paths in R. The adjacency matrix M_{R^+} is called the *connectivity matrix*. Note that $M_{R^+}[i,j] = 1$ if and only if the ith vertex and the jth vertex are connected by some path in R; otherwise, $M_{R^+}[i,j] = 0$.

In general, it is rather tedious to calculate R^+. One way is to find R^2, $R^3, \ldots, R^n$ to obtain $R^+ = R^1 \cup R^2 \cup \cdots \cup R^n$. Another way to find R^+ is to use the equation $M_{R^+} = M_R \oplus (M_R)^2 \oplus \cdots \oplus (M_R)^n$.

Warshall's Algorithm

Warshall's algorithm is another well-known algorithm for calculating the connectivity matrix M_{R^+}.

The vertices $x_1, x_2, \ldots, x_{k-1}$, in a path $(x_0, x_1, \ldots, x_k)$ from vertex x_0 to vertex x_k, are called **interior vertices** of the path.

Let R be a relation on a set $V = \{v_1, v_2, \ldots, v_n\}$. In what follows, we shall construct a sequence of matrices $P_0, P_1, P_2, \ldots, P_n$. The last matrix P_n will be the connectivity matrix M_{R^+}.

In the following, $P_k[i,j]$ stands for the (i,j) entry of the matrix P_k, where $1 \leq i \leq n, 1 \leq j \leq n$, and $0 \leq k \leq n$. Warshall's algorithm begins with a matrix $P_0 = M_R$. Each of the matrices $P_1, P_2, \ldots, P_n$ is computed as follows:

P_0 $P_0 = M_R$

P_1 $P_1[i,j] = 1$ if there is a path in R from v_i to v_j with interior vertices, if any, from the set $\{v_1\}$

         $P_1[i,j] = 0$ otherwise

P_2 $P_2[i,j] = 1$ if there is a path in R from v_i to v_j with interior vertices, if any, from the set $\{v_1, v_2\}$

         $P_2[i,j] = 0$ otherwise

2.5 REACHABILITY AND WARSHALL'S ALGORITHM

$\vdots$

P_k $P_k[i,j] = 1$ if there is a path in R from v_i to v_j with interior vertices, if any, from the set $\{v_1, v_2, \ldots, v_k\}$
$P_k[i,j] = 0$ otherwise

$\vdots$

P_n $P_n[i,j] = 1$ if there is a path in R from v_i to v_j with interior vertices, if any, from the set $\{v_1, v_2, \ldots, v_n\}$
$P_n[i,j] = 0$ otherwise

Note: $P_0[i,j] = 1$ if there is a path from v_i to v_j with no interior vertices; $P_0[i,j] = 0$ otherwise. Hence, the definition of P_0 is consistent with the definition of P_k for $1 \leq k \leq n$.

Since any path in R must have interior vertices, if any, from the set $V = \{v_1, v_2, \ldots, v_n\}$, we see that $P_n = M_{R^+}$.

PRACTICE PROBLEM 1   In Example 2.4.10, $V = \{v_1, v_2, v_3, v_4\}$ and $R = \{(v_1, v_2), (v_1, v_4), (v_2, v_1), (v_2, v_2), (v_3, v_3), (v_4, v_1), (v_4, v_2), (v_4, v_3)\}$. Find P_2 and note that $P_2 \neq (M_R)^2$.

In the computation of $P_1, P_2, \ldots, P_n$, we note that, if $P_{k-1}[i,j] = 1$, then $P_k[i,j] = 1$. In other words, if there is a path from v_i to v_j with interior vertices from the set $\{v_1, v_2, \ldots, v_{k-1}\}$, then there is a path from v_i to v_j with interior vertices from the set $\{v_1, v_2, \ldots, v_{k-1}, v_k\}$. Hence, if $P_{k-1}[i,j] = 1$, then $P_k[i,j] = 1$ for $k = 1, \ldots, n$.

We further note that, for $k = 1, \ldots, n$, we have $P_k[i,j] = 1$ if both $P_{k-1}[i,k] = 1$ and $P_{k-1}[k,j] = 1$. In other words, there is a path from v_i to v_j with interior vertices from the set $\{v_1, v_2, \ldots, v_k\}$ if there are paths from v_i to v_k and from v_k to v_j both with interior vertices from the set $\{v_1, v_2, \ldots, v_{k-1}\}$. This is the basis of Warshall's algorithm. We give a proof of this important result.

THEOREM 2.5.1 $P_k[i,j] = 1$ if and only if

$$P_{k-1}[i,j] = 1 \text{ or } (P_{k-1}[i,k] = 1 \text{ and } P_{k-1}[k,j] = 1), \quad \text{for } k = 1, \ldots, n$$

Proof   ($\rightarrow$) First assume $P_k[i,j] = 1$. This means there exists a path in R from v_i to v_j with interior vertices, if any, from the set $\{v_1, v_2, \ldots, v_k\}$. We may assume, without loss of generality, that the interior vertices of this path are distinct since, if there is a path from v_i to v_j, then there is a path from v_i to v_j with no repeated interior vertices. If this path does not contain v_k as an interior vertex, then $P_{k-1}[i,j] = 1$. If v_k is an interior vertex of this path, then there is a path from v_i to v_k with interior vertices from the set $\{v_1, v_2, \ldots, v_{k-1}\}$ and there is a path from v_k to v_j with interior vertices from the set $\{v_1, v_2, \ldots, v_{k-1}\}$. See Figure 2.5.1. Therefore, $P_{k-1}[i,k] = 1$ and $P_{k-1}[k,j] = 1$.

($\leftarrow$) To prove the converse, assume that $P_{k-1}[i,j] = 1$ or that both $P_{k-1}[i,k] = 1$ and $P_{k-1}[k,j] = 1$. Clearly, if $P_{k-1}[i,j] = 1$, then $P_k[i,j] = 1$.

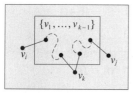

FIGURE 2.5.1

If $P_{k-1}[i, k] = 1$ and $P_{k-1}[k, j] = 1$, then there is a path from v_i to v_k with interior vertices from the set $\{v_1, v_2, \ldots, v_{k-1}\}$ and a path from v_k to v_j with interior vertices from the set $\{v_1, v_2, \ldots, v_{k-1}\}$. Combining these two paths, we can form a path from v_i to v_j with interior vertices from the set $\{v_1, v_2, \ldots, v_{k-1}, v_k\}$. ∎

We summarize **Warshall's algorithm** in concise form for easy reference.

ALGORITHM 2.5.2 **WARSHALL'S ALGORITHM FOR COMPUTING M_{R^+}**

$$P_0 = M_R;$$
For $k = 1, \ldots, n,$
$$P_k[i, j] = P_{k-1}[i, j] + (P_{k-1}[i, k] \cdot P_{k-1}[k, j]),$$
where $1 \leq i \leq n, 1 \leq j \leq n$ \hfill (1)

We will use Warshall's algorithm to compute the transitive closure R^+ of a relation R in the next example. Since this is our first example using Warshall's algorithm, we include some details.

EXAMPLE 2.5.3 Let $V = \{v_1, v_2, v_3\}$ and $R = \{(v_1, v_2), (v_2, v_1), (v_2, v_3)\}$. Find the connectivity relation R^+.

Solution   Since $n = 3$, the connectivity matrix M_{R^+} is the matrix P_3. First, let

$$P_0 = M_R = \begin{bmatrix} 0 & 1 & 0 \\ 1 & 0 & 1 \\ 0 & 0 & 0 \end{bmatrix}$$

For $k = 1$, we find the nine entries of P_1 using Formula (1), as follows:

$P_1[1, 1] = P_0[1, 1] + (P_0[1, 1] \cdot P_0[1, 1]) = 0 + (0 \cdot 0) = 0$
$P_1[1, 2] = P_0[1, 2] + (P_0[1, 1] \cdot P_0[1, 2]) = 1 + (0 \cdot 1) = 1 + 0 = 1$
$P_1[1, 3] = P_0[1, 3] + (P_0[1, 1] \cdot P_0[1, 3]) = 0 + (0 \cdot 0) = 0$
$P_1[2, 1] = P_0[2, 1] + (P_0[2, 1] \cdot P_0[1, 1]) = 1 + (1 \cdot 0) = 1 + 0 = 1$
$P_1[2, 2] = P_0[2, 2] + (P_0[2, 1] \cdot P_0[1, 2]) = 0 + (1 \cdot 1) = 0 + 1 = 1$
$P_1[2, 3] = P_0[2, 3] + (P_0[2, 1] \cdot P_0[1, 3]) = 1 + (1 \cdot 0) = 1 + 0 = 1$
$P_1[3, 1] = P_0[3, 1] + (P_0[3, 1] \cdot P_0[1, 1]) = 0 + (0 \cdot 0) = 0$
$P_1[3, 2] = P_0[3, 2] + (P_0[3, 1] \cdot P_0[1, 2]) = 0 + (0 \cdot 1) = 0 + 0 = 0$
$P_1[3, 3] = P_0[3, 3] + (P_0[3, 1] \cdot P_0[1, 3]) = 0 + (0 \cdot 0) = 0$

Hence,

$$P_1 = \begin{bmatrix} 0 & 1 & 0 \\ 1 & 1 & 1 \\ 0 & 0 & 0 \end{bmatrix}$$

2.5 REACHABILITY AND WARSHALL'S ALGORITHM

Note that $P_1[1, 2] = 1$, $P_1[2, 1] = 1$, and $P_1[2, 3] = 1$ because the entry of P_0 at the same location is 1. We could have skipped the computations for these entries of P_1.

For $k = 2$, we have $P_2[1, 2] = 1$, $P_2[2, 1] = 1$, $P_2[2, 2] = 1$, and $P_2[2, 3] = 1$ since P_1 has entry 1 at the same location. We find the remaining entries of P_2 using Formula (1):

$$P_2[1, 1] = P_1[1, 1] + (P_1[1, 2] \cdot P_1[2, 1]) = 0 + (1 \cdot 1) = 0 + 1 = 1$$
$$P_2[1, 3] = P_1[1, 3] + (P_1[1, 2] \cdot P_1[2, 3]) = 0 + (1 \cdot 1) = 1$$
$$P_2[3, 1] = P_1[3, 1] + (P_1[3, 2] \cdot P_1[2, 1]) = 0 + (0 \cdot 1) = 0$$
$$P_2[3, 2] = P_1[3, 2] + (P_1[3, 2] \cdot P_1[2, 2]) = 0 + (0 \cdot 1) = 0$$
$$P_2[3, 3] = P_1[3, 3] + (P_1[3, 2] \cdot P_1[2, 3]) = 0 + (0 \cdot 1) = 0$$

Hence,

$$P_2 = \begin{bmatrix} 1 & 1 & 1 \\ 1 & 1 & 1 \\ 0 & 0 & 0 \end{bmatrix}$$

For $k = 3$, each entry of the first two rows of P_3 is 1, since each entry of the first two rows of P_2 is 1. The remaining entries of P_3 are computed as follows:

$$P_3[3, 1] = P_2[3, 1] + (P_2[3, 3] \cdot P_2[3, 1]) = 0 + (0 \cdot 0) = 0$$
$$P_3[3, 2] = P_2[3, 2] + (P_2[3, 3] \cdot P_2[3, 2]) = 0 + (0 \cdot 0) = 0$$
$$P_3[3, 3] = P_2[3, 3] + (P_2[3, 3] \cdot P_2[3, 3]) = 0 + (0 \cdot 0) = 0$$

Hence,

$$P_3 = \begin{bmatrix} 1 & 1 & 1 \\ 1 & 1 & 1 \\ 0 & 0 & 0 \end{bmatrix}$$

P_3 is the matrix at the end of execution of Warshall's algorithm, and $P_3 = M_{R^+}$. From the matrix M_{R^+}, we obtain $R^+ = \{(v_1, v_1), (v_1, v_2), (v_1, v_3), (v_2, v_1), (v_2, v_2), (v_2, v_3)\}$. ∎

Although the computation of R^+ appears lengthy, it can be performed quickly, especially after a bit of practice.

We emphasize the fact that the matrices $P_1, P_2, \ldots, P_n$ are, in general, different from the powers of M_R. This difference results in significant savings of computation. In Example 2.5.3,

$$(M_R)^2 = \begin{bmatrix} 0 & 1 & 0 \\ 1 & 0 & 1 \\ 0 & 0 & 0 \end{bmatrix} \otimes \begin{bmatrix} 0 & 1 & 0 \\ 1 & 0 & 1 \\ 0 & 0 & 0 \end{bmatrix} = \begin{bmatrix} 1 & 0 & 1 \\ 0 & 1 & 0 \\ 0 & 0 & 0 \end{bmatrix}$$

which is different from

$$P_1 = \begin{bmatrix} 0 & 1 & 0 \\ 1 & 1 & 1 \\ 0 & 0 & 0 \end{bmatrix}$$

and different from

$$P_2 = \begin{bmatrix} 1 & 1 & 1 \\ 1 & 1 & 1 \\ 0 & 0 & 0 \end{bmatrix}$$

PRACTICE PROBLEM 2 Let $V = \{v_1, v_2, v_3, v_4\}$, and let $R = \{(v_1, v_2), (v_2, v_1), (v_3, v_2), (v_4, v_3), (v_4, v_4)\}$ be a relation on V. Use Warshall's algorithm to find the connectivity matrix M_{R^+}.

EXAMPLE 2.5.4 Let $A = \{1, 2, 3, 4, 5\}$ and $R = \{(i, j) : j = i + 1\}$. Suppose we have found

$$P_2 = \begin{bmatrix} 0 & 1 & 1 & 0 & 0 \\ 0 & 0 & 1 & 0 & 0 \\ 0 & 0 & 0 & 1 & 0 \\ 0 & 0 & 0 & 0 & 1 \\ 0 & 0 & 0 & 0 & 0 \end{bmatrix}$$

Then, $P_3[2, 4]$ and $P_3[3, 5]$ can be computed as follows:

$$P_3[2, 4] = P_2[2, 4] + (P_2[2, 3] \cdot P_2[3, 4]) = 0 + (1 \cdot 1) = 1$$
$$P_3[3, 5] = P_2[3, 5] + (P_2[3, 3] \cdot P_2[3, 5]) = 0 + (0 \cdot 0) = 0$$

PRACTICE PROBLEM 3 In Example 2.5.4, find the entries $P_3[1, 3]$ and $P_3[4, 5]$.

Warshall's Algorithm in Pseudocode

M_{R^+} is easily calculated using a digital computer. We can readily rewrite the following pseudocode version of Warshall's algorithm (Algorithm 2.5.2) in any programming language. The matrix M_R is identified as MATRIX and the matrices $P_0, P_1, \ldots, P_n$ are all identified as PATH; both MATRIX and PATH are two-dimensional Boolean arrays.

ALGORITHM 2.5.5 **PATH**

PATH := MATRIX
For k = 1 to n do
  For i = 1 to n do
    For j = 1 to n do
      PATH[i, j] := PATH[i, j] + (PATH[i, k] · PATH[k, j])
        (*Where + and · are Boolean addition and
        Boolean multiplication, respectively*)
Return PATH.

After this algorithm is executed, the final matrix PATH is the connectivity matrix M_{R^+}.

Now we introduce a concept that is closely related to the connectivity relation. Suppose $V = \{v_1, v_2, \ldots, v_n\}$. Given a relation R on V, the **reachability relation** $R^* = \{(v_i, v_j) : v_i = v_j \text{ or there is a path in } R \text{ from } v_i \text{ to } v_j\}$. The adjacency matrix M_{R^*} corresponding to R^* is called the **reachability matrix** of R.

Let $D = \{(v, v) : v \in V\}$, the **diagonal relation** on V. Note that the $n \times n$ identity matrix I is the adjacency matrix for D.

By definition, $R^* = D \cup R^+$. Hence, $M_{R^*} = M_D \oplus M_{R^+} = I \oplus M_{R^+}$.

The reachability relation R^* is both reflexive and transitive and is also called the **reflexive transitive closure** of R.

The reachability relation R^* specifies which vertices can be reached from other vertices. Each vertex can reach itself, and a vertex v can reach another vertex w if there is a path from v to w. Hence, R^* gives the whole picture of reachability. The question "Can I get there from here?" can be readily answered by finding R^*.

In order to apply Warshall's algorithm to compute the reachability matrix M_{R^*}, we need the following theorem. The proof of Theorem 2.5.6 is left as an exercise. (See Exercise 12, Exercise Set 2.5.)

THEOREM 2.5.6 $R^* = (D \cup R)^+$.

Warshall's algorithm for computing M_{R^+} provides an efficient way of computing M_{R^*} from the adjacency matrix M_R and the identity matrix I. By Theorem 2.5.6 we can obtain M_{R^*} by applying Warshall's algorithm beginning with the matrix $I \oplus M_R$.

PRACTICE PROBLEM 4 Make appropriate changes in Algorithm 2.5.5 (Path) to obtain Algorithm Reach, which will compute M_{R^*}. In the algorithm, the reachability matrix should be identified as REACH, the adjacency matrix M_R is identified as MATRIX, and the $n \times n$ identity matrix I is identified as IDENTITY.

Warshall's algorithm (Algorithm 2.5.2) is straightforward and systematic. However, slight modification of this method will greatly increase the efficiency of the algorithm. As in Formula (1), we begin with $P_0 = M_R$. From Formula (1) we have

$$P_k[i, j] = P_{k-1}[i, j] + (P_{k-1}[i, k] \cdot P_{k-1}[k, j])$$

Using the preceding equation, we see the following:

If $P_{k-1}[i, k] = 0$, then $P_k[i, j] = P_{k-1}[i, j]$.
If $P_{k-1}[i, k] = 1$, then $P_k[i, j] = P_{k-1}[i, j] + P_{k-1}[k, j]$.

We incorporate these shortcuts into the following algorithm.

ALGORITHM 2.5.7 NEWPATH

>PATH := MATRIX
>for k = 1 to n do
> for i = 1 to n do
>   if PATH[i, k] = 1 then
>   for j = 1 to n do
>   PATH[i, j] := PATH[i, j] + PATH[k, j]   (*Where + is Boolean addition*)
>Return PATH.

Algorithm 2.5.7 (Newpath) is significantly faster than Algorithm 2.5.5 (Path) for many relations since it does not execute the inner loop when PATH$[i, k] = 0$. The initial matrix is M_R, which usually has many zero entries. For these zero entries, the inner loop is not executed. The authors conducted a simple experiment on a microcomputer with a 16×16 adjacency matrix M_R. Using Algorithm 2.5.5 (Path) it took more than 43 seconds to find M_{R^+}, whereas Algorithm 2.5.7 (Newpath) performed the same task in less than 6 seconds.

Algorithm Newpath is stated in concise form below for easy reference.

ALGORITHM 2.5.7 NEWPATH (Concise Form)

$$P_0 = M_R;$$
$$\text{For } k = 1, \ldots, n,$$
$$P_k[i,j] = P_{k-1}[i,j] \text{ if } P_{k-1}[i,k] = 0,$$
$$P_k[i,j] = P_{k-1}[i,j] + P_{k-1}[k,j] \text{ if } P_{k-1}[i,k] = 1, \quad (2)$$
$$\text{where } 1 \leq i \leq n,\ 1 \leq j \leq n$$

PRACTICE PROBLEM 5 Let $V = \{v_1, v_2, v_3\}$ and $R = \{(v_1, v_2), (v_2, v_1), (v_2, v_3)\}$. Use Formula (2) to find M_{R^+}. Compare the computation with that in Example 2.5.3.

■ **PROGRAMMING NOTES**

Algorithm 2.5.7 (Newpath) as a Pascal Procedure As before, we are assuming MATRIX and PATH are $n \times n$ Boolean arrays.

```
procedure NewPath;
   var
      i, j, k : integer;
   begin
      PATH := MATRIX;
      for k := 1 to n do
         for i := 1 to n do
            if PATH[i, k] then
               for j := 1 to n do
                  PATH[i, j] := PATH[i, j] or PATH[k, j]
   end;
```

Floyd's Shortest Path Algorithm

In automobile travel, finding the shortest route from one city to another is often desirable. This problem is similar to finding the shortest path in a digraph between two vertices. In 1962, R. W. Floyd constructed a **shortest path algorithm**, which we will call Algorithm Short and will implement as procedure Short.

Let $V = \{v_1, v_2, \ldots, v_n\}$, and let R be a relation on V. In procedure Short, MATRIX represents the adjacency matrix M_R, and SHORT represents an $n \times n$ matrix with integer entries. Both MATRIX and SHORT are $n \times n$ arrays of integers. We assume a previously defined function min, which outputs the minimum of two integers. Now, we state Algorithm Short and procedure Short.

ALGORITHM 2.5.8 **SHORT** This algorithm finds the length of the shortest path between two vertices if there is a path between the vertices. If there is no path between two vertices, the result so indicates.

The initial matrix S_0 is M_R with some modifications:

$S_0[i, i] = 0 \quad \text{for } i = 1, 2, \ldots, n$

$S_0[i, j] = 1 \quad \text{if } M_R[i, j] = 1 \text{ and } i \neq j$

$S_0[i, j] = b \quad \text{if } M_R[i, j] = 0 \text{ and } i \neq j, \text{ where } b > n$

The matrix S_k, for $1 \leq k \leq n$, is computed as follows:

$S_k[i, j] = \min(S_{k-1}[i, j], S_{k-1}[i, k] + S_{k-1}[k, j]), \quad \text{for } 1 \leq i \leq n, 1 \leq j \leq n$

The matrix S_n is the shortest path matrix.

```
procedure Short;
  var
    i, j, k : integer;
  begin
        for i := 1 to n do                  (*Initialize SHORT*)
          for j := 1 to n do
            if MATRIX[i, j] = 1 then
              SHORT[i, j] := 1
            else
              SHORT[i, j] := 300;           (*Use a large number, at least*)
                                            (*n + 1, to indicate there is no*)
        for i := 1 to n do                  (*edge (i, j).*)
          SHORT[i, i] := 0;                 (*Do not count the edge (i, i).*)
         for k := 1 to n do                 (*Generate SHORT*)
           for i := 1 to n do
             for j := 1 to n do
               SHORT[i, j] := min(SHORT[i, j], SHORT[i, k] + SHORT[k, j])
                                            (*Where + is ordinary integer*)
                                            (*addition and min stands for*)
                                            (*minimum.*)
  end;
```

After execution of procedure Short, we obtain the **shortest path matrix** SHORT, whose (i, j) entry is the number of edges in the shortest path from v_i to v_j if the entry is less than 300. If the (i, j) entry is 300, there is no path from v_i to v_j. The proof that Algorithm 2.5.8 (Short) is correct is similar to the proof (in Theorem 2.5.1) that Warshall's algorithm is correct.

Now we apply Algorithm 2.5.8 (Short) in the following example. For simplicity, let S_0 denote the matrix SHORT after initialization, and let S_k denote the matrix SHORT for $k = 1, 2, \ldots, n$. The matrix S_n is the shortest path matrix at the end of Algorithm Short, where n is the number of vertices in the set V.

EXAMPLE 2.5.9 Let $V = \{v_1, v_2, v_3\}$ and $R = \{(v_1, v_2), (v_2, v_1), (v_1, v_3), (v_3, v_2), (v_3, v_3)\}$. Use Algorithm Short to find the shortest path matrix S_3.

Solution The adjacency matrix is

$$M_R = \begin{bmatrix} 0 & 1 & 1 \\ 1 & 0 & 0 \\ 0 & 1 & 1 \end{bmatrix}$$

We initialize to obtain

$$S_0 = \begin{bmatrix} 0 & 1 & 1 \\ 1 & 0 & 300 \\ 300 & 1 & 0 \end{bmatrix}$$

For $k = 1$, we use the following formula from Algorithm Short:

$$S_1[i, j] = \min(S_0[i, j], S_0[i, 1] + S_0[1, j]), \quad \text{for } 1 \le i \le 3,\ 1 \le j \le 3$$

We find

$S_1[1, 1] = \min(S_0[1, 1], S_0[1, 1] + S_0[1, 1]) = 0$
$S_1[1, 2] = \min(S_0[1, 2], S_0[1, 1] + S_0[1, 2]) = 1$
$S_1[1, 3] = 1$
$S_1[2, 1] = 1$
$S_1[2, 2] = 0$
$S_1[2, 3] = 2$
$S_1[3, 1] = 300$
$S_1[3, 2] = 1$
$S_1[3, 3] = 0$

Hence,

$$S_1 = \begin{bmatrix} 0 & 1 & 1 \\ 1 & 0 & 2 \\ 300 & 1 & 0 \end{bmatrix}$$

Similarly, we find

$$S_2 = \begin{bmatrix} 0 & 1 & 1 \\ 1 & 0 & 2 \\ 2 & 1 & 0 \end{bmatrix} \quad \text{and} \quad S_3 = \begin{bmatrix} 0 & 1 & 1 \\ 1 & 0 & 2 \\ 2 & 1 & 0 \end{bmatrix}$$

The shortest path matrix S_3 shows the number of edges in the shortest path between vertex v_i and vertex v_j. See Figure 2.5.2. Note that the (i, i) entry is 0 since the number of edges in the shortest path from v_i to v_i is 0. Since there is no entry 300 in S_3, every vertex v_j is reachable from every vertex v_i. ∎

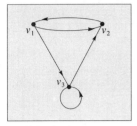

FIGURE 2.5.2

We have noted that Floyd's algorithm (Algorithm Short) yields the *length* of the shortest path, not the shortest path itself. It is possible to modify Floyd's algorithm so that it determines the shortest path, if one exists, between any two vertices in a digraph. (See *Data Structures and Algorithms* by Aho, Hopcroft, and Ullman, listed in the References.)

EXERCISE SET 2.5

1. Let $V = \{v_1, v_2\}$, and let R be a relation on V defined by $R = \{(v_1, v_2), (v_2, v_1)\}$.
   a. Find M_R.
   b. Find P_0, P_1, and P_2 using Warshall's algorithm (Algorithm 2.5.2).
   c. Find $(M_R)^2$.
   d. Compare P_1 and P_2 with $(M_R)^2$.
   e. Find the connectivity matrix M_{R^+} and the connectivity relation R^+ by using $M_{R^+} = P_2$.
   f. Find M_{R^+} and R^+ by using $M_{R^+} = M_R \oplus (M_R)^2$. Compare with your results from part e.

2. Let $V = \{v_1, v_2, v_3, v_4\}$ and $R = \{(v_1, v_2), (v_2, v_3), (v_3, v_4), (v_2, v_1)\}$.
   a. Find R^+ by finding P_4.
   b. Find R^+ by finding $M_{R^+} = M_R \oplus (M_R)^2 \oplus (M_R)^3 \oplus (M_R)^4$.

3. Refer to Exercise 1.
   a. Find the reachability matrix M_{R^*}.   b. Find the reachability relation R^*.

4. Refer to Exercise 2.
   a. Find M_{R^*}.   b. Find R^*.

5. Let $V = \{v_1, v_2, v_3, v_4\}$ and $R = \{(v_2, v_1), (v_2, v_3), (v_3, v_2), (v_3, v_4), (v_4, v_4), (v_4, v_3), (v_4, v_1)\}$.
   a. Draw a digraph to represent R.
   b. Determine $R \circ R$ from the digraph. Compare $R \circ R$ with $(M_R)^2$.
   c. Find M_{R^*} from $M_{R^*} = I \oplus M_R \oplus (M_R)^2 \oplus (M_R)^3 \oplus (M_R)^4$.
   d. Find M_{R^*} by Warshall's algorithm (Algorithm 2.5.2).

6. Let $V = \{1, 2, 3, 4\}$. Find a relation R on V such that $R^+ \neq R^*$.

7. Let $V = \{v_1, v_2, v_3, v_4\}$ and

$$P_2 = \begin{bmatrix} 1 & 0 & 1 & 0 \\ 0 & 0 & 0 & 1 \\ 0 & 1 & 0 & 1 \\ 0 & 0 & 1 & 0 \end{bmatrix}$$

Find:
a. $P_3[2,4]$ b. $P_3[1,2]$ c. $P_3[2,1]$
d. $P_3[4,2]$ e. $P_3[2,2]$ f. $P_4[2,2]$

8. Let $V = \{v_1, v_2, v_3, v_4, v_5\}$ and

$$P_3 = \begin{bmatrix} 0 & 1 & 1 & 1 & 1 \\ 0 & 0 & 0 & 0 & 1 \\ 0 & 1 & 0 & 1 & 1 \\ 1 & 1 & 1 & 1 & 1 \\ 0 & 0 & 0 & 0 & 0 \end{bmatrix}$$

Find:
a. $P_4[3,4]$ b. $P_4[3,2]$ c. $P_4[3,3]$
d. $P_4[2,3]$ e. $P_5[2,2]$ f. $P_5[3,1]$

Exercises 9 and 10 require an understanding of Floyd's shortest path algorithm, described in the Programming Notes on page 85.

9. Let $V = \{v_1, v_2, v_3\}$ and $R = \{(v_1, v_1), (v_1, v_2), (v_1, v_3), (v_2, v_2), (v_2, v_3), (v_3, v_1)\}$.
   a. Draw a digraph to represent R and find M_R.
   b. Find S_0, S_1, S_2, and the shortest path matrix S_3.

10. Let $V = \{v_1, v_2, v_3\}$ and $R = \{(v_2, v_1), (v_2, v_2), (v_3, v_1), (v_3, v_2)\}$.
    a. Draw a digraph to represent R and find M_R.
    b. Find S_0, S_1, S_2, and the shortest path matrix S_3.

11. Algorithm 2.5.5 (Path) was modified to obtain the more efficient Algorithm 2.5.7 (Newpath). A similar modification can be made on Algorithm Reach (from Practice Problem 4) to achieve higher efficiency. We state the new algorithm as Algorithm Newreach and express it here in mathematical notation:

$$\left. \begin{array}{ll} P_0 = M & \\ P_k[i,j] = P_{k-1}[i,j] & \text{if } i = k \text{ or } P_{k-1}[i,k] = 0 \\ P_k[i,j] = P_{k-1}[i,j] + P_{k-1}[k,j] & \text{if } i \neq k \text{ and } P_{k-1}[i,j] = 1 \end{array} \right\} \quad (3)$$

Write Algorithm Newreach in pseudocode.

12. Prove Theorem 2.5.6. In other words, prove $(D \cup R)^+ = D \cup R^+$. [*Hint:* It is clear that D is a subset of both $(D \cup R)^+$ and $D \cup R^+$. So we need prove only that, if there is a path from x to y in R, there is also a path from x to y in $D \cup R$; and, if there is a path from x to y in $D \cup R$, there is a path from x to y in R.]

13. Assume R is a transitive relation on $\{1, 2, \ldots, n\}$. What must be true about the matrices $P_1, P_2, \ldots, P_n$? Justify your answer.

14. Assume R is a relation on $\{1, 2, \ldots, n\}$ and assume $P_n = M_R$. What must be true about R? Justify your answer.

15. Let $A = \{1, 2, 3, 4\}$, and define R on A by $x \, R \, y$ if $x = y + 1$.
    a. Find the transitive closure R^+.
    b. Show that $x \, R^* \, y$ if and only if $y \leq x$.

16. Let $A = \{0, 1, 2, 3, 4, 5, 6\}$, and define R on A by $x \, R \, y$ if $|x - y| = 3$.
    a. Find the transitive closure R^+.
    b. Show that $x \, R^* \, y$ if and only if $x - y$ is divisible by 3.

■ PROGRAMMING EXERCISES

17. Refer to Exercise 11. Write a procedure in Pascal to implement Algorithm Newreach.
18. Write a computer program that includes procedure Short and that will find the shortest path matrix for a given relation R.

2.6
APPLICATION: DEADLOCK DETECTION

The history of telecommunications network development is one of decreasing message transmission cost and increasing speed and quality of communication. Some classical problems in network design and analysis concern network load, message transit time, and reliability issues. Pioneer work on these problems was accomplished during the ARPANET development, which began in 1969. Algorithms and protocols were invented to prevent network congestion and deadlocks. The problem of detecting deadlocks in a network continues to receive considerable attention in computer science research. How do we detect possible deadlocks? Is there a method to identify nodes in a given network where deadlocks could occur?

A **buffer** is a temporary data storage device in a network. A buffer is free if it is able to receive and store more messages. **Buffer deadlock** occurs when two or more nodes have no free buffers, and each node is waiting for some other node to provide a buffer.

As an example, we consider a simple telecommunications network with nodes A, B, C, D, E, and F (see Figure 2.6.1). The directed edges connecting the nodes indicate the communications links.

For simplicity, let us assume nodes C, D, and E have no free buffers. If messages in each node are waiting for buffers in one or more nodes, then this network is in a buffer deadlock. Specifically, assume that node C requires assignment of a buffer in node E before a message can be transmitted from C to A. If messages in node E are also waiting for a buffer in node D, and those in node D are waiting for a buffer in C, then nodes E, D, and C are in a buffer deadlock.

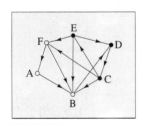

FIGURE 2.6.1

The network in Figure 2.6.1 is a digraph with vertices A, B, C, D, E, and F. For example, (C, E, D, C) and (F, B, D, C, F) are closed paths. Every closed path in a network is a potential deadlock. Since for every closed path there is a corresponding cycle, it follows that we can detect potential deadlocks by finding the cycles.

In this example of a very small network, all the cycles may be easily found. For networks with a large number of nodes, finding all cycles is very time-consuming. However, we can use Warshall's algorithm (Algorithm 2.5.2) to find the connectivity matrix for the adjacency matrix corresponding to the network. A cycle beginning and ending at a node in the network is indicated by a 1 on the main diagonal of the connectivity matrix. Hence, we can identify the nodes where cycles, and therefore potential deadlocks, occur. By detecting potential deadlocks at design time, we can modify the network design to decrease the frequency of deadlocks.

EXERCISE SET 2.6

1. Find the adjacency matrix of the digraph in Figure 2.6.1.
2. From the adjacency matrix in Exercise 1, use Warshall's algorithm (Algorithm 2.5.2) to find the connectivity matrix.
3. Identify all the cycles at each node of the digraph in Figure 2.6.1.

KEY TERMS

Directed graph (digraph)
Vertices (nodes)
Edges (arcs)
Initial vertex of an edge
Terminal vertex of an edge
Digraph
Path
Initial vertex of a path
Terminal vertex of a path
Closed path at v_0
Length of path
Simple path
Circuit
Cycle
Self-loop
Reachability problem
Reachable vertex
Project network
Binary relation (relation)
n-tuple
Coordinates

Product (Cartesian product)
n-ary relation
Degree of relation
Unary relation
Ternary relation
Table
Projection
Reflexive property
Symmetric property
Antisymmetric property
Transitive property
Transitive closure
Acquaintance relation
Correspondence between digraphs and binary relations
Connectivity relation
Composition of binary relations
Pigeonhole principle
Matrix
Row index
Column index
Square matrix of order n
Main diagonal

Adjacency matrix for a digraph
Boolean addition and multiplication for $\{0, 1\}$
Boolean addition and multiplication for matrices
Boolean sum
Boolean product
Zero matrix
Identity matrix
Adjacency matrix for a relation
Connectivity matrix
Interior vertices
Warshall's algorithm
Reachability relation
Reachability matrix
Diagonal relation
Reflexive transitive closure
Floyd's shortest path algorithm
Shortest path matrix
Buffer
Buffer deadlock

REVIEW EXERCISES

1. For each of the following, find a set A and a relation R on A satisfying the given conditions.
   a. R is reflexive but neither symmetric nor transitive.
   b. R is neither reflexive nor symmetric but is transitive.
   c. R is not symmetric but is reflexive and transitive.

2. In each of the following relations R, determine if R is reflexive, symmetric, transitive, antisymmetric. If not, give a counterexample.
   a. Let A be a group of five siblings, with three brothers and two sisters, all having the same parents. Define $R \subseteq A \times A$ by $R = \{(x, y): x$ is the sister of $y\}$.
   b. Let $A = \mathbb{N} = \{0, 1, 2, 3, \ldots\}$. R is defined by $x \mathrel{R} y$ whenever $x \cdot y$ is even for all $x, y \in \mathbb{N}$, where $x \cdot y$ is ordinary multiplication.

3. Determine if the following relation R on $A = \{a, b, c, d\}$ is reflexive, symmetric, transitive, antisymmetric. Let $R = \{(a, a), (a, d), (b, a), (b, b), (c, a), (c, b), (c, c), (c, d), (d, b), (d, d)\}$.

4. Let $A = \{1, 2, \ldots, 9\}$, and define R on A by $R = \{(i, j):(j = i^2)$ or $(i = j + 1)\}$. Find:
   a. $R \circ R$ b. R^+

5. Let $A = \{1, 2, 3, 4, 5\}$ and $R = \{(i, j): j = i + 1\}$. Find:
   a. M_R b. $(M_R)^2$ c. P_3 d. R^+

6. Let R be a relation on a finite set A. Justify your answer to each of the following questions.
   a. Suppose R is reflexive. Is R^2 also reflexive?
   b. Suppose R is symmetric. Is R^2 also symmetric?
   c. Suppose R is transitive. Is R^2 also transitive?

7. Let R and S be relations on a finite set A. Justify your answer to each of the following questions.
   a. Suppose R and S are reflexive. Is $R \circ S$ also reflexive?
   b. Suppose R and S are symmetric. Is $R \circ S$ also symmetric?
   c. Suppose R and S are transitive. Is $R \circ S$ also transitive?

For Exercises 8 and 9 you may assume the following: A relation R is transitive if and only if $R^2 \subseteq R$.

8. Let R be a reflexive and transitive relation on a finite set A. Are M_{R^2} and M_R equal? Justify your answer.

9. Let R be a relation on a finite set A. Prove: R is transitive if and only if $M_R = M_R \oplus M_{R^2}$.

10. Let $A = \{1, 2, 3, 4, 5\}$, and define S on A by $x \mathrel{S} y$ if $x + 1 = y$.
    a. Find the transitive closure S^+.
    b. Show that $x \mathrel{S^*} y$ if and only if $x \leq y$.

11. Let $A = \{1, 2, 3, 4, 5\}$, and define T on A by $x \mathrel{T} y$ if $|x - y| = 2$.
    a. Find the transitive closure T^+.
    b. Show that $x \mathrel{T^*} y$ if and only if $x - y$ is divisible by 2.

PROGRAMMING EXERCISES

12. Write a computer program to implement Warshall's algorithm (Algorithm 2.5.2). Use your program to find R^+ in Exercises 4 and 5 above.

13. Use Exercise 9 to write a computer program that takes the adjacency matrix of a relation as input and determines whether the relation is transitive.

REFERENCES

Aho, A. V., J. E. Hopcroft, and J. D. Ullman. *Data Structures and Algorithms.* Reading, MA: Addison-Wesley, 1983.

Ahuja, V. *Design and Analysis of Computer Communication Networks.* New York: McGraw-Hill, 1982.

Codd, E. F. "A Relational Model of Data for Large Shared Data Banks." *Communication of ACM*, 13 (1970): 377–387.

Date, C. J. *An Introduction to Database Systems.* Reading, MA: Addison-Wesley, 1975.

Floyd, R. W. "Algorithm 97: Shortest Path." *Communication of ACM*, 5:6 (1962): 345.

Frank, H., and W. Chou. "Network Properties of the ARPA Computer Network." *Networks*, 4 (1974): 213–239.

Gries, D. *Compiler Construction for Digital Computers.* New York: Wiley, 1971.

Harary, F., R. Z. Norman, and D. Cartwright. *Structural Models: An Introduction to the Theory of Directed Graphs.* New York: Wiley, 1965.

Warshall, S. "A Theorem on Boolean Matrices." *Journal of ACM*, 9 (1962): 11–12.

CHAPTER 3

RELATIONS AND FUNCTIONS

We continue the study of relations begun in Chapter 2. In Section 3.1 we define and discuss equivalence relations and partial orderings. A good working knowledge of these two important concepts is fundamental for any mathematician or computer scientist.

Functions are surely even more important. In the last two sections of this chapter, we define functions and consider several types with special properties. A good understanding of functions will be gained by working the many examples presented in this chapter.

APPLICATION Many different algorithms can be devised to execute the same task. Selecting one over the others is a significant issue. Thus, an important aspect of algorithm design and analysis is the determination of execution costs for a given algorithm. One type of cost is computer time to execute an algorithm. Execution time, or time complexity, of a given algorithm is of great interest to computer scientists. In Section 3.5, we will use functions to analyze the time complexity of algorithms.

3.1
EQUIVALENCE RELATIONS AND PARTIAL ORDERINGS

In Chapter 2, we introduced the reflexive, symmetric, antisymmetric, and transitive properties for a relation R defined on a set A. Here are two examples to illustrate the four properties.

EXAMPLE 3.1.1 Let $R = \{(m, n): m \leq n\}$ be a relation on the set $\mathbb{Z}$ of integers.

    a. R is reflexive since $m \leq m$ holds for each $m \in \mathbb{Z}$; that is, $(m, m) \in R$ for each $m \in \mathbb{Z}$.
    b. R is not symmetric since $(-3, 8) \in R$ but $(8, -3) \notin R$.
    c. R is antisymmetric because, if $m \neq n$, then $m > n$ or $n > m$. Equivalently, if $m \leq n$ and $n \leq m$, then $m = n$.
    d. R is transitive for, if both $m \leq n$ and $n \leq p$, then $m \leq p$. In other words, if (m, n) and (n, p) are in R, then (m, p) is in R, for all m, n, p in $\mathbb{Z}$. ∎

EXAMPLE 3.1.2 Let $R = \{(x, y): y = x + 2\}$ be a relation on the set $\mathbb{R}$ of real numbers.

    a. R is not reflexive since $(9, 9) \notin R$.
    b. R is not symmetric since $(2, 4) \in R$ but $(4, 2) \notin R$.
    c. R is antisymmetric because, if $(a, b) \in R$ and $(b, a) \in R$, then $b = a + 2$ and $a = b + 2$. But this means $b = (b + 2) + 2$; so $0 = 4$, an impossibility. Hence, the antisymmetric condition is vacuously true.
    d. R is not transitive since $(2, 4) \in R$ and $(4, 6) \in R$ but $(2, 6) \notin R$. ∎

Equivalence Relations

DEFINITION   A relation R on a set A that is reflexive, symmetric, and transitive is called an **equivalence relation** on A.

EXAMPLE 3.1.3 Let

$$A = \{\text{John Smith, Amy Lee, Peter Lucas, David Lucas, Martha Lucas,}$$
$$\text{James Lee, Charles Smith}\}$$

and define R on A by $(x, y) \in R$ if x and y have the same last name. R is an equivalence relation. ∎

PRACTICE PROBLEM 1 Refer to Example 3.1.3. Verify that R is an equivalence relation on A.

EXAMPLE 3.1.4 Consider the set T of all triangles in the plane. Triangles T_1 and T_2 are said to be **similar** if the corresponding angles of T_1 have the same measure as those of T_2. We can see that similarity on T is an equivalence relation. ∎

DEFINITION

> Let R be an equivalence relation on A, and let $a \in A$. The **equivalence class** of the element a is $[a] = \{x : x \in A \text{ and } x \, R \, a\}$.

In Example 3.1.3, the equivalence relation R determines three equivalence classes:

{John Smith, Charles Smith}
{James Lee, Amy Lee}
{Peter Lucas, David Lucas, Martha Lucas}

The elements in each equivalence class are related to one another by the relation R. Also, the equivalence class [John Smith] is identical with the equivalence class [Charles Smith]; they are both equal to the set {John Smith, Charles Smith}. In general, as stated in Lemma 3.1.7 following, the equivalence classes $[x]$ and $[y]$ are identical if and only if $x \, R \, y$.

Let us list several facts about the equivalence classes of Example 3.1.3:

1. Every equivalence class is nonempty.
2. The equivalence classes are **pairwise disjoint**; that is, for every $x, y \in A$, if $[x] \cap [y] \neq \emptyset$, then $[x] = [y]$.
3. The union of all the equivalence classes is equal to A.

DEFINITION

> A **partition** of a set A is a collection of pairwise disjoint, nonempty subsets of A such that A is the union of these subsets.

So the collection of equivalence classes in Example 3.1.3 forms a partition of A.

Let us study two more examples before looking at the general situation concerning the connection between equivalence relations and partitions.

EXAMPLE 3.1.5 Let $A = \{1, 2, 3, 4, 5, 6, 7\}$. Define a relation $\equiv$ on A by $m \equiv n$ if $n - m = 2k$ for some $k \in \mathbb{Z}$. That is, $m \equiv n$ if $n - m$ is divisible by 2. Show that $\equiv$ is an equivalence relation, and find all the equivalence classes.

Solution First we show that $\equiv$ is an equivalence relation.

   a. The relation $\equiv$ is reflexive since $m - m = 0 = 2 \cdot 0$ and $0 \in \mathbb{Z}$.
   b. The relation $\equiv$ is symmetric for, if $n - m = 2k$, then $m - n = 2(-k)$.
   c. To show that $\equiv$ is transitive, we first assume $m \equiv n$ and $n \equiv p$. We need to prove $m \equiv p$. By assumption, $n - m = 2k$ and $p - n = 2j$ for some $k, j \in \mathbb{Z}$. But $p - m = (p - n) + (n - m) = 2j + 2k = 2(j + k)$. Hence, $p - m$ is divisible by 2 and, therefore, $m \equiv p$.

Therefore, $\equiv$ is an equivalence relation.

The equivalence classes are $[1] = \{1, 3, 5, 7\}$ and $[2] = \{2, 4, 6\}$. So this equivalence relation $\equiv$ determines the partition of A consisting of the set of odd integers in A and the set of even integers in A. ∎

EXAMPLE 3.1.6 Let $A = \{1, 2, 3, 4, 5, 6, 7\}$. Define a relation $\equiv$ on A by $a \equiv b$ if $b - a$ is divisible by 3. Show that $\equiv$ is an equivalence relation, and find all the equivalence classes.

Solution The proof that $\equiv$ is an equivalence relation is similar to the proof given in Example 3.1.5 above. This equivalence relation $\equiv$ determines a partition of the set A into three disjoint subsets $[1] = \{1, 4, 7\}$, $[2] = \{2, 5\}$, and $[3] = \{3, 6\}$. Clearly, $A = \{1, 4, 7\} \cup \{2, 5\} \cup \{3, 6\} = [1] \cup [2] \cup [3]$. ∎

Note that different equivalence relations in the two preceding examples yield different partitions of the same set A.

PRACTICE PROBLEM 2 Define $\equiv$ on $\{1, 2, 3, 4, 5, 6, 7, 8, 9, 10\}$ by $x \equiv y$ if $x - y = 4k$ for some $k \in \mathbb{Z}$. Verify that $\equiv$ is an equivalence relation, and find all the equivalence classes.

We have illustrated how an equivalence relation on a set A determines a partition of A. Before proving that every equivalence relation determines a partition, we state the following lemma.

LEMMA 3.1.7 Let R be an equivalence relation on the set A.

For $x, y \in A$, $[x] = [y]$ if and only if $x\ R\ y$.

The proof of Lemma 3.1.7 will be postponed until Chapter 5.

THEOREM 3.1.8 Let R be an equivalence relation on the set A. The collection of equivalence classes $\{[z] : z \in A\}$ is a partition of A.

Proof We need to prove: 1. $[z] \neq \emptyset$ for all $z \in A$. 2. The equivalence classes are pairwise disjoint. 3. The union of the equivalence classes is A.

1. By definition of $[z]$, $z \in [z]$ since $z\ R\ z$. Hence, $[z] \neq \emptyset$ for all $z \in A$.
2. Assume $[x] \cap [y] \neq \emptyset$. Then there is some $z \in [x] \cap [y]$. By definition, $z\ R\ x$ and $z\ R\ y$. Hence, $x\ R\ z$ and $z\ R\ y$. So $x\ R\ y$. By Lemma 3.1.7 we conclude $[x] = [y]$.
3. Since $[x] \subseteq A$ for each $x \in A$, the union of all the equivalence classes of A is a subset of A. Next, we prove that A is a subset of the union of all the equivalence classes. Let $y \in A$. Then $y \in [y]$. Hence, every element of A is an element of some equivalence class. Therefore, A is a subset of the union of the equivalence classes. ∎

PRACTICE PROBLEM 3 Specify where the reflexive, symmetric, and transitive properties for R were used in the proof of Theorem 3.1.8.

Posets

DEFINITION

> A relation R on a set A that is reflexive, antisymmetric, and transitive is called a **partial ordering** on A (or **partial order** of A). The set A is called a **partially ordered set** (or **poset**) under the partial ordering R, and we may write (A, R) is a poset.

EXAMPLE 3.1.9 Let $A = \{1, 2, 3, 4\}$. The relation $R = \{(a, b): a \leq b\}$ on A is a partial ordering on A. Thus, A is a partially ordered set under R. We note that $R = \{(1, 1), (1, 2), (1, 3), (1, 4), (2, 2), (2, 3), (2, 4), (3, 3), (3, 4), (4, 4)\}$. ∎

EXAMPLE 3.1.10 Refer to Example 3.1.1. The relation $R = \{(m, n): m \leq n\}$ is a partial ordering on $\mathbb{Z}$. Therefore, $(\mathbb{Z}, R)$ is a poset. ∎

DEFINITION

> Let (A, R) be a poset. Elements $x, y \in A$ are **comparable** if
>
> $(x, y) \in R$ or $(y, x) \in R$

In Example 3.1.10 any two elements of $\mathbb{Z}$ are comparable.

DEFINITION

> A **linear ordering** R on A (or **linear order** of A) is a partial ordering R on A such that any two elements of A are comparable.

Both partial orderings defined in Examples 3.1.9 and 3.1.10 are also linear orderings.

In the next example we give a very important partial ordering that is not a linear ordering.

EXAMPLE 3.1.11 **SET INCLUSION IS A PARTIAL ORDER** Let $P(\mathbb{Z})$ be the set of all subsets of $\mathbb{Z}$, the power set of $\mathbb{Z}$. Define a relation R on $P(\mathbb{Z})$ as follows: For A, B in $P(\mathbb{Z})$, $(A, B) \in R$ if $A \subseteq B$. Clearly, $P(\mathbb{Z})$ is a partially ordered set under R. Note that the subsets $\{1, 2\}$ and $\{0, 1\}$ are not comparable members of $P(\mathbb{Z})$ under R. Thus, the poset $(P(\mathbb{Z}), R)$ is not a linear ordering. ∎

PRACTICE PROBLEM 4 Refer to Example 3.1.11.

a. Find two more members of $P(\mathbb{Z})$ that are not comparable.
b. Find a member S of $P(\mathbb{Z})$ that is comparable to every other member of $P(\mathbb{Z})$. How many such members are there in $P(\mathbb{Z})$?

We have used R to stand for a generic partial ordering and have given several examples with $\leq$ used to denote a specific partial ordering. From now on we will use $\preceq$ to denote any partial ordering. When A is a set with a partial order $\preceq$, we may write $(A, \preceq)$ is a poset.

Since a partial order of a finite set is a relation, it can be represented by a digraph, as shown in the next example.

EXAMPLE 3.1.12 Define a partial order of $A = \{1, 2, 3, 4, 8\}$ by $x \preceq y$ if x is a factor of y. Hence, $a \preceq a$ and $1 \preceq a$ for all $a \in A$. Also, $2 \preceq 4$, $2 \preceq 8$, and $4 \preceq 8$. The digraph for this poset is shown in Figure 3.1.1. ∎

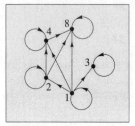

FIGURE 3.1.1

We can simplify digraphs for partial orders by eliminating edges corresponding to the reflexive property or the transitive property. In other words, we eliminate all edges (x, x), and any edge (x, z) whenever (x, y) and (y, z) are edges. We eliminate arrowheads from the edges by always drawing the directed edges so they point upwards. With these conventions, the digraph for a partial order is called a **Hasse diagram**, named after the mathematician Helmut Hasse (1898–1979).

The partial order of A pictured in Figure 3.1.1 is represented by the Hasse diagram in Figure 3.1.2.

Let $U = \{a, b\}$ and recall that set inclusion $\subseteq$ is a partial order of the power set $P(U)$. Figure 3.1.3 shows the Hasse diagram for the set inclusion partial order of $P(U)$.

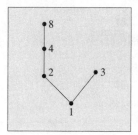

FIGURE 3.1.2

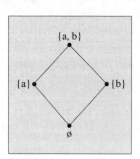

FIGURE 3.1.3

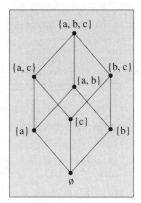

FIGURE 3.1.4

The Hasse diagram shown in Figure 3.1.4 represents the set inclusion partial order of $P(U)$ where $U = \{a, b, c\}$.

For a finite set A, the Hasse diagram is enough to specify the partial ordering on A. Two substantially different partial orderings on the same set are specified in the next example.

3.1 EQUIVALENCE RELATIONS AND PARTIAL ORDERINGS

EXAMPLE 3.1.13 Let $A = \{0, 1, 2, 3, 4\}$.

a. Figure 3.1.5 shows the Hasse diagram for the standard linear ordering $\leq$ on A.

b. Figure 3.1.6 shows the Hasse diagram for the partial ordering defined by $x \preceq y$ if $x = 0$ or x is a factor of y. ∎

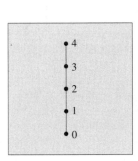

FIGTRE 3.1.5

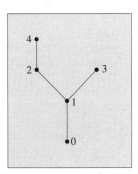

FIGURE 3.1.6

EXERCISE SET 3.1

1. Let $D_{35} = \{1, 5, 7, 35\}$, and define a partial order by $x \preceq y$ if y is divisible by x. Draw the Hasse diagram for the poset $(D_{35}, \preceq)$. Compare this Hasse diagram with that of Figure 3.1.3.

2. Let $D_{30} = \{1, 2, 3, 5, 6, 10, 15, 30\}$, and define a partial order by $x \preceq y$ if x is a factor of y. Draw the Hasse diagram for the poset $(D_{30}, \preceq)$. Compare this Hasse diagram with that of Figure 3.1.4.

3. Let C be the set of all computer programs in Pascal. Define a relation R on C by the following:

    For any two programs x and y in C, $x\,R\,y$ if x and y produce the same output from the same input.

    Show that R is an equivalence relation on C.

4. Let P be the set of all people and R be a relation defined on P as described in parts a, b, c, and d below. Determine if R has the reflexive, symmetric, antisymmetric, or transitive properties. Assume x and y are arbitrary members of P.
    a. $x\,R\,y$ if x is the sister of y.
    b. $x\,R\,y$ if x and y have the same father.
    c. $x\,R\,y$ if x and y have the same mother.
    d. $x\,R\,y$ if x and y have the same paternal grandfather.

5. Let $A = \{2, 3, 6, 12\}$, and define a partial order by $x \preceq y$ if x is a factor of y. Draw the Hasse diagram for the poset $(A, \preceq)$.

6. Let $A = \{1, 2, 4, 5, 10\}$, and define a partial order by $x \preceq y$ if x is a factor of y. Draw the Hasse diagram for the poset $(A, \preceq)$.

7. Let $A = \{a, b, c, d\}$, and let $P(A)$ be the power set of A.
   a. List all the members (subsets of A) in $P(A)$.
   b. Define a relation R on $P(A)$ by $x \, R \, y$ if $x \subseteq y$. Show that $P(A)$ is a poset under R.

8. Refer to Exercise 7.
   a. Draw the Hasse diagram for the poset $(P(A), R)$.
   b. Find two members in $P(A)$ that are comparable to every other member in $P(A)$.

9. Let L be the set of all lines in the plane. Lines l and m in L are said to be **perpendicular**, written $l \perp m$, if l and m intersect at right angles. Is $\perp$ a transitive relation on L? If not, give a counterexample.

10. Let T be the set of all triangles in the plane. Triangles T_1 and T_2 are said to be **congruent**, written $T_1 \approx T_2$, if their corresponding sides have the same length and their corresponding angles have the same measure. Verify that the relation $\approx$ on T is an equivalence relation.

11. Define a relation $\equiv$ on $\mathbb{Z}$ by $m \equiv n$ if $m - n$ is divisible by 3.
    a. Show that $\equiv$ is an equivalence relation.
    b. Determine all the equivalence classes under $\equiv$. How many are there?
    c. Verify that the equivalence classes in part b form a partition of $\mathbb{Z}$.

12. Let d be a fixed positive integer. Define a relation $\equiv$ on $\mathbb{Z}$ by $m \equiv n$ whenever $m - n$ is divisible by d.
    a. Show that $\equiv$ is an equivalence relation.
    b. Determine all the equivalence classes under $\equiv$. How many are there?
    The equivalence relation in Exercise 12 is called **congruence modulo d** and is frequently written as $m \equiv n (\mod d)$ or simply $\equiv$ modulo d. Congruence modulo d is discussed in Chapter 6.

13. Let M be a 4×4 array where
    $$M = \begin{bmatrix} a & b & c & d \\ e & f & g & h \\ i & j & k & l \\ m & n & o & p \end{bmatrix}$$

    Define R on the set $A = \{a, b, c, d, e, f, g, h, i, j, k, l, m, n, o, p\}$ by $x \, R \, y$ if x and y are in the same row of the array M.
    a. Show that R is an equivalence relation.
    b. How many equivalence classes are there?

14. Repeat Exercise 13 with R on A defined by $x \, R \, y$ if x and y are in the same column of the array M.

15. Let A be a set of identifiers in the computer programming language Pascal. Define R on A by $x \, R \, y$ if the two identifiers x and y have at least one character in common.
    a. Show that R is reflexive and symmetric.
    b. Show that R is not an equivalence relation.

16. Let A be a set of identifiers, each consisting of at least three characters, in the computer programming language Pascal. Define R on A by $x \, R \, y$ if the two identifiers x and y have the first three characters in common.
    a. Show that R is an equivalence relation.
    b. What is in the equivalence class [ab4uq]?

17. Let D be the set of digits $\{0, 1, 2, \ldots, 9\}$. Define a relation R on the power set $P(D)$ by $A\ R\ B$ if $\text{Card}(A) = \text{Card}(B)$.
   a. Show that R is an equivalence relation.
   b. List the elements in the equivalence class $[\varnothing]$ and in $[D]$.
   c. List the elements in the equivalence class $[\{4\}]$.
   d. Describe the elements in the equivalence class $[\{4, 7\}]$.

18. Let E be the set of English words. Give a precise definition of the lexicographical (or dictionary) order R on the set E. Show that (E, R) is a poset.

19. Consider the set $A = \{(m, n) : m, n \in \mathbb{Z} \text{ and } n \neq 0\}$. Define the relation R on the set A by $(m, n)\ R\ (p, q)$ if $mq = np$.
   a. Verify that R is an equivalence relation on A.
   b. Find three members other than $(2, 5)$ in the equivalence class $[(2, 5)]$.
   c. How many members are in the equivalence class $[(2, 5)]$?
   d. Let Q be the set of all equivalence classes $[(m, n)]$, where $(m, n) \in A$. Explain why Q is like the set of all rational numbers.

20. Let $A = \mathbb{Z}$, and define S on A by $x\ S\ y$ if $x + 1 = y$.
   a. Find the transitive closure S^+.
   b. What well-known partial order relation is the reflexive transitive closure S^*?

21. Let $A = \mathbb{Z}$, and define T on A by $x\ T\ y$ if $|x - y| = 2$.
   a. Find the transitive closure T^+.
   b. What well-known equivalence relation is T^*?

■ **PROGRAMMING EXERCISES**

22. Write a computer program to determine whether a given relation on a finite set is an equivalence relation.

23. Write a computer program to determine whether a given relation on a finite set is a partial order.

3.2
EXTREMAL ELEMENTS IN A PARTIALLY ORDERED SET

Certain elements in a poset, called **extremal elements**, have properties that make them especially useful. Later in this section we will discuss a sorting method that makes use of a certain type of extremal element. The first two types of extremal elements we discuss are least elements and greatest elements.

DEFINITION

In a poset $(A, \leq)$,

v is a **least element** of A if $v \in A$ and $v \leq x$ for all $x \in A$;
u is a **greatest element** of A if $u \in A$ and $x \leq u$ for all $x \in A$.

In Example 3.1.13a, the least element is 0 and the greatest element is 4. In Example 3.1.13b, the least element is 0, but there is no greatest element. For example, 4 is not a greatest element since 3 is not a factor of 4. A given partially ordered set may possess neither a least element nor a greatest element.

PRACTICE PROBLEM 1 Draw the Hasse diagram for a poset that has neither a least element nor a greatest element.

If a least element does exist for a poset, it must be unique. Similarly for the greatest element. We state this fact as a theorem and leave the proof as an exercise. (See Exercise 9, Exercise Set 3.2.)

THEOREM 3.2.1 There is at most one least element and at most one greatest element for a given poset.

We write $a \prec b$ when $a \preceq b$ and $a \neq b$. When $a \prec b$, we say a is **strictly less than** b or b is **strictly greater than** a.

DEFINITION

In a poset $(A, \preceq)$,

m is a **minimal element** of A if there is no $x \in A$ such that $x \prec m$;
q is a **maximal element** of A if there is no $x \in A$ such that $q \prec x$.

Note that, in Example 3.1.13b, both 3 and 4 are maximal elements. Unlike least and greatest elements, minimal and maximal elements are not necessarily unique. We illustrate this fact in the next example.

EXAMPLE 3.2.2 Let $A = \{a, b, c, d, e, f, g\}$ and the partial order $\preceq$ be defined by the Hasse diagram pictured in Figure 3.2.1. The elements a, b, and c are all minimal elements; f and g are maximal elements.

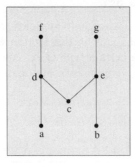

FIGURE 3.2.1

It is also possible for a poset not to have a minimal or maximal element. But finite, nonempty posets always contain maximal and minimal elements.

THEOREM 3.2.3 If A is a finite, nonempty set and $(A, \preceq)$ is a poset, then there is at least one minimal element and at least one maximal element.

Proof We will prove there is at least one minimal element. The proof that there is at least one maximal element is similar.

Pick any element $a_1 \in A$. If a_1 is minimal, we are done. If not, we can find a_2 such that $a_2 \prec a_1$. If a_2 is minimal, we are done. If not, we can find a_3 such that $a_3 \prec a_2 \prec a_1$. Similarly, we can find a_4, and so on. This process must terminate, because A is finite. The final a_k we pick yields a linearly ordered subset $\{a_k, a_{k-1}, \ldots, a_2, a_1\}$ of A such that $a_k \prec \cdots \prec a_3 \prec a_2 \prec a_1$. Clearly, a_k is a minimal element. ∎

PRACTICE PROBLEM 2 Find a poset that has neither a minimal element nor a maximal element. [*Hint:* Refer to Theorem 3.2.3.]

The existence of maximal and minimal elements is crucial for some applications. For example, we will make use of minimal elements in the following sorting method.

Topological Sorting

In Section 2.1 we discussed project networks. Certain tasks in a project network must be finished before others can begin. In other words, the relationship among tasks is a **precedence relationship**. A project network can be considered as a partially ordered set of tasks.

Suppose that we have a partial order R describing the precedence relationships among certain tasks. Further suppose there is only one team of people to carry out these tasks. How do we order these tasks so they can be done one at a time? What we need is a linear ordering L of the tasks that maintains the given precedence in R among tasks. In other words, if (x, y) is in R, then (x, y) must be in L. So we need to find a linear order L so that $R \subseteq L$.

Given a poset (A, R), how do we find a linear order L of A including R? The method for accomplishing this is called **topological sorting**, which we illustrate in the next example.

EXAMPLE 3.2.4 Let the partial order R on $\{a, b, c, d, e, f, g\}$ be as shown by the Hasse diagram in Figure 3.2.2(a), page 104. Find a linear order L including R.

Solution First, place a minimal element, say e, into the linear order L; delete the vertex e and any edges adjacent to it from the Hasse diagram. Repeat for the minimal element a. At this stage we have $e \prec a$ in the linear order L we are building. The resulting Hasse diagram now has minimal elements b and c. Repeat the process for c, then for b. Now we have $e \prec a \prec c \prec b$, and the

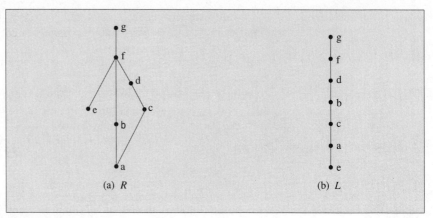

FIGURE 3.2.2

Hasse diagram has minimal element d. Repeat until we have a linear order L, where $e \prec a \prec c \prec b \prec d \prec f \prec g$ [see Figure 3.22(b)]. ∎

PRACTICE PROBLEM 3 Refer to Example 3.2.4. Find two other linear orders L_1 and L_2 such that $R \subseteq L_1 \cap L_2$.

A topological sorting algorithm is stated in the programming notes at the end of this section.

DEFINITION

In a poset $(A, \preceq)$, where $B \subseteq A$,

$v \in A$ is a **lower bound** of B if $v \preceq b$ for all $b \in B$.
$u \in A$ is an **upper bound** of B if $u \succeq b$ for all $b \in B$.

Note that lower bounds and upper bounds need not be elements of the given set.

For the poset $(A, \preceq)$ defined in Example 3.2.2, the set $\{a, c\}$ has no lower bounds but it has upper bounds d and f. The element d (or f) is an upper bound of $\{a, c\}$ because d (or f) is above both a and c in the Hasse diagram. There are no lower bounds because no elements are below both a and c in the Hasse diagram. For a given poset, it is not difficult to find lower and upper bounds of subsets by examining the Hasse diagram.

PRACTICE PROBLEM 4 Refer to Example 3.2.2.

a. Find all lower bounds and upper bounds of $B = \{d, f\}$.
b. Find all lower bounds and upper bounds of $B = \{c, d, e\}$.
c. Find all lower bounds and upper bounds of $B = \{a, b\}$.

3.2 EXTREMAL ELEMENTS IN A PARTIALLY ORDERED SET

DEFINITION

Let $(A, \preceq)$ be a poset and $B \subseteq A$.

g is a **greatest lower bound** of B, written $g = \text{glb}(B)$, if g is a lower bound of B and $g \geq v$ for all lower bounds v of B.

h is a **least upper bound** of B, written $h = \text{lub}(B)$, if h is an upper bound of B and $h \leq u$ for all upper bounds u of B.

For the poset $(A, \preceq)$ defined in Example 3.2.2, the set $\{a, c\}$ has no greatest lower bound since it has no lower bounds. The element d is the least upper bound of $\{a, c\}$; that is, $\text{lub}(\{a, c\}) = d$.

PRACTICE PROBLEM 5   Refer to Example 3.2.2.

a. Find all greatest lower bounds and least upper bounds of $B = \{d, f\}$.
b. Find all greatest lower bounds and least upper bounds of $B = \{c, d, e\}$.
c. Find all greatest lower bounds and least upper bounds of $B = \{a, b\}$.

For a poset $(A, \preceq)$ and a subset $B \subseteq A$, there may be no $\text{lub}(B)$ or $\text{glb}(B)$. However, when a least upper bound or a greatest lower bound does exist, then it is unique. We state this fact as a theorem.

THEOREM 3.2.5   Given a poset $(A, \preceq)$ and a subset $B \subseteq A$, there is at most one $\text{lub}(B)$ and at most one $\text{glb}(B)$.

Proof   We will prove there is at most one $\text{lub}(B)$. Suppose $h = \text{lub}(B)$ and $k = \text{lub}(B)$. Then both h and k are upper bounds of B. Since h is a least upper bound of B, we have $h \preceq k$. Similarly, $k \preceq h$. Since $\preceq$ is antisymmetric, we have $h = k$. The proof for the greatest lower bound is similar and is left as an exercise. (See Exercise 8, Exercise Set 3.2.) ∎

▮ PROGRAMMING NOTES

ALGORITHM 3.2.6   **TOPOLOGICAL SORTING**   Given a partial order R on a set A with n elements, this algorithm constructs a linear ordering L of A including R.

```
Begin
    H₁ := R.                                           (*Initialize*)
    For i = 1 to n do
        V(i) := a minimal element of Hᵢ.
        Remove V(i) and all edges adjacent to V(i) from Hᵢ.
        Name the resulting relation Hᵢ₊₁.              (*New relation*)

    Return the linear order V(1) ≺ V(2) ≺ ··· ≺ V(n).
End.
```

EXERCISE SET 3.2

1. A partial order $\preceq$ is defined on $\{a, b, c, d, e\}$ by the Hasse diagram in Figure 3.2.3.
   a. Find any maximal or minimal elements.
   b. Find any greatest or least elements.

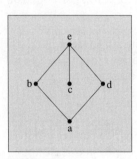

FIGURE 3.2.3 **FIGURE 3.2.4**

2. A partial order $\preceq$ is defined on $\{a, b, c, d, e, f, g\}$ by the Hasse diagram in Figure 3.2.4.
   a. Find any maximal or minimal elements.
   b. Find any greatest or least elements.

3. Define a partial order $\preceq$ on $\{2, 3, 6, 12\}$ by $x \preceq y$ if x is a factor of y.
   a. Find any maximal or minimal elements.
   b. Find any greatest or least elements.

4. Define a partial order $\preceq$ on $\{1, 2, 4, 5, 10\}$ by $x \preceq y$ if x is a factor of y.
   a. Find any maximal or minimal elements.
   b. Find any greatest or least elements.

5. Consider the set inclusion partial order on $P(\{a, b, c\})$.
   a. Find any maximal or minimal elements.
   b. Find any greatest or least elements.

6. Let $\leq$ be the usual order on the set $\mathbb{R}$ of real numbers. Consider the set $A = \{x : x \in \mathbb{R} \text{ and } x < \sqrt{2}\}$.
   a. Verify that $(A, \leq)$ is a poset.
   b. Find the greatest lower bound (if any) of A.
   c. Find the least upper bound (if any) of A.

7. Repeat Exercise 6 for the set $B = \{y : y \in \mathbb{R}^+ \text{ and } 5 \leq y^2 < 9\}$.

8. Let $(A, \preceq)$ be a poset and assume $B \subseteq A$. Prove: There is at most one glb(B).

9. Let $(A, \preceq)$ be a poset. Prove: There is at most one least element for $(A, \preceq)$.

10. Use topological sorting to find a linear ordering including the partial order defined in Exercise 1.

11. Use topological sorting to find a linear ordering including the partial order defined in Exercise 2.

12. Let $D_{28} = \{1, 2, 4, 7, 14, 28\}$, and define a partial order by $x \preceq y$ if y is divisible by x. Use topological sorting to find a linear ordering including this partial order.

13. Let $D_{30} = \{1, 2, 3, 5, 6, 10, 15, 30\}$, and define a partial order by $x \preceq y$ if x is a factor of y. Use topological sorting to find a linear ordering including this partial order.

14. Let R be a partial ordering on $\{v_1, v_2, \ldots, v_n\}$. Suppose v_1 is a maximal element. What conditions must be satisfied by certain entries of the adjacency matrix M_R?

15. Repeat Exercise 14 with the assumption that v_3 is a minimal element.

16. Let R be a partial ordering on $\{v_1, v_2, \ldots, v_n\}$. Suppose v_1 is a least element. What conditions must be satisfied by certain entries of the adjacency matrix M_R?

17. Let R be a partial ordering on $\{v_1, v_2, \ldots, v_n\}$. Suppose v_1 is the least upper bound of $\{v_2, v_3\}$. What conditions must be satisfied by certain entries of the adjacency matrix M_R?

∎ PROGRAMMING EXERCISE

18. Write a computer program to implement the topological sorting algorithm.

3.3 FUNCTIONS

The concept of a function plays a central role in the disciplines of computer science, mathematics, and many related fields. Computer languages have many standard functions. Most languages allow the user to define functions. Functions make mathematical concepts easier to describe and understand. They make computer programs easier to write, more structured, and easier to read. We will discuss important applications of functions throughout the book.

A function is a special type of a relation. A function can be considered as a rule that takes input and produces unique output. We define a function as follows.

DEFINITION

Given nonempty sets D and C, a **function** f from D to C, denoted by $f: D \to C$, is a relation from D to C satisfying the property

For every $x \in D$, there is a unique $y \in C$ so that the ordered pair (x, y) belongs to the relation f.

This modern definition of *function* was first proposed in 1837 by Dirichlet. The set D is called the **domain** of f, and the set C is called the **codomain** of f.

Let $f: D \to C$ be a function. By definition, for each $x \in D$, there is a unique element y in the codomain assigned to x by the function f. We write $y = f(x)$ to denote the element y assigned to x. We say y is the **image** of x, and x is the **preimage** of y. By definition, every element of the domain has a unique image. In other words, if $(x, y) \in f$ and $(x, z) \in f$, then $y = z$.

In order to specify a function, we must give the ordered pairs, the domain, and the codomain. Instead of listing ordered pairs, we will often use the more common method of writing $f(x) = y$ for $(x, y) \in f$.

Every function with a finite domain can be pictured as a digraph. We illustrate this in the next example.

EXAMPLE 3.3.1 Let $D = \{a, b, c\}$ and $C = \{1, 2, 3, 4\}$ be the domain and codomain of the function f given by $f(a) = 1$, $f(b) = 3$, and $f(c) = 3$. Thus, $f = \{(a, 1), (b, 3), (c, 3)\}$ when expressed as a set of ordered pairs. The function f is pictured as a digraph in Figure 3.3.1. Note that 3 is the image of both b and c; in other words, 3 has preimages b and c. ∎

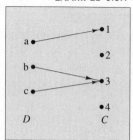

FIGURE 3.3.1

Another useful method for representing a function is to use a table. The function described in Example 3.3.1 is also defined by Table 3.3.1.

TABLE 3.3.1

x	a	b	c
$f(x)$	1	3	3

Our next example illustrates the two main ways a relation can fail to be a function.

EXAMPLE 3.3.2 Let R and S be relations from the set $D = \{a, b, c\}$ to the set $C = \{1, 2, 3, 4\}$ given by $R = \{(a, 1), (c, 3)\}$ and $S = \{(a, 1), (b, 3), (c, 3), (c, 4)\}$. See Figure 3.3.2. The relations R and S are not functions from the set $D = \{a, b, c\}$ to the set $C = \{1, 2, 3, 4\}$.

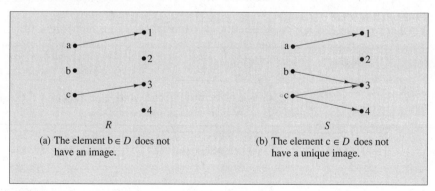

(a) The element $b \in D$ does not have an image.

(b) The element $c \in D$ does not have a unique image.

FIGURE 3.3.2 ∎

PRACTICE PROBLEM 1 Let $D = \{a, b, c, d\}$ and $C = \{1, 2, 3\}$. Let f and g be relations defined by $f = \{(a, 1), (b, 3)\}$ and $g = \{(a, 1), (b, 1), (c, 2), (d, 3)\}$. Represent f and g as digraphs, and determine whether f and g are functions from D to C. Give reasons for your answers.

Many useful functions have domains with an infinite number of elements.

EXAMPLE 3.3.3 **a.** Let $D = \mathbb{N} = \{0, 1, 2, \ldots\} = C$. Define $f : D \to C$ by $f(n) = 2^n$. So, $f(0) = 1$, $f(1) = 2$, $f(2) = 4$, and so on.

3.3 FUNCTIONS

b. Let $D = \mathbb{R}^+$ and $C = \mathbb{R}$. Define $g:D \to C$ by $g(x) = \log_2 x$, where $\log_2$ is the logarithm base 2 function. So, $g(1) = 0$, $g(2) = 1$, $g(4) = 2$, and so on. Also, $g(\frac{1}{2}) = -1$. ∎

Not all functions can be defined with a single formula.

EXAMPLE 3.3.4 Suppose we want to define a function g from the nonnegative integers to the integers that assigns to each even integer a value so that $g(0) = 0$, $g(2) = 1$, $g(4) = 2, \ldots$; and assigns to each odd integer a value so that $g(1) = -1$, $g(3) = -2$, $g(5) = -3, \ldots$. We can give a two-case definition for g as follows:

$$g(n) = \begin{cases} \dfrac{n}{2} & \text{if } n \text{ is even} \\ \dfrac{-(n+1)}{2} & \text{if } n \text{ is odd} \end{cases}$$

∎

Sequences

DEFINITION

> A **sequence** is any function with domain $\mathbb{N}$ or a subset of $\mathbb{N}$ of the form $\{m, m+1, \ldots\} \subseteq \mathbb{N}$.

We usually think of a sequence as a list, in order, of specified elements. For example, the list $0, 1, 4, 9, \ldots, n^2, \ldots$ is a sequence. This sequence is formally defined as the function $T: \mathbb{N} \to \mathbb{Z}$, where $T(n) = n^2$. It is common practice to describe a sequence such as T by using a lowercase letter t and designating its values as $t_n = n^2$.

In order to distinguish sequences (where order is important) from sets (where order does not matter), we will frequently use parentheses. So we write a sequence as $(t_n)_{n=0,1,\ldots}$ or $(t_0, t_1, \ldots)$. For the sequence with values $t_n = n^2$, we can write $(0, 1, 4, 9, \ldots, n^2, \ldots)$.

EXAMPLE 3.3.5 Let $s_1 = 1$, $s_2 = \frac{1}{2}$, $s_3 = \frac{1}{3}$, $s_4 = \frac{1}{4}, \ldots$. We designate this sequence by $(s_n)_{n=1,2,\ldots}$, where $s_n = 1/n$, or by $(1, \frac{1}{2}, \frac{1}{3}, \ldots, 1/n, \ldots)$. We can also formally define the sequence as a function $S:D \to \mathbb{R}$, where $D = \{1, 2, 3, \ldots\}$ and $S(n) = 1/n$. ∎

Sequences will be pivotal in our discussion of recurrence relations in Chapter 9. Here is a sequence useful for algorithm analysis.

EXAMPLE 3.3.6 We define the sequence $(s_n)_{n=1,2,\ldots}$ by

s_n = The sum of the first n positive integers

So, $s_n = 1 + 2 + \cdots + n$.

We derive an algebraic formula for s_n. The famous mathematician Carl Friedrich Gauss (1777–1855) is said to have invented this derivation while still in grade school. We write the summands of $s_n = 1 + 2 + 3 + \cdots + (n-1) + n$ in reverse order and add the resulting version of s_n to the original version, column by column, as follows:

$$\begin{array}{rcccccccc}
s_n & = & 1 & + & 2 & + & 3 & + \cdots + (n-1) + & n \\
s_n & = & n & + & (n-1) & + & (n-2) & + \cdots + 2 + & 1 \\
\hline
2s_n & = & (n+1) & + & (n+1) & + & (n+1) & + \cdots + (n+1) + & (n+1)
\end{array}$$

The right-hand side of the resulting sum has n terms. Hence, $2s_n = n(n+1)$. Therefore,

$$s_n = \frac{n(n+1)}{2}$$

Many applications require finite lists of elements. A **finite sequence** is a function with domain a finite subset of $\mathbb{N}$. Usually the domain of a finite sequence will be a set such as $\{1, 2, \ldots, m\}$. For example, the finite sequence (2, 4, 8, 16, 32) can be defined by $s_n = 2^n$ with domain $\{1, 2, 3, 4, 5\}$.

It will be convenient in what follows to write finite sequences without using parentheses or commas. For example, (a, b, b, a, b) is a finite sequence of the symbols a and b that we may write more simply as abbab. A finite sequence of symbols written adjacent to one another without parentheses or commas is called a **string** of symbols. We will be particularly interested in strings of the symbols 0 and 1. Strings of 0's and 1's are called **bit strings**.

Boolean-Valued Functions

DEFINITION

Any function with codomain $C = \{0, 1\}$ is called a **Boolean-valued function**.

The set of Boolean-valued functions includes the characteristic functions, to be introduced shortly, and the Boolean functions to be studied in Chapter 7.

EXAMPLE 3.3.7   Let $D = \{a, b, c\}$ and $C = \{0, 1\}$. Define $f: D \to C$ by $f(a) = 1$, $f(b) = 0$, and $f(c) = 1$. This is conveniently shown as Table 3.3.2.

TABLE 3.3.2

x	a	b	c
$f(x)$	1	0	1

The second row in Table 3.3.2 is a bit string. It will be convenient to think of bit strings as corresponding to subsets of a given universal set. In Example 3.3.7 assume the given set D contains the elements in the order listed: a, b, c. The bit string corresponding to the subset $\{a, c\}$ is $101 = f(a)f(b)f(c)$; $f(b) = 0$ because $b \notin \{a, c\}$.

To make this idea more precise, we introduce characteristic functions.

3.3 FUNCTIONS

DEFINITION

Let $U = \{u_1, u_2, \ldots, u_n\}$ be a set and $A \subseteq U$ be some subset. The **characteristic function** of the set A is a function $\chi_A: U \to \{0, 1\}$ defined by

$$\chi_A(x) = \begin{cases} 1 & \text{if } x \in A \\ 0 & \text{if } x \notin A \end{cases}$$

The function f of Example 3.3.7 is the characteristic function of the set $A = \{a, c\}$ with domain $U = \{a, b, c\}$. Hence, $f = \chi_A : \{a, b, c\} \to \{0, 1\}$.

For any finite set $U = \{u_1, u_2, \ldots, u_n\}$ and subset $A \subseteq U$, the *bit string corresponding to the set A* is equal to $\chi_A(u_1)\chi_A(u_2)\ldots\chi_A(u_n)$, where $\chi_A: U \to \{0, 1\}$ is the characteristic function of the set A. Note that the bit string depends on the order of the elements in the set U. We will assume a fixed order for the elements in U.

EXAMPLE 3.3.8 Let $U = \{u_1, u_2, u_3, u_4, u_5\}$.

a. Let $A = \{u_2, u_3, u_5\}$. The table for χ_A is as shown in Table 3.3.3. The bit string for A is $\chi_A(u_1)\chi_A(u_2)\chi_A(u_3)\chi_A(u_4)\chi_A(u_5) = 01101$.

b. Let $B = \{u_1, u_3, u_5\}$. The table for χ_B is as shown in Table 3.3.4. The bit string for B is 10101.

TABLE 3.3.3

x	u_1	u_2	u_3	u_4	u_5
$\chi_A(x)$	0	1	1	0	1

TABLE 3.3.4

x	u_1	u_2	u_3	u_4	u_5
$\chi_B(x)$	1	0	1	0	1

PRACTICE PROBLEM 2 Refer to Example 3.3.8.

a. Find the table for $\chi_{A \cup B}$ and the bit string for $A \cup B$. Compare with χ_A and χ_B.

b. Find the table for $\chi_{A \cap B}$ and the bit string for $A \cap B$. Compare with χ_A and χ_B.

c. Find the tables for $\chi_{A'}$ and $\chi_{B'}$ and the corresponding bit strings. Compare with χ_A and χ_B.

The Boolean addition ($+$), multiplication ($\cdot$), and complement ($'$) operations are defined on $\{0, 1\}$ by Tables 3.3.5–3.3.7.

TABLE 3.3.5

+	0	1
0	0	1
1	1	1

TABLE 3.3.6

$\cdot$	0	1
0	0	0
1	0	1

TABLE 3.3.7

x	x'
0	1
1	0

Let $U = \{u_1, u_2, \ldots, u_n\}$, $A \subseteq U$, and $B \subseteq U$. It is left as an exercise to verify the following rules for characteristic functions. (See Exercise 14, Exercise Set 3.3.) They will be discussed again in the programming notes of this section.

> **RULES FOR CHARACTERISTIC FUNCTIONS**
> **a.** $\chi_{A'}(u) = (\chi_A(u))'$ for all $u \in U$, where $A' = U - A$, $(\chi_A(u))' = 0$ if $\chi_A(u) = 1$, and $(\chi_A(u))' = 1$ if $\chi_A(u) = 0$.
> **b.** $\chi_{A \cup B}(u) = \chi_A(u) + \chi_B(u)$ for all $u \in U$, where the $+$ on the right is Boolean addition.
> **c.** $\chi_{A \cap B}(u) = \chi_A(u) \cdot \chi_B(u)$ for all $u \in U$, where the $\cdot$ on the right is Boolean multiplication.

Partial Functions

It is often useful to consider relations called *partial functions*, which are "almost" functions. Partial functions are especially useful in the study of nondeterministic finite state machines and Turing machines, which are discussed in any course on automata theory.

Partial functions occur often in algebra. Functions are frequently specified by giving a rule of association. For example, a definition such as: "Let f be the function defined by $f(x) = 1/(1 - x^2)$" is very common. This sentence alone is not enough to specify a function. We still need to know the domain and the codomain. In this case the codomain is usually assumed to be the real numbers, and the domain is some subset of the real numbers. Since $f(x)$ is not defined at 1 and -1 but is defined for all $x \neq \pm 1$, we take the domain to be $\mathbb{R} - \{-1, 1\}$. We assumed the domain to be $D \subseteq \mathbb{R}$ and then found D so that $f(x)$ is defined for every $x \in D$ and undefined for every $x \in \mathbb{R} - D$. We say f is a partial function from the real numbers to the real numbers.

DEFINITION

> A **partial function** f from A to C is a relation from A to C such that $f: D \to C$ is a function, where $D \subseteq A$ and, for all $x \in A - D$, there does not exist a unique $y \in C$ so that $(x, y) \in f$. For $x \in A - D$, f is said to be **undefined** at x. The set D is called the **domain of definition** for the partial function f.

EXAMPLE 3.3.9
a. The relations of Example 3.3.2 are both partial functions.
b. The relation g from $\mathbb{R}$ to $\mathbb{R}$ defined by $g(x) = \log_2 x$ is a partial function. The domain of definition for g is $\mathbb{R}^+$.
c. The relation f from $\mathbb{N}$ to $\mathbb{R}$ defined by $f(n) = 1/[n(n - 3)]$ is a partial function. Here the domain of definition for f is $D = \mathbb{N} - \{0, 3\}$. ∎

■ **PROGRAMMING NOTES**

Standard Functions in Pascal There are many standard functions in Pascal. We will give just two examples here.

The **trunc** function is the truncating function with domain $\mathbb{R}$ and codomain $\mathbb{Z}$. For any $x \in \mathbb{R}$, trunc(x) represents the integer part of x. Any fractional part of x is cut off. For example,

$$\text{trunc}(5.8) = 5$$
$$\text{trunc}(2.0) = 2$$
$$\text{trunc}(-3.2) = -3$$
$$\text{trunc}(-7.8) = -7$$
$$\text{trunc}(0.0) = 0$$

The **abs** function is the absolute value function with domain = $\mathbb{R}$ and codomain = $\mathbb{R}^+ \cup \{0\}$. For $x \in \mathbb{R}$, abs(x) = $|x|$, where $|x|$ is the mathematical notation for the absolute value of x. For example,

$$\text{abs}(-12.0) = 12.0$$
$$\text{abs}(2.6) = 2.6$$
$$\text{abs}(0.0) = 0.0$$
$$\text{abs}(-0.37) = 0.37$$

User-Defined Functions in Pascal The square function **sqr** is a standard function in Pascal but the cube function is not. We illustrate how to define a function in Pascal by defining the cube function using multiplication * and the square function sqr:

```
function Cube (x : real) : real;
   begin
      Cube := x*sqr(x)
   end;
```

In Pascal, writing "Cube (x:real)" corresponds in mathematical terminology to requiring that the variable x be from the domain $\mathbb{R}$, or at most those real numbers that are available on the computer running the program. Similarly, the ": real" following corresponds to declaring the codomain as the set of reals available on the computer that runs the program. Note that the last assignment, "Cube := x*sqr(x)," assigns the function value to be a value of type real. A function in Pascal always returns a value from its codomain. So, a single variable function $f:D \to C$ is written in Pascal as "function f (x:D):C."

Here is a function that takes a finite set as input and returns the cardinality of the set. Assume Universe is a previously defined set type. For specificity, we will declare Universe in the following way:

```
type
    Universe = set of 'A' .. 'z';

function Card (Aset : Universe) : integer;
    var
        Count : integer;
        Element : char;
    begin
        Count := 0;
        for Element := 'A' to 'z' do
            if Element in Aset then
                Count := Count + 1;
        Card := Count
    end;
```

The function declaration for Card specifies that Aset is of type Universe. The function Card can be used in a Pascal program along with the procedure Getletters, discussed in Chapter 1, to count the number of distinct letters in a line of text.

Bit String Implementation of Sets in Pascal Characteristic functions are helpful for understanding how sets are implemented in the programming language Pascal. When sets are used in a Pascal program, a universal set must be declared first. This is necessary so that a location in memory can be reserved for the corresponding bit strings. Every subset of a universal set with n elements corresponds to a bit string with some combination of n 0's and 1's. Implementation of sets in Pascal makes use of the corresponding bit strings for each declared set.

The rules for characteristic functions (see page 112) have great significance for the bit string implementation of sets. Rule a applied to bit strings says: Apply the complement operation to all coordinates of the bit string for A to get the bit string for A'. Rule b says: Add the bit string for A to the bit string for B, coordinate by coordinate, to get the bit string for $A \cup B$. Rule c says: Multiply the bit string for A by the bit string for B, coordinate by coordinate, to get the bit string for $A \cap B$.

EXERCISE SET 3.3

1. Given a relation f from the set $A = \{a, b, c, d\}$ to the set $B = \{a, b, c\}$, where $f = \{(a, b), (b, b), (c, b), (d, c)\}$, determine whether f is a function from domain A to codomain B.

2. Given a relation g from the set $D = \{1, 2, 3, 4, 5\}$ to the set $C = \{a, b, c, d\}$, where $g = \{(1, a), (2, c), (3, d), (4, c)\}$, determine whether g is a function from domain D to codomain C.

3. Find the domain of definition for each of the given partial functions.
   a. f from $\mathbb{R}$ to $\mathbb{R}$ defined by $f(x) = 1/(x + 3)$
   b. g from $\mathbb{R}$ to $\mathbb{R}$ defined by $g(x) = \sqrt{x + 8}$

4. Find the domain of definition for each of the given partial functions.
   a. f from $\mathbb{R}$ to $\mathbb{R}$ defined by $f(x) = 1/(x^2 - 9)$
   b. g from $\mathbb{Z}$ to $\mathbb{R}$ defined by $g(z) = \sqrt{z - 2}$

5. Let $U = \{u_1, u_2, u_3, u_4\}$. Write the characteristic function and the bit string for each given subset of U.
   a. $A = \{u_1, u_2\}$ b. $A = \{u_3\}$ c. $A = U$ d. $A = \emptyset$

6. Let $U = \{u_1, u_2, u_3, u_4, u_5, u_6\}$. Write the characteristic function and the bit string for each given subset of U.
   a. $A = \{u_1, u_2\}$ b. $A = \{u_3, u_5, u_6\}$ c. $A = U$ d. $A = \emptyset$

7. Define a sequence s_n by $(s_0, s_1, s_2, \ldots) = (0, 1, 2, \frac{1}{3}, 4, \frac{1}{5}, \ldots)$. Give a two-case definition for s_n.

8. Define a sequence t_n by $(t_0, t_1, t_2, \ldots) = (0, 1, \frac{1}{2}, 9, 4, \frac{1}{5}, \ldots)$. Give a three-case definition for t_n.

9. Recall the equivalence relation $\equiv$ defined on $\mathbb{Z}$ by $m \equiv n$ whenever $m - n$ is divisible by 3. Let C be the set $\{[0], [1], [2]\}$ of equivalence classes. Define $g: \mathbb{Z} \to C$ by $g(x) = [x]$.
   a. Find $g(1)$, $g(-1)$, and $g(18)$.
   b. Write the set $\{x : g(x) = [0]\}$ as a subset of $\mathbb{Z}$.

10. Recall the equivalence relation $\equiv$ defined on $\mathbb{Z}$ by $m \equiv n$ whenever $m - n$ is divisible by 4. Let C be the set $\{[0], [1], [2], [3]\}$ of equivalence classes. Define $g: \mathbb{Z} \to C$ by $g(x) = [x]$.
    a. Find $g(1)$, $g(-2)$, and $g(19)$.
    b. Write the set $\{x : g(x) = [4]\}$ as a subset of $\mathbb{Z}$.

11. Define the **floor function** or **greatest integer function** $L(x) = \lfloor x \rfloor$ by $L: \mathbb{R} \to \mathbb{Z}$, where $L(x) = n$ for $n \leq x < n + 1$. Note that $L(x)$ is the largest integer less than or equal to x. For example, $L(5.4) = \lfloor 5.4 \rfloor = 5$. Find:
    a. $\lfloor 2.3 \rfloor$ b. $\lfloor 3.0 \rfloor + \lfloor 8.6 \rfloor$ c. $\lfloor -1.7 \rfloor$ d. $\lfloor -1.3 \rfloor - \lfloor -2.5 \rfloor$

12. We define the **ceiling function** $H(x) = \lceil x \rceil$ by $H: \mathbb{R} \to \mathbb{Z}$, where $H(x) = n$ for $n - 1 < x \leq n$. Note that $H(x)$ is the smallest integer greater than or equal to x. For example, $\lceil 6.9 \rceil = 7$. Find:
    a. $\lceil 3.7 \rceil$ b. $\lceil 5.0 \rceil - \lceil 8.3 \rceil$ c. $\lceil -6.3 \rceil$ d. $\lceil -6.37 \rceil - \lceil -12.47 \rceil$

13. Let G be the set of all characteristic functions from domain $S = \{a, b, c\}$ to codomain $\{0, 1\}$. For each subset $A \subseteq S$, the function χ_A is the corresponding characteristic function of A. Let $G = \{\chi_A : A \subseteq S\}$. Define a relation $\preceq$ on G by:

    $\chi_A \preceq \chi_B$ if $A \subseteq B$

    a. Verify that the relation $\preceq$ on G is a partial ordering.
    b. Draw a Hasse diagram for the poset $(G, \preceq)$.
    c. What can you say about the Hasse diagram of part b and that of Figure 3.1.4 (page 98)?

14. Let $U = \{u_1, u_2, \ldots, u_n\}$, and let A and B be subsets of U. Verify each of the following rules for characteristic functions.
    a. $\chi_{A'}(u) = (\chi_A(u))'$ b. $\chi_{A \cup B}(u) = \chi_A(u) + \chi_B(u)$ c. $\chi_{A \cap B}(u) = \chi_A(u) \cdot \chi_B(u)$

15. Let F be the set of all functions from domain $\mathbb{R}$ to codomain $\mathbb{R}$. Define a relation $\preceq$ on F by:

    $f \preceq g$ if $f(x) \leq g(x)$ for every $x \in R$

    a. Verify that $(F, \preceq)$ is a poset.
    b. Is $\preceq$ on F an equivalence relation? If not, give a counterexample.

16. Consider $\mathbb{N}$ with the standard order $\leq$, and let D be the set of all nonempty subsets of $\mathbb{N}$. Define $g: D \to \mathbb{N}$ by $g(A) = \text{glb}(A)$.
    a. Explain why g is a well-defined function, and find $g(\{x: x^3 > 8\})$.
    b. Explain why g would be a partial function if $\mathbb{N}$ were replaced by $\mathbb{Z}$ in the definition.

17. Let f be a function with domain D and codomain C. Define a relation R on D by $x \, R \, z$ if $f(x) = f(z)$. Show that R is an equivalence relation on D.

18. Let g be a function with domain $\mathbb{Z}$ and codomain $\mathbb{Z}$. Suppose g is defined by $g(x) = x^2$ for $x \in \mathbb{Z}$. Define a relation R on $\mathbb{Z}$ by $m \, R \, n$ if $g(m) = g(n)$. By Exercise 17, we know R is an equivalence relation on $\mathbb{Z}$.
    a. Find the elements in the equivalence class $[7]$.
    b. Find the elements in the equivalence class $[0]$.
    c. Find the set of all equivalence classes on $\mathbb{Z}$.

19. Let $(A, \preceq)$ be a poset. Define a function $f: A \to P(A)$ by $f(x) = \{a \in A : a \prec x\}$. What can you say about x when $f(x)$ is the empty set?

20. Define a sequence $(t_1, t_2, \ldots)$ by $t_n =$ The number of bit strings corresponding to the set $\{a_0, a_1, \ldots, a_n\}$ with exactly two 1's.
    a. Find t_1.
    b. Find t_5.
    c. Test the conjecture $t_m = (m + 1)m/2$ for several values of m.

PROGRAMMING EXERCISES

21. Write a function in Pascal corresponding to the function $f: \mathbb{N} \to \mathbb{N}$ defined by $f(n) = 2^n$.

22. Write a function in Pascal corresponding to the function $g: \mathbb{R} \to \mathbb{R}$ defined by $g(x) = x^5$.

23. Use the trunc function to define a function in Pascal corresponding to the floor function.

24. Consider a function given by $y = F(x)$, where $F:[a, b] \to \mathbb{R}^+$ and $[a, b]$ is a closed interval in $\mathbb{R}$. Let A denote the area bounded by the graph of $y = F(x)$, the x-axis, and the lines $x = a$ and $x = b$. Write a function in Pascal to approximate the area A. Hint: Let $h = (b - a)/n$. The area A is approximated by the sum I of areas of n rectangles, where

    $$I = h \cdot F(a) + h \cdot F(a + h) + h \cdot F(a + 2h) + h \cdot F(a + 3h) + \cdots + h \cdot F(a + (n - 1)h)$$

25. Let $K:[0, 1] \to \mathbb{R}^+$ be a function given by $K(x) = x^2$. Choose $n = 40$ in the function from Exercise 24 to approximate the area bounded by the graph of $y = K(x)$, the x-axis, and the lines $x = 0$ and $x = 1$.

3.4
SPECIAL FUNCTIONS

In this section we consider functions with special and useful properties, and then discuss functions composed of other functions.

3.4 SPECIAL FUNCTIONS

One-to-One Functions

DEFINITION

A function $f:D \to C$ is **one-to-one** (or **1–1**) if

$f(x) = f(z)$ implies $x = z$ for all $x, z \in D$.

In other words, every element of the codomain of a 1–1 function has at most one preimage. A one-to-one function is also called an **injective function**.

EXAMPLE 3.4.1

a. Let $D = \{a, b, c\}$ and $C = \{1, 2, 3, 4\}$, with $f:D \to C$ defined by $f(a) = 2$, $f(b) = 3$, and $f(c) = 2$. The function f is not 1–1 since $f(a) = f(c)$ and $a \neq c$.

b. The function of Example 3.3.1 (page 108) is shown in Figure 3.4.1. This function is not 1–1 because 3 has two preimages.

c. Both functions of Example 3.3.3 (page 108) are 1–1. To verify that $f: \mathbb{N} \to \mathbb{N}$ defined by $f(n) = 2^n$ is 1–1 we assume $f(m) = f(k)$. Then $2^m = 2^k$. Hence, by taking the logarithm base 2 of both sides of the equation $2^m = 2^k$, we obtain $m = \log_2 2^m = \log_2 2^k = k$. Therefore, $m = k$. ∎

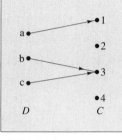

FIGURE 3.4.1

PRACTICE PROBLEM 1

Consider $g: \mathbb{R}^+ \to \mathbb{R}$ defined by $g(x) = \log_2 x$, from Example 3.3.3b (page 109). Verify that g is 1–1.

Here is the pigeonhole principle, used in Chapter 2, stated in terms of a 1–1 function.

PIGEONHOLE PRINCIPLE

Assume C and D are finite sets.

If $f:D \to C$ is a 1–1 function, then $\text{Card}(D) \leq \text{Card}(C)$.

For this statement of the pigeonhole principle, we imagine D as a set of pigeons, C as a set of pigeonholes, and f as the function placing each pigeon in a pigeonhole. We interpret "f is 1–1" as "no pigeonhole contains more than one pigeon."

Onto Functions

DEFINITION

A function $f:D \to C$ is **onto** (or **surjective**) if,

for every $y \in C$, there is at least one $x \in D$ so that $f(x) = y$.

In other words, every element of the codomain of an onto function has at least one preimage.

EXAMPLE 3.4.2
 a. Let $D = \{a, b, c\}$ and $C = \{1, 2\}$, with $f:D \to C$ defined by $f(a) = 1$, $f(b) = 1$, and $f(c) = 2$. Then f is onto.
 b. The function shown in Figure 3.4.1 is not onto since 2 and 4 do not have preimages.
 c. The function $g:\mathbb{R}^+ \to \mathbb{R}$ defined by $g(x) = \log_2 x$ is onto. To verify that g is onto, let us assume $y \in \mathbb{R}$. Then, $2^y \in \mathbb{R}^+$ and $\log_2 2^y = y$. Hence, $\log_2 x = y$ for $x \in D = \mathbb{R}^+$. ∎

PRACTICE PROBLEM 2
 a. Verify that $f:\mathbb{R} \to \mathbb{R}^+$ defined by $f(x) = 2^x$ is onto.
 b. Show that $h:\mathbb{N} \to \mathbb{N}$ defined by $h(n) = 2^n$ is not onto.

Bijective Functions

DEFINITION

> A function f is **bijective** if f is both 1–1 and onto.

A bijective function is also called a **bijection** or a **1–1 onto function**. Every element of the codomain of a bijective function has exactly one preimage.

EXAMPLE 3.4.3 The functions $f:\mathbb{R} \to \mathbb{R}^+$ defined by $f(x) = 2^x$ and $g:\mathbb{R}^+ \to \mathbb{R}$ defined by $g(x) = \log_2 x$ are both bijective. ∎

DEFINITION

> Let A be any set. The **identity function** $1_A : A \to A$ is defined by
>
> $1_A(x) = x$ for all $x \in A$

Every identity function is bijective.

Permutation Functions

The bijective functions from a finite set onto itself constitute a very useful class of functions.

DEFINITION

> Let A be a finite set. A bijective function from A to A is called a **permutation of** A.

Permutations will be discussed more fully in Section 4.2.

EXAMPLE 3.4.4 Let $A = \{1, 2, 3\}$.

3.4 SPECIAL FUNCTIONS

TABLE 3.4.1

x	1	2	3
f(x)	2	3	1

a. The function 1_A is a permutation of A.
b. The function $f: A \to A$ defined by Table 3.4.1 is also a permutation. ∎

PRACTICE PROBLEM 3 Refer to Example 3.4.4. There are four other permutations of $\{1, 2, 3\}$. Find them and write the table for each.

Writing a permutation in its tabular form, as in Example 3.4.4, suggests another notation for permutations. Let $A = \{1, 2, \ldots, n\}$, and assume $f: A \to A$ is a permutation. We may write f as a two-dimensional array, as follows:

$$f = \begin{pmatrix} 1 & 2 & 3 & \cdots & n \\ f(1) & f(2) & f(3) & \cdots & f(n) \end{pmatrix}$$

So the array form for f in Example 3.4.4b is

$$f = \begin{pmatrix} 1 & 2 & 3 \\ 2 & 3 & 1 \end{pmatrix}$$

Compositions of Functions

Many functions we study are composed of other simpler functions.

DEFINITION

Let $f: D \to A$ and $g: B \to C$, where $A \subseteq B$. Define the **composition** $g \circ f: D \to C$ by

$(g \circ f)(x) = g(f(x))$ for all $x \in D$

Because the codomain A of f is a subset of the domain B of g, $g(f(x))$ is a well-defined element of C for all possible values $f(x)$ in A.

The diagram in Figure 3.4.2 gives a pictorial representation of $g \circ f$. It is clear from the diagram that evaluating $(g \circ f)(x)$ involves evaluating $f(x)$ first and then evaluating $g(f(x))$.

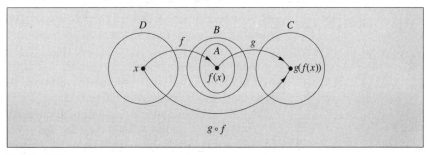

FIGURE 3.4.2

EXAMPLE 3.4.5 Let $f:\mathbb{R} \to [1, \infty)$ be defined by $f(x) = x^2 + 1$ and $g:\mathbb{R}^+ \to \mathbb{R}$ be defined by $g(x) = \log_2 x$. Then, $(g \circ f)(x) = g(f(x)) = g(x^2 + 1) = \log_2(x^2 + 1)$. ∎

EXAMPLE 3.4.6 Let

$$f = \begin{pmatrix} 1 & 2 & 3 \\ 2 & 3 & 1 \end{pmatrix} \quad \text{and} \quad g = \begin{pmatrix} 1 & 2 & 3 \\ 2 & 1 & 3 \end{pmatrix}$$

Then,

$$g \circ f = \begin{pmatrix} 1 & 2 & 3 \\ 1 & 3 & 2 \end{pmatrix}$$

since $(g \circ f)(1) = g(f(1)) = g(2) = 1$, $(g \circ f)(2) = g(f(2)) = g(3) = 3$, and $(g \circ f)(3) = g(f(3)) = g(1) = 2$. ∎

PRACTICE PROBLEM 4 Verify

$$f \circ g = \begin{pmatrix} 1 & 2 & 3 \\ 3 & 2 & 1 \end{pmatrix}$$

where f and g are as given in Example 3.4.6.

EXAMPLE 3.4.7 Let $f:\mathbb{R} \to \mathbb{R}^+$ be defined by $f(x) = 2^x$ and $g:\mathbb{R}^+ \to \mathbb{R}$ be defined by $g(x) = \log_2 x$. So, $g \circ f:\mathbb{R} \to \mathbb{R}$ and $f \circ g:\mathbb{R}^+ \to \mathbb{R}^+$ are well-defined compositions. Further, $(g \circ f)(x) = g(f(x)) = g(2^x) = x$ and $(f \circ g)(x) = f(\log_2 x) = 2^{\log_2 x} = x$. Hence, $g \circ f = 1_{\mathbb{R}}$ and $f \circ g = 1_{\mathbb{R}^+}$, where $1_{\mathbb{R}}$ and $1_{\mathbb{R}^+}$ are the identity functions on $\mathbb{R}$ and $\mathbb{R}^+$, respectively. ∎

The functions f and g of Example 3.4.7 are said to be inverses of each other.

DEFINITION Let $f:D \to C$ be a function. Then, $g:C \to D$ is an **inverse** of f if $g \circ f = 1_D$ and $f \circ g = 1_C$. When g is the inverse of f, we write $g = f^{-1}$.

Note that f^{-1} does *not* denote the reciprocal $1/f$.

PRACTICE PROBLEM 5 Let $f:\mathbb{R} \to \mathbb{R}$ be a function defined by $f(x) = 8x + 1$ for each $x \in \mathbb{R}$. Let $g:\mathbb{R} \to \mathbb{R}$ be a function defined by $g(x) = (x - 1)/8$ for each $x \in \mathbb{R}$. Convince yourself that f and g are inverses of each other.

A function has an inverse if and only if it is bijective. If an inverse for a given function exists, it is unique. In addition, the inverse is also bijective. These facts will be proved in Chapter 5.

Let $f:D \to C$ be a bijective function, and assume $v = f(u)$. Then, $f^{-1}(v) = f^{-1}(f(u)) = (f^{-1} \circ f)(u) = 1_D(u) = u$. So, the inverse f^{-1} maps the image v of u back to the preimage u of v. Figure 3.4.3 gives a pictorial representation.

The idea discussed above and illustrated in Figure 3.4.3 can be exploited to find some inverse functions.

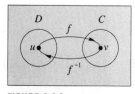

FIGURE 3.4.3

EXAMPLE 3.4.8 Define $f: \mathbb{R} \to \mathbb{R}$ by $f(x) = 3x + 5$. Any inverse function f^{-1} must have the property $f^{-1}(v) = u$ if $f(u) = v$.

To find f^{-1}, we set $v = f(u) = 3u + 5$. Solve for u to get $u = (v-5)/3$. But $u = f^{-1}(v)$, so $f^{-1}(v) = (v-5)/3$. Hence, $f^{-1}(x) = (x-5)/3$. ∎

PRACTICE PROBLEM 6 Refer to Example 3.4.8. Verify that $(f \circ f^{-1})(x) = x = (f^{-1} \circ f)(x)$.

Since any function that has an inverse must be bijective, Practice Problem 6 enables us to conclude that the function f of Example 3.4.8 is bijective.

EXAMPLE 3.4.9 **a.** Let $f = \begin{pmatrix} 1 & 2 & 3 \\ 2 & 3 & 1 \end{pmatrix}$. Then $f^{-1} = \begin{pmatrix} 1 & 2 & 3 \\ 3 & 1 & 2 \end{pmatrix}$.

b. Let $g = \begin{pmatrix} 1 & 2 & 3 \\ 2 & 1 & 3 \end{pmatrix}$. Then $g^{-1} = \begin{pmatrix} 1 & 2 & 3 \\ 2 & 1 & 3 \end{pmatrix} = g$. ∎

Permutations always have inverses. (Why?) Furthermore, the inverse can be found by exchanging the top row and the bottom row in the array form of the function and then rearranging the columns so the top row is in the natural order. (Why?) For

$$f = \begin{pmatrix} 1 & 2 & 3 \\ 2 & 3 & 1 \end{pmatrix}$$

exchange rows to get

$$\begin{pmatrix} 2 & 3 & 1 \\ 1 & 2 & 3 \end{pmatrix}$$

and then rearrange columns so that the top row is in order to obtain the inverse

$$f^{-1} = \begin{pmatrix} 1 & 2 & 3 \\ 3 & 1 & 2 \end{pmatrix}$$

PRACTICE PROBLEM 7 Verify that

$$f \circ f^{-1} = \begin{pmatrix} 1 & 2 & 3 \\ 1 & 2 & 3 \end{pmatrix} = g^{-1} \circ g$$

where f and g are as defined in Example 3.4.9.

Operations

We have defined three operations on the set $\{0, 1\}$. The Boolean complement operation is a unary operation on $\{0, 1\}$; the Boolean addition and multiplication operations are binary operations on $\{0, 1\}$.

DEFINITION

A **unary operation** on a nonempty set A is a function

$h: A \to A$

A **binary operation** on a nonempty set A is a function

$k: A \times A \to A$

EXAMPLE 3.4.10   **a.** Let $A = \mathbb{Z}$ and $h(x) = -x$. Then h is a unary operation on $\mathbb{Z}$.
b. Let $A = \mathbb{N}$ and $h(x) = -x$. Then h is *not* a unary operation on $\mathbb{N}$ since $h(2) = -2$ and $-2 \notin \mathbb{N}$.
c. Let B be a set, and let $A = P(B)$, the power set of B. Let $h(Y) = B - Y$. Then h is a unary operation on $P(B)$. ∎

EXAMPLE 3.4.11   **a.** Let $A = \mathbb{Z}$ and $k(x, y) = x - y$. Then k is a binary operation on $\mathbb{Z}$.
b. Let $A = \mathbb{N}$ and $k(x, y) = x - y$. Then k is *not* a binary operation on $\mathbb{N}$ since $k(3, 8) = 3 - 8 = -5$ and $-5 \notin \mathbb{N}$.
c. Let B be a set, and let $A = P(B)$, the power set of B. Let $k(X, Y) = X \cup Y$. Then k is a binary operation on $P(B)$. ∎

■ **PROGRAMMING NOTES**

Bijective Functions in Pascal   Characters read into a computer are encoded as bit strings. Each bit string is the binary representation of an integer. We will discuss binary numbers and the conversion from binary to decimal representation in Chapter 6. The ASCII (American Standard Code for Information Interchange) encoding scheme is very common. The ASCII character set includes 128 characters, and each character corresponds to the decimal equivalent of its encoded bit string.

The Pascal function **ord** maps characters to their corresponding integer equivalents. The function **chr** maps an integer to its corresponding character. Let D be the character set available on a given computer. For specificity we assume this is the ASCII character set. Let C be the set $\{0, 1, 2, \ldots, 127\}$. Hence, ord: $D \to C$ and chr: $C \to D$. For example, ord('E') = 69 and chr(69) = 'E'. Clearly, ord(chr(69)) = ord('E') = 69 and chr(ord('E')) = chr(69) = 'E'. The functions ord and chr are bijective and are inverses of each other.

Another pair of bijective Pascal functions are the predecessor and successor functions. The functions **pred** and **succ** are defined on linearly ordered sets. The ASCII character set is linearly ordered. For this set pred('q') = 'p' and succ('p') = 'q'. Let D be the ASCII character set without the first character, and let C be the ASCII character set without the last character. Then pred: $D \to C$ and succ: $C \to D$. The predecessor function and the successor function are inverses of each other.

Operations in Pascal   Several standard Pascal functions are unary operations. The functions abs, sqr, exp [where $\exp(x) = e^x$], and the logical operator NOT are four examples. Binary operations in Pascal include set union and intersection, and the logical operators AND and OR.

3.4 SPECIAL FUNCTIONS

EXERCISE SET 3.4

1. Given a relation f from the set $A = \{a, b, c, d\}$ to the set $B = \{a, b, c\}$, where $f = \{(a, b), (b, b), (c, b), (d, c)\}$, determine whether f is a function from domain A to codomain B. If it is a function, determine whether f is injective.

2. Given a relation g from the set $D = \{1, 2, 3, 4\}$ to the set $C = \{a, b, c, d\}$, where $g = \{(1, a), (2, c), (3, d), (4, b)\}$, determine whether g is a function from domain D to codomain C. If it is a function, determine whether g is injective.

3. Let
$$f = \begin{pmatrix} 1 & 2 & 3 & 4 \\ 2 & 1 & 4 & 3 \end{pmatrix} \quad \text{and} \quad g = \begin{pmatrix} 1 & 2 & 3 & 4 \\ 1 & 3 & 4 & 2 \end{pmatrix}$$
be two permutation functions on $\{1, 2, 3, 4\}$.
   a. Find $f \circ g$ and $g \circ f$.
   b. Find f^{-1} and verify that $f \circ f^{-1} = \begin{pmatrix} 1 & 2 & 3 & 4 \\ 1 & 2 & 3 & 4 \end{pmatrix} = f^{-1} \circ f$.
   c. Find g^{-1} and verify that $g \circ g^{-1} = \begin{pmatrix} 1 & 2 & 3 & 4 \\ 1 & 2 & 3 & 4 \end{pmatrix} = g^{-1} \circ g$.

4. Let
$$f = \begin{pmatrix} 1 & 2 & 3 & 4 \\ 2 & 1 & 3 & 4 \end{pmatrix} \quad \text{and} \quad g = \begin{pmatrix} 1 & 2 & 3 & 4 \\ 3 & 1 & 4 & 2 \end{pmatrix}$$
be two permutation functions on $\{1, 2, 3, 4\}$.
   a. Find $f \circ g$ and $g \circ f$.
   b. Find f^{-1} and verify that $f \circ f^{-1} = \begin{pmatrix} 1 & 2 & 3 & 4 \\ 1 & 2 & 3 & 4 \end{pmatrix} = f^{-1} \circ f$.
   c. Find g^{-1} and verify that $g \circ g^{-1} = \begin{pmatrix} 1 & 2 & 3 & 4 \\ 1 & 2 & 3 & 4 \end{pmatrix} = g^{-1} \circ g$.

5. Let $f: \mathbb{R} \to \mathbb{R}$ be defined by $f(x) = 2 - 5x$. Assume f is bijective and use the method of Example 3.4.8 to find f^{-1}. Verify that $f \circ f^{-1} = 1_{\mathbb{R}} = f^{-1} \circ f$.

6. Let $f: \mathbb{R} \to \mathbb{R}$ be defined by $f(x) = 1 + x^3$. Assume f is bijective and use the method of Example 3.4.8 to find f^{-1}. Verify that $f \circ f^{-1} = 1_{\mathbb{R}} = f^{-1} \circ f$.

For each of the following pairs of functions f and g in Exercises 7–12 determine whether $f \circ g$ is well-defined. Write $(f \circ g)(x)$ explicitly if it is well-defined. Also determine whether $g \circ f$ is well-defined. Write $(g \circ f)(x)$ explicitly if it is well-defined.

7. $f: \mathbb{Z} \to \mathbb{Z}$ defined by $f(x) = 3x - 2$ $g: \mathbb{N} \to \mathbb{Z}$ defined by $g(n) = 2 - n$
8. $f: \mathbb{N} \to \mathbb{R}$ defined by $f(n) = 2 - \sqrt{n}$ $g: \mathbb{Z} \to \mathbb{Z}$ defined by $g(x) = x + 7$
9. $f: \mathbb{N} \to \mathbb{R}$ defined by $f(n) = \log_2(n + 1)$ $g: \mathbb{N} \to \mathbb{N}$ defined by $g(n) = 2^n - 1$
10. $f: \mathbb{Z} \to \mathbb{N}$ defined by $f(x) = 3x^2 + 1$ $g: \mathbb{N} \to \mathbb{Z}$ defined by $g(n) = 2 - n$
11. $f: \mathbb{N} \to \mathbb{Z}$ defined by $f(n) = 2 - n^2$ $g: \mathbb{Z} \to \mathbb{Z}$ defined by $g(x) = x + 7$
12. $f: \mathbb{N} \to \mathbb{R}$ defined by $f(n) = \log_3(n + 1)$ $g: \mathbb{N} \to \mathbb{N}$ defined by $g(n) = 3^n - 1$

13. We define a predecessor function and a successor function on $\mathbb{Z}$ as follows: Define $p: \mathbb{Z} \to \mathbb{Z}$ by $p(x) = x - 1$ and $s: \mathbb{Z} \to \mathbb{Z}$ by $s(x) = x + 1$. Verify that p and s are inverses of each other.

14. We define a predecessor function and a successor function on $\mathbb{N}$ as follows: Define $p: D \to \mathbb{Z}$ by $p(x) = x - 1$ and $s: \mathbb{N} \to C$ by $s(x) = x + 1$. Specify the sets D and C, and verify that p and s are inverses of each other.

15. Determine whether each of the following is a unary operation on the given set.
    a. $h(x) = 2^x$ on $\mathbb{N}$ **b.** $h(x) = \log_2(x + 1)$ on $\mathbb{N}$

16. Determine whether each of the following is a unary operation on the given set.
    a. $h(x) = \dfrac{1}{x}$ on $\mathbb{N}$ **b.** $h(x) = \dfrac{1}{x}$ on $\mathbb{R}$ **c.** $h(x) = \dfrac{1}{x}$ on $\mathbb{R} - \{0\}$

17. Determine whether each of the following is a binary operation on the given set.
    a. $k(x, y) = \dfrac{x}{y + 1}$ on $\mathbb{N}$ **b.** $k(x, y) = x^2 + y$ on $\mathbb{N}$
    c. $k(x, y) = x^2 + y^2 - xy$ on $\mathbb{N}$

18. Determine whether each of the following is a binary operation on the given set.
    a. $k(T, S) = T \cap S$ on $P(U)$ (in other words, $T, S \subseteq U$)
    b. $k(T, S) = T \cup S$ on $P(U)$

19. Let $D = \{1, 2, 3\} = C$. Either define functions $f: D \to C$ with the following properties or assert that such a function does not exist.
    a. f is bijective. **b.** f is onto but not 1–1.
    c. f is 1–1 but not onto. **d.** f is neither 1–1 nor onto.

20. Either define functions $f: \mathbb{N} \to \mathbb{N}$ with the following properties or assert that such a function does not exist.
    a. f is bijective. **b.** f is onto but not 1–1.
    c. f is 1–1 but not onto. **d.** f is neither 1–1 nor onto.

21. Either define functions $f: \mathbb{N} \to \mathbb{Z}$ with the following properties or assert that such a function does not exist.
    a. f is bijective. [*Hint:* Consider Example 3.3.4 (page 109).]
    b. f is 1–1 but not onto.
    c. f is onto but not 1–1.
    d. f is neither 1–1 nor onto.

22. Show that, if m pigeons occupy n pigeonholes, then at least one pigeonhole will have more than $\lfloor (m - 1)/n \rfloor$ pigeons. [*Hint:* Recall that $\lfloor x \rfloor$ is the largest integer less than or equal to x. Now use algebra and an inequality to do the exercise.]

23. Suppose a computer's memory, with seven locations, has a capacity averaging 1024 bits. Use the result of Exercise 22 to show that there is a location with at least 1024 bits.

24. Suppose 27 students are placed at random in chairs that are arranged in an 8 column by 5 row array. Use the result of Exercise 22 to show the following:
    a. There are at least 6 students in some row.
    b. There are at least 4 students in some column.

■ PROGRAMMING EXERCISE

25. User-defined data types in Pascal are linearly ordered. Consider the following defined data type:

    type
        months = (Jan, Feb, March, April, May, June, July, Aug, Sept, Oct, Nov, Dec);

The functions ord, pred, and succ are all defined on months. For example, ord(Jan) = 0 and pred(March) = Feb.
a. Find ord(June) and ord(Sept).
b. Designate the domain and codomain of the functions pred and succ, and verify that pred and succ are inverses of each other.

3.5
APPLICATION: TIME-COMPLEXITY FUNCTIONS

In order to compare one algorithm with another for relative efficiency, a measure of the algorithms' cost must be defined. The cost of a given algorithm is an important aspect of its design and analysis. One measure of this cost is the time required for execution. The execution time, or time complexity, of a given algorithm is of great importance in computer science.

When choosing an algorithm to solve a problem, we need to know how fast the computer program implementing the algorithm will run. What factors affect the run time of a program? We usually consider the following four factors:

1. The input to the program
2. The quality of code generated by the compiler used to create the program
3. The nature and speed of the instructions on the machine used to execute the program
4. The time complexity of the algorithm underlying the program

As a result of factor 1, the run time of a program is defined as a function of the input. Generally, the run time depends on the size of the input (for example, the number of items in a list). So, we use a **time-complexity function** T, where the domain of T is the nonnegative integers $\mathbb{N}$ and the codomain is the nonnegative reals. For example, when a program has time complexity proportional to n^2, we write $T(n) = cn^2$ for some constant c, called the **constant of proportionality**.

Because of factors 2 and 3, it is not possible to express $T(n)$ in some standard time unit such as seconds. Instead we can only hope to conclude that some given algorithm has time complexity (for example) proportional to n^2. The constant of proportionality will be unspecified because of the unknown factors such as the compiler and the hardware used to execute the program.

Factor 4, the time complexity of the algorithm, is measured by the function value $T(n)$. Even though the units of $T(n)$ are unspecified, we can think of $T(n)$ as the number of instructions on a computer, the number of comparisons in a sorting routine, the number of swaps in a searching algorithm, or some other appropriate measure of the cost of the algorithm in question.

Growth Rates of Time-Complexity Functions

Table 3.5.1 (page 126) illustrates the rate of growth for various time-complexity functions.

TABLE 3.5.1 COMPARISON OF SEVERAL TIME-COMPLEXITY FUNCTIONS

n	$\log_2 n$	n	$n \log_2 n$	n^2	n^3	2^n	$n!$
2	1	2	2	4	8	4	2
10	3.32	10	33.2	100	1000	1024	$3.63 \cdot 10^6$
32	5	32	160	1024	$3.28 \cdot 10^4$	$4.29 \cdot 10^9$	$2.63 \cdot 10^{35}$
64	6	64	384	4096	$2.62 \cdot 10^5$	$1.84 \cdot 10^{19}$	$1.27 \cdot 10^{89}$
100	6.64	100	664	10^4	10^6	$1.27 \cdot 10^{30}$	$9.33 \cdot 10^{177}$

Computer speeds are rated in millions of machine instructions executed per second, or MIPS. A certain computer rated as a 1000 MIPS machine executes 1 billion (10^9) instructions per second. Recall that 1 μs = 1 microsecond = 10^{-6} second. An algorithm of time complexity n^2 will take $100/10^9 = 10^{-7}$ second, or $10^{-7}/10^{-6} = 10^{-1} = 0.10$ μs when $n = 100$. An algorithm of complexity 2^n will require approximately 1 μs when $n = 10$ and approximately $4 \cdot 10^{13}$ years when $n = 100$. Note that, even if a computer could execute 1 trillion (10^{12}) instructions per second, nearly $4 \cdot 10^{10}$, or 40 billion years, are required to execute 2^n instructions when $n = 100$.

Clearly, for large n, say $n > 60$, algorithms with time complexity proportional to 2^n or $n!$ are not feasible. On the other hand, algorithms with time complexity proportional to $\log_2 n$, n, $n \log_2 n$, n^2, or n^3 are considered to be relatively efficient. Table 3.5.1 shows $\log_2 n$, n, $n \log_2 n$, n^2, and n^3 to be in order of decreasing efficiency. For example, an algorithm with time complexity proportional to $n \log_2 n$ is preferable to one proportional to n^2.

Counting and Algorithm Analysis

The number of steps taken by an algorithm usually is a function of its input. This property is particularly evident in the case of loops. So, we examine algorithm efficiency using loops as examples.

EXAMPLE 3.5.1 Consider the following loops.

a. In the loop, P is some executable statement.

    for i = 1 to 10 do
      P

It is clear that P will be executed 10 times in this loop.

b. It is clear that P will be executed n times in the loop

    for i = 1 to n do
      P

∎

EXAMPLE 3.5.2 Consider the following nested loops.

a. In the loop, the inner loop (the j-loop) will execute n times for each execution of the outer loop (the i-loop).

3.5 APPLICATION: TIME-COMPLEXITY FUNCTIONS

    for i = 1 to n do
        for j = 1 to n do
            P

Since the outer loop will execute n times, we see that P will execute $n \cdot n = n^2$ times.

b. Similarly, P will execute n^3 times in

    for i = 1 to n do
        for j = 1 to n do
            for k = 1 to n do
                P

Recall that Warshall's algorithm had the same form as the nested loop in Example 3.5.2b and, hence, has time complexity n^3.

EXAMPLE 3.5.3 Consider the nested loop

    for i = 1 to n do
        for j = 1 to i do
            P

This situation requires more thought, for now the inner loop depends on the value of the outer-loop control variable, i. When $i = 1$, the j-loop will execute once. When $i = 2$, the j-loop will execute twice. In general, we see that, when $i = m$, the j-loop will execute m times. Hence, P will execute $1 + 2 + \cdots + n$ times in this nested loop.

In Example 3.3.6 (page 109) we proved

$$1 + 2 + \cdots + n = \frac{n(n+1)}{2} = \frac{n^2}{2} + \frac{n}{2} \tag{1}$$

Hence, the nested loop above has time complexity $n^2/2 + n/2$.

EXERCISE SET 3.5

In Exercises 1–4 suppose we have a computer that executes 1 billion instructions per second. Use Table 3.5.1 to find the amount of time it takes to run an algorithm with the given time complexity for each value of n.

1. Time complexity $n \log_2 n$
   a. $n = 10$ **b.** $n = 32$ **c.** $n = 100$

2. Time complexity $\log_2 n$
   a. $n = 10$ **b.** $n = 32$ **c.** $n = 100$

3. Time complexity n^2
   a. $n = 10$ **b.** $n = 32$ **c.** $n = 100$

4. Time complexity $n!$
   a. $n = 10$ **b.** $n = 32$ **c.** $n = 100$

How many times will P execute in the nested loops of Exercises 5–7?

5. for i := 1 to n do
    for j := i to n do
      P

6. for i := 1 to n do
    for j := 1 to i do
      for k := 1 to n do
        P

7. for i := 1 to n − 1 do
    for j := i + 1 to n do
      P

KEY TERMS

Equivalence relation
Equivalence class
Pairwise disjoint
Partition
Partial ordering
Partial order
Partially ordered set (poset)
Comparable elements
Linear ordering
Hasse diagram
Extremal elements
Least element
Greatest element
Strictly less than
Strictly greater than
Minimal element
Maximal element
Precedence relationship

Topological sorting
Lower bound
Upper bound
Greatest lower bound
Least upper bound
Function
Domain
Codomain
Image
Preimage
Sequence
Finite sequence
String
Bit string
Boolean-valued function
Characteristic function
Partial function

Domain of definition
One-to-one (1–1) function
Injective function
Onto (or surjective) function
Bijective function (bijection, 1–1 onto function)
Identity function
Permutation of A
Composition of functions
Inverse function
Unary operation
Binary operation
Time-complexity function

REVIEW EXERCISES

1. Let $U = \mathbb{Z}$, $A = \{-3, 7, 9\}$, $B = \{x : x \leq 8\}$, and $D = \{1, 3, 5, 7, 9\}$. Find:
  a. B' **b.** $A \cup D$ **c.** $A \cap D$ **d.** $A \cup B$
  e. $D - B$ **f.** $A \cap B$ **g.** $A \cap (B \cup D)$ **h.** $(A \cap B) \cup D$
  i. $P(A)$ **j.** $A \times \emptyset$ **k.** $A \times D$

2. Let $U = \mathbb{N}$, $A = \{1, 3, 5\}$, $B = \{x : x^2 + 6x - 7 = 0\}$, and $D = \{x : x < 5\}$. Find:
  a. D' **b.** $A \cup D'$ **c.** $A \cap D'$ **d.** $A \cap B$ **e.** $D - A$ **f.** $A - D$
  g. Card(B)   **h.** $A \times B$ **i.** $P(A)$ **j.** $A \times A$ **k.** $B \times B$

3. Decide whether each of the following assertions about sets is true or false. If the statement is false, give a specific example showing it is false.
  a. $D \cap (A \cup B) = D \cup (A \cap B)$ **b.** $D \cap (A \cup B) \subseteq D \cup (A \cap B)$
  c. $(A - B) - C = A - (B - C)$

4. Decide whether each of the following assertions about sets is true or false. If the statement is false, give a specific example showing it is false.
  a. $A \cap B = B - (B - A)$ **b.** If $A \cap B = \emptyset$, then $P(A) \cap P(B) = \emptyset$.
  c. If $A \cup B = A \cup C$, then $B = C$.

REVIEW EXERCISES

5. Which of the following functions are 1–1, onto, bijective? All the functions have $\mathbb{R}$ as domain and codomain.
   a. $f(x) = x^3 + 1$ **b.** $g(x) = x^4 + 1$ **c.** $j(x) = \cos x$ **d.** $k(x) = x + \sin x$

6. Recall the floor function $G(x) = \lfloor x \rfloor$ defined by $G : \mathbb{R} \to \mathbb{Z}$, where $G(x) = n$ for $n \leq x < n + 1$. So, for example, $G(2.3) = \lfloor 2.3 \rfloor = 2$, $G(3.0) = 3$, and $\lfloor -1.3 \rfloor = -2$. In the following, specify the domain and codomain for both f and h.
   a. Find a function f such that $G \circ f = 1_{\mathbb{Z}}$, or explain why no such function can be found.
   b. Find a function h such that $h \circ G = 1_{\mathbb{R}}$, or explain why no such function can be found.

7. Let $D_{42} = \{1, 2, 3, 6, 7, 14, 21, 42\}$, and define a partial order by $x \preceq y$ if y is divisible by x.
   a. Draw the Hasse diagram for the poset $(D_{42}, \preceq)$.
   b. Find any maximal or minimal elements.
   c. Find any greatest or least elements.

8. Let $\mathbb{Z}$ be the set of all integers. Define a relation $\equiv$ on $\mathbb{Z}$ by $m \equiv n$ if $m - n$ is divisible by 5.
   a. Show that $\equiv$ is an equivalence relation.
   b. Determine all the equivalence classes under $\equiv$. How many are there?
   c. Verify that the equivalence classes in part b form a partition of $\mathbb{Z}$.

9. Given a relation f from the set $A = \{a, b, c, d, e\}$ to the set $B = \{1, 2, 3\}$, where $f = \{(a, 2), (b, 2), (c, 2), (e, 3)\}$, determine whether f is a function. If f is not a function, determine whether f is a partial function. If so, what is its domain of definition?

10. Let $U = \{u_1, u_2, u_3, u_4, u_5, u_6, u_7\}$. Write the characteristic function and the bit string for each given subset of U.
    a. $A = \{u_1, u_2, u_7\}$ **b.** $A = \{u_3, u_5, u_6\}$
    c. $A = U - \{u_1, u_3\}$ **d.** $A = \{u_1, u_2, u_7\}'$

11. Let $f : \mathbb{R} \to \mathbb{R}$ be defined by $f(x) = 3 - x^5$. Assume f is bijective, and find f^{-1}. Verify $f \circ f^{-1} = 1_{\mathbb{R}} = f^{-1} \circ f$.

12. For the following pair of functions f and g, determine whether $g \circ f$ is well-defined. Write $(g \circ f)(x)$ explicitly if it is well-defined. Do the same for $f \circ g$.

    $f : \mathbb{N} \to \mathbb{R}$ defined by $f(n) = 2 + \sqrt{n + 7}$
    $g : \mathbb{N} \to \mathbb{Z}$ defined by $g(x) = x - 7$

13. Determine whether each of the following is a unary operation on the given set.
    a. $h(x) = 2^{-x}$ on $\mathbb{N}$ **b.** $h(x) = \log_2 4^x$ on $\mathbb{N}$

14. Determine whether each of the following is a binary operation on the given set.
    a. $k(x, y) = (x - 1)^2 y$ on $\mathbb{N}$ **b.** $k(x, y) = x^2 y - 1$ on $\mathbb{N}$

15. Let R be a relation on a set A. Define a relation S on A by $x\, S\, y$ if $y\, R\, x$. Prove:
    a. R is symmetric if $R = S$.
    b. If R is a partial order, then S is a partial order.

16. Let A be an infinite set and $P(A)$ be its power set. Define a relation R on $P(A)$ by $X\, R\, Y$ if $(X - Y) \cup (Y - X)$ is a finite set.
    a. Show that R is an equivalence relation.
    b. Describe the members in the equivalence class containing the empty set.
    c. Describe the members in the equivalence class $[A]$.

17. Define a function f with domain $\mathbb{Z}$ and codomain $A = \{-1, 1\}$, which is a subset of $\mathbb{Z}$, by

$$f(x) = \begin{cases} 1 & \text{if } x \text{ is even} \\ -1 & \text{if } x \text{ is odd} \end{cases}$$

Prove $f(x + y) = f(x) \cdot f(y)$ for all x, y in $\mathbb{Z}$.

REFERENCES

Aho, A., J. Hopcroft, and J. Ullman. *Data Structures and Algorithms.* Reading, MA: Addison-Wesley, 1983.

Knuth, D. *The Art of Computer Programming: Volume 1, Fundamental Algorithms,* 2nd ed. Reading, MA: Addison-Wesley, 1973.

CHAPTER 4

COMBINATORICS AND FINITE PROBABILITY

Combinatorics, or combinatorial mathematics, is the branch of mathematics that studies and analyzes the techniques of counting. Gottfried Wilhelm Leibniz (1646–1716), who invented calculus independently of Isaac Newton, first used the term *combinatorial* in his *Dissertation on the Combinatorial Art* in 1666. In that treatise Leibniz considered problems related to the factorial and the binomial coefficients. He foresaw the many applications of combinatorial mathematics when he observed that combinatorics has "applications to the whole sphere of sciences." Indeed, combinatorial reasoning is necessary for the study of computer systems analysis, operations research, and finite probability, among many other fields of applied mathematical research.

In this chapter and in Chapter 9, we will discuss many basic techniques of counting and show how combinatorics is applied to a wide variety of problems. In our discussion of probability in Section 4.4, we apply concepts introduced in Sections 4.1–4.3.

APPLICATION   In the application of Section 4.5, we will solve an interesting combinatorial problem posed by the mathematician and computer scientist Donald Knuth in Volume 1 of his classic series of books *The Art of Computer Programming*. The problem and its solution involve permutations, bijective functions, binomial coefficients, and well-balanced parentheses.

4.1
BASIC COUNTING TECHNIQUES

To solve many problems in computer science, we can use techniques of counting. For example, to estimate the relative efficiency of two computer programs, we need to count the number of times certain steps are executed in each program. The problem of counting is not always easy. Some counting problems are extremely difficult despite the availability of high-speed computers.

All the sets under discussion will be assumed to be finite. If A is a set, recall that Card(A) denotes the number of elements in the set A.

EXAMPLE 4.1.1 Consider the sets

$A = \{1, 2, 3, 4, 5\}$
$E = \{a, b, c, d, e, \ldots, x, y, z\}$
$D = \{0, 1, 2, 3, 4, 5, 6, 7, 8, 9\}$
$G = \emptyset$

We see that Card(A) = 5, Card(E) = 26, Card(D) = 10, and Card(G) = 0. ∎

A few simple rules of combinatorics will make counting easier. The first of these rules gives us a method of counting the number of elements in a union of two sets.

SUM RULE

Card($A \cup B$) = Card(A) + Card(B) − Card($A \cap B$)
Card($A \cup B$) = Card(A) + Card(B), if $A \cap B = \emptyset$

The **sum rule** is also called the **law of inclusion and exclusion**.

We illustrate the sum rule for the case $A \cap B = \emptyset$ and the case $A \cap B \neq \emptyset$ in Figures 4.1.1(a) and 4.1.1(b). In Figure 4.1.1(b), we have $A \cap B \subseteq A$ and $A \cap B \subseteq B$. The elements in $A \cap B$ are counted once in Card(A) and again

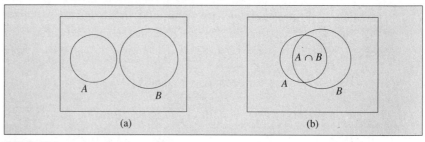

FIGURE 4.1.1

4.1 BASIC COUNTING TECHNIQUES

in Card(B). Therefore, in order to obtain Card($A \cup B$), we need to subtract Card($A \cap B$) once from the sum Card(A) + Card(B). Thus, Card($A \cup B$) = Card(A) + Card(B) − Card($A \cap B$).

EXAMPLE 4.1.2 Consider the sets

$A = \{a, d, e, g, k\}$
$B = \{1, 2, 3, 5\}$
$C = \{e, g, 4, 7, 8\}$

We have $A \cup B = \{a, d, e, g, k, 1, 2, 3, 5\}$, $A \cap B = \emptyset$, $A \cup C = \{a, d, e, g, k, 4, 7, 8\}$, and $A \cap C = \{e, g\}$. We have Card($A \cup B$) = 9, Card(A) = 5, and Card(B) = 4. This illustrates the sum rule: Card($A \cup B$) = 9 = 5 + 4 = Card(A) + Card(B) when $A \cap B = \emptyset$.

Since Card($A \cup C$) = 8, Card(A) = 5, Card(C) = 5, and Card($A \cap C$) = 2, we have Card($A \cup C$) = 8 = 5 + 5 − 2 = Card(A) + Card(C) − Card($A \cap C$). ∎

PRACTICE PROBLEM 1 Let

$A = \{1, 2, 3, 4, 5, 6, 7\}$
$B = \{4, 7, 10, 12, 13\}$
$C = \{2, 4, 6\}$

a. Find Card($A \cup B$), Card($A \cup C$), and Card($B \cup C$).
b. Find Card($A \cap B$), Card($A \cap C$), and Card($B \cap C$).
c. Verify that Card($A \cup B$) = Card(A) + Card(B) − Card($A \cap B$).
d. Verify that Card($A \cup C$) = Card(A) + Card(C) − Card($A \cap C$).
e. Verify that Card($B \cup C$) = Card(B) + Card(C) − Card($B \cap C$).

EXAMPLE 4.1.3 A car manufacturer finds that the most common production defects are faulty brakes and broken headlights. In testing a sample of 80 cars, the manufacturer recorded the following data:

20 cars have faulty brakes
15 cars have broken headlights
10 cars have both defects

a. How many cars in the sample have at least one of these two defects?
b. How many cars in this sample have neither of these two defects?

Solution a. Let B be the set of cars in this sample with faulty brakes, and let H be the set if cars in this sample with broken headlights. Using the sum rule, we have

$$\text{Card}(B \cup H) = \text{Card}(B) + \text{Card}(H) - \text{Card}(B \cap H)$$
$$= 20 + 15 - 10 = 25$$

There are 25 cars in this sample with at least one of these two defects.

b. If a car has neither of the defects, then it does not have at least one of the defects. Since $80 - 25 = 55$, there are 55 cars in this sample with neither of these two defects. ∎

PRACTICE PROBLEM 2 In surveying a sample of 1000 college juniors, we find that 800 of them can write computer programs in Pascal, 500 of them can write computer programs in FORTRAN, and 325 of them can write computer programs in both Pascal and FORTRAN. Find the number of juniors in this sample who can write computer programs in neither Pascal nor FORTRAN.

Next we recall a formula from Chapter 1. It is called the product rule in counting.

PRODUCT RULE

$\text{Card}(A \times B) = \text{Card}(A) \cdot \text{Card}(B)$ where A and B are nonempty sets

This **product rule** holds whether $A \cap B \neq \emptyset$ or $A \cap B = \emptyset$.

We illustrate the product rule by the next example.

EXAMPLE 4.1.4 Let $A = \{a, b, c\}$ and $B = \{b, 1, 2, d\}$. Find $A \times B$ and verify the product rule with A and B.

Solution For $a \in A$ in the first position, we can have each of b, 1, 2, and d in B in the second position. The same situation holds for b and c in A. We show this in Figure 4.1.2. Hence, $\text{Card}(A \times B) = 12 = 3 \cdot 4 = \text{Card}(A) \cdot \text{Card}(B)$.

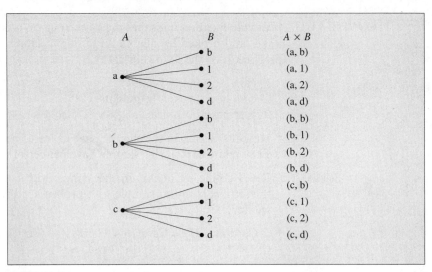

FIGURE 4.1.2

We often need to apply the sum rule and product rule in the same problem, as illustrated by the next example.

EXAMPLE 4.1.5 A variable name in the programming language BASIC is either a single (English) alphabetic letter or a single alphabetic letter followed by a single decimal digit. How many possible BASIC variable names are there?

Solution   Let A denote the set of 26 alphabetic letters, and let D denote the set of decimal digits $\{0, 1, 2, 3, 4, 5, 6, 7, 8, 9\}$. Let V be the set of all possible BASIC variable names. Therefore, V is represented by the set $A \cup (A \times D)$. Since $A \cap (A \times D) = \emptyset$, we can use the sum rule to obtain

$$\text{Card}(V) = \text{Card}(A) + \text{Card}(A \times D)$$

Applying the product rule to $A \times D$ yields

$$\text{Card}(V) = \text{Card}(A) + \text{Card}(A) \cdot \text{Card}(D)$$
$$= 26 + 26 \cdot 10 = 26 + 260 = 286$$

Hence, there are 286 possible BASIC variable names. ∎

The product rule can be extended to more than two sets:

$$\text{Card}(A_1 \times A_2 \times \cdots \times A_n) = \text{Card}(A_1) \cdot \text{Card}(A_2) \cdot \cdots \cdot \text{Card}(A_n)$$

EXAMPLE 4.1.6 In years past, telephone numbers began with two alphabetic letters followed by five digits. How many different telephone numbers were possible? (The letter "Q" was not used in the telephone number prefixes.)

Solution   Let $\Omega = \{A, B, C, D, \ldots, X, Y, Z\} - \{Q\}$ and $D = \{0, 1, 2, \ldots, 8, 9\}$. The set of all possible telephone numbers is $\Omega \times \Omega \times D \times D \times D \times D \times D$, and the number of different telephone numbers is $\text{Card}(\Omega \times \Omega \times D \times D \times D \times D \times D) = 25 \cdot 25 \cdot 10 \cdot 10 \cdot 10 \cdot 10 \cdot 10 = 62{,}500{,}000$. ∎

PRACTICE PROBLEM 3   The dialing mechanism of a telephone has either ten buttons or ten finger holes in the dial. As a result, the number of seven-character numbers is the same as the number of seven-digit telephone numbers. How many seven-digit telephone numbers are there? (Note that the first digit of a telephone number cannot be a "0" or a "1".)

The sum rule also can be extended to more than two sets:

$$\text{Card}(A_1 \cup A_2 \cup \cdots \cup A_n) = \text{Card}(A_1) + \text{Card}(A_2) + \cdots + \text{Card}(A_n),$$
when $A_1, \ldots, A_n$ are pairwise disjoint

The sum rule on $A_1, \ldots, A_n$, when the sets are not pairwise disjoint, can be developed for $n > 2$. For illustration, let us consider the case of $n = 3$. First we write $A_1 \cup A_2 \cup A_3 = (A_1 \cup A_2) \cup A_3$, then we apply the sum rule for the sets $(A_1 \cup A_2)$ and A_3.

$$\begin{aligned}
\text{Card}(A_1 \cup A_2 \cup A_3) &= \text{Card}((A_1 \cup A_2) \cup A_3) \\
&= [\text{Card}(A_1 \cup A_2) + \text{Card}(A_3)] - \text{Card}((A_1 \cup A_2) \cap A_3) \\
&= [\text{Card}(A_1) + \text{Card}(A_2) - \text{Card}(A_1 \cap A_2) + \text{Card}(A_3)] \\
&\quad - \text{Card}[(A_1 \cap A_3) \cup (A_2 \cap A_3)] \\
&= [\text{Card}(A_1) + \text{Card}(A_2) - \text{Card}(A_1 \cap A_2) + \text{Card}(A_3)] \\
&\quad - [\text{Card}(A_1 \cap A_3) + \text{Card}(A_2 \cap A_3) - \text{Card}[(A_1 \cap A_3) \cap (A_2 \cap A_3)]] \\
&= [\text{Card}(A_1) + \text{Card}(A_2) - \text{Card}(A_1 \cap A_2) + \text{Card}(A_3)] \\
&\quad - [\text{Card}(A_1 \cap A_3) + \text{Card}(A_2 \cap A_3) - \text{Card}(A_1 \cap A_2 \cap A_3)] \\
&= \text{Card}(A_1) + \text{Card}(A_2) + \text{Card}(A_3) - \text{Card}(A_1 \cap A_2) - \text{Card}(A_1 \cap A_3) \\
&\quad - \text{Card}(A_2 \cap A_3) + \text{Card}(A_1 \cap A_2 \cap A_3)
\end{aligned}$$

The sum rule for four sets A_1, A_2, A_3, and A_4 is stated below. We leave the derivation as an exercise. (See Exercise 18, Exercise Set 4.1.)

$$\begin{aligned}
\text{Card}(A_1 \cup A_2 \cup A_3 \cup A_4) &= \text{Card}(A_1) + \text{Card}(A_2) + \text{Card}(A_3) + \text{Card}(A_4) - \text{Card}(A_1 \cap A_2) \\
&\quad - \text{Card}(A_1 \cap A_3) - \text{Card}(A_1 \cap A_4) - \text{Card}(A_2 \cap A_3) \\
&\quad - \text{Card}(A_2 \cap A_4) - \text{Card}(A_3 \cap A_4) + \text{Card}(A_1 \cap A_2 \cap A_3) \\
&\quad + \text{Card}(A_1 \cap A_2 \cap A_4) + \text{Card}(A_1 \cap A_3 \cap A_4) \\
&\quad + \text{Card}(A_2 \cap A_3 \cap A_4) - \text{Card}(A_1 \cap A_2 \cap A_3 \cap A_4)
\end{aligned}$$

EXAMPLE 4.1.7 Let $F = \{n : 1000 \leq n \leq 9999\}$ be the set of all four-digit positive integers. Find the number of all integers in F with at least one kth digit equal to k for $k = 1, 2, 3, 4$.

Solution Let B_k be the set of all integers in F with kth digit equal to k for $k = 1, 2, 3, 4$. For example, $1479 \in B_1$, $4237 \in B_2$, and $9238 \in B_2 \cap B_3$. We seek to find $\text{Card}(B_1 \cup B_2 \cup B_3 \cup B_4)$ using the sum rule for four sets.

$$\begin{aligned}
\text{Card}(B_1 \cup B_2 \cup B_3 \cup B_4) &= 1000 + 900 + 900 + 900 - 100 - 100 \\
&\quad - 100 - 90 - 90 - 90 + 10 + 10 + 10 + 9 - 1 = 3168 \quad \blacksquare
\end{aligned}$$

PRACTICE PROBLEM 4 Let $A = \{a, b, c, d, e, f, g\}$, $B = \{a, b, c, h, i, j, k, l\}$, $C = \{b, c, d, m, n, o\}$, and $D = \{b, d, e, h, m, q, r, s, t\}$.

    a. Find the cardinalities of A, B, C, D, $A \cap B$, $A \cap C$, $A \cap D$, $B \cap C$, $B \cap D$, $C \cap D$, $A \cap B \cap C$, $A \cap B \cap D$, $A \cap C \cap D$, $B \cap C \cap D$, and $A \cap B \cap C \cap D$.

    b. Find $\text{Card}(A \cup B \cup C \cup D)$ using the sum rule for four sets.

Counting techniques can also be applied to sets of functions. Theorem 4.1.8 motivates a very useful and concise notation for the set of all functions from a specified domain to a given codomain.

THEOREM 4.1.8 If $\text{Card}(D) = m \geq 0$ and $\text{Card}(C) = n \geq 1$, then the number of all functions $f : D \rightarrow C$ is n^m.

4.1 BASIC COUNTING TECHNIQUES

Proof If $m = 0$, then $D = \emptyset$. There is exactly one function $f: \emptyset \to C$ since there is only one way to assign no element from D to elements of C. This is the "do nothing" function. In this case $n^m = n^0 = 1$.

Assume $m \neq 0$, and let $D = \{a_1, \ldots, a_m\}$ and $C = \{b_1, \ldots, b_n\}$. Each $a_i \in D$ can be assigned under f to any of the n elements of C. There are n possible images for a_1, there are n possible images for $a_2, \ldots,$ and there are n possible images for a_m. Hence, there are $n \cdot n \cdots \cdot n = n^m$ possible ways of assigning elements from D to elements of C. In other words, there are n^m functions from D to C. ∎

It is common for C^D to denote the set of all functions from D to C. With this notation, Theorem 4.1.8 states that $\text{Card}(C^D) = \text{Card}(C)^{\text{Card}(D)}$.

EXAMPLE 4.1.9 How many four-digit binary numbers (numbers of base 2) are there?

Solution We use the two digits in $C = \{0, 1\}$ to represent numbers in base 2. Each four-digit number in base 2 corresponds to a function

$$f: \{1, 2, 3, 4\} \to \{0, 1\}$$

(Why?) By Theorem 4.1.8, there are 2^4 such functions. Hence, the number of four-digit binary numbers is 2^4. ∎

The question in Example 4.1.9 could have been phrased: "How many bit strings of length 4 are there?" Looking at the question this way makes it clear we are seeking the number of finite sequences with domain $\{1, 2, 3, 4\}$ and codomain $\{0, 1\}$. This approach also leads directly to the answer 2^4.

The next example poses another question about finite sequences or strings.

EXAMPLE 4.1.10 How many eight-digit numbers in base 6 are there?

Solution We use the six digits in $D = \{0, 1, 2, 3, 4, 5\}$ to represent numbers in base 6. Each eight-digit number in base 6 corresponds to a function (finite sequence)

$$f: \{1, 2, 3, 4, 5, 6, 7, 8\} \to \{0, 1, 2, 3, 4, 5\}$$

By Theorem 4.1.8, there are 6^8 such functions. Hence, the number of eight-digit numbers in base 6 is 6^8. ∎

The following is an immediate result of Theorem 4.1.8.

COROLLARY 4.1.11 If A is a set with n elements, then there are 2^n subsets of A. In other words, $\text{Card}(P(A)) = 2^{\text{Card}(A)}$.

Proof Let B be any subset of A. Let χ_B be the characteristic function on A, where $\chi_B: A \to \{0, 1\}$ is defined by

$$\chi_B(x) = \begin{cases} 1 & \text{if } x \in B \\ 0 & \text{if } x \in A - B \end{cases}$$

Clearly, $\chi_B = \chi_C$ if and only if $B = C$, where B and C are subsets of A. So, the number of all characteristic functions on A is equal to the number of subsets of A. The set of all functions $f: A \to \{0, 1\}$ is the same as the set of all characteristic functions on A. Hence, the number of all subsets of A is $\text{Card}(\{0, 1\})^{\text{Card}(A)} = 2^n$. ∎

EXAMPLE 4.1.12 Let $A = \{a, b, c, d\}$. Find:

a. $\text{Card}(A \times A)$ b. $\text{Card}(P(A))$ c. $P(A)$

Solution
a. $\text{Card}(A \times A) = (\text{Card } A) \cdot (\text{Card } A) = 4 \cdot 4 = 16$, by the product rule
b. $\text{Card}(P(A)) = 2^{\text{Card}(A)} = 2^4 = 16$, by Corollary 4.1.11
c. $P(A) = \{\emptyset, \{a\}, \{b\}, \{c\}, \{d\}, \{a, b\}, \{a, c\}, \{a, d\}, \{b, c\}, \{b, d\}, \{c, d\},$
$\{a, b, c\}, \{a, b, d\}, \{a, c, d\}, \{b, c, d\}, A\}$ ∎

PRACTICE PROBLEM 5 Apply the product rule and Corollary 4.1.11 to find $\text{Card}(S)$, $\text{Card}(P(S))$, and $\text{Card}(S \times S)$ when S is $\emptyset$, $\{a\}$, $\{a, b\}$, or $\{a, b, c\}$, respectively.

EXERCISE SET 4.1

1. There are 45 students in a statistics class and 62 students in a calculus class. A delegation of 2 members is being formed to present a petition to the department chair. In how many ways can a delegation of 1 statistics student and 1 calculus student be formed?

2. How many five-digit numbers of base 3 are there?

3. A 16-bit string is a sequence of 16 symbols, each symbol a 0 or a 1. How many 16-bit strings are there?

4. Find the number of bit strings with length greater than 2 and less than or equal to 8.

5. A local service club must elect first a president and then a vice president from among its membership of 30 people. In how many ways can these officers be elected?

6. A local restaurant serves a four-course meal. The customer may choose any one of 4 items for the first course, any one of 7 items for the second course, any one of 11 items for the third course, and any one of 3 items for the fourth course. Assuming that the customer does not skip a course, how many choices of meals are there?

7. Refer to Exercise 6. Assume that the customer may skip one course. How many choices of meals are there?

8. A FORTRAN identifier is a string of from one to six characters. The first character must be alphabetic, and the remaining characters may be alphanumeric (i.e., alphabetic or numeric). Find the number of FORTRAN identifiers.

9. Let $A = \{1, 2, 3, 4, 5, 6\}$ and $B = \{a, b, c\}$. Find:
   a. $\text{Card}(A \times B)$ b. $\text{Card}(A \times A)$ c. $\text{Card}(A^B)$ d. $\text{Card}(P(A))$
   e. $\text{Card}(A \cup B)$

10. Let $A = \{1, 2, 3, 4\}$ and $B = \{a, b, c, 1, 2\}$. Find:
    a. $\text{Card}(A \times B)$ b. $\text{Card}(A \times A)$ c. $\text{Card}(A^B)$ d. $\text{Card}(P(B))$
    e. $\text{Card}(A \cup B)$

11. Suppose a questionnaire has nine questions. In how many ways can the questionnaire be answered if:
    a. Each question requires a yes or no response?
    b. Each question is multiple-choice with five choices?

12. Suppose a questionnaire has five questions. In how many ways can the questionnaire be answered if:
    a. Each question requires a yes or no response?
    b. Each question requires a yes or no response, but the respondent may choose to skip one of the questions?

13. Suppose a questionnaire has 11 questions. In how many ways can the questionnaire be answered if each question requires a yes or no response followed by a 4-part multiple-choice.

14. A small ice cream shop sells seven flavors of ice cream. It also offers any combination of the following toppings: nuts, hot fudge, butterscotch. How many different single-scoop ice cream cones can be ordered?

15. A restaurant offers a basic cheese pizza with a choice of any combination of the extra ingredients: sausage, pepperoni, tomatoes, black olives, anchovies, white onions, bacon, green peppers. How many different varieties of pizza can a customer order?

16. A survey of 100 college students revealed the following data:

    18 like to eat chicken
    40 like to eat beef
    20 like to eat lamb
    12 like to eat chicken and beef
    5 like to eat chicken and lamb
    4 like to eat beef and lamb
    3 like to eat all three

    We will classify the student who does not like to eat any of the three kinds of meat (i.e., chicken, beef, and lamb) as a non–meat eater.
    a. How many students in our sample like to eat at least one of the three kinds of meat?
    b. How many non–meat eaters are in our sample?
    c. How many students in our sample like to eat only lamb?

17. From a survey of 120 people, the following data were obtained:

    90 owned a car
    40 owned a home
    35 owned a computer
    32 owned a car and a home
    21 owned a home and a computer
    26 owned a car and a computer
    17 owned all three

    a. How many people in this survey sample owned either a car, a home, or a computer?
    b. How many people owned neither a car, nor a home, nor a computer?
    c. How many people owned only a computer?

18. Derive a formula for Card($A_1 \cup A_2 \cup A_3 \cup A_4$). [*Hint:* Write $A_1 \cup A_2 \cup A_3 \cup A_4$ as $(A_1 \cup A_2 \cup A_3) \cup A_4$, and apply the sum rule.]

19. How many integers between 1 and 500, inclusive, are:
    a. Not divisible by 2?
    b. Divisible by either 2 or 3?
    c. Divisible by either 2 or 3 or 7?
    d. Divisible by 2 or 3 or 5 or 7?

20. Find the number of six-digit strings in which each of the digits 2, 5, and 7 appears at least once.

21. Find the number of integers in the set $\{n: 100 \leq n \leq 999\}$ with at least one kth digit equal to k for $k = 1, 2, 3$.

22. Find the number of integers in the set $\{n: 1000 \leq n \leq 9999\}$ with at least one kth digit not equal to k for $k = 1, 2, 3, 4$.

23. Let A and B be finite sets. Show that $\text{Card}(A) = \text{Card}(A - B) + \text{Card}(A \cap B)$.

24. Let A and B be finite sets, with $A - B = \{a, b, c, d, e, f, g\}$ and $A \cap B = \{h, i, j, k, l, m\}$. Find $\text{Card}(A)$.

25. Let $B = \{0, 1\}$ and $A = B \times \cdots \times B = B^n$. The set $S = B^A$ of all functions $f: A \to \{0, 1\}$ is called the set of all Boolean functions of n variables. Find $\text{Card}(S)$.

26. Let $A = \{0, 1, 2, \ldots, 9\}$.
    a. Find the number of functions $f: A \to A$.
    b. Find the number of binary relations on A.

27. Let $D = \{1, 2, \ldots, n\}$ and $C = \{1, 2, \ldots, m\}$.
    a. Find the number of functions $f: D \to C$.
    b. Find the number of binary relations from D to C.
    c. Find the number of functions $g: C \to D$.

28. Let $A = \{1, 2, \ldots, n\}$.
    a. Find the number of binary relations on A.
    b. Find the number of binary relations on A that are reflexive.

29. Let $A = \{1, 2, \ldots, n\}$. Find the number of binary relations R on A with the property: for every $m \in A$, there is an element $k \in A$ so that $m \, R \, k$.

4.2
PERMUTATIONS

We now discuss the problem of counting arrangements of the elements of a set. If we have two elements a and b, it is easy to see that there are two arrangements—namely, ab and ba. If we have a set $A = \{a, b, c\}$, how many arrangements are there of the elements in the set A? With a little trial and error, we find six arrangements:

abc, acb, bac, bca, cab, cba

In Example 4.2.1, we will use a systematic method to find these arrangements. Each of these arrangements is an example of a permutation of the set A. We recall from Section 3.4 that a permutation of a set A with n elements is a bijective function from A onto A. A permutation of a set A with n elements corresponds to an arrangement of the n elements.

For example, acb is one of the six arrangements of the set $\{a, b, c\}$. This arrangement corresponds to a bijective function $f: A \to A$, where $f(a) = a$,

4.2 PERMUTATIONS

$f(b) = c$, and $f(c) = b$. In other words, acb $= f(a)f(b)f(c)$. Similarly, the arrangement cba corresponds to another bijective function, $g:A \to A$, where $g(a) = c$, $g(b) = b$, and $g(c) = a$. Conversely, every bijective function $h:A \to A$ corresponds to an arrangement of A—namely, the arrangement $h(a)h(b)h(c)$, where $h(a)$, $h(b)$, and $h(c)$ are all the elements of A. For example, if $h(a) = b$, $h(b) = c$, and $h(c) = a$, the arrangement is bca. Note that we assume the elements of the set A are in a fixed order—in this case, a, b, c. Therefore, the number of arrangements of a set A equals the number of bijective functions from A to A.

PRACTICE PROBLEM 1 Consider the set $A = \{a, b, c, d, e\}$ and the bijective function $f:A \to A$, with $f(a) = b$, $f(b) = c$, $f(c) = d$, $f(d) = e$, $f(e) = a$. Find the arrangement of A corresponding to the bijective function f.

In a permutation, it is assumed that no element is repeated, and the order of the arrangement of the elements is important. Sometimes, a permutation is called an ordering or order.

The next example shows a simple method of finding all the permutations of a set with three elements.

EXAMPLE 4.2.1 Let $A = \{a, b, c\}$. Find all the permutations of A.

Solution We can find all the permutations of A by a tree diagram, as shown in Figure 4.2.1.

Each permutation of A has three elements since Card(A) = 3. The six permutations of A are displayed in the right-hand column of Figure 4.2.1.

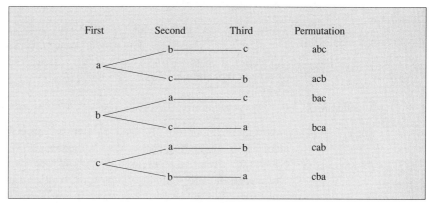

FIGURE 4.2.1

It is always possible to use a tree diagram to find all the permutations of a finite set A. However, the process is tedious when Card(A) is greater than 4. To find the number of permutations of a set A, we can use the following, extremely useful **fundamental counting principle**.

> **FUNDAMENTAL COUNTING PRINCIPLE**
> If an arrangement can be constructed in k successive steps, and if the first step can be performed in n_1 ways, the second step can be performed in n_2 ways, ..., and the kth step can be performed in n_k ways, then the total number of arrangements is the product $n_1 \cdot n_2 \cdots n_k$.

Now let us apply the fundamental counting principle to Example 4.2.1. Since there are three choices for the first element in the arrangement, two choices for the second element, and one choice for the third element, there are $3 \cdot 2 \cdot 1 = 6$ arrangements in all.

The fundamental counting principle will be applied again in the next example.

EXAMPLE 4.2.2 A student is enrolled in four courses, CS 150, Math 242, Art 120, Chinese 160. In how many different orders can her final examinations in these four courses be scheduled? Assume there are no schedule conflicts.

Solution By the fundamental counting principle, we see that there are $4 \cdot 3 \cdot 2 \cdot 1 = 24$ different orders in which the student's final examinations can be scheduled.

Factorial Notation

The product of the first n consecutive integers is

$$n! = n(n-1)(n-2) \cdots 3 \cdot 2 \cdot 1$$

where $n!$ is read **n factorial**. Thus, $1! = 1$, $2! = 2 \cdot 1 = 2$, $3! = 3 \cdot 2 \cdot 1 = 6$, $4! = 4 \cdot 3 \cdot 2 \cdot 1 = 24$, etc. The special case of $0!$ is defined by $0! = 1$.

The next theorem gives a general formula for the number of permutations of a set A with n elements.

THEOREM 4.2.3 If A has n elements, then the number of permutations of A is $n!$.

Proof Every arrangement of A has n elements. There are n choices for the first element in an arrangement, $n-1$ choices for the second element, $n-2$ choices for the third element, ..., 2 choices for the $(n-1)$th element, and 1 choice for the nth element. Therefore, by the fundamental counting principle, there are $n(n-1)(n-2) \cdots 2 \cdot 1 = n!$ permutations of A.

PRACTICE PROBLEM 2 How many permutations can be formed from the letters in the word NUMBER?

EXAMPLE 4.2.4 Let $A = \{a, b, c, d, e\}$. Find the number of permutations of A beginning with c and ending with d.

4.2 PERMUTATIONS

Solution For the first element of the arrangement, there is only 1 choice (namely, c). For the second element, there are 3 remaining choices since c is the first element and d is the last element. For the third element, there are 2 remaining choices. For the fourth element, there is 1 remaining choice. For the fifth element, there is only 1 choice (namely, d). Hence, by the product rule, there are $1 \cdot 3 \cdot 2 \cdot 1 \cdot 1 = 6$ permutations of A beginning with c and ending with d.

r-Permutations

DEFINITION

Let A be a set with n elements, and let $0 \leq r \leq n$. An **r-permutation** of A is an arrangement of r elements of A.

An n-permutation of A is a permutation of A. A 0-permutation of a set is the empty arrangement of the set. There is exactly one empty arrangement of a set.

EXAMPLE 4.2.5 Consider a set $A = \{a, b, c, d\}$. How many 2-permutations of A are there?

Solution Every 2-permutation of A has a first element followed by a second element. We make use of a tree diagram, shown in Figure 4.2.2. There are twelve 2-permutations of A, and they are listed in the right-hand column of Figure 4.2.2.

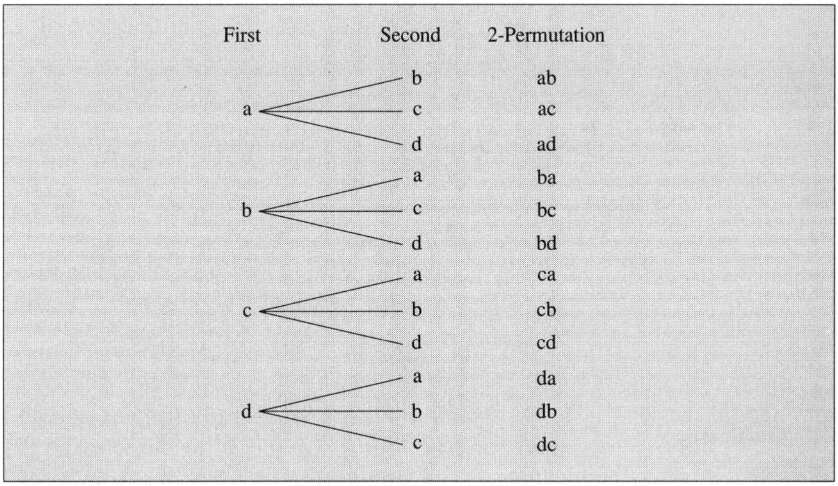

First	Second	2-Permutation
a	b	ab
	c	ac
	d	ad
b	a	ba
	c	bc
	d	bd
c	a	ca
	b	cb
	d	cd
d	a	da
	b	db
	c	dc

FIGURE 4.2.2

This problem can also be solved by applying the fundamental counting principle. In making a 2-permutation, there are four ways the first element can be selected from the set A. Having chosen the first element, there are three ways the second element can be selected. Hence, there are twelve 2-permutations. ∎

PRACTICE PROBLEM 3 How many 3-permutations of {a, b, c, d} are there?

If a set A has n elements, we can find the number of r-permutations of A, where $0 < r \leq n$, by the fundamental counting principle. The number of r-permutations of a set with n elements is usually denoted by **$P(n, r)$**.

THEOREM 4.2.6 The number of r-permutations of a set with n elements is

$$P(n, r) = n(n - 1)(n - 2) \cdots [n - (r - 1)] \qquad \text{where } 0 < r \leq n$$

Proof There are r elements in an r-permutation of A. The first element can be selected in n ways; the second element can be selected in $n - 1$ ways, because one of the n elements has been selected; the third element can be selected in $n - 2$ ways. This process continues until the rth element, which can be selected in $n - (r - 1)$ ways. By the fundamental counting principle, $P(n, r) = n(n - 1)(n - 2) \cdots [n - (r - 1)]$. ∎

We note that, when $r = n$, Theorem 4.2.6 says that the number of n-permutations of A is

$$P(n, n) = n(n - 1)(n - 2) \cdots [n - (n - 1)]$$
$$= n(n - 1)(n - 2) \cdots 1 = n!$$

Since an n-permutation of A is a permutation of A, Theorem 4.2.3 is a special case of Theorem 4.2.6.

EXAMPLE 4.2.7 The Winsome baseball team has 25 members. What is the number of all possible 9-person batting orders, assuming no substitutes?

Solution Since 9 members will be playing, we seek the number of 9-permutations of a set of 25 elements, or $P(25, 9)$.

We apply Theorem 4.2.6 with $n = 25$ and $r = 9$. Since $n - (r - 1) = 25 - (9 - 1) = 25 - 8 = 17$, the number of batting orders is

$$P(25, 9) = 25 \cdot 24 \cdot 23 \cdots 17 \qquad \text{9 factors} \qquad ∎$$

In Example 4.2.7, a 9-person batting order can be considered to be a 1–1 function from $D = \{1, 2, \ldots, 9\}$ to the set of 25 players available. In general, $P(n, r)$ is the number of 1–1 functions from a set D with r elements to a set C with n elements.

PRACTICE PROBLEM 4 How many different 3-letter initials, with distinct letters, can be constructed from the English alphabet? How many 1–1 functions are there from {0, 1, 2} to the set of 26 letters of the alphabet?

4.2 PERMUTATIONS

Permutations with Repetition

In our discussion of permutations and r-permutations, no repeated symbols are allowed. For example, abba is not considered to be a 4-permutation. However, in some applications, it makes sense to use repeated symbols. For example, 3545555 is a perfectly acceptable telephone number, as is 3331000.

EXAMPLE 4.2.8 How many seven-digit strings can be constructed from the set $D = \{0, 1, 2, 3, 4, 5, 6, 7, 8, 9\}$? How many seven-digit telephone numbers can be constructed from the set D if 0 or 1 is not allowed as a first digit in any telephone number? Assume repeated digits can be used.

Solution By the fundamental counting principle, the number of seven-digit strings with repeated digits is $10 \cdot 10 \cdot 10 \cdot 10 \cdot 10 \cdot 10 \cdot 10 = 10^7 = 10{,}000{,}000$. If 0 or 1 cannot be used as a first digit in a telephone number, then the number of such seven-digit telephone numbers is $8 \cdot 10 \cdot 10 \cdot 10 \cdot 10 \cdot 10 \cdot 10 = 8 \cdot 10^6 = 8{,}000{,}000$. ∎

PRACTICE PROBLEM 5 Some banks design four-digit passwords for their Automatic Teller Machine (ATM) customers by the following specifications: The first digit and the fourth digit must be nonzero; the second digit and the third digit must be identical but different from the first digit and different from the fourth digit; the fourth digit must be different from the first digit. How many such four-digit passwords are possible?

DEFINITION

> Let A be a set with n elements and r be *any* nonnegative integer. An **r-permutation with replacement** (or **with repetition**) is an arrangement of r elements from A, where each of the r elements can be repeated.

THEOREM 4.2.9 The number of r-permutations with replacement of a set with n elements is n^r, where r is any nonnegative integer.

Proof First assume $r > 0$. For each of the r elements in the r-permutation with replacement, there are n choices of elements since repetition is allowed. By the fundamental counting principle, the number of r-permutations with replacement is $n \cdot n \cdots \cdot n = n^r$.

For $r = 0$, there is one 0-permutation, the empty permutation. Hence, the number of 0-permutations is $n^0 = 1$. ∎

We give another example to illustrate r-permutations without and with replacement.

EXAMPLE 4.2.10 Suppose a college dormitory telephone extension has four digits.

a. Find the number of different telephone extensions if no repeated digits are allowed.

b. Find the number of different telephone extensions if 0 cannot be the first digit and repeated digits can be used.

Solution The set of digits $D = \{0, 1, 2, 3, 4, 5, 6, 7, 8, 9\}$ has ten elements.

a. If no repeated digits are allowed, a four-digit telephone extension is the same as a 4-permutation of D. By Theorem 4.2.6, the number of four-digit telephone extensions is $P(10, 4) = 10 \cdot 9 \cdot 8 \cdot 7 = 5040$.

b. For the first digit, there are nine choices of digits since 0 cannot be used. For each of the remaining three digits in the telephone extension, there are ten choices since repeated digits can be used. Hence, the number of four-digit telephone extensions with replacement is $9 \cdot 10 \cdot 10 \cdot 10 = 9000$.

PRACTICE PROBLEM 6 Refer to Example 4.2.10. Find the number of seven-digit strings with replacement such that every digit in the string is an odd integer.

The next example involves both the fundamental counting principle and permutations without repetition.

EXAMPLE 4.2.11 Find the number of permutations of the letters in COMPUTER, such that the letters in MUTE are together in any order.

Solution COPRMUTE is one such permutation, as is COPRMTEU. In order to keep the letters in MUTE together, let Ω denote the set $\{M, U, T, E\}$. We consider Ω as a symbol in a permutation. The number of permutations of C, O, P, R, Ω is 5!. In each of these 5! permutations, the presence of Ω will keep the letters in MUTE together. Now, for each of these 5! permutations, there are 4! permutations of the letters in Ω. Hence, the number of permutations of COMPUTER that contain the letters in MUTE together in any order is $5!4! = 2880$.

PRACTICE PROBLEM 7 Refer to Example 4.2.11. Find the number of permutations of the letters in COMPUTER such that the letters in PUT are together in any order.

We summarize the discussion of r-permutations (without replacement) and r-permutations with replacement of a set with n elements in Table 4.2.1.

TABLE 4.2.1

NAME OF SAMPLE	ORDER COUNTS?	REPETITIONS ALLOWED?	NUMBER OF WAYS TO SELECT THE SAMPLE
r-Permutation	Yes	No	$P(n, r) = n(n - 1) \cdots [n - (r - 1)]$
r-Permutation with replacement	Yes	Yes	n^r

PROGRAMMING NOTES

The Number of Possible Words with *r* Bits Some computers have 16-bit words, whereas others have 8-bit words, 32-bit words, or 60-bit words. As an example, a 16-bit word is the same as a 16-permutation with replacement of the set $B = \{0, 1\}$. The string 0110111001101111 is an example of a 16-bit word. Recall that a string of elements from $B = \{0, 1\}$ is commonly called a bit string. By Theorem 4.2.9, with $n = 2$ and $r = 16$, the number of all 16-bit words is 2^{16}. Similarly, the number of all 8-bit words is 2^8, the number of all 32-bit words is 2^{32}, and the number of all 60-bit words is 2^{60}.

A Function in Pascal for Calculating *P(n, r)* Recall that $P(n, r)$ is the number of *r*-permutations of a set with *n* elements. In the following function, we denote $P(n, r)$ by RPermutation.

```
function RPermutation (n, r : integer) : integer;
    var
        i, P : integer;
    begin
        if r <= n then
            begin
                P := 1;
                for i := n downto n - r + 1 do
                    P := i * P;
                RPermutation := P
            end
    end;
```

EXERCISE SET 4.2

1. Use a tree diagram to find all permutations of the set $A = \{1, 2, 3\}$.
2. Use a tree diagram to find all permutations of the set $B = \{a, b, c, d\}$.
3. Use Theorem 4.2.3 to determine the number of all permutations of the set $S = \{a, b, c, d, e\}$.
4. Use Theorem 4.2.3 to find the number of all permutations of the set $C = \{1, 2, 3, 4, 5, 6\}$.
5. How many 3-permutations of the set $S = \{a, b, c, d, e\}$ are there?
6. How many 2-permutations of the set $C = \{1, 2, 3, 4, 5, 6\}$ are there?
7. Five airplanes—a 747, a DC10, an L1011, an A300, and a BA146—are waiting for takeoff. There is only one runway available.
    a. If three airplanes can be accommodated in one hour, in how many different orders can three of the five airplanes take off in one hour?
    b. If the runway can accommodate all five airplanes in one hour, in how many different orders can they take off in one hour?
8. In standard Pascal an identifier must begin with a letter followed by any number (or none) of letters and digits. How many different Pascal identifiers are there having four or fewer characters?
9. How many different Pascal identifiers are there having six or fewer characters?

10. Find the number of 5-permutations with replacement of the set $D = \{1, 2, 3, 4\}$.
11. How many 6-letter strings of the set $S = \{a, b, c, d, e, f, g, h\}$ are there if repeated symbols can be used?
12. Find the number of all bit strings of length 5.
13. Since the ASCII system uses 7 bits to represent each character in a computer, the total number n of characters that can be encoded is the number of bit strings of length 7. Find the number n.
14. Determine the number of bit strings of length 8 that begin and end with 1.
15. Determine the number of bit strings of length 10 that begin with 0 and end with 1.
16. Suppose you are going to use bit strings of a fixed length to encode 40 different characters.
    a. What is the minimum number of bits needed?
    b. If the minimum number of bits is used, how many bit strings are not being used to encode characters?
17. Suppose you are going to use bit strings of a fixed length to encode 96 different characters.
    a. What is the minimum number of bits needed?
    b. If the minimum number of bits is used, how many bit strings are not being used to encode characters?
18. a. If we draw 6 cards from a deck of 52 cards, one at a time without replacement, in how many different orders can the 6 cards be drawn?
    b. Suppose we draw 6 cards from a deck of 52 cards, replacing each card in the deck each time. In how many different orders can the 6 cards be drawn?
19. A computer center has eight different programs to run. Three of them are written in Pascal, two in FORTRAN, and three in Ada. Find the number of different orders for running these eight programs if:
    a. They can be run in any order.
    b. The programs in the same language must be run consecutively.
20. The local city council is composed of three Democrats and two Republicans. In how many ways may they be seated in a row behind the council chambers table if:
    a. They can sit in any order.
    b. Members of the same party must sit together.
    c. No two members of the same party may sit together.
21. The local city council is composed of three Democrats, three Republicans, and an independent. In how many ways may they be seated in a row behind the council chambers table if:
    a. They can sit in any order.
    b. Members of the same party must sit together.
    c. No two members of the same party may sit together.
22. Find the number of permutations of the letters in PROBLEMS, such that the letters in PROBE are together in any order.
23. Find the number of permutations of the letters in SPECIAL, such that the letters in AEI are together in any order.
24. Let $A = \{a, b, c, d\}$ and $B = \{1, 2, 3, 4, 5, 6, 7\}$.
    a. Find the number of 1–1 functions $h: B \to B$.

4.3 COMBINATIONS   **149**

       b. Find the number of 1–1 functions $f: A \to B$.
       c. Find the number of 1–1 functions $g: B \to A$.

25. Let $A = \{a, b, c, d, e\}$ and $B = \{1, 2, 3\}$.
       a. Find the number of 1–1 functions $h: A \to A$.
       b. Find the number of 1–1 functions $f: A \to B$.
       c. Find the number of 1–1 functions $g: B \to A$.

26. Determine the number of functions from a three-element set to a five-element set that are *not* 1–1.

27. Determine the number of functions from a five-element set to an eight-element set that are *not* 1–1.

28. Suppose each of four tasks will be assigned to one of nine people. In how many ways may this be done if:
       a. There are no restrictions on the assignments.
       b. No person is to do more than one task.

4.3 COMBINATIONS

A permutation of a set involves arrangement in which order counts. We introduce the concept of **combinations** of a set, in which the order does not count.

DEFINITION

> Let A be a set with n elements and $0 \leq r \leq n$. An **r-combination** of A is a subset of A that contains r elements. The number of r-combinations in A is denoted by $\boldsymbol{C(n, r)}$, read "n choose r."

An r-combination of a set A with n elements is an unordered selection of r elements from A. There is no repetition among the r elements. When $r = n$, we have $C(n, r) = C(n, n) = 1$, since there is precisely one n-combination of a set with n elements.

The number $C(n, r)$ when $r = 0$ deserves a special note. We can interpret $C(n, 0)$ as the number of ways of selecting 0-combinations from a set A of n elements. Since there is exactly one way of selecting an empty subset (with 0 elements) from the set A, we have $C(n, 0) = 1$.

EXAMPLE 4.3.1   Find all 3-combinations and 3-permutations of the set $A = \{a, b, c\}$.

       Solution   There is one 3-combination of A, namely, the set A itself. The number of 3-permutations of A is 3!. The six 3-permutations of A are abc, acb, bac, bca, cab, and cba. Since order counts in 3-permutations, every 3-combination gives rise to $3! = 3 \cdot 2 \cdot 1 = 6$ 3-permutations. ∎

EXAMPLE 4.3.2   Find all 2-combinations of the set $A = \{a, b, c\}$.

       Solution   There are three 2-combinations of A—namely, $\{a, b\}, \{a, c\}$, and $\{b, c\}$. The order of the elements is not important. For example, $\{a, b\}$ and $\{b, a\}$ are considered to be the same 2-combination. ∎

PRACTICE PROBLEM 1   Find all 3-combinations of the set $\{1, 2, 3, 4\}$.

We now develop a formula for finding $C(n, r)$, the number of r-combinations of a set with n elements.

THEOREM 4.3.3 $$C(n, r) = \frac{n(n-1)(n-2) \cdots [n-(r-1)]}{r!} = \frac{P(n, r)}{r!} \quad \text{for } 0 < r \leq n$$

Proof   We first find a relationship between the number of r-permutations and the number of r-combinations of a set A. For every r-combination, there are $r!$ r-permutations since there are $r!$ arrangements of r elements. By the fundamental counting principle, the number of r-permutations is $C(n, r) \cdot r!$. By Theorem 4.2.6 (page 144), the number of r-permutations of A is $P(n, r) = n(n-1)(n-2) \cdots [n-(r-1)]$.

Hence, $C(n, r) \cdot r! = n(n-1)(n-2) \cdots [n-(r-1)]$. Dividing both sides of the last equation by $r!$ yields

$$C(n, r) = \frac{n(n-1)(n-2) \cdots [n-(r-1)]}{r!} = \frac{P(n, r)}{r!} \quad \text{for } 0 < r \leq n \quad \blacksquare$$

The next example is an application of Theorem 4.3.3.

EXAMPLE 4.3.4   A mathematics class has 30 students. A committee of 5 students is being selected to present a petition to the Chair for a facility to hold a geometry festival. In how many ways can a committee of 5 students be formed?

Solution   A committee of 5 students is a 5-combination of the set of 30 students. By Theorem 4.3.3, the number of 5-combinations is

$$C(30, 5) = \frac{30 \cdot 29 \cdot 28 \cdot 27 \cdot 26}{5!} = 142{,}506 \quad \blacksquare$$

From Theorem 4.3.3, we have, for $0 < r \leq n$,

$$C(n, r) = \frac{n(n-1)(n-2) \cdots [n-(r-1)]}{r!} = \frac{P(n, r)}{r!}$$

This formula can be rewritten in concise form using only factorial notation. First we use Theorem 4.2.6 to rewrite $P(n, r)$, for $0 < r \leq n$, as follows:

$$\begin{aligned} P(n, r) &= n(n-1)(n-2) \cdots [(n-(r-1)] \\ &= n(n-1)(n-2) \cdots (n-r+1) \\ &= n(n-1)(n-2) \cdots (n-r+1) \cdot \frac{(n-r)!}{(n-r)!} \\ &= \frac{n(n-1)(n-2) \cdots (n-r+1)[(n-r)!]}{(n-r)!} = \frac{n!}{(n-r)!} \end{aligned}$$

4.3 COMBINATIONS

Hence,

$$P(n, r) = \frac{n!}{(n-r)!}$$

Now,

$$C(n, r) = \frac{P(n, r)}{r!} = \frac{n!}{r!(n-r)!}$$

The concise formulas just derived also apply in the case of $r = 0$. Recall that $C(n, 0) = 1$, $C(n, n) = 1$, $P(n, 0) = 1$, and $P(n, n) = 1$. Hence,

$$P(n, r) = \frac{n!}{(n-r)!} \quad \text{for } 0 \leq r \leq n$$

$$C(n, r) = \frac{n!}{r!(n-r)!} \quad \text{for } 0 \leq r \leq n$$

We need to read problems carefully to decide whether to use methods of permutations, combinations, or the fundamental counting principle. Sometimes, we must use more than one method, as illustrated by the next example.

EXAMPLE 4.3.5 A discrete mathematics class consists of 13 sophomores and 7 juniors.

  a. How many 5-person committees can be formed from this class?
  b. How many 5-person committees with 3 sophomores and 2 juniors can be formed from this class?

Solution a. The number of 5-person committees is

$$C(20, 5) = \frac{20!}{5!(20-5)!} = \frac{20!}{5!15!}$$
$$= \frac{20 \cdot 19 \cdot 18 \cdot 17 \cdot 16 \cdot 15!}{5 \cdot 4 \cdot 3 \cdot 2 \cdot 1 \cdot 15!}$$
$$= \frac{20 \cdot 19 \cdot 18 \cdot 17 \cdot 16}{5 \cdot 4 \cdot 3 \cdot 2 \cdot 1} = 15{,}504$$

  b. The number of ways of selecting 3 sophomores from 13 sophomores is $C(13, 3)$. The number of ways of selecting 2 juniors from 7 juniors is $C(7, 2)$. By the fundamental counting principle, the number of 5-person committees with 3 sophomores and 2 juniors is the product

$$C(13, 3) \cdot C(7, 2) = \frac{13!}{3!(13-3)!} \cdot \frac{7!}{2!(7-2)!} = \frac{13 \cdot 12 \cdot 11 \cdot 10!}{3 \cdot 2 \cdot 1 \cdot 10!} \cdot \frac{7 \cdot 6 \cdot 5!}{2 \cdot 1 \cdot 5!}$$
$$= \frac{13 \cdot 12 \cdot 11}{3 \cdot 2 \cdot 1} \cdot \frac{7 \cdot 6}{2 \cdot 1} = 6006$$

PRACTICE PROBLEM 2 A computer software company employs 25 women and 20 men. An executive board of 5 directors is to be composed of 3 women and 2 men. In how many ways can such a board of directors be selected?

An interesting result follows from the definition of $C(n, r)$.

COROLLARY 4.3.6 Let A be a set of n elements and $0 \leq r \leq n$. Then

$$C(n, r) = C(n, n - r)$$

Proof By definition, $C(n, r)$ is the number of r-combinations of A, and $C(n, n - r)$ is the number of $(n - r)$-combinations of A. Each time we select an r-combination of A, we leave behind an $(n - r)$-combination of A. Hence, there is a one-to-one correspondence between $(n - r)$-combinations and r-combinations of A. Therefore, $C(n, r) = C(n, n - r)$.

We can also use algebra to verify the equation $C(n, r) = C(n, n - r)$. This is left as an exercise. (See Exercise 27a, Exercise Set 4.3.)

The next example is a reminder of how the numbers $C(n, r)$ are related to sets and their subsets.

EXAMPLE 4.3.7 Let $A = \{a, b, c, d\}$. Find $C(n, r)$ for $r = 0, 1, 2, 3, 4$.

Solution
$$C(4, 0) = 1$$
$$C(4, 1) = \frac{4!}{1!(4-1)!} = \frac{4!}{1!3!} = 4$$
$$C(4, 2) = \frac{4!}{2!(4-2)!} = \frac{4!}{2!2!} = 6$$
$$C(4, 3) = \frac{4!}{3!(4-3)!} = \frac{4!}{3!1!} = 4$$
$$C(4, 4) = \frac{4!}{4!(4-4)!} = \frac{4!}{4!0!} = 1$$

Hence, we have one 0-element subset of A, four 1-element subsets of A, six 2-element subsets of A, four 3-element subsets of A, and one 4-element subset of A. ∎

The numbers $C(n, r)$ appear as coefficients in the expansion of the nth power of a binomial such as $(x + y)^n$. For this reason, these numbers are called the **binomial coefficients**.

EXAMPLE 4.3.8 Expand $(x + y)^4$. Show that the coefficients in the expansion of $(x + y)^4$ are $C(4, r)$ for $r = 0, 1, 2, 3, 4$.

4.3 COMBINATIONS

Solution Using algebra, we find

$$(x + y)^4 = (x + y)^2 \cdot (x + y)^2$$
$$= (x^2 + 2xy + y^2) \cdot (x^2 + 2xy + y^2)$$
$$= x^4 + 4x^3y + 6x^2y^2 + 4xy^3 + y^4$$
$$= 1x^4 + 4x^3y + 6x^2y^2 + 4xy^3 + 1y^4$$

The coefficients in the expansion of $(x + y)^4$ are 1, 4, 6, 4, 1, which are precisely $C(4, 0)$, $C(4, 1)$, $C(4, 2)$, $C(4, 3)$, $C(4, 4)$ found in Example 4.3.7. ∎

We rewrite $(x + y)^4$ in order to study the patterns in the coefficients and the exponents of each term:

$(x + y)^4 = 1x^4 + 4x^3y + 6x^2y^2 + 4xy^3 + 1y^4$ or
$(x + y)^4 = C(4, 0)x^4 + C(4, 1)x^3y + C(4, 2)x^2y^2 + C(4, 3)xy^3 + C(4, 4)y^4$ or
$(x + y)^4 = C(4, 0)x^4y^0 + C(4, 1)x^3y^1 + C(4, 2)x^2y^2 + C(4, 3)x^1y^3 + C(4, 4)x^0y^4$
$$\tag{1}$$

Let us examine the pattern in the coefficients and the exponents in Equation (1). We note that the sum of the exponent of x and the exponent of y in Equation (1) is always 4. The exponent of x decreases from 4 to 0, whereas the exponent of y increases from 0 to 4, as we read Equation (1) left to right. The five coefficients are $C(4, 0)$, $C(4, 1)$, $C(4, 2)$, $C(4, 3)$, and $C(4, 4)$.

Now, we follow the example of $(x + y)^4$ and write the general expansion of n powers of a binomial expression $x + y$. We note that, in each term, the sum of the exponent of x and the exponent of y is n. In each coefficient $C(n, r)$, r is the exponent of y and $n - r$ is the exponent of x. We have

$$(x + y)^n = C(n, 0)x^ny^0 + C(n, 1)x^{n-1}y^1 + C(n, 2)x^{n-2}y^2 + \cdots$$
$$+ C(n, r)x^{n-r}y^r + \cdots + C(n, n - 1)x^1y^{n-1} + C(n, n)x^0y^n$$

This formula is expressed concisely, in summation notation, in the **Binomial Theorem**.

THEOREM 4.3.9 **THE BINOMIAL THEOREM**

$$(x + y)^n = \sum_{r=0}^{n} C(n, r)x^{n-r}y^r$$

Proof The expression $(x + y)^n$ is the product of $x + y$ multiplied by itself n times. We multiply these n factors by choosing a term, x or y, from each factor, multiplying the terms together, combining like terms, and adding the resulting expressions. The first term $C(n, 0)x^ny^0$ is the result of having chosen x from all n factors $x + y$ and y from no factors. The next term $C(n, 1)x^{n-1}y$ comes from having chosen x from $n - 1$ factors and y from the remaining one factor in all possible ways and adding the resulting terms. These choices can be made in $n = C(n, 1)$ ways.

In general, we choose x from $n - r$ factors and y from the remaining r factors in all possible ways. To see that there are $C(n, r)$ ways of doing this, we consider the n factors as n tags placed in a row. For any given choice of x's and y's, we label each tag with a "1" if x is chosen from the corresponding factor and each tag with a "0" if y is chosen from the corresponding factor. But each of these resulting bit strings corresponds to an r-element subset of a set with n elements. Hence, there are exactly as many choices as there are r-element subsets of a set with n elements. The result follows. ∎

The technique, used in the proof of Theorem 4.3.9, of restating the problem as a problem about bit strings and then using our knowledge of the correspondence between bit strings and sets to solve the problem is very important and will be used again. It is a technique worth a little extra time and study.

Theorem 4.3.9 is very useful for expanding expressions such as $(x + y)^5$. When finding the numbers $C(n, r)$, we should make use of the fact that $C(n, r) = C(n, n - r)$.

EXAMPLE 4.3.10 Use the Binomial Theorem to expand $(x + y)^5$.

Solution First we need to find $C(5, 0)$, $C(5, 1)$, $C(5, 2)$, $C(5, 3)$, $C(5, 4)$, $C(5, 5)$:

$C(5, 0) = 1$

$C(5, 1) = \dfrac{5!}{1!4!} = 5$

$C(5, 2) = \dfrac{5!}{2!3!} = 10$

$C(5, 3) = C(5, 2) = 10 \qquad C(5, 3) = C(5, 5 - 2) = C(5, 2)$, because $C(n, n - r) = C(n, r)$ where $n = 5$ and $r = 2$

$C(5, 4) = C(5, 1) = 5$

$C(5, 5) = C(5, 0) = 1$

The coefficients 1, 5, 10, 10, 5, 1 are symmetric due to the equation $C(n, r) = C(n, n - r)$. Hence,

$(x + y)^5 = x^5 + 5x^4y + 10x^3y^2 + 10x^2y^3 + 5xy^4 + y^5$ ∎

PRACTICE PROBLEM 3 Use the method of Example 4.3.10 to find the expansion $(a + b)^6$.

PRACTICE PROBLEM 4 Express the sum $C(5, 0)3^5 + C(5, 1)3^44 + C(5, 2)3^34^2 + C(5, 3)3^24^3 + C(5, 4)3 \cdot 4^4 + C(5, 5)4^5$ as one number in exponential form.

The next theorem gives an interesting and useful relationship among the numbers $C(n, r)$, $C(n - 1, r - 1)$, and $C(n - 1, r)$.

THEOREM 4.3.11 $C(n, r) = C(n - 1, r - 1) + C(n - 1, r)$, for $0 < r \leq n$.

Proof We give an algebraic proof here:

$$C(n-1, r-1) + C(n-1, r) = \frac{(n-1)!}{(r-1)![(n-1)-(r-1)]!} + \frac{(n-1)!}{r![(n-1)-r]!}$$

$$= \frac{(n-1)!}{(r-1)!(n-r)!} + \frac{(n-1)!}{r![(n-1)-r]!}$$

$$= \frac{(n-1)!}{(r-1)!(n-r)!} \cdot \frac{r}{r} + \frac{(n-1)!}{r![(n-1)-r]!} \cdot \frac{n-r}{n-r}$$

$$= \frac{r \cdot (n-1)!}{r!(n-r)!} + \frac{(n-r) \cdot (n-1)!}{r!(n-r)!}$$

$$= \frac{r \cdot (n-1)! + (n-r) \cdot (n-1)!}{r!(n-r)!}$$

$$= \frac{(r+n-r) \cdot (n-1)!}{r!(n-r)!} = \frac{n!}{r!(n-r)!} = C(n, r) \quad \blacksquare$$

We will encounter the equation of Theorem 4.3.11 again when we study recurrence relations in Chapter 9. It is also useful for computing binomial coefficients, as we now illustrate.

The binomial coefficients are displayed in the well-known **Pascal's triangle**. See Figure 4.3.1. The top row ($n = 0$) has the lone coefficient 1 in the binomial expansion $(x + y)^0 = 1$. The second row ($n = 1$) has the two coefficients in the binomial expansion $x + y = 1x + 1y$. The third row ($n = 2$) has the three coefficients in the binomial expansion $(x + y)^2 = 1x^2 + 2xy + 1y^2$. The fourth row ($n = 3$) has the four coefficients in the binomial expansion $(x + y)^3 = 1x^3 + 3x^2y + 3xy^2 + 1y^3$, and so on.

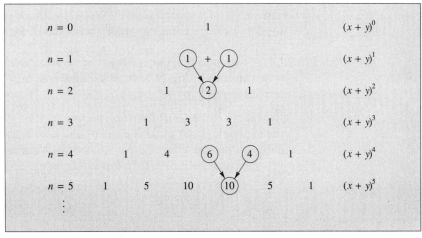

FIGURE 4.3.1 Pascal's triangle

Pascal's triangle can be constructed easily. The very top row is 1, add a 1 on the left side and right side of each subsequent row. The middle term of the $n = 2$ row is obtained by adding two terms from the $n = 1$ row, namely $2 = 1 + 1$. Please refer to Figure 4.3.1.

The theory behind the construction of Pascal's triangle is the equation $C(n, r) = C(n - 1, r - 1) + C(n - 1, r)$ of Theorem 4.3.11. For example, since $C(5, 3) = 10$, $C(4, 2) = 6$, and $C(4, 3) = 4$, we see that $10 = 6 + 4$ is $C(5, 3) = C(4, 2) + C(4, 3)$ where $n = 5$ and $r = 3$. See the $n = 4$ and $n = 5$ rows in Figure 4.3.1.

Pascal's triangle provides a nice way to find the binomial coefficients. In addition, it has many fascinating properties, one of which is found by doing the next practice problem.

PRACTICE PROBLEM 5 Find the sum of the binomial coefficients in each of the first six rows of Pascal's triangle. Write a general expression for the sum in terms of n.

PRACTICE PROBLEM 6 Write the expansion of $(x + y)^4$ using the binomial coefficients from Pascal's triangle.

r-Combination with Replacement

In the above discussion, an *r*-combination of a set has no repeated elements. An *r*-combination of a set with n elements, when repetition is allowed, is called an **r-combination with replacement** (or **with repetition**). The number of *r*-combinations of a set with n elements with replacement is

$$C(n - 1 + r, r) \tag{2}$$

The methods introduced in Example 4.3.12 indicate how to prove that expression (2) is the correct formula for calculating the number of *r*-combinations with replacement of a set with n elements. However, we omit the details. For a complete proof, see *Applied Combinatorics* by F. S. Roberts.

EXAMPLE 4.3.12 Given two brands of mineral water, if we want to select four bottles of mineral water and repetition of brands is allowed, in how many ways can we make our selection?

Solution Let the two brands be denoted by a and b. After a little thinking, we see that the 4-combinations of the 2-element set {a, b} with repetition are aaaa, aaab, aabb, abbb, and bbbb. So the number of 4-combinations of the set of two brands with repetition is 5. In this example, $n = 2$, $r = 4$, and $C(n - 1 + r, r) = C(2 - 1 + 4, 4) = C(5, 4) = 5$.

To get an insight into why the formula holds, let us look at the example from a slightly different point of view. What distinguishes the combinations aabb and aaab is the place at which the a's stop and the b's begin. We can just as well use a marker, say "0", to denote this. So, the combination aabb

4.3 COMBINATIONS

corresponds to 11011 and the combination aaab corresponds to 11101. We use "1" to stand for either letter a or b because the placement of the marker is what matters. The five combinations above correspond, respectively, to 11110, 11101, 11011, 10111, 01111. But these strings are the bit strings that correspond to all the 4-element subsets of a set with 5 elements. And there are $C(5, 4)$ such sets! Note that $C(5, 4)$ is $C(n - 1 + r, r)$ for $n = 2$ and $r = 4$. We see that $n - 1 = 2 - 1$ is the number of markers used, and r is the number of bottles selected.

Now suppose we want to choose from three brands. It will then be necessary to use two markers by the above analysis. In general, we need one less marker than the number of brands. It is not hard to see that, if we want to select seven bottles and there are three brands, then we need to use two markers. A bit string like 110111101 will correspond to two of the first brand, four of the second brand, and one of the third brand. This bit string corresponds to a 7-element subset of a set with 9 elements. Hence, the number of 7-combinations with repetition of a set with 3 elements is $C(n - 1 + r, 7) = C(3 - 1 + 7, 7) = C(9, 7)$. The number 9 is the sum of the number of markers required and the number of bottles selected.

PRACTICE PROBLEM 7 There are four different varieties of rose bushes available at a local nursery. In how many ways can we select eight rose bushes, with repetition allowed, for our garden?

We summarize the discussion of r-combinations of a set with n elements in Table 4.3.1.

TABLE 4.3.1

NAME OF SAMPLE	ORDER COUNTS?	REPETITION ALLOWED?	NUMBER OF WAYS TO SELECT THE SAMPLE
r-Combination	No	No	$C(n, r) = \dfrac{n!}{r!(n - r)!}$
r-Combination with replacement	No	Yes	$C(n - 1 + r, r) = \dfrac{(n + r - 1)!}{r!(n - 1)!}$

PROGRAMMING NOTES

The Eight-Queens Problem A classic problem of computer science is to write an algorithm that solves the eight-queens problem from chess. Recall that a chessboard is an 8×8 matrix of alternating white and black squares. The problem is to find all arrangements of eight queens on the board so that no queen can eliminate any other queen. By the rules of chess, the queen can eliminate another piece whenever that piece occupies the same row, same column, or same diagonal as the queen if there is no other piece between

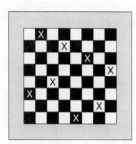

FIGURE 4.3.2

the queen and that piece. The eight-queens problem is nontrivial; a full solution eluded the brilliant mathematician Gauss who worked on it in 1850. Figure 4.3.2 shows one configuration of eight queens such that no queen can eliminate any other queen.

As a first attempt at finding an algorithm, we might try placing the eight queens on the board in eight squares (called an **eight-queens configuration**) then test whether the configuration is a solution. We call a configuration in which no queen can eliminate any other queen an **admissible configuration**. So, assuming we can represent each eight-queens configuration, we can write a computer program that generates all eight-queens configurations and that tests each to check whether it is an admissible configuration.

Is this an efficient strategy? To see how many eight-queens configurations are possible, we can apply our knowledge of combinations. For each eight-queens configuration, we need to choose 8 squares from a possible 64. Hence, the number of configurations is $C(64, 8)$. But $C(64, 8) = 4{,}426{,}165{,}368$, or approximately 4.5 billion configurations.

We can do much better than this. In an admissible configuration there must be exactly one queen per row. Since there are eight rows and each row has eight squares, there are $8^8 = 16{,}777{,}216$ configurations in which a single queen occupies each row. Note that we may assume each queen is in a specific row—first queen in row one, second queen in row two, etc.

We can reduce the number of configurations to be tested still further because we also know there must be exactly one queen per column. Furthermore, every square on the board lies at the intersection of a row and a column. Hence, when the first queen is placed in a square of row one, there are seven squares remaining in row two for the second queen, then there are six squares remaining in row three for the third queen, etc. Hence, there are $8! = 40{,}320$ configurations in which each row and each column is occupied by exactly one queen.

We have reduced the number of configurations to be tested from 4.5 billion to approximately 40 thousand. Testing 40 thousand configurations in a computer program is feasible. So, even though we have not yet solved the eight-queens problem, we have made significant progress by using our knowledge of combinatorics. For a recursive algorithm that solves the eight-queens problem, please see *Data Structures and Program Design* by R. L. Kruse. That algorithm tests fewer than 40 thousand configurations by using the fact that two queens cannot occupy the same diagonal. The lesson of this programming note is that basic counting is very useful for analyzing and simplifying difficult problems.

A Computer Program to Generate Pascal's Triangle The following program will generate and print the first 12 rows (rows 0 through 11) in Pascal's triangle. The program generates the binomial coefficients by the rule $C(n, r) = C(n-1, r-1) + C(n-1, r)$ of Theorem 4.3.11. We are using "Binomial[n,r]" to stand for $C(n, r)$. In the program each binomial coefficient is implemented as an entry in a two-dimensional array.

```
program PascalTriangle (input, output);
   const  Max = 11;
   type  CombCoeff = array[0..Max, 0..Max] of integer;
   var   Binomial : CombCoeff;

   procedure GenerateBinomial;
      var
         r, n : integer;
      begin
         for n := 0 to Max do
            begin
               Binomial[n, 0] := 1;
               Binomial[n, n] := 1;
               for r := 1 to n - 1 do
                  Binomial[n, r] := Binomial[n - 1, r - 1] + Binomial[n - 1, r]
            end
      end;

   procedure PrintBinomial;
      var
         r, n : integer;
      begin
         for n := 0 to Max do
            begin
               for r := 0 to n do
                  write(Binomial[n, r]);
               writeln
            end
      end;

   begin                                          (*Main Program*)
      GenerateBinomial;
      PrintBinomial
   end.
```

EXERCISE SET 4.3

1. Let $A = \{a, b, c, d\}$. Find the number of 3-combinations of A, where no symbol is repeated.

2. Let $A = \{a, b, c, d\}$. Find the number of 3-combinations of A, where repetition of symbols is allowed.

3. Let $A = \{a, b, c, d, e, f, g, h\}$. Find the number of 5-combinations of A where:
   a. No symbol is repeated.   **b.** Repetition of symbols is allowed.

4. Suppose that a 4-letter password consists of letters in the set $\{a, b, c, d, o\}$ and cannot begin with o. How many such passwords are there if repetition is allowed?

5. Let A be a set with 5 elements. If replacement is not allowed, find the number of
   a. 3-permutations of A **b.** 3-combinations of A

6. Let A be a set with 7 elements. If replacement is not allowed, find the number of
   a. 4-permutations of A b. 4-combinations of A
7. How many different bit strings of length 7 with two 1's and five 0's are there?
8. How many different bit strings of length 9 with three 1's and six 0's are there?
9. Suppose a certain organization of 23 people includes as officers a president, a vice president, and a secretary.
   a. In how many ways can the officers be selected from the organization?
   b. How many committees of 3 people can be formed from this group?
10. Suppose a certain organization of 31 people includes as officers a president, a vice president, a secretary, and a treasurer.
    a. In how many ways can the officers be selected from the organization?
    b. How many committees of 4 people can be formed from this group?
11. Suppose a certain organization of 28 people includes a president and a vice president.
    a. How many committees of 5 people can be formed from this group?
    b. How many committees of 5 people, including neither the president nor the vice president, can be formed from this group?
    c. How many committees of 5 people, including the president or the vice president or both, can be formed from this group?
12. Suppose a certain organization of 39 people includes as officers a president, a vice president, and a secretary.
    a. How many committees of 7 people can be formed from this group?
    b. How many committees of 7 people, including all the officers, can be formed from this group?
    c. How many committees of 7 people, including no more than one of the officers, can be formed from this group?
13. Suppose we are to construct a path in a digraph with vertices as shown in Figure 4.3.3. The path must go from the lower left vertex to the upper right vertex and must include only edges that point up or to the right. We call such a path an *up/right path*. A sample up/right path is shown. Note that every such path must have 11 edges, with 4 pointing up and 7 pointing right. Find the number of such paths.

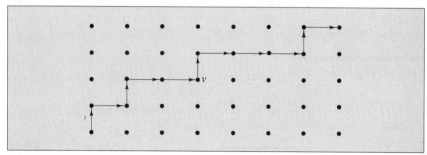

FIGURE 4.3.3

14. Refer to Figure 4.3.3.
    a. How many up/right paths go through vertex v?
    b. How many up/right paths do not go through vertex v?

4.3 COMBINATIONS

For Exercises 15–19 order does not matter.

15. In how many ways can we select seven cases of beer from three brands? Assume each case contains bottles of the same brand.
16. In how many ways can we select ten microcomputers if there are three different models available?
17. In how many ways can we select seven boxes of storage disks if there are four brands available?
18. Suppose the local computer store stocks eight different models of microcomputers. In how many ways can we select five microcomputers if:
    a. Each microcomputer selected is a different model.
    b. There is no restriction on the models selected.
19. Suppose the local candy store stocks 11 different kinds of candy. In how many ways can we select a bag of six pieces of candy if:
    a. Each piece selected is of a different kind.
    b. There is no restriction on the kinds of candy in the selection.
20. a. List the entries $C(8, 0), C(8, 1), \ldots, C(8, 8)$ of the $n = 8$ row in Pascal's triangle.
    b. Expand $(a + b)^8$.
    c. Simplify $C(8, 0) + C(8, 1) + \cdots + C(8, 8)$.
21. a. List the entries $C(7, 0), C(7, 1), \ldots, C(7, 7)$ of the $n = 7$ row in Pascal's triangle.
    b. Expand $(a + b)^7$.
    c. Simplify $C(7, 0) + C(7, 1) + \cdots + C(7, 7)$.
22. Use the Binomial Theorem to simplify each of the following.
    a. $C(37, 0)2^{37} + C(37, 1)2^{36}3 + \cdots + C(37, 36)2 \cdot 3^{36} + C(37, 37)3^{37}$
    b. $C(37, 0)5^{37} + C(37, 1)5^{36} + \cdots + C(37, 36)5 + C(37, 37)$
23. Use the Binomial Theorem to simplify each of the following.
    a. $C(23, 0)3^{23} + C(23, 1)3^{22}4 + \cdots + C(23, 22)3 \cdot 4^{22} + C(23, 23)4^{23}$
    b. $C(23, 0)7^{23} + C(23, 1)7^{22} + \cdots + C(23, 22)7 + C(23, 23)$
24. a. List the first three entries $C(5, 5), C(6, 5),$ and $C(7, 5)$ of the $r = 5$ diagonal in Pascal's triangle.
    b. Verify that $C(5, 5) + C(6, 5) + C(7, 5) = C(8, 6)$.
25. Verify each step of the following proof that $C(5, 5) + C(6, 5) + C(7, 5) = C(8, 6)$.
    Proof
    $$C(5, 5) + C(6, 5) + C(7, 5) = C(6, 6) + C(6, 5) + C(7, 5)$$
    $$= C(7, 6) + C(7, 5) = C(8, 6)$$
26. Simplify each of the following:
    a. $C(7, 7) + C(8, 7) + \cdots + C(53, 7)$ [*Hint:* Consider Exercise 25 and use the equation $C(n, r) = C(n - 1, r - 1) + C(n - 1, r)$.]
    b. $C(7, 0) + C(8, 1) + \cdots + C(53, 46)$ [*Hint:* Use the equation $C(n, r) = C(n, n - r)$ and part a.]
27. Use algebra to verify the following:
    a. $C(n, r) = C(n, n - r)$, where $0 \leq r \leq n$
    b. $C(n, r) = (n/r) \cdot C(n - 1, r - 1)$, where $1 \leq r \leq n$
28. Use the Binomial Theorem to show
    $$C(n, 0) + C(n, 1) + C(n, 2) + \cdots + C(n, n - 1) + C(n, n) = 2^n \quad \text{for all } n \in \mathbb{N}$$

29. Use Exercises 27b and 28 to show
$$C(n, 1) + 2 \cdot C(n, 2) + 3 \cdot C(n, 3) + \cdots + n \cdot C(n, n) = n \cdot 2^{n-1} \quad \text{for all } n \in \mathbb{N}$$

30. Here is the outline of a combinatorial proof that
$$C(n, 0)^2 + C(n, 1)^2 + C(n, 2)^2 + \cdots + C(n, n-1)^2 + C(n, n)^2 = C(2n, n) \quad \text{for all } n \in \mathbb{N}$$
Justify each step.

1. $C(2n, n)$ is the number of n-combinations of the set $S = \{1, 2, \ldots, 2n\}$.
2. Let $A = \{1, 2, \ldots, n\}$ and $B = \{n + 1, n + 2, \ldots, 2n\}$. Every n-combination of S is the union of a j-combination of A and an $n - j$ combination of B for some $j = 0, 1, \ldots, n$.
3. By step 2, $C(2n, n) = C(n, 0) \cdot C(n, n) + C(n, 1) \cdot C(n, n-1) + C(n, 2) \cdot C(n, n-2) + \cdots + C(n, n) \cdot C(n, 0)$.
4. The result follows from step 3.

4.4
INTRODUCTION TO FINITE PROBABILITY

The history of probability is believed to owe its start to problems arising from games of chance. Archaeologists have found gambling artifacts in Egypt dating from as early as 3500 B.C. The earliest dice, from Iraq and India, date from about 3000 B.C. Dice playing was popular for several thousand years before the 16th century.

Mathematician–physician–astrologer Gerolamo Cardano is believed to have written the first mathematical analysis of the cast of dice in 1526. Cardano proposed a probability theory and issued advice to his students on how to wager. Galileo took up the analysis of dice playing in the late 16th century, as did Pascal in 1654—both at the request of gamesters who had been troubled by disappointing and costly experiences with dice. Galileo's calculations are like those used by mathematicians today. Modern probability theory was on its way at last.

The greatest advance in probability theory came in the middle of the 17th century with a manuscript by Christian Huygens (1629–1695), *De Ratiociniis in Ludao Aleae* ("On Ratiocination in Dice Games"). Since that time, probability theory has developed into a branch of theoretical mathematics with wide application.

The theory of probability is stated in the language of set theory. Probability problems are formulated with respect to a universal set called the *sample space*. Loosely speaking, the elements in the sample space are possible outcomes of an experiment.

An **experiment** is a measurement or observation of some process. By assumption, an experiment results in exactly one outcome from among several possible outcomes. A **sample point** is an outcome of the experiment. A **sample space** is the set of all sample points.

4.4 INTRODUCTION TO FINITE PROBABILITY

EXAMPLE 4.4.1 Rolling a die and observing the upper face value yields one number, an integer between 1 and 6. There are six possible outcomes of this experiment—namely, observing a 1, 2, 3, 4, 5, or 6. The set $S = \{1, 2, 3, 4, 5, 6\}$ is the sample space for this experiment. ∎

An **event** is a subset of the sample space S. If an event contains exactly one sample point, it is called a **simple event**. If an event is the empty set, it is called the **impossible event**.

We say an event E **occurs** whenever the outcome is an element of E. For a given event E, the sample points in E are called **favorable outcomes**. Two events A and B in a sample space are **mutually exclusive** if $A \cap B \neq \emptyset$.

In Example 4.4.1, the singleton set $\{2\}$ is the simple event of observing a 2. The set $A = \{1, 3, 5\}$ is the event of observing an odd number, and the set $B = \{1, 2, 3, 4\}$ is the event of observing a number less than 5. Since A and B have some common sample points, events A and B are not mutually exclusive. The events $C = \{2, 4, 6\}$ and $A = \{1, 3, 5\}$ are mutually exclusive. See Figure 4.4.1.

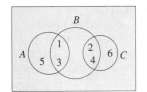

FIGURE 4.4.1

If a balanced die is rolled, what is the chance of observing a 3? Suppose we roll a balanced die 6 million times. We expect to observe a 3 approximately 1 million times. Hence, we expect the chance of observing a 3 to be approximately $\frac{1,000,000}{6,000,000}$, or $\frac{1}{6}$.

This intuitive sense of what is expected is formalized as part of the assumptions for a theory of probability. The theoretical foundation makes precise analysis of the problems possible.

In Example 4.4.1, the sample space S has 6 sample points. Assuming each simple event $S_i = \{i\}$ is equally likely for $1 \leq i \leq 6$, we assign a probability of $\frac{1}{6}$ to each event S_i, denoted by $\Pr(S_i)$. Let $P(S)$ be the power set of S. We can now define a probability function $\Pr: P(S) \to [0, 1]$ using $\Pr(S_i) = \frac{1}{6}$ as follows. If E is any event, then $\Pr(E)$ is equal to the sum of $\Pr(S_i)$ for each $S_i \subseteq E$. For instance, if $E = S_1 \cup S_2 \cup S_6$, then $\Pr(E) = \Pr(S_1) + \Pr(S_2) + \Pr(S_6) = \frac{1}{6} + \frac{1}{6} + \frac{1}{6} = \frac{1}{2}$. It follows that $\Pr(S) = \Pr(S_1) + \Pr(S_2) + \Pr(S_3) + \Pr(S_4) + \Pr(S_5) + \Pr(S_6) = 1$.

Defining the probability function after its values on simple events have been assigned is made explicit in the following algorithm.

ALGORITHM 4.4.2 **DEFINING A PROBABILITY FUNCTION ON A FINITE SAMPLE SPACE**

1. Define the sample space S.
2. Assign probabilities to the simple events so that, if $S_1, S_2, \ldots, S_n$ are the simple events, then

$$\sum_{i=1}^{n} \Pr(S_i) = 1$$

3. For each event $E \subseteq S$, define $\Pr(E)$ to be the sum of the probabilities for the simple events corresponding to the sample points of E.

Note that steps 1 and 2 above will always involve some assumptions about the experiment. For the problems we will solve in this text, those assumptions will be explicitly stated. Many errors made while solving problems in probability result from not stating explicitly how the sample space is chosen for a given experiment. Several exercises in Exercise Set 4.4 will ask you to state how you have chosen the sample space.

An **equiprobable space** is a sample space in which equal probabilities are assigned to all simple events. We assumed above that the sample space in Example 4.4.1 is an equiprobable space.

PRACTICE PROBLEM 1 Let $S = \{a, b, c, d, e\}$ be an equiprobable space. Find:

a. $\Pr(\{a\})$ **b.** $\Pr(\{a, b\})$ **c.** $\Pr(\{c, d, e\})$

In Practice Problem 1 we saw that the probability of an event E is equal to the cardinality of the set E divided by the cardinality of the sample space. That is, the probability of an event is the number of favorable outcomes divided by the total number of outcomes. This is always true for equiprobable spaces.

THEOREM 4.4.3 Let S be a finite equiprobable sample space. For a given event E,

$$\Pr(E) = \frac{\text{Card}(E)}{\text{Card}(S)}$$

Proof We will use Algorithm 4.4.2. Suppose S has n sample points and the event E has m sample points. Since S is an equiprobable space, $\Pr(\{x\}) = 1/n$ for each sample point x. Hence, $\Pr(E) = m(1/n) = m/n = \text{Card}(E)/\text{Card}(S)$. ∎

Example 4.4.4 illustrates Theorem 4.4.3. This example also shows the possibility of choosing more than one sample space to describe an experiment, and the desirability of choosing a sample space that makes the assignment of probabilities convenient.

EXAMPLE 4.4.4 **ROLLING TWO DICE** If we roll two balanced dice, the sum of the upper face values yields one number, an integer between 2 and 12. We might choose the sample space so that there are 11 possible outcomes of this experiment—namely, observing a 2, 3, ..., or 12. Instead, we prefer to let the sample space $S = D \times D$, where $D = \{1, 2, 3, 4, 5, 6\}$. So,

$$S = \{(1, 1), (1, 2), (1, 3), \ldots, (1, 6), (2, 1), (2, 2), (2, 3), \ldots, (2, 6), \ldots, (6, 1), \\ (6, 2), (6, 3), \ldots, (6, 6)\}$$

The sample points are all possible pairs of values on the upper faces of the dice, and the total number of possible outcomes is 36. We obtain an equiprobable space with this sample space.

4.4 INTRODUCTION TO FINITE PROBABILITY

The events of interest are those subsets of S that include sample points that have coordinates adding to a given value. Let $E_i = \{(m, n) : m + n = i\}$, where $2 \leq i \leq 12$. For example, $E_5 = \{(1, 4), (2, 3), (3, 2), (4, 1)\}$. Figure 4.4.2, which shows the sums corresponding to each of the 36 pairs, is useful for calculating the probabilities $\Pr(E_i)$. These probabilities are shown in Table 4.4.1. For example, $\Pr(E_5) = \frac{4}{36} = \frac{1}{9}$ because there are four pairs of numbers with coordinates adding to 5.

	1	2	3	4	5	6
1	2	3	4	5	6	7
2	3	4	5	6	7	8
3	4	5	6	7	8	9
4	5	6	7	8	9	10
5	6	7	8	9	10	11
6	7	8	9	10	11	12

FIGURE 4.4.2

TABLE 4.4.1

SUM	PROBABILITY
2	$\frac{1}{36}$
3	$\frac{2}{36}$
4	$\frac{3}{36}$
5	$\frac{4}{36}$
6	$\frac{5}{36}$
7	$\frac{6}{36}$
8	$\frac{5}{36}$
9	$\frac{4}{36}$
10	$\frac{3}{36}$
11	$\frac{2}{36}$
12	$\frac{1}{36}$

Choosing the sample space to be the 11 outcomes 2 through 12 would not have resulted in an equiprobable space, because, for example, one is more likely to observe the sum of 7 than the sum of 12. ∎

PRACTICE PROBLEM 2 Refer to Example 4.4.4. Find the events E_3, E_7, and E_9 and their respective probabilities.

Let us summarize what we know about a probability function as defined by Algorithm 4.4.2. It is left as an exercise to show that the **probability function** $\Pr : P(S) \to [0, 1]$ satisfies the following properties. (See Exercise 25, Exercise Set 4.4.)

PROPERTIES OF THE PROBABILITY FUNCTION $\Pr : P(S) \to [0, 1]$

1. $\Pr(S) = 1$
2. $0 \leq \Pr(E) \leq 1$ for any event E in S
3. If A and B are mutually exclusive events, then $\Pr(A \cup B) = \Pr(A) + \Pr(B)$.

Property 3 above is referred to as the **additive law for mutually exclusive events**.

The complement A' of a set A is the event containing all sample points in S and not in A. So $\Pr(A')$ is the probability that an event A will *not* occur. The probability that an event A will not occur is equal to 1 minus the probability that the event will occur, as proved in the next theorem.

THEOREM 4.4.5 $\quad \Pr(A') = 1 - \Pr(A)$

Proof Clearly, $S = A \cup A'$, and A and A' are mutually exclusive events. By the additive law for mutually exclusive events, $\Pr(A) + \Pr(A') = \Pr(A \cup A') = \Pr(S) = 1$ and the theorem follows. ∎

Notice the fact that

$$\Pr(A) = 1 - \Pr(A') \quad \text{and} \quad \Pr(\emptyset) = 0$$

are corollaries to Theorem 4.4.5.

PRACTICE PROBLEM 3 Prove that the probability of an impossible event is zero—that is, $\Pr(\emptyset) = 0$.

To compute the probability that an event E will occur, it is frequently easier to compute the probability of the event E' and then find $\Pr(E)$ by using $\Pr(E) = 1 - \Pr(E')$.

Example 4.4.6 illustrates the utility of Theorem 4.4.5.

EXAMPLE 4.4.6 **TOSSING A FAIR COIN** Suppose we toss a fair coin, with a head (H) on one side and a tail (T) on the other side, three times. By *fair coin* we mean it is assumed to be equally likely that either a head or a tail will come up on each toss. The sample space is $S = \{$HHH, HHT, HTH, HTT, THH, THT, TTH, TTT$\}$. This is an equiprobable space.

We can find all sample points in the sample space from the tree diagram in Figure 4.4.3. For example, tracing the top branch gives the sample point HHH, tracing the second from the top branch gives the sample point HHT, and so on.

What is the probability of obtaining at least one head after three tosses of the coin? Let A be the event of tossing the coin three times and obtaining at least one head. Then A' is the event of obtaining three tails in three tosses. But the probability of obtaining exactly three tails is $\frac{1}{8}$ since there are eight equally likely simple events. Hence, $\Pr(A) = 1 - \frac{1}{8} = \frac{7}{8}$. ∎

PRACTICE PROBLEM 4 Refer to Example 4.4.6. Find the probability of obtaining at least one tail on the last two tosses.

4.4 INTRODUCTION TO FINITE PROBABILITY

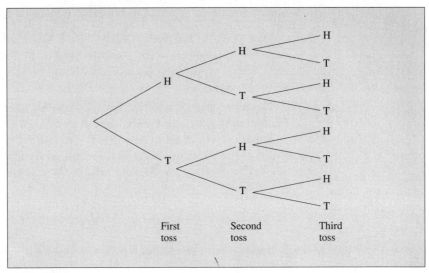

FIGURE 4.4.3

The Additive Law of Probability

We know that $\Pr(A \cup B) = \Pr(A) + \Pr(B)$ if A and B are mutually exclusive events. This law, generalized to events that are not necessarily mutually exclusive, is called the **additive law of probability**.

THEOREM 4.4.7 **THE ADDITIVE LAW OF PROBABILITY**

$$\Pr(A \cup B) = \Pr(A) + \Pr(B) - \Pr(A \cap B)$$

Proof First we write $A \cup B = (A - B) \cup (A \cap B) \cup (B - A)$. We note that the events $A - B$, $A \cap B$, and $B - A$ are mutually exclusive. Applying the additive law for mutually exclusive events twice, we obtain

$$\Pr(A \cup B) = \Pr((A - B) \cup (A \cap B)) + \Pr(B - A)$$
$$= \Pr(A - B) + \Pr(A \cap B) + \Pr(B - A) \quad (1)$$

Now we write $A = (A - B) \cup (A \cap B)$ and $B = (B - A) \cup (A \cap B)$, where $(A - B) \cap (A \cap B) = \emptyset$ and $(B - A) \cap (A \cap B) = \emptyset$. Again by the additive law for mutually exclusive events, we have

$$\Pr(A) = \Pr(A - B) + \Pr(A \cap B) \quad \text{and} \quad \Pr(B) = \Pr(B - A) + \Pr(A \cap B)$$

From the last two equations, we obtain

$$\Pr(A - B) = \Pr(A) - \Pr(A \cap B) \quad (2)$$

and

$$\Pr(B - A) = \Pr(B) - \Pr(A \cap B) \quad (3)$$

Using Equations (2) and (3), Equation (1) can be rewritten as

$$\Pr(A \cup B) = \Pr(A - B) + \Pr(A \cap B) + \Pr(B - A)$$
$$= \Pr(A) - \Pr(A \cap B) + \Pr(A \cap B) + \Pr(B) - \Pr(A \cap B)$$
$$= \Pr(A) + \Pr(B) - \Pr(A \cap B) \qquad \blacksquare$$

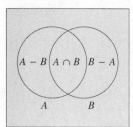

FIGURE 4.4.4 $A \cup B =$ $(A - B) \cup (A \cap B) \cup (B - A)$

The more general additive law of probability reduces to the additive law for mutually exclusive events when $A \cap B = \emptyset$, since $\Pr(\emptyset) = 0$.

Equation (1), $\Pr(A \cup B) = \Pr(A - B) + \Pr(A \cap B) + \Pr(B - A)$, in the proof of the additive law, is best illustrated by the Venn diagram in Figure 4.4.4. The diagram shows very clearly the relation between the set $A \cup B$ and the disjoint sets $A - B$, $A \cap B$, and $B - A$.

The Venn diagram in Figure 4.4.4 is also useful for finding probabilities of certain events when the probabilities of other events are given.

EXAMPLE 4.4.8 Suppose we are given that $\Pr(A) = .4$ and $\Pr(B - A) = .3$. Then we can see that $\Pr(A \cup B) = .4 + .3 = .7$. However, there is not enough information to find $\Pr(A \cap B)$. This is true because the values $\Pr(A \cap B) = .1$ and $\Pr(A - B) = .3$, as well as $\Pr(A \cap B) = .2$ and $\Pr(A - B) = .2$ and other combinations adding to .4, are all consistent with the given information.

Suppose, on the other hand, we are given that $\Pr(A) = .4$, $\Pr(B') = .5$, and $\Pr(B - A) = .3$. Then we can deduce that $\Pr(B) = .5$. Hence, $\Pr(A \cap B) = \Pr(B) - \Pr(B - A) = .5 - .3 = .2$. $\blacksquare$

EXAMPLE 4.4.9 Refer to Example 4.4.4. What is the probability of throwing an odd number or a number less than 8? Let A be the event of throwing the dice and observing a sum that is odd, and let B be the event of throwing a number less than 8. We seek $\Pr(A \cup B)$. Using Table 4.4.1 to calculate the probabilities $\Pr(A)$ and $\Pr(B)$, we obtain

$$\Pr(A) = \tfrac{2}{36} + \tfrac{4}{36} + \tfrac{6}{36} + \tfrac{4}{36} + \tfrac{2}{36} = \tfrac{1}{2}$$
$$\Pr(B) = \tfrac{1}{36} + \tfrac{2}{36} + \tfrac{3}{36} + \tfrac{4}{36} + \tfrac{5}{36} + \tfrac{6}{36} = \tfrac{7}{12}$$

Similarly, we calculate $\Pr(A \cap B) = \tfrac{2}{36} + \tfrac{4}{36} + \tfrac{6}{36} = \tfrac{1}{3}$. Therefore,

$$\Pr(A \cup B) = \tfrac{1}{2} + \tfrac{7}{12} - \tfrac{1}{3} = \tfrac{3}{4}$$

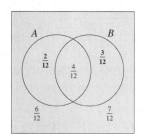

FIGURE 4.4.5

A Venn diagram (see Figure 4.4.5) can help clarify this solution and also give us more information. The numbers in boldface type are the probabilities $\Pr(A - B) = \tfrac{2}{12}$ and $\Pr(B - A) = \tfrac{3}{12}$, which can also be deduced from the given probabilities. $\blacksquare$

Binomial Coefficients

The next three examples illustrate how the binomial coefficients arise in the solutions of some common probability problems. Let us look again at the coin tossing experiment considered in Example 4.4.6.

4.4 INTRODUCTION TO FINITE PROBABILITY

EXAMPLE 4.4.10 Recall that the equiprobable sample space is $S = \{\text{HHH, HHT, HTH, HTT, THH, THT, TTH, TTT}\}$. If we had used the labels "1" for H and "0" for T, this sample space would be the set of all bit strings of length 3. With that analogy in mind, we can apply our knowledge of the binomial coefficients to answer many probability questions. For example, the probability of tossing exactly two heads is $C(3, 2)/8$ since there are exactly $C(3, 2)$ bit strings of length 3 with two 1's.

Of course, the same reasoning applies if we toss a coin, say, seven times. The probability of obtaining exactly two heads is $C(7, 2)/2^7$. What is the probability of selecting, at random, a 2-element subset from a set with 7 elements? The answer is precisely the same; it is $C(7, 2)/2^7$. ∎

There is another kind of question that has a solution dependent on the binomial coefficients.

EXAMPLE 4.4.11 Suppose your 23-person programming class has been preparing for a programming contest. The instructor decides to pick a three-person team at random, so every possible three-person team can be assumed to be an equally likely choice. What is the probability that you will be chosen for the contest team?

Solution We choose the sample space to be the set of all 3-element subsets of a 23-element set. From our previous work in Section 4.3, there are $C(23, 3)$ possible teams or outcomes. How many favorable outcomes are there? If we assume you are a member of a team, there are $C(23, 2)$ ways to choose the remainder of the team. Hence, there are $C(23, 2)$ favorable outcomes. Therefore, the probability that you will be chosen for the team is $C(23, 2)/C(23, 3) = \frac{1}{7}$. ∎

Similar problems arise when analyzing card games. In a standard 52-card deck, there are four suits (spades, hearts, diamonds, and clubs) with 13 cards in each suit. There are 13 denominations of cards (aces, kings, queens, jacks, 10's, 9's, ..., 2's), and each denomination has a card in each suit (for example, ace of spades, ace of hearts, ace of diamonds, and ace of clubs).

In the game of draw poker, the initial hand dealt consists of 5 cards. Since order does not matter, we can take the sample space to be all the possible 5-element subsets of the 52-card deck. Hence, the total number of outcomes for the initial hand dealt in draw poker is $C(52, 5)$.

EXAMPLE 4.4.12 **DRAW POKER**

a. What is probability of drawing four aces in a poker hand?
b. A full house is any hand with three of a kind and a pair. What is the probability of drawing a full house consisting of three aces and two kings?
c. What is the probability of drawing a hand with at least one ace?

Solution a. Since the fifth card in the hand can be any of the other 48 cards, there are 48 favorable outcomes. Therefore, the probability is $48/C(52, 5) = .000905$.
b. There are $C(4, 3)$ ways to get the aces and $C(4, 2)$ ways to get the kings. Hence, there are $C(4, 3) \cdot C(4, 2)$ favorable outcomes, and the probability is equal to $C(4, 3) \cdot C(4, 2)/C(52, 5) = .000452$.
c. Let A be the event of drawing at least one ace. It is easier to calculate the probability of drawing no aces, $\Pr(A')$, and then use the equation $\Pr(A) = 1 - \Pr(A')$. Since 48 cards in the deck are not aces, there are $C(48, 5)$ poker hands with no aces. Hence, $\Pr(A') = C(48, 5)/C(52, 5)$. Therefore, the probability of drawing a hand with at least one ace is $1 - C(48, 5)/C(52, 5) = 341$. ∎

The Birthday Problem

Imagine yourself at a party with 35 people. In such a situation you might wonder whether there are two people at the party with the same birthday. What is the probability of this happening? We are assuming there are 365 possible birthdays. Before going on to read the solution to this problem, make a guess. What do you think is the probability that two or more people in a group of 35 will have the same birthday?

To simplify the analysis, we will first solve this problem for a smaller number of people. Then we will use the same reasoning process to solve a more general problem.

EXAMPLE 4.4.13 **THE BIRTHDAY PROBLEM** For a randomly chosen group of seven people, what is the probability of two or more people having the same birthday (same day and month but not necessarily the same year)? We will assume 365 days in a year. So we are asking how probable is it, among seven people, that two or more of them have birthdays falling on the same one of a possible 365 days.

Solution Let E be the event: Two or more people have a birthday on the same day of the year. It will be easier to calculate $\Pr(E')$ in this problem. The probability that no two people have the same birthday is $\Pr(E')$. Now let us think of the seven people in the class as a 7-tuple. The coordinates of this 7-tuple are assigned the numbers $1, 2, \ldots, 365$ for the possible birthdates. So the sample space consists of 7-tuples with entries from $1, 2, \ldots, 365$. The total number of possible outcomes is 365^7. A favorable outcome for E' is a 7-tuple with no matching entries. The number of favorable outcomes is calculated by noting that there are 365 possible entries for the first coordinate, 364 possible entries for the second coordinate, 363 possible entries for the third coordinate, etc. Hence,

$$\Pr(E') = \frac{365 \cdot 364 \cdot 363 \cdot 362 \cdot 361 \cdot 360 \cdot 359}{365^7} \cong .9438$$

Therefore, $\Pr(E) = 1 - \Pr(E') \cong .0562$. So the probability of having two people in a randomly chosen group of seven with the same birthday is a little greater than 1 in 20. ∎

Now we will solve the birthday problem for a group of n people. Let E_n be the event: In a group of n people, two or more people have a birthday on the same day of the year. We found above that $\Pr(E'_7) \cong .9438$. Applying the techniques used to calculate $\Pr(E'_7)$, we obtain

$$\Pr(E'_n) = \frac{365 \cdot 364 \cdot 363 \cdots (366 - n)}{365^n}$$

Table 4.4.2 gives several values of $\Pr(E_n) = 1 - \Pr(E'_n)$. These results are really quite amazing. It takes a group of only 23 to make the probability of a duplication greater than .50. Moreover, a duplication of birthdays is practically a certainty in a group of 55.

What was your guess for $\Pr(E_{35})$? We have shown that, in the hypothetical gathering mentioned at the beginning of this example, the probability of a duplication of birthdays is greater than $\frac{4}{5}$. Using the terminology of odds, we say there is greater than a 4-to-1 chance of a duplication.

If $\Pr(E) = p$, we say the **odds in favor** of E are p to $1 - p$. For example, if $\Pr(E) = \frac{1}{3}$, the odds in favor of E are $\frac{1}{3}$ to $\frac{2}{3}$, or 1 to 2. Conversely, if the odds in favor of an event E are a to b, then $\Pr(E) = a/(a + b)$. So, if the odds in favor of E are 3 to 2, then $\Pr(E) = \frac{3}{5}$. When the probability is equal to 50, we say the **odds are even** of E occurring. For a randomly chosen group of 23 people, the odds are slightly greater than even that there will be a duplication of birthdays.

TABLE 4.4.2

n	$\Pr(E_n)$
5	.0245
15	.2509
20	.4098
21	.4422
22	.4743
23	.5059
24	.5371
25	.5675
35	.8139
45	.9408
55	.9862

EXERCISE SET 4.4

1. Let $S = \{x_0, x_1, x_2, x_3, x_4\}$ be a sample space.
   a. List two simple events for S.
   b. List an event for S that is not simple.
   c. Assume S is an equiprobable space with probability function Pr. If S_i is a simple event in S, then find $\Pr(S_i)$. Also find $\Pr(\{x_0, x_2, x_4\})$.

2. Let $S = \{x_0, x_1, x_2, x_3, x_4, x_5, x_6\}$ be a sample space.
   a. List two simple events for S.
   b. List an event for S that is not simple.
   c. Assume S is an equiprobable space with probability function Pr. If S_i is a simple event in S, then find $\Pr(S_i)$. Also find $\Pr(\{x_0, x_2, x_5\})$.

3. Let $S = \{x_0, x_1, x_2, x_3, x_4\}$ be a sample space with probability function Pr. Let $E_i = \{x_i\}$. Assume that $\Pr(E_1) = \Pr(E_3) = 2\Pr(E_0)$ and $\Pr(E_0) = \Pr(E_2) = \Pr(E_4)$. Find:
   a. $\Pr(E_i)$ for all i b. $\Pr(\{x_0, x_2\})$ c. $\Pr(\{x_1, x_2, x_3\})$

4. Let $S = \{x_0, x_1, x_2, x_3, x_4, x_5\}$ be a sample space with probability function Pr. Let $E_i = \{x_i\}$. Assume that $\Pr(E_0) = \Pr(E_1) = 2\Pr(E_2)$ and $\Pr(E_2) = \Pr(E_3) = \Pr(E_4) = \Pr(E_5)$. Find:
   a. $\Pr(E_i)$ for all i b. $\Pr(\{x_0, x_2\})$ c. $\Pr(\{x_1, x_2, x_3\})$

5. Suppose we toss a fair coin, with a head (H) on one side and a tail (T) on the other side, four times.
   a. Specify the sample space.
   b. Find the probability of obtaining exactly four heads.
   c. Find the probability of obtaining at least three heads.

6. Suppose we toss a fair coin, with a head (H) on one side and a tail (T) on the other side, seven times.
   a. Specify the sample space.
   b. Find the probability of obtaining exactly five heads.
   c. Find the probability of obtaining at least five heads.

7. Suppose a computer program is designed to select at random a 3-permutation of the set $\{a, b, c, d, e\}$. Assume the set of all 3-permutations of $\{a, b, c, d, e\}$ is an equiprobable sample space.
   a. List several sample points.
   b. Find the probability of selecting any given 3-permutation.
   c. Find the probability of selecting a 3-permutation starting with the letter b.
   d. Find the probability of selecting a 3-permutation containing the letter b.

8. Suppose the letters in the word PASCAL are rearranged at random. Find the probability that the two A's will be adjacent in the new arrangement. First specify the sample space.

9. Let A and B be two events in a sample space. Suppose $\Pr(A) = .5$, $\Pr(B) = .3$, and $\Pr(A \cap B) = .2$. Find:
   a. $\Pr(A \cup B)$ b. $\Pr(B')$ c. $\Pr(A - B)$ d. $\Pr(B - A)$

10. Let A and B be two events in a sample space S. Suppose $\Pr(A) = .6$, $\Pr(B - A) = .4$, and $\Pr(B') = .3$. Find:
    a. $\Pr(B)$ b. $\Pr(A \cap B)$ c. $\Pr(A \cup B)$

11. Let A and B be two events in a sample space S. Is it possible to assign the probabilities $\Pr(B') = .6$, $\Pr(A \cap B) = .2$, and $\Pr(B - A) = .3$? Explain.

12. Suppose your 15-member computer club is selecting, at random, a 5-person committee to plan the end-of-year party. What is the probability that a given member of the club will be chosen for the committee? First specify the sample space.

13. It is known that a shipment of computer disk drives has 18 disk drives with no defects, 6 with minor defects, and 2 with major defects. The inspector will select 2 disk drives at random for inspection.
    a. What is the probability that both disk drives will be found to have minor defects?
    b. What is the probability that both will be found to be nondefective?
    c. What is the probability that both will be found to have major defects?

14. Suppose a computer program is designed to select at random a subset of cardinality 4 from the set $\{a, b, c, d, e, f, g\}$. Assume the power set of $\{a, b, c, d, e, f, g\}$ is an equiprobable space.
    a. List several sample points.
    b. Find the probability of selecting any given subset of cardinality 4.
    c. Find the probability of selecting a subset of cardinality 4 that contains the letter b.

15. Suppose a computer program is designed to select at random a subset of $\{a, b, c, d, e\}$. Assume the power set of $\{a, b, c, d, e\}$ is an equiprobable space.

4.4 INTRODUCTION TO FINITE PROBABILITY

    a. List several sample points.
    b. Find the probability of selecting any given subset.
    c. Find the probability of selecting a subset of cardinality 3.
    d. Find the probability of selecting a subset A where $\text{Card}(A) \geq 3$.

16. Suppose a single card is drawn at random from a standard 52-card deck. Find each of the following probabilities.
    a. Probability of drawing an ace of spades
    b. Probability of drawing an ace
    c. Probability of drawing an ace or any spade

17. Suppose two cards are drawn without replacement from a standard 52-card deck and order is irrelevant. Find each of the following.
    a. The probability of drawing an ace and a king
    b. The probability of drawing two hearts
    c. The probability of drawing two hearts or an ace and a king

18. Suppose five cards are drawn without replacement from a standard 52-card deck and order is irrelevant. Find each of the following.
    a. The probability of drawing four aces and a king
    b. The probability of drawing any four of a kind

19. Suppose five cards are drawn without replacement from a standard 52-card deck and order is irrelevant. Find each of the following.
    a. The probability of drawing any two aces and two kings plus any other card
    b. The probability of drawing any two pairs

20. Suppose five cards are drawn without replacement from a standard 52-card deck and order is irrelevant. Find each of the following.
    a. The probability of drawing a royal flush (ace, king, queen, jack, and 10 of a single suit)
    b. The probability of drawing a straight flush (A straight flush is any sequence of five cards in a single suit that is not a royal flush. Since an ace can count as a 1 in straights, the hand ace, 2, 3, 4, 5 of a single suit is one example of a straight flush.)

21. Let E_n be the event: At least two people in a randomly chosen group of n people have birthdays in the same month. Find a formula for $\Pr(E_n)$.

22. Suppose you are in a group of 20 people. What is the probability that at least one other person in the group has the same birthday as you? Find a formula for the probability that another person in a group of n people will have the same birthday as you.

23. If five people choose, at random, a digit from among the ten digits, what is the probability that each will choose a different digit?

24. If 10 people choose, at random, a letter from the English alphabet of 26 letters, what is the probability that each will choose a different letter?

25. Let S be a sample space. Prove that a probability function $\Pr: P(S) \to [0, 1]$ satisfies each of the following properties.
    a. $\Pr(S) = 1$
    b. $0 \leq \Pr(E) \leq 1$ for any event in S
    c. If A and B are mutually exclusive events, then

$$\Pr(A \cup B) = \Pr(A) + \Pr(B)$$

26. Let A and B be events in a sample space with $A \subseteq B$. Prove each of the following.
   a. $\Pr(B - A) = \Pr(B) - \Pr(A)$ **b.** $\Pr(A) \leq \Pr(B)$

■ PROGRAMMING EXERCISE

27. Refer to the birthday problem. Write a computer program to calculate the probabilities $\Pr(E_n)$ for $n = 10, 15, 20, \ldots, 65$.

4.5

APPLICATION: STACK PERMUTATIONS AND CATALAN NUMBERS

In *Fundamental Algorithms*, volume 1 of his classic series of books, *The Art of Computer Programming*, Donald Knuth posed what we will call the stack permutation problem. A **stack** is a linearly ordered list for which all insertions and deletions are made from one end, called the **top**, of the list. Stacks are often likened to a pile of dinner plates. When the plates are individually placed on and taken from the pile, the last one placed on the pile will be the first one taken from the pile. A stack, like the dinner-plate pile, is a **last-in-first-out (lifo)** structure. An insertion to the top of a stack is called a **push**; a deletion is called a **pop**.

Knuth pointed out that a stack can be represented by the railway switching network shown in Figure 4.5.1. Imagine three railcars, labeled 1, 2, 3 left to right, on the input side of the track shown in Figure 4.5.1. This is the standard arrangement 123 of the three symbols 1, 2, and 3. Now suppose we perform the following operations on the three cars:

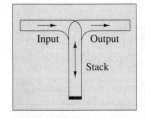

FIGURE 4.5.1

1. Move car 3 onto the stack.   Push
2. Move car 3 onto the output. Pop
3. Move car 2 onto the stack.   Push
4. Move car 1 onto the stack.   Push
5. Move car 1 onto the output. Pop
6. Move car 2 onto the output. Pop

The cars are now on the output side in the arrangement 213. We have a new permutation of the numbers obtained by a sequence of pushes and pops.

Intuitively, an admissible sequence of pushes and pops is any sequence of pushes from the input and pops to the output that results in an empty input and an empty stack and that can be carried out in the railway switching network.

A sequence of pushes and pops is an **admissible sequence** if the sequence contains an equal total number of pushes and pops and if at no time during the process are there more pops than pushes. There cannot be more pops than pushes during the process since we cannot pop an empty stack. Any permutation of the numbers $N_n = \{1, 2, 3, \ldots, n\}$ obtained on the output with an admissible sequence of pushes and pops applied to the standard arrangement $123 \ldots n$ placed on the input is called a **stack permutation**.

There are six permutations of $\{1, 2, 3\}$. How many of those six are stack permutations? It is easy to see that 123 and 321 are both stack permutations. Furthermore, we can verify that 132 and 312 are both stack permutations. But 231 is not a stack permutation. To see this, note that any sequence of pushes and pops leading to 1 on the right of an arrangement must at some point have had a stack with 1 on the top of the stack, with the other numbers below it. After 1 is popped, 2 must be on top and 3 underneath, since 3 must have gone on the stack before 2. Therefore, 3 cannot be adjacent to 1. Hence, there are five stack permutations of N_3.

The Stack Permutation Problem

Find a formula for the number of stack permutations of N_n. To solve this stack permutation problem, we will transform it into an equivalent problem involving strings of symbols. Let us use the symbols U (for push) and O (for pop) to encode a sequence of pushes and pops as a string of U's and O's. For example, the string that corresponds to the sequence leading to 213 is written UOUUOO. A string of U's and O's corresponding to an admissible sequence is an **admissible string**. Note that an admissible string has exactly as many U's as O's. When reading the string left to right, there are always at least as many U's as O's at any time.

The stack permutation problem is equivalent to the problem of finding a formula for the number of admissible strings with n U's and n O's. A bijective function and the relationship between binomial coefficients and bit strings can now be used to solve this problem. We first prove a lemma.

LEMMA 4.5.1 There is a bijective function from the set of strings with n U's and n O's that are not admissible to the set of all strings with $n + 1$ O's and $n - 1$ U's.

Proof First we find a function and then prove it is bijective by finding its inverse. Consider any string of n U's and n O's that is not admissible. At some place in the string, as it is read left to right, there are more O's than U's. Let k be the first position where the number of O's exceeds the number of U's. To the right of the kth position, there is one more U than there are O's. To obtain a string with $n + 1$ O's and $n - 1$ U's, we replace every U with an O and every O with a U in the positions of the string to the right of the kth position.

To find the inverse of this function, we consider any string with $n + 1$ O's and $n - 1$ U's. Again, let k be the first position of the string where the number of O's exceeds the number of U's. To the right of the kth position, the number of O's exceeds the number of U's by 1 since the number of O's originally exceeded the number of U's by 2. To obtain a string with n U's and n O's, we replace every U with an O and every O with a U to the right of the kth position. The string so obtained is not admissible because, at the kth position, the number of O's exceeds the number of U's. ∎

Since two finite sets have an equal number of elements if there is a bijective function from one set to the other, we can find the number of strings with n U's and n O's that are not admissible by finding the number of strings with $n + 1$ O's and $n - 1$ U's. Since strings of U's and O's are just bit strings written with different symbols, the number of strings with $n + 1$ O's and $n - 1$ U's is $C(2n, n - 1)$.

THEOREM 4.5.2 The number of admissible strings with n U's and n O's is

$$C(2n, n) - C(2n, n - 1)$$

Proof The total number of strings with n U's and n O's is $C(2n, n)$. By the work above, the number of strings with n U's and n O's that are not admissible is $C(2n, n - 1)$. The result follows. ∎

Catalan Numbers

The number $\text{Cat}(n) = C(2n, n) - C(2n, n - 1)$ is called the nth **Catalan number**. We have thus proven that the nth Catalan number is the solution to the stack permutation problem. That is, there are $C(2n, n) - C(2n, n - 1)$ stack permutations of the set $N_n = \{1, 2, 3, \ldots, n\}$. Recall that we found five stack permutations of $\{1, 2, 3\}$, and note that $C(6, 3) - C(6, 2) = 20 - 15 = 5$.

A string of parentheses is **well-balanced** if it has exactly as many left parentheses as right parentheses and if, when the string is read left to right, there are always at least as many left parentheses as right parentheses at any time. Hence, each string of well-balanced parentheses corresponds to an admissible string of U's and O's. Therefore $\text{Cat}(n)$ is the number of well-balanced strings of parentheses with n left parentheses and n right parentheses.

We have seen that Catalan numbers were used to solve problems regarding the number of stack permutations of a set and the number of strings of well-balanced parentheses. It may be surprising to learn that neither of these problems motivated the discovery of Catalan numbers.

Catalan numbers were originally discovered as a means to solve a geometry problem posed by Euler in 1759. The problem is to find a formula for the number of ways to divide a convex polygon with $n + 2$ sides into triangles with $n - 1$ nonintersecting diagonals. This problem was solved by J. A. von Segner in 1759. Later, in 1838, the problem was also solved by E. Catalan. He proved that there are $\text{Cat}(n)$ ways of triangulating a convex polygon with $n - 1$ diagonals. We show the solution for $n = 3$ (pentagons) in Figure 4.5.2.

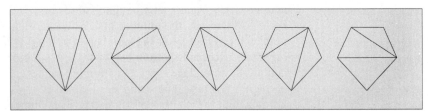

FIGURE 4.5.2 Triangulations of a pentagon

EXERCISE SET 4.5

1. We have shown that Cat(3) = 5. Find Cat(0), Cat(1), Cat(2), Cat(4).
2. **a.** Find the number of stack permutations of $\{1, 2, 3, 4, 5\}$.
   b. Find the number of strings of well-balanced parentheses with five left parentheses and five right parentheses.
3. Test whether each of the following permutations of $\{1, 2, 3, 4\}$ is a stack permutation. If it is a stack permutation, give the admissible sequence of pushes and pops.
   a. 2143   **b.** 3214   **c.** 2431
4. Test whether each of the following permutations of $\{1, 2, 3, 4, 5\}$ is a stack permutation. If it is a stack permutation, give the admissible sequence of pushes and pops.
   a. 12435   **b.** 32154   **c.** 24135
5. **a.** Find the number of triangulations of a hexagon.
   b. Draw all the triangulations of a hexagon.
6. Prove $\mathrm{Cat}(n) = (1/n) \cdot C(2n, n-1)$.
7. Prove $\mathrm{Cat}(n-1) = (1/n) \cdot C(2n-2, n-1)$.

KEY TERMS

Sum rule
Law of inclusion and exclusion
Product rule
Fundamental counting principle
Factorial notation, n!
r-Permutation
P(n, r)
r-Permutation with replacement (or with repetition)
r-Combination
C(n, r)
Binomial coefficients
Binomial Theorem
Pascal's triangle
r-Combination with replacement (or with repetition)
Experiment
Sample point
Sample space
Event
Simple event
Impossible event
Occurrence of event
Favorable outcome
Mutually exclusive events
Equiprobable space
Probability function
Additive law for mutually exclusive events
Additive law of probability
Odds
Stack
Top
Last-in-first-out (lifo)
Push, pop
Admissible sequence
Stack permutation
Catalan number
Well-balanced parentheses

REVIEW EXERCISES

1. **a.** How many decimal numbers have five or fewer digits?
   b. How many subsets of $\{0, 1, \ldots, 9\}$ are there with five or fewer elements?
2. Suppose a computer club has 73 members (38 females and 35 males).
   a. In how many ways can a president, a vice president, and a secretary be chosen?
   b. In how many ways can a committee of five be chosen?
   c. In how many ways can a committee of seven be chosen if it is to be composed of three females and four males?

3. The local computer store carries five different types of microcomputers.
   a. Suppose the local college wants to buy 20 of these micros and does not specify how many of each it wants. In how many ways can the store choose this set of 20? Assume the store has at least 20 of each type and order does not matter.
   b. In how many ways can the store choose this set of 20 if the college specifies that exactly 10 must be Macintoshes?
4. a. How many bit strings of length 7 have exactly 3 0's?
   b. How many bit strings of length n have exactly k 0's?
   c. How many bit strings of length $n + m$ have exactly k 1's among the first n bits and exactly j 1's among the last m bits?
5. Let $D = \{1, 2, \ldots, n\} = C$. Find the number of onto functions $f:D \to C$.
6. Let $D = \{1, 2, \ldots, n, n + 1\}$ and $C = \{1, 2, \ldots, n\}$. Find the number of onto functions $f:D \to C$.
7. Let $D = \{1, 2, \ldots, n\}$ and $C = \{0, 1\}$. Find the number of onto Boolean functions $f:D \to C$.
8. Let A and B be two events in a sample space. Suppose $\Pr(A) = .7$, $\Pr(B) = .5$, and $\Pr(A \cap B) = .3$. Find:
   a. $\Pr(A \cup B)$ b. $\Pr(B')$ c. $\Pr(A - B)$ d. $\Pr(B - A)$
9. Suppose five cards are drawn without replacement from a standard 52-card deck and order is irrelevant. Find each of the following.
   a. The probability of drawing five hearts
   b. The probability of drawing a flush (a flush is any five cards of the same suit.)
10. Suppose 13 cards are drawn without replacement from a standard 52-card deck and order is irrelevant. Find each of the following.
    a. The probability of drawing 13 hearts
    b. The probability of drawing 7 hearts and 6 spades
11. It is known that a shipment of computer disk drives has 45 disk drives with no defects, 3 with minor defects, and 2 with major defects. The inspector will select 2 disk drives at random for inspection.
    a. What is the probability that both disk drives will have minor defects?
    b. What is the probability that both will be nondefective?
    c. What is the probability that both will have major defects?
    d. What is the probability that one will be defective and one will have no defects?
12. How many partitions of $\{1, 2, 3\}$ are there?
13. How many equivalence relations can be defined on $\{1, 2, 3\}$?
14. Let $V = \{1, 2, \ldots, n\}$.
    a. How many digraphs $G = (V, E)$ are there?
    b. How many digraphs $G = (V, E)$ are there with exactly k edges?
    c. How many symmetric digraphs $G = (V, E)$ with no self-loops are there?
15. Suppose you are to place square tiles in a 2×4 rectangular pattern on a bathroom wall. How many different patterns can you make if:
    a. You have 15 distinct tiles to use?
    b. You have 5 boxes of tiles and tiles in different boxes are different colors? Each box has at least 8 tiles of the same color. Tiles of the same color are indistinguishable from one another.

16. Suppose you have 5 distinct boxes and 12 objects. In how many ways can the objects be partitioned among the boxes if:
    a. The objects are distinct?   **b.** The objects are indistinguishable?

17. Suppose you have 11 distinct boxes and 7 objects. In how many ways can the objects be partitioned among the boxes if:
    a. The objects are distinct?   **b.** The objects are indistinguishable?

Let $D = \{d_1, d_2, \ldots, d_n\}$ and $C = \{c_1, c_2, \ldots, c_m\}$. Think of the elements of D as objects, the elements of C as containers, and a function $f: D \to C$ as a rule for placing each object of D in a container of C. When the objects and the containers are both distinct, we know the number of functions in C^D is m^n. Verify the facts given in Exercises 18–21.

18. When both the objects and the containers are distinct, the number of 1–1 functions $f: D \to C$ is $m(m-1) \cdots (m-n+1)$.

19. When the objects are indistinguishable and the containers are distinct, the number of functions $f: D \to C$ is $C(m+n-1, n)$.

20. When the objects are indistinguishable and the containers are distinct, the number of 1–1 functions $f: D \to C$ is $C(m, n)$.

21. When the objects are indistinguishable and the containers are distinct, the number of onto functions $f: D \to C$ is $C(n-1, m-1)$.

REFERENCES

Bogart, K. *Introductory Combinatorics.* Marshfield, MA: Pitman, 1983.

Knuth, D. E. *The Art of Computer Programming: Volume 1, Fundamental Algorithms,* 2nd ed. Reading, MA: Addison-Wesley, 1973.

Kruse, R. L. *Data Structures and Program Design,* 2nd ed. Englewood Cliffs, NJ: Prentice-Hall, 1987.

Roberts, F. S. *Applied Combinatorics.* Englewood Cliffs, NJ: Prentice-Hall, 1984.

Snell, J. L. *Introduction to Probability Theory with Computing.* Englewood Cliffs, NJ: Prentice-Hall, 1975.

Tucker, A. *Applied Combinatorics.* New York: Wiley, 1980.

CHAPTER 5

LOGIC AND PROOF

Everyone must be able to recognize logical arguments. For computer scientists, mathematicians, engineers, and scientists, the requirement is compelling. The close relationship between logic and computer science is not accidental. To understand computer science and mathematics, we must be able to prove or disprove statements and to verify the correctness of algorithms.

Symbolic logic can be described as the analytical study of the art of reasoning. In this chapter we will discuss the two most important branches of symbolic logic—propositional logic and predicate logic.

Our objective is to define and study several common proof techniques and to justify them by principles of logic. These techniques are powerful tools for the analysis of proofs, algorithms, and computer programs. We will use these techniques to construct rigorous proofs for many true statements.

Mathematical induction is one of the most useful proof techniques in mathematics and computer science. We will study many examples of proof by mathematical induction. Induction and other techniques will be used to prove several theorems stated and used in earlier chapters. Finally, we will discuss predicate logic.

APPLICATION	**ARTIFICIAL INTELLIGENCE—AUTOMATED THEOREM PROVING**
	Artificial intelligence (AI) is an important and lively branch of computer science. Ever since the invention of the digital computer, mathematicians, computer scientists, psychologists, and philosophers have been investigating the relationship between human thought processes and the theory of machine (automata) processes.

AI researchers see potential applications for computers that go far beyond routine numerical calculation and text processing. Some see the computer as a paradigm for the study of human processes such as memory, vision, speech, and hearing. Others within AI believe higher-order activities of the brain, not physical mechanisms, are most important. This latter group of researchers is interested in programming machines to emulate logical reasoning, to play games such as chess, and to understand natural languages such as English. For them, whether or not the machine performs its task by emulating the human brain is unimportant; it is results that are paramount.

Much early enthusiasm for a mathematical theory closely relating the brain and machine has waned because of the overwhelming complexity of the brain compared to a digital computer. Nevertheless, the pursuit of a mechanical explanation for mental processes goes on, but more progress has been made in emulating mental processes. In particular, automating logical and mathematical thought processes has been a fruitful line of research. The application in Section 5.6 discusses the role of propositional logic in automated theorem proving.

5.1
PROPOSITIONAL LOGIC

In this section we continue the study of propositional logic, begun in Section 1.5. Propositional logic is an interesting area of study in its own right. An extensive study of logic would take us far beyond our immediate purpose, however. Our point of view is that logic is important because it forms the basis for proof techniques, and therefore has special utility for mathematics and computer science.

The AND ($\wedge$), OR ($\vee$), and NOT ($\neg$) connectives are defined by truth tables in Section 1.5. The design of logic networks for electronic devices depends on three basic logic gates, the AND gate, the OR gate, and the NOT gate. These are important analogs of the three basic statement connectives AND, OR, and NOT. Logic networks and logic gates will be studied in Chapter 7.

Recall that the IMPLIES ($\rightarrow$) and the IF AND ONLY IF ($\leftrightarrow$) connectives are defined by the truth tables given in Tables 5.1.1 and 5.1.2, respectively.

TABLE 5.1.1

p	q	$p \rightarrow q$
T	T	T
T	F	F
F	T	T
F	F	T

TABLE 5.1.2

p	q	$p \leftrightarrow q$
T	T	T
T	F	F
F	T	F
F	F	T

The next example gives several statements with their truth values. These statements concern a binary relation R on a set A.

EXAMPLE 5.1.1 Let $A = \{a, b, c\}$ and $R = \{(a, b), (b, c), (a, a), (c, b), (c, c)\}$.

STATEMENT	TRUTH VALUE
$(b, c) \in R \rightarrow (c, b) \in R$	True
$(b, a) \in R \rightarrow (a, b) \in R$	(Vacuously) True
$(a, b) \in R \rightarrow (b, a) \in R$	False
$[(a, b) \in R \wedge (b, c) \in R] \rightarrow (a, c) \in R$	False
$[(a, c) \in R \wedge (c, b) \in R] \rightarrow (a, b) \in R$	(Vacuously) True
$(a, b) \in R \leftrightarrow a \in A$	True

PRACTICE PROBLEM 1 Refer to Example 5.1.1. Supply the truth value for each of the following statements.

a. $[(b, c) \in R \wedge (c, b) \in R] \rightarrow b = c$ **b.** $d \in A \rightarrow (d, d) \in R$

5.1 PROPOSITIONAL LOGIC

Remarks on Punctuation and Parentheses

The meaning of a phrase in English can differ wildly depending on punctuation. Consider the following two phrases:

What do you think? You got an A on the final!
What? Do you think you got an A on the final?

For symbolized statements, punctuation is accomplished by using parentheses. To keep statements from looking cluttered, we will use certain conventions for leaving out some parentheses. The most important conventions are illustrated in the next example.

EXAMPLE 5.1.2 The NOT ($\neg$), AND ($\wedge$), and OR ($\vee$) symbols always apply to as little as possible. The IMPLIES symbol ($\rightarrow$) is applied only after other symbols along with the three basic connectives have been evaluated. The IF AND ONLY IF symbol ($\leftrightarrow$) is applied last.

a. $\neg p \wedge q$ stands for $(\neg p) \wedge q$, not for $\neg(p \wedge q)$
b. $\neg p \rightarrow q$ stands for $(\neg p) \rightarrow q$, not for $\neg(p \rightarrow q)$
c. $\neg p \vee \neg q$ stands for $(\neg p) \vee (\neg q)$, not for $\neg[p \vee (\neg q)]$
d. $p \wedge q \rightarrow r$ stands for $(p \wedge q) \rightarrow r$, not for $p \wedge (q \rightarrow r)$
e. $\neg p \vee q \rightarrow r \wedge q$ stands for $[(\neg p) \vee q] \rightarrow (r \wedge q)$
f. $\neg p \leftrightarrow p \rightarrow q \wedge r$ stands for $(\neg p) \leftrightarrow [p \rightarrow (q \wedge r)]$ ∎

Use parentheses for clarity and to avoid confusion. The hierarchical order of the connectives is NOT, then OR and AND at the same level, followed by the IMPLIES connective and the IF AND ONLY IF connective. Parenthesized expressions are always performed first. This order is summarized as follows:

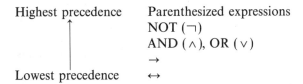

Highest precedence — Parenthesized expressions
NOT ($\neg$)
AND ($\wedge$), OR ($\vee$)
$\rightarrow$
Lowest precedence — $\leftrightarrow$

Caution: In programming languages such as Pascal, where NOT, AND, and OR are Boolean operators, AND is given precedence over OR.

Constructing Truth Tables

A basic principle of propositional logic is that the truth values of a statement composed of other statements and the four basic connectives $\neg$, $\vee$, $\wedge$, and $\rightarrow$ are determined by the truth values of the constituent statements and by the way in which those statements are combined with the connectives. We have already shown, by means of truth tables, how the truth values of the statements $\neg p$, $p \vee q$, $p \wedge q$, and $p \rightarrow q$ are determined by the truth values of p and q. Those truth tables serve as definitions of the connectives $\neg$, $\vee$, $\wedge$, and $\rightarrow$.

To exhibit the truth values of any statement composed of the four basic connectives and other statements, we will construct truth tables based on the truth tables for $\neg$, $\vee$, $\wedge$, and $\rightarrow$. The construction of truth tables involves a step-by-step calculation of the truth values of the constituent statements of the given statement. This is illustrated in the next example.

EXAMPLE 5.1.3 Construct the truth table for $(p \vee q) \wedge (\neg p \rightarrow q)$.

Solution The truth values are calculated in the following order.

1. We give the possible truth values for the statements p and q.
2. The truth values for $p \vee q$ and $\neg p$ are then calculated.
3. Next we calculate the truth values for $\neg p \rightarrow q$.
4. Finally, the truth values for $(p \vee q) \wedge (\neg p \rightarrow q)$ can be calculated.

$$(p \vee q) \wedge (\neg p \rightarrow q)$$
Steps 2   4   2   3

It is helpful to number the columns of the truth table showing the order in which truth values are calculated. (See Table 5.1.3.)

TABLE 5.1.3

	p	q	$(p \vee q) \wedge (\neg p \rightarrow q)$
	T	T	T T F T
	T	F	T T F T
	F	T	T T T T
	F	F	F F T F
Steps	1	1	2 4 2 3

The column with the highest number contains the truth values for the truth table being constructed. The truth table is shown in Table 5.1.4.

TABLE 5.1.4

p	q	$(p \vee q) \wedge (\neg p \rightarrow q)$
T	T	T
T	F	T
F	T	T
F	F	F

PRACTICE PROBLEM 2 Consider $(p \wedge \neg q) \rightarrow (p \vee q)$.

a. Place numbers, starting with 2, under the connectives to specify the order for evaluating truth values of constituent statements.
b. Construct the truth table with the columns numbered in order.

EXAMPLE 5.1.4 Construct the truth table for $(p \wedge \neg q) \leftrightarrow \neg(p \rightarrow q)$.

Solution The truth table is given in Table 5.1.5.

TABLE 5.1.5

p	q	$(p \wedge \neg q) \leftrightarrow \neg(p \rightarrow q)$
T	T	F F T F T
T	F	T T T T F
F	T	F F T F T
F	F	F T T F T
Steps 1	1	3 2 6 5 4

The Exclusive OR

The OR connective is defined so that a statement such as "Bill will play tennis or Jean will go dancing" is true when Bill plays tennis and Jean goes dancing (when both constituent statements are true). For that reason the OR connective is sometimes called the "inclusive OR." A statement such as "She will wear a jacket or she will catch cold," involving *or* used in ordinary discourse, is frequently considered to be false when both constituent statements are true. This interpretation of the meaning of *or* is used to define the "exclusive OR" logical connective, abbreviated $\veebar$.

The connective $\veebar$ is called the **exclusive OR** connective, and it is defined by the truth table in Table 5.1.6.

Asserting that $p \veebar q$ is true means that either p or q is true but it is not the case that p and q are both true. A simple example will show that the OR connective and the exclusive OR connective are not to be confused.

TABLE 5.1.6

p	q	$p \veebar q$
T	T	F
T	F	T
F	T	T
F	F	F

EXAMPLE 5.1.5 Construct the truth tables for $p \vee \neg q$ and $p \veebar \neg q$.

Solution The truth tables are given in Tables 5.1.7 and 5.1.8, respectively.

TABLE 5.1.7

p	q	$p \vee \neg q$
T	T	T
T	F	T
F	T	F
F	F	T

TABLE 5.1.8

p	q	$p \veebar \neg q$
T	T	T
T	F	F
F	T	F
F	F	T

EXERCISE SET 5.1

In Exercises 1–4, let p stand for "The irises are blooming" and let q stand for "It is winter." Write each of the given pairs of statements in English and determine truth values for statements p and q for which the given statements have different truth values.

1. $\neg p \to q$; $\neg p \leftrightarrow q$
2. $\neg p \land q$; $\neg p \veebar q$
3. $p \to p \lor q$; $p \to p \land q$
4. If q, then $\neg p \lor q$; if q, then $\neg q \land p$.

In Exercises 5 and 6, use the following symbols:

p: Orchids are blooming.
q: Swimming is good exercise.
r: Dancing is fun.

5. Translate each of the following statements into symbolic notation.
   a. Dancing is fun, and swimming is good exercise or orchids are blooming.
   b. Swimming is good exercise only if orchids are not blooming or dancing is not fun.
6. Translate each of the following statements into symbolic notation.
   a. If swimming is not good exercise, then dancing is fun or orchids are not blooming.
   b. Dancing is not fun or swimming is not good exercise or orchids are not blooming.

For the statements given in Exercises 7–12, place numbers, starting with 2, under the connectives to specify the order for evaluating the truth values of constituent statements.

7. $(p \to \neg q) \land r \to q$
8. $p \to \neg q \land r$
9. $(p \to q) \to [(p \land r) \to (q \land r)]$
10. $(p \to q) \to [(q \to r) \to (p \to r)]$
11. $[(p \to q) \land (r \to s)] \to [(p \lor r) \to (q \lor s)]$
12. $[(p \to q) \land (r \to s)] \to [(\neg q \lor \neg s) \to (\neg p \lor \neg r)]$

Construct truth tables for the statements in Exercises 13–24.

13. a. $p \land q \to p$
    b. $p \to p \lor q$
14. a. $p \land (p \to q) \to q$
    b. $q \land (p \to q) \to p$
15. a. $p \land (\neg p \lor q) \to q$
    b. $\neg p \land (p \to q) \to q$
16. $(p \to q) \land (q \to r) \to (p \to r)$
17. $(p \to q) \to [(p \lor r) \to (q \lor r)]$
18. $(p \to q) \to [(p \land r) \to (q \land r)]$
19. $(p \lor q \to r) \to (p \to r) \land (q \to r)$
20. $(p \land q \to r) \to (p \to r) \land (q \to r)$
21. $(p \land q \to r) \to (p \to r) \lor (q \to r)$
22. $[(p \lor q) \land r] \to (\neg p \lor q)$
23. $\neg(p \land \neg q) \lor (r \to q)$
24. $[\neg p \lor (p \to \neg r)] \land r$
25. Construct the truth tables for the following.
    a. $p \veebar p$ b. $p \veebar \neg p$
26. Construct the truth tables for the following.
    a. $p \veebar (q \veebar r)$ b. $(p \veebar q) \veebar r$
27. Construct the truth tables for the following.
    a. $p \veebar (q \land r)$ b. $(p \veebar q) \land (p \veebar r)$
28. Construct the truth tables for the following.
    a. $p \land (q \veebar r)$ b. $(p \land q) \veebar (p \land r)$
29. Write a statement composed of p and q and the connectives $\neg$, $\lor$, and $\land$ that is true when exactly one of p or q is true and false otherwise.

30. Write a statement composed of p, q, and r and the connectives $\neg$, $\vee$, and $\wedge$ that is true when exactly one of p, q, or r is true and false otherwise.

31. In a remote village, every villager is either a liar or a truth teller but not both. A liar always lies; a truth teller always tells the truth. Suppose you encounter two villagers, Amy and James. James says "If I am a truth teller, then Amy is a truth teller." Is Amy a liar or a truth teller? What about James?

5.2
LOGICAL EQUIVALENCE AND TAUTOLOGIES

In algebra two expressions that may not look alike, such as $(x-2)(x+1)$ and $x^2 - x - 2$, are said to be equivalent because they take on the same value for every possible real value x. There is a similar relationship, called logical equivalence, between statements in propositional logic.

Logical Equivalence

By examining the four combinations of truth values for p and q, we can see that $p \rightarrow q$ and $\neg p \vee q$ have the same truth tables. See Tables 5.2.1 and 5.2.2. We say that $p \rightarrow q$ and $\neg p \vee q$ are logically equivalent, and we write $(p \rightarrow q) \equiv (\neg p \vee q)$. In general, two statements are **logically equivalent** if they have the same truth values whenever all constituent statements in one have the same values as the corresponding statements in the other. It is left as an exercise to show that $p \equiv q$ if and only if $p \leftrightarrow q$ is a tautology. (See Exercise 22, Exercise Set 5.2.) For deleting parentheses, the logical equivalence $\equiv$ has the lowest precedence.

TABLE 5.2.1

p	q	$p \rightarrow q$
T	T	T
T	F	F
F	T	T
F	F	T

TABLE 5.2.2

p	q	$\neg p \vee q$
T	T	T
T	F	F
F	T	T
F	F	T

Several logical equivalences, including the one just introduced, are stated below. They are important, and we will use each of them later.

SOME LAWS OF LOGIC

a. $\neg(\neg p) \equiv p$ — **Rule of double negation**
b. $(p \rightarrow q) \equiv (\neg p \vee q)$ — **OR form** of an implication
c. $(p \rightarrow q) \equiv (\neg q \rightarrow \neg p)$ — **Contrapositive** of an implication
d. $\neg(p \rightarrow q) \equiv (p \wedge \neg q)$ — **Negation** of an implication
e. $\neg(p \vee q) \equiv (\neg p \wedge \neg q)$ — **De Morgan's law**
f. $\neg(p \wedge q) \equiv (\neg p \vee \neg q)$ — **De Morgan's law**
g. $(p \wedge r \rightarrow q) \equiv [r \rightarrow (p \rightarrow q)]$ — **Rule for direct proof**
h. $(p \wedge \neg q \rightarrow 0) \equiv (p \rightarrow q)$ — **Rule for proof by contradiction**

The symbol O used in Law h stands for any statement that is always false. (See the discussion of contradictions that follows.) The OR form and the contrapositive are useful alternative ways of writing an implication. Roughly speaking, De Morgan's laws can be paraphrased: Negating an OR makes it an AND, whereas negating an AND makes it an OR. Both direct proof and proof by contradiction are introduced in Section 5.3.

PRACTICE PROBLEM 1 Compute the truth tables for $\neg(p \vee q)$ and $\neg p \wedge \neg q$ to verify De Morgan's Law e above.

The truth tables for $p \vee \neg p$ and $q \rightarrow p \vee q$ are given in Tables 5.2.3 and 5.2.4, respectively. The statements $p \vee \neg p$ and $q \rightarrow p \vee q$ are two examples of tautologies. Any statement with a T in every line of its truth table is a **tautology**. The symbol I will be used for any tautology. Hence, $p \vee \neg p \equiv I$ and $(q \rightarrow p \vee q) \equiv I$.

TABLE 5.2.3

p	$p \vee \neg p$
T	T
F	T

TABLE 5.2.4

p	q	$q \rightarrow p \vee q$
T	T	T
T	F	T
F	T	T
F	F	T

The statement $[p \wedge (p \rightarrow q)] \rightarrow q$ is a very useful tautology called **modus ponens**.

PRACTICE PROBLEM 2 Compute the truth table for $[p \wedge (p \rightarrow q)] \rightarrow q$ and thus verify that modus ponens is a tautology.

The truth tables for $p \wedge \neg p$ and $(\neg p \vee q) \wedge (p \wedge \neg q)$ are given in Tables 5.2.5 and 5.2.6, respectively. The statements $p \wedge \neg p$ and $(\neg p \vee q) \wedge (p \wedge \neg q)$ are examples of contradictions. Any statement with an F in every line of its truth table is a **contradiction**. The symbol O will be used for any contradiction. Hence, $p \wedge \neg p \equiv O$ and $(\neg p \vee q) \wedge (p \wedge \neg q) \equiv O$.

TABLE 5.2.5

p	$p \wedge \neg p$
T	F
F	F

TABLE 5.2.6

p	q	$(\neg p \vee q) \wedge (p \wedge \neg q)$
T	T	F
T	F	F
F	T	F
F	F	F

Let us summarize the three concepts just introduced.

5.2 LOGICAL EQUIVALENCE AND TAUTOLOGIES

Two statements p and q are *logically equivalent* (written $p \equiv q$) if p and q have the same truth values whenever all constituent statements in p have the same values as the corresponding statements in q.

A statement p is a *tautology* (written $p \equiv I$) if the truth table for p has a T in every line.

A statement q is a *contradiction* (written $q \equiv O$) if the truth table for q has an F in every line.

Here are some important rules called the **Boolean laws of logic**.

BOOLEAN LAWS OF LOGIC

1. **a.** $p \vee q \equiv q \vee p$ **b.** $p \wedge q \equiv q \wedge p$ Commutative
2. **a.** $(p \vee q) \vee r$ **b.** $(p \wedge q) \wedge r$ Associative
   $\equiv p \vee (q \vee r)$ $\equiv p \wedge (q \wedge r)$
3. **a.** $p \vee (q \wedge r)$ **b.** $p \wedge (q \vee r)$ Distributive
   $\equiv (p \vee q) \wedge (p \vee r)$ $\equiv (p \wedge q) \vee (p \wedge r)$
4. **a.** $p \vee O \equiv p$ **b.** $p \wedge I \equiv p$
5. **a.** $p \vee \neg p \equiv I$ **b.** $p \wedge \neg p \equiv O$

Each law can be verified by a calculation of the truth tables for the two statements in each logical equivalence. Please see Section 1.2 for the analogous Boolean laws for set theory.

EXAMPLE 5.2.1   Table 5.2.7 gives a verification of Law 3b since $p \wedge (q \vee r)$ and $(p \wedge q) \vee (p \wedge r)$ yield the same truth values at every line.

TABLE 5.2.7

p	q	r	$p \wedge (q \vee r)$	$(p \wedge q) \vee (p \wedge r)$
T	T	T	T	T
T	T	F	T	T
T	F	T	T	T
T	F	F	F	F
F	T	T	F	F
F	T	F	F	F
F	F	T	F	F
F	F	F	F	F

By the commutative laws (Laws 1a and 1b of the Boolean laws of logic), we see that $p \vee q$ is logically equivalent to $q \vee p$ and that $p \wedge q$ is logically equivalent to $q \wedge p$. So, order of the constituent statements is not important in either an OR statement or an AND statement. By the associative laws (Laws 2a and 2b), we can write $p \vee q \vee r$ for either $(p \vee q) \vee r$ or $p \vee (q \vee r)$ and $p \wedge q \wedge r$ for either $(p \wedge q) \wedge r$ or $p \wedge (q \wedge r)$ without fear of ambiguity.

Is order important for an implication? In other words, is $q \to p$ logically equivalent to its converse $p \to q$? It is left as an exercise (see Exercise 1a, Exercise Set 5.2) to construct a truth table for $q \to p$ and compare it to that of $p \to q$. It turns out that the truth tables are not the same, so $p \to q$ is not logically equivalent to $q \to p$. So, the converse of an implication is not logically equivalent to the implication.

An example from algebra will serve to illustrate the distinction between an implication and its converse.

EXAMPLE 5.2.2 The implication "If an integer greater than 2 is prime, then the integer is odd" is true. However, its converse "If an integer greater than 2 is odd, then it is prime" is false. ∎

By calculating the truth table for $(p \to q) \land (q \to p)$, we see that the biconditional $p \leftrightarrow q$ is logically equivalent to $(p \to q) \land (q \to p)$. Hence, the biconditional $p \leftrightarrow q$ is true whenever the implication $p \to q$ and its converse are both true.

PRACTICE PROBLEM 3 Calculate the truth table for $(p \to q) \land (q \to p)$, and verify that $p \leftrightarrow q$ is true precisely when the truth value of p is the same as the truth value of q.

For the implication $p \to q$, we say p is a **sufficient condition** for q or that q is a **necessary condition** for p. Hence, for the biconditional $p \leftrightarrow q$, we say q is a **necessary and sufficient condition** for p.

Counterexamples

By definition, a statement is not a tautology whenever at least one line of its truth table has a truth value F on the right. To show that a statement is not a tautology, a line in the truth table for that statement must be found with a truth value F. The combination of truth values in any line that produces a truth value F for the statement is called a **counterexample**.

In Practice Problem 2 it was shown that $[p \land (p \to q)] \to q$ (modus ponens) is a tautology.

EXAMPLE 5.2.3 The truth table in Table 5.2.8 shows that $[q \land (p \to q)] \to p$ is not a tautology.

TABLE 5.2.8

p	q	$[q \land (p \to q)] \to p$	
T	T	T	T
T	F	F	T
F	T	T	F
F	F	F	T

The combination of truth values on the left in line 3 of Table 5.2.8 is a counterexample. ∎

Functionally Complete Sets of Logical Connectives

Although we will not prove it here, all statements in propositional logic can be written using only the three basic connectives AND, OR, and NOT. It is possible and frequently useful to write all statements using only one or two connectives. For example, all statements can be written in terms of the connectives OR and NOT. To see this we need only write AND in terms of OR and NOT. This can be done using one of De Morgan's laws and the rule of double negation to obtain $(p \land q) \equiv \neg(\neg p \lor \neg q)$. (See the laws of logic given on page 187.)

A set of logical connectives is **functionally complete** whenever all statements in propositional logic can be written in terms of the connectives in the set. The set of the three basic connectives AND, OR, and NOT is functionally complete. We can show that any set A of connectives is logically complete if we show that the three basic connectives can all be written in terms of the connectives in A. We have just shown that the set {NOT, OR} is functionally complete using that method.

PRACTICE PROBLEM 4 Use De Morgan's law to show that the set of connectives {NOT, AND} is functionally complete.

Our discussion of functionally complete sets applies to logic networks, to be studied in Chapter 7. Each logical connective has an analogous logic gate. As we will see in Chapter 7, all logic networks can be designed using only the three basic AND, OR, and NOT gates. Because {NOT, OR} is functionally complete, it is possible to design every logic network using only NOT gates and OR gates. Similarly, it is possible to design every logic network using only NOT gates and AND gates.

Can a single logic gate be designed so that all logic networks can be written in terms of that logic gate? In other words, is there a single logical connective such that all statements can be written in terms of that logical connective? We will see in the following example and the exercises that there are such connectives.

If the functionally complete set is a singleton set, we say that the connective in the set is functionally complete. Example 5.2.4 gives a functionally complete logical connective.

EXAMPLE 5.2.4 The NOR Connective   Define $p \downarrow q$ by the truth table given in Table 5.2.9. It is clear that $p \downarrow q \equiv \neg(p \lor q)$. The logical connective $\downarrow$ is called a **NOR** (not or) connective. To show that $\downarrow$ is functionally complete, we will write each of the three basic connectives in terms of $\downarrow$. Since {NOT, AND} is functionally complete, it is sufficient to write NOT and AND in terms of $\downarrow$.

TABLE 5.2.9

p	q	NOR $p \downarrow q$
T	T	F
T	F	F
F	T	F
F	F	T

a. $\neg p \equiv \neg(p \lor p) \equiv p \downarrow p$
b. $p \land q \equiv \neg(\neg p) \land \neg(\neg q) \equiv \neg(\neg p \lor \neg q) \equiv \neg p \downarrow \neg q \equiv (p \downarrow p) \downarrow (q \downarrow q)$ ∎

The logic gate for $p \downarrow q$ is called a NOR gate. So, any logic network can be written entirely in terms of NOR gates. There is also a **NAND** (not and)

PROGRAMMING NOTES

logical connective, written with the logic symbol ↑. You are asked to show that ↑ is functionally complete in Exercise Set 5.2 (see Exercise 23). The analogous logic gate is the NAND gate; any logic network can be designed with only NAND gates.

Propositional Logic in Pascal   Propositional logic is implemented in Pascal with the built-in data type **Boolean** and its operations. A variable of data type Boolean takes on only the values true and false. When the declaration

   var
      Flag:Boolean

is made, the variable Flag can have either the value true or the value false. The variable Flag can be assigned a specific value using an assignment such as:

   Flag := false

In Pascal, the type Boolean is a predefined, linearly ordered type with the following implicit definition:

   type Boolean = (false, true)

The Pascal ord function, which returns the ordinal position of a type value, yields ord(false) = 0 and ord(true) = 1.

Statement variables $p, q, r, \ldots$ are implemented in Pascal as Boolean variables. The three basic logical connectives $\neg$, $\wedge$, and $\vee$ are implemented in Pascal as the Boolean operations NOT, AND, and OR. What is the Pascal version of the implication connective?

Since Boolean is an enumerated type, all of the Pascal comparison operations (=, <> (not equal), >, <, >=, and <=) are defined for the Boolean type. Now consider the truth table for the Pascal inequality $p <= q$ restricted to the Boolean values true and false. See Table 5.2.10. Of course, this is exactly the same as the truth table for the implication connective →. Hence, the Pascal equivalent for the implication connective is the Pascal inequality <=.

Similarly, we see that the Pascal counterpart of the biconditional connective ↔ is the equality = restricted to the type Boolean.

It is a simple matter to write a Pascal program that determines the truth value of any meaningful combination of statements and connectives, no matter how complicated, provided that we supply the truth values of the statement variables occurring in the expression. For example,

   var p, q, r, value:Boolean;
     p := true;
     q := false;
     r := true;
     value := ((p AND q) OR (p AND NOT r)) <= NOT(r OR q)

can be used to evaluate the logical expression $[(p \wedge q) \vee (p \wedge \neg r)] \rightarrow \neg(r \vee q)$ for one particular combination of truth values for p, q, and r.

TABLE 5.2.10

p	q	$p <= q$
True	True	True
True	False	False
False	True	True
False	False	True

5.2 LOGICAL EQUIVALENCE AND TAUTOLOGIES

A more challenging problem is to write a program that will determine and print the truth table for a given expression of propositional logic. Another challenging problem is to write a computer program that will determine whether a given statement is a tautology and, if not, print a counterexample. (See Exercises 29 and 30, Exercise Set 5.2.)

EXERCISE SET 5.2

1. Compute the appropriate truth tables to verify each of the following.
   a. $q \to p$ is not logically equivalent to $p \to q$
   b. $\neg q \to \neg p$ is logically equivalent to $p \to q$

   In Exercises 2–9 verify each of the logical equivalences by calculating the appropriate truth tables.

2. $\neg(\neg p) \equiv p$
3. $\neg(p \wedge q) \equiv \neg p \vee \neg q$
4. $p \vee (q \wedge r) \equiv (p \vee q) \wedge (p \vee r)$
5. $p \vee \neg p \equiv I$
6. $p \wedge \neg p \equiv O$
7. $(p \wedge r \to q) \equiv [p \to (r \to q)]$
8. $(p \wedge \neg q \to O) \equiv (p \to q)$
9. $(p \vee q \to r) \equiv (p \to r) \wedge (q \to r)$

   Verify that each statement given in Exercises 10–17 is a tautology by constructing the appropriate truth table.

10. $[(p \to q) \wedge (q \to r)] \to (p \to r)$
11. $(p \to q) \to [(p \vee r) \to (q \vee r)]$
12. $(p \to q) \to [(p \wedge r) \to (q \wedge r)]$
13. $(p \to q) \to [(q \to r) \to (p \to r)]$
14. $[(p \to q) \wedge (r \to s)] \to [(p \vee r) \to (q \vee s)]$
15. $[(p \to q) \wedge (r \to s)] \to [(p \wedge r) \to (q \wedge s)]$
16. $[(p \to q) \wedge (r \to s)] \to [(\neg q \vee \neg s) \to (\neg p \vee \neg r)]$
17. $[(p \to q) \wedge (r \to s)] \to [(\neg q \wedge \neg s) \to (\neg p \wedge \neg r)]$

    In Exercises 18b and 19b you may use the following **substitution principle** of propositional logic. If $r \equiv s$, then r may be substituted for s, at any occurrence of s, in a statement P to obtain another statement Q that is logically equivalent to P. For example, $(p \wedge q) \vee q$ is logically equivalent to $(p \wedge q) \vee \neg(\neg q)$ since $\neg(\neg q) \equiv q$.

18. Verify $\neg(p \to q) \equiv p \wedge \neg q$ in two ways.
    a. Calculate the appropriate truth tables.
    b. Use the rule of double negation, the OR form of an implication, and one of De Morgan's laws.

19. Verify that $p \wedge (\neg p \vee q) \to q$ is a tautology in two ways.
    a. Calculate the appropriate truth table.
    b. Use the OR form of an implication and modus ponens.

20. Neither of the following is a tautology. Find a counterexample in each case.
    a. $p \vee \neg q \vee (q \to p)$ b. $[(p \wedge q) \vee r] \to [(p \vee q) \wedge r]$

21. Neither of the following is a tautology. Find a counterexample in each case.
    a. $(p \wedge q \to r) \leftrightarrow [(p \to r) \wedge (q \to r)]$ b. $[(p \wedge q) \vee r] \to [p \wedge (q \vee r)]$

22. Show that $p \equiv q$ is true if and only if $p \leftrightarrow q$ is a tautology.

TABLE 5.2.11

p	q	$p \uparrow q$
T	T	F
T	F	T
F	T	T
F	F	T

23. Define $p \uparrow q$ by the truth table given in Table 5.2.11. It is clear that $p \uparrow q \equiv \neg(p \wedge q)$. The logic gate for $p \uparrow q$ is called a NAND (not and) gate. We call $\uparrow$ the **Sheffer stroke**. Prove that the Sheffer stroke is functionally complete by showing each of the following.
    a. $p \uparrow p \equiv \neg p$ **b.** $(p \uparrow p) \uparrow (q \uparrow q) \equiv p \vee q$ **c.** $\neg(\neg p \vee \neg q) \equiv p \wedge q$

24. The exclusive OR connective $\underline{\vee}$ was defined in Section 5.1. It is clear from the truth table defining $\underline{\vee}$ that $p \underline{\vee} q \equiv (p \wedge \neg q) \vee (\neg p \wedge q)$. Show that each of the following holds.
    a. $p \underline{\vee} p \equiv O$ **b.** $p \underline{\vee} \neg p \equiv I$ **c.** $\neg(p \underline{\vee} q) \equiv \neg p \underline{\vee} q \equiv p \underline{\vee} \neg q$

25. Refer to the exclusive OR connective. Show that each of the following holds.
    a. $p \underline{\vee} q \equiv \neg(p \leftrightarrow q)$ **b.** $p \underline{\vee} (q \underline{\vee} r) \equiv (p \underline{\vee} q) \underline{\vee} r$

    Exercises 26–28 involve the If...then...else connective. Recall that If...then...else is a Boolean control statement in Pascal and other programming languages. But If...then...else is also a connective in propositional logic.

    If...then...else is a 3-ary connective and is defined by:

    1. When p is true: If p then q else r has the same truth value as q.
    2. When p is false: If p then q else r has the same truth value as r.

26. Give the truth table for the If...then...else connective.

27. Prove:
    a. (If p then $\neg p$ else p) $\equiv O$ **b.** (If p then q else I) $\equiv (p \rightarrow q)$

28. Prove:
    a. (If p then q else O) $\equiv p \wedge q$ **b.** (If p then O else I) $\equiv \neg p$
    c. (If $\neg p$ then q else I) $\equiv p \vee q$
    By proving these three facts you have proved that the connective If...then...else is functionally complete.

■ PROGRAMMING EXERCISES

29. Write a computer program that will construct the truth table for
    $$(p \rightarrow \neg q) \wedge (q \vee r) \rightarrow (p \vee r)$$

30. Write a computer program that will determine whether a given statement is a tautology and, if the statement is not a tautology, the program will print a counterexample.

5.3

PROOF TECHNIQUES

In previous chapters, we have delayed the proofs of some facts until this chapter. Before presenting those proofs, and many others, we introduce the most basic and useful proof techniques. First we discuss what constitutes a valid mathematical argument in propositional logic.

A mathematical argument is a set of statements called premises together with a statement called the conclusion. The goal of a mathematical argument is to deduce logically the conclusion from the premises. Premises are statements assumed to be true in the context of the argument. They can include tautologies, definitions, axioms, previously proved theorems, and statements assumed to be true for the given argument. An argument has a finite number of premises.

Recall that a tautology is a statement that is always true. An argument with three premises P_1, P_2, and P_3 and a conclusion q is a **valid argument** in propositional logic if the implication $(P_1 \land P_2 \land P_3) \to q$ is a tautology. Whenever this is the case, q is said to **follow logically** from the premises P_1, P_2, and P_3. From the truth table for an implication and the definition of tautology, we see that the conclusion of a valid argument must be true whenever all the premises are true.

An argument with two premises P_1 and P_2 and a conclusion q is an **invalid argument** in propositional logic if the implication $(P_1 \land P_2) \to q$ is not a tautology. So, a counterexample for an invalid argument with premises P_1 and P_2 and conclusion q is any combination of truth values that makes P_1 true, P_2 true, and q false.

The definitions given above for valid and invalid arguments generalize to arguments with any finite set of premises $P_1, P_2, \ldots, P_n$.

A **proof** is a step-by-step demonstration that a given argument is valid. Mathematicians use logical rules of procedure based on known tautologies to construct proofs. In what follows we analyze several important proof techniques. A few rules of procedure will be made explicit and their use illustrated by examples. The goal is to understand how and when the rules are applied.

We reproduce here for easy reference several important laws of logic stated earlier (in Section 5.2). Each of them will be useful for analyzing and constructing proofs.

LAWS OF LOGIC FOR PROOF TECHNIQUES

a. $\neg(\neg p) \equiv p$ — Rule of double negation
b. $(p \to q) \equiv (\neg p \lor q)$ — OR form of an implication
c. $(p \to q) \equiv (\neg q \to \neg p)$ — Contrapositive of an implication
d. $\neg(p \lor q) \equiv (\neg p \land \neg q)$ — De Morgan's law
e. $\neg(p \land q) \equiv (\neg p \lor \neg q)$ — De Morgan's law
f. $(p \land r \to q) \equiv [r \to (p \to q)]$ — Rule for direct proof
g. $(p \land \neg q \to O) \equiv (p \to q)$ — Rule for proof by contradiction
h. $(p \lor r \to q) \equiv [(p \to q) \land (r \to q)]$ — Rule for proof by cases
i. $[p \land (p \to q) \to q] \equiv I$ — Modus ponens

The rule for proof by cases (Law h) was not stated earlier, but it will be useful in what follows.

For many arguments the conclusion is itself an implication. The following is a rule of procedure for the **direct proof** of an implication.

DIRECT PROOF

To prove an implication $p \to q$, assume p and use p along with other known true statements to deduce q. This is based on the tautology

$(p \land r \to q) \equiv [r \to (p \to q)]$

where r stands for the collection of known true statements.

The first example illustrates a direct proof of an implication in propositional logic.

EXAMPLE 5.3.1   Premises $\neg(p \wedge n)$, $\neg q \to n$
Prove $p \to q$

Proof
1. p Assumption
2. $\neg(p \wedge n)$ Premise
3. $\neg p \vee \neg n$ Step 2 and De Morgan's law
4. $p \to \neg n$ Step 3, OR form, and double negation
5. $\neg n$ Steps 1 and 4 and modus ponens
6. $\neg q \to n$ Premise
7. $\neg n \to q$ Step 6, contrapositive, and double negation
8. q Steps 5 and 7 and modus ponens

Let r stand for the two given premises and the tautologies used in the proof. The form of the proof is $p \wedge r \to q$. Hence, $p \to q$ follows logically from r by the rule of direct proof. ∎

In steps 3 and 7 above, we substituted a statement that is logically equivalent to the statement in the previous step. This is justified by the **substitution principle** for propositional logic. If $r \equiv s$, then r may be substituted for s, at *any* occurrence of s, in a statement P to obtain another statement Q that is logically equivalent to P. This substitution principle was used in Exercise Set 5.2.

We cannot write the proofs of most mathematical statements so that they consist of only premises and tautologies. So the rules of procedure for propositional logic, such as the direct proof rule, are not sufficient to justify most mathematical proofs. Nevertheless, the tautologies and rules of procedure we discuss provide valuable insight into the logical format of mathematical proofs. For each mathematical argument there is a rule of procedure that is a paradigm for the logical format of the proof.

The next example illustrates how the rule of direct proof, used as a model, can illuminate the logical format of a proof.

EXAMPLE 5.3.2   Prove that the square of every even integer is even. In other words, prove: If x is an even integer, then x^2 is also an even integer. First, note that an integer m is even if $m = 2j$ for some integer j.

Proof
1. Let x be an even integer.   Assumption
2. $x = 2k$, for some integer k Step 1, definition of even
3. $x^2 = (2k)(2k)$ Step 2, rule of algebra
4. $x^2 = 2(2kk)$ Step 3, rule of algebra
5. Hence, x^2 is an even integer.   Step 4, $2kk$ is an integer

Let r stand for the reasons given in steps 2–5, p stand for "x is an even integer," and q stand for "x^2 is an even integer." The form of the proof is $(p \wedge r) \to q$. Hence, $p \to q$ follows logically from r. ∎

It is important to note that a proof of an implication $p \to q$ is not a proof of q. For example, we cannot conclude from Example 5.3.2 that the square of any integer is even. Indeed, that statement is false since $3^2 = 9$ and 9 is not even.

For some statements it is easier to give an **indirect proof**. An indirect proof can take several forms. The most useful is a **proof by contradiction**.

PROOF BY CONTRADICTION

To prove an implication $p \to q$, assume p and assume the negation of q, then deduce any contradiction. This is based on the tautology

$$(p \wedge \neg q \to O) \equiv (p \to q)$$

where O stands for any contradiction.

In the proof that follows, we assume the following facts of arithmetic: An integer is either even or odd; if an integer is not odd, then it is even, and conversely.

EXAMPLE 5.3.3 Prove that, if the square of an integer is odd, then the integer is also odd. Here is a proof by contradiction.

Proof
1. Let x^2 be odd and x not be odd. Assumption
2. x is even. Fact of arithmetic
3. x^2 is even. By Example 5.3.2
4. x^2 is not odd. Fact of arithmetic
5. x^2 is odd and not odd. Steps 1 and 4, contradiction

Therefore, if x^2 is odd, then x is also odd. ∎

So far we have written proofs in a formal step-by-step fashion. From here on we employ a more commonly used informal style.

In what follows we will need the basic **additive rule for inequalities** of real numbers: $(a \leq c$ and $b \leq d) \to a + b \leq c + d$.

EXAMPLE 5.3.4 Let $m, n \in \mathbb{N}$. Prove: If $m + n < 89$, then $m < 45$ or $n < 45$. Here is a proof by contradiction.

Proof Assume $m + n < 89$ and suppose it is not the case that $m < 45$ or $n < 45$. By De Morgan's law, $m \geq 45$ and $n \geq 45$. Hence, $m + n \geq 90$ by the additive rule of inequalities. This contradicts our hypothesis, so the proof is complete. ∎

Another form of indirect proof is a **proof by contrapositive**.

> **PROOF BY CONTRAPOSITIVE**
> Proof of an implication $p \to q$ is equivalent to a proof that $\neg q \to \neg p$. This is based on the tautology
> $$(p \to q) \equiv (\neg q \to \neg p)$$

EXAMPLE 5.3.5 Let $m, n \in \mathbb{N}$. Prove: If $m + n > 50$, then $m > 25$ or $n > 25$. The contrapositive of this statement is: If it is not the case that $m > 25$ or $n > 25$, then it is not the case that $m + n > 50$. By De Morgan's law, we must prove: If $m \leq 25$ and $n \leq 25$, then $m + n \leq 50$. But this follows from the additive rule for inequalities. ∎

Proofs by contrapositive and by contradiction for an implication $p \to q$ are very similar, and the choice of one over the other is frequently a matter of style. A proof by contradiction is often easier, since more is assumed to be true; we are able to assume both $\neg q$ and p. On the other hand, a proof by contradiction is likely to be less elegant than one using the contrapositive. In any case, a direct proof is usually easier to understand and should be chosen over an indirect proof whenever possible.

PRACTICE PROBLEM 1 Let $m, n \in \mathbb{N}$. Prove: If $mn < 100$, then $m < 10$ or $n < 10$.

a. Use proof by contradiction.
b. Use proof by contrapositive.

[*Hint:* You may assume that, if $a \leq b$, then $ac \leq bc$ for $a, b, c \in \mathbb{N}$.]

Many statements needing proof are biconditionals, or "if and only if" statements. Because $(p \leftrightarrow q) \equiv [(p \to q) \wedge (q \to p)]$, we need to prove two implications in order to prove a statement of the form $p \leftrightarrow q$. Recall from Section 1.5 that, in proofs of $p \leftrightarrow q$, the proof of $p \to q$ is labeled ($\to$) whereas the proof of $q \to p$ is labeled ($\leftarrow$).

EXAMPLE 5.3.6 Prove: An integer is even if and only if its square is even.

Proof ($\to$) This part was proved in Example 5.3.2.

($\leftarrow$) We do a proof by contradiction. Suppose x is an integer with x^2 even but x is not even. Since x is not even, $x = 2k + 1$ for some integer k. Hence, $x^2 = (2k + 1)^2 = 2(2k^2 + 2k) + 1$. So x^2 is not even. We have a contradiction. ∎

PRACTICE PROBLEM 2 Prove: The product of two integers is even if and only if at least one of the integers is even. [*Hint:* For ($\to$), do a proof by contradiction. Suppose x

and y are two integers with xy even, x odd, and y odd. For ($\leftarrow$), do a direct proof. Assume either one of the integers is even.]

The following is a rule of procedure for **proof by cases** of an implication.

PROOF BY CASES To prove the statement $p \vee r \rightarrow q$, it is necessary and sufficient to consider two cases: Prove q logically follows from p, and prove q also logically follows from r. This proof technique is based on the law of logic $(p \vee r \rightarrow q) \equiv [(p \rightarrow q) \wedge (r \rightarrow q)]$.

EXAMPLE 5.3.7 Prove that $n^2 + n$ is even for all positive integers n.

Proof First note that $n^2 + n = n(n + 1)$. By Practice Problem 2, the product of two integers is even if at least one of the integers is even.

Case 1. Assume n is even. Then we are done.
Case 2. Assume n is odd. Then $n + 1$ is even and again we are done. ∎

In Section 1.2 several theorems of set theory were stated and the proofs delayed until this chapter. Before giving some of those proofs, we state the most important method of proof for set theory.

Following is the **pick-a-point method** for proving $A \subseteq B$:

PROOF BY THE PICK-A-POINT METHOD

Take an arbitrary element u in the universal set U and assume $u \in A$. Then use any known true statements, including properties of A and B, to prove $u \in B$.

This method of proof is based on the definition: $A \subseteq B$ if $u \in A$ implies $u \in B$ for all $u \in U$. The pick-a-point method will be discussed again in Section 5.5 when we discuss predicate logic.

EXAMPLE 5.3.8 Following are proofs of some statements from Chapter 1 using the pick-a-point method.

a. Prove: $A \cap B \subseteq A$.
 Proof Assume $u \in A \cap B$. Then $u \in A$ and $u \in B$, by definition of $A \cap B$. Hence, $u \in A$, since $p \wedge q \rightarrow p$ is a tautology. Done!
b. Prove: $\emptyset \subseteq A$.
 Proof We do a proof by contradiction. Suppose $\emptyset \nsubseteq A$. Hence, there is an element u such that $u \in \emptyset$ and $u \notin A$, by definition of $\subseteq$. But $u \notin \emptyset$, since $\emptyset$ is the empty set. Contradiction!
c. Prove: If $A \subseteq B$, then $B' \subseteq A'$.

Proof Assume $A \subseteq B$. To prove $B' \subseteq A'$ using the pick-a-point method, we let $u \in B'$. Hence, $u \notin B$, by definition of B'. Since $A \subseteq B$, $u \in A$ implies $u \in B$. Hence, $u \notin A$, since $\neg q \wedge (p \rightarrow q) \rightarrow \neg p$ is a tautology. Thus, $u \in A'$, and we have proved $B' \subseteq A'$. ∎

Every Boolean law for set theory can be proved using the corresponding Boolean law of logic and the pick-a-point method. The Boolean laws of logic in Section 5.2 and the Boolean laws for set theory in Section 1.2 are restated here for easy reference.

BOOLEAN LAWS OF LOGIC

1. **a.** $p \vee q \equiv q \vee p$ **b.** $p \wedge q \equiv q \wedge p$
2. **a.** $(p \vee q) \vee r \equiv p \vee (q \vee r)$ **b.** $(p \wedge q) \wedge r \equiv p \wedge (q \wedge r)$
3. **a.** $p \vee (q \wedge r) \equiv (p \vee q) \wedge (p \vee r)$ **b.** $p \wedge (q \vee r) \equiv (p \wedge q) \vee (p \wedge r)$
4. **a.** $p \vee O \equiv p$ **b.** $p \wedge I \equiv p$
5. **a.** $p \vee \neg p \equiv I$ **b.** $p \wedge \neg p \equiv O$

BOOLEAN LAWS FOR SET THEORY

1. **a.** $A \cup B = B \cup A$ **b.** $A \cap B = B \cap A$
2. **a.** $(A \cup B) \cup C = A \cup (B \cup C)$ **b.** $(A \cap B) \cap C = A \cap (B \cap C)$
3. **a.** $A \cup (B \cap C) = (A \cup B) \cap (A \cup C)$ **b.** $A \cap (B \cup C) = (A \cap B) \cup (A \cap C)$
4. **a.** $A \cup \emptyset = A$ **b.** $A \cap U = A$
5. **a.** $A \cup A' = U$ **b.** $A \cap A' = \emptyset$

Recall that, for any two sets A and B, $A = B$ if and only if A and B have exactly the same elements. Hence, to prove $A = B$, we prove both $A \subseteq B$ and $B \subseteq A$.

EXAMPLE 5.3.9 Here are proofs of some Boolean laws for set theory. The reasons given are from the list of Boolean laws of logic.

a. Prove: $A \cup \emptyset = A$.
Proof Let u be an arbitrary element of the universal set U. We must prove that $u \in A \cup \emptyset$ if and only if $u \in A$.

($\rightarrow$) Assume $u \in A \cup \emptyset$. Then, $u \in A$ or $u \in \emptyset$. By Boolean law of logic 4a, $u \in A$ or $u \in \emptyset$ if and only if $u \in A$. Therefore, $u \in A$.

($\leftarrow$) The converse is proved by reversing the steps above.

Hence, $u \in A \cup \emptyset$ if and only if $u \in A$.

b. Prove: $A \cup (B \cap C) = (A \cup B) \cap (A \cup C)$.
Proof Let u be an arbitrary element of the universal set U. We must prove that $u \in A \cup (B \cap C)$ if and only if $u \in (A \cup B) \cap (A \cup C)$.

($\rightarrow$) Assume $u \in A \cup (B \cap C)$. Then, $u \in A$ or ($u \in B$ and $u \in C$) by definition of $\cup$ and $\cap$. By Boolean law of logic 3a, $u \in A$ or ($u \in B$ and $u \in C$) if and only if ($u \in A$ or $u \in B$) and ($u \in A$ or $u \in C$). Hence, $u \in (A \cup B) \cap (A \cup C)$.

($\leftarrow$) The converse is proved by reversing the steps above. ∎

Proofs of other statements from Section 1.2 are left as exercises. See Exercise Set 5.3.

It is also possible to prove statements about sets using the Boolean laws for set theory directly, without invoking a pick-a-point proof.

EXAMPLE 5.3.10 Prove $(A - D) \cup (B - D) = (A \cup B) - D$.

Proof
$$(A - D) \cup (B - D) = (A \cap D') \cup (B \cap D')$$
$$= (D' \cap A) \cup (D' \cap B)$$
$$= D' \cap (A \cup B)$$
$$= (A \cup B) \cap D'$$
$$= (A \cup B) - D \quad ∎$$

PRACTICE PROBLEM 3 Supply a reason for each step of the proof in Example 5.3.10.

In Sections 3.3 and 3.4 we discussed functions. Recall that a function is bijective if and only if it is both 1–1 and onto. The definitions of both a 1–1 function and an onto function are given in Section 3.4. Proofs of three theorems first stated in Chapter 3 are included in the next example.

EXAMPLE 5.3.11 **a.** Prove: If $f:D \to C$ has an inverse f^{-1}, then f is bijective.
Proof Assume $f:D \to C$ has an inverse $f^{-1}:C \to D$. Recall that $f \circ f^{-1} = 1_C$ and $f^{-1} \circ f = 1_D$. To prove f is 1–1, we show that $f(x) = f(z)$ implies $x = z$. Assume $f(x) = f(z)$. Hence, $f^{-1}(f(x)) = f^{-1}(f(z))$ since f^{-1} is a function. So, $(f^{-1} \circ f)(x) = (f^{-1} \circ f)(z)$ by definition of composition. Since $f^{-1} \circ f = 1_D$, we have $x = z$. Therefore, f is 1–1.

To prove f is onto, we show that, for any $y \in C$, there is an element $w \in D$ such that $f(w) = y$. Assume $y \in C$. Since $f \circ f^{-1} = 1_C$, we have $(f \circ f^{-1})(y) = y$. By definition of composition, $f(f^{-1}(y)) = y$. But $f^{-1}(y) \in D$, since $f^{-1}:C \to D$. Hence, $y = f(f^{-1}(y))$, and $f^{-1}(y) \in D$. Therefore, f is onto.

b. Prove: If $g:D \to A$ is onto and $f:A \to C$ is onto, then $f \circ g:D \to C$ is onto.
Proof Assume $y \in C$. Then, $y = f(z)$ for some $z \in A$, since f is onto. Hence, $z = g(x)$ for some $x \in D$, since g is onto. By definition of $f \circ g$, we have $y = f(g(x)) = (f \circ g)(x)$ for some $x \in D$. Therefore, $f \circ g$ is onto.

c. Prove: If $g:D \to A$ is 1–1 and $f:A \to C$ is 1–1, then $f \circ g:D \to C$ is 1–1.
Proof Assume $(f \circ g)(x) = (f \circ g)(z)$. Then, $f(g(x)) = f(g(z))$. Since f is 1–1, we have $g(x) = g(z)$. Since g is 1–1, we have $x = z$. Therefore, $f \circ g$ is 1–1. ∎

See Exercise Set 5.3 for more proofs of theorems pertaining to functions.

In Exercise 12, Exercise Set 3.1, we defined a relation $\equiv \pmod{d}$ on $\mathbb{Z}$ as follows: Assume d is a positive integer. Then $x \equiv y \pmod{d}$ whenever $x - y$ is a multiple of d. We now prove $\equiv \pmod{d}$ is an equivalence relation.

THEOREM 5.3.12 Let d be a fixed positive integer. The relation $\equiv \pmod{d}$ is an equivalence relation.

Proof The relation $\equiv \pmod{d}$ is reflexive because $x - x = 0 \cdot d$, and hence, $x \equiv x \pmod{d}$.

To prove $\equiv \pmod{d}$ is symmetric, we assume $x \equiv y \pmod{d}$. So there is an integer k such that $x - y = kd$. Hence, $y - x = (-k)d$. Therefore, $y - x$ is a multiple of d, and $y \equiv x \pmod{d}$.

To prove $\equiv \pmod{d}$ is transitive, we assume $x \equiv y \pmod{d}$ and $y \equiv z \pmod{d}$. Hence, $x - y = jd$ and $y - z = hd$ for some integers j and h. Adding, we obtain $x - z = (x - y) + (y - z) = (j + h)d$. Therefore, $x - z$ is a multiple of d, and $x \equiv z \pmod{d}$. ∎

We asserted in Section 3.1 that the equivalence classes for an equivalence relation on a set partitioned the set. We gave a partial proof and omitted the following theorem needed to complete the proof. Recall that we denote an equivalence class containing $x \in A$ by $[x] = \{z : z \in A \land z R x\}$.

THEOREM 5.3.13 Let R be an equivalence relation on the set A. For all a, b in A, $[a] = [b]$ if and only if $a R b$.

Proof ($\rightarrow$) Assume $[a] = [b]$. Then, $a \in [b]$ since $a \in [a]$. Hence, $a R b$.

($\leftarrow$) Assume $a R b$ and prove $[a] = [b]$. We will prove $[a] \subseteq [b]$ using the pick-a-point method. Let $x \in [a]$. Then $x R a$. So $x R a$ and $a R b$. Hence, $x R b$ since R is transitive. Therefore, $x \in [b]$. The proof that $[b] \subseteq [a]$ is similar. Therefore, $[a] = [b]$. ∎

EXERCISE SET 5.3

In Exercises 1–6 prove each of the given theorems of propositional logic.

1. Premises: $p \rightarrow \neg q, s \lor q$; conclusion: $p \rightarrow s$
2. Premises: $p \rightarrow m, m \rightarrow t, m$; conclusion: $t \lor \neg p$
3. Premises: $p \rightarrow q, \neg p \land r \rightarrow s, \neg q$; conclusion: $s \lor \neg r$
4. Premises: $p \rightarrow q, \neg p \land r \rightarrow s, \neg q$; conclusion: $p \lor r \rightarrow s$
5. Premises: $p \rightarrow \neg n, \neg p \rightarrow \neg s, \neg s \rightarrow q, r \rightarrow n$; conclusion: $r \rightarrow \neg s$
6. Premises: $p \rightarrow \neg n, \neg p \rightarrow \neg s, s, r \rightarrow n$; conclusion: $\neg r$

7. Prove each of the following with a direct proof.
   a. If x and y are even integers, then xy is even.
   b. If x and y are odd integers, then $x + y$ is even.

8. Prove each of the following using a proof by contradiction.
   a. If xy is odd, then x and y are both odd integers.
   b. If $x + y$ is even and x is odd, then y is odd.
9. Prove each of the following theorems of set theory.
   a. $B \subseteq A \cup B$ b. $(A')' = A$ c. Boolean law 1b for set theory
10. Prove each of the following theorems of set theory.
    a. Boolean law 3b for set theory   b. Boolean law 4b for set theory
    c. Boolean law 5b for set theory
11. Prove the following facts about Cartesian products of sets.
    a. $(A \cup B) \times C = (A \times C) \cup (B \times C)$ b. $(A \cap B) \times C = (A \times C) \cap (B \times C)$
12. Prove the following facts about Cartesian products of sets.
    a. $A \times (B \cup C) = (A \times B) \cup (A \times C)$ b. $A \times (B \cap C) = (A \times B) \cap (A \times C)$
13. a. Prove $(A \times B) \cup (C \times D) \subseteq (A \cup C) \times (B \cup D)$.
    b. Find an example with $(A \times B) \cup (C \times D) \neq (A \cup C) \times (B \cup D)$.
14. Let $f : D \to C$ be a function.
    a. Prove $f \circ 1_D = f$.   b. Prove $1_C \circ f = f$.
15. Prove: If $f : D \to C$ is a bijective function, then f^{-1} is a bijective function.
16. Prove that the inverse of a bijective function is unique. That is, prove: If $f : D \to C$ is a bijective function and $f \circ g = 1_C$ and $g \circ f = 1_D$, then $g = f^{-1}$.
17. a. Use a pick-a-point proof to prove $D - (A \cap B) = (D - A) \cup (D - B)$. By substituting $D = U$ in the result above we get the corollary $(A \cap B)' = A' \cup B'$.
    b. Use $(A \cap B)' = A' \cup B'$ and the result $(C')' = C$ in Exercise 9b to prove that $D - (A \cup B) = (D - A) \cap (D - B)$. Of course, this law has the corollary $(A \cup B)' = A' \cap B'$.
    The four laws proved here are called **De Morgan's laws for sets**.
18. Prove, without using the pick-a-point method, that $(A \cap B) \cup (A - B) = A$.
19. Prove, without using the pick-a-point method, that $(A \cap B) \cup (C - A) \cup (B \cap C) = (A \cap B) \cup (C - A)$.
20. Assume $g : D \to A$ and $f : A \to C$ are functions. In Example 5.3.11b we proved: If $g : D \to A$ is onto and $f : A \to C$ is onto, then $f \circ g : D \to C$ is onto. The converse of this theorem is: If $f \circ g : D \to C$ is onto, then $g : D \to A$ and $f : A \to C$ are both onto. The following counterexample shows that the converse is false. Let $D = \{1\} = C$ and $A = \{a, b\}$. Define $g : D \to A$ by $g(1) = a$ and $f : A \to C$ by $f(a) = 1 = f(b)$. *Note:* In the example, $f \circ g$ is both onto and 1–1, whereas g is not onto but is 1–1 and f is onto but not 1–1. We also proved, in Example 5.3.11c that: If $g : D \to A$ is 1–1 and $f : A \to C$ is 1–1, then $f \circ g : D \to C$ is 1–1. Our example is also a counterexample to the converse of this theorem. However, it is possible to prove a partial converse to each of these theorems.
    a. Prove: If $f \circ g$ is onto, then f is onto.
    b. Prove: If $f \circ g$ is 1–1, then g is 1–1.
21. Let $f : A \to A$ be a function. Recall that any function is also a relation. In this case, $f = \{(x, y) : y = f(x)\}$ is a relation on A. Prove that the only function $f : A \to A$ that is also an equivalence relation is the identity function 1_A.
22. Let R be a symmetric and transitive binary relation on a set A. Show that, if for every $x \in A$ there exists $y \in A$ such that (x, y) is in R, then R is an equivalence relation on A.

23. Prove that the following statements are logically equivalent.
   i. $A \subseteq B$ ii. $A' \cup B = U$ iii. $A \cap B' = \emptyset$

24. Prove: If $m = ab$, where a and b are positive integers, then $a \leq \sqrt{m}$ or $b \leq \sqrt{m}$.

25. Suppose that the circumference of a circular wheel is divided into ten sectors. The digits $0, 1, 2, \ldots, 9$ are randomly assigned to the ten sectors. Prove that there are two consecutive sectors such that the sum of the digits from those sectors is at least 8.

26. A microcomputer is used for 90 hours over a period of 14 days. Prove that for two consecutive days the microcomputer was used for at least 12 hours.

5.4
INTRODUCTION TO MATHEMATICAL INDUCTION

If consecutive odd integers starting with 1 are added, a nice pattern emerges. We get

$(n = 1) \quad 1 = 1$
$(n = 2) \quad 1 + 3 = 4$
$(n = 3) \quad 1 + 3 + 5 = 9$
$(n = 4) \quad 1 + 3 + 5 + 7 = 16$
$\quad \vdots \qquad\qquad \vdots$

where n is the number of odd integers to be added. It appears that the sum of the first n odd integers is always equal to the square of n. How do we prove such a statement? Verifying an infinite sequence of statements is out of the question. The **principle of mathematical induction** is useful in such cases.

PRINCIPLE OF MATHEMATICAL INDUCTION

To prove a statement consisting of an infinite sequence of statements $p(1), p(2), p(3), \ldots$, apply the following steps:

Basis Step   Prove $p(1)$.
Induction Step   Prove the implication $p(k) \rightarrow p(k + 1)$ for $k = 1, 2, 3, \ldots$.

The statement $p(k)$ in the induction step is called the **induction hypothesis**.

It may look like nothing is gained by invoking mathematical induction, since the induction step is also an infinite sequence of statements. But in practice the induction hypothesis is assumed to be true for an arbitrary integer k and a proof of $p(k + 1)$ is given. Because the integer k is arbitrary and the proof is one that would hold for any k, the induction step is done only once.

5.4 INTRODUCTION TO MATHEMATICAL INDUCTION

So mathematical induction reduces the original infinite number of verifications to two proofs. How does this work?

EXAMPLE 5.4.1 Consider an infinite loop in a computer program that outputs numbers in the following pattern:

1. The first number output is 1.
2. If the kth number is k^2, then the $(k+1)$th number is $(k+1)^2$.

The output must be:

$(n=1)$	1	By step 1
$(n=2)$	4	By step 2 and $n=1$ output
$(n=3)$	9	By step 2 and $n=2$ output
$(n=4)$	16	By step 2 and $n=3$ output
$\vdots$	$\vdots$	$\vdots$

Conjecture: On the nth level of the loop, output must be n^2. Because of the principle of mathematical induction, this conjecture must be true. We are ignoring the possibility of an overflow resulting when n^2 exceeds the capacity of the computer to represent integers. ∎

Following is the proof of a mathematical statement, and the reasoning is identical to that in Example 5.4.1.

EXAMPLE 5.4.2 Prove that the sum of the first n odd integers is equal to the square of n for $n = 1, 2, 3, \ldots$.

Proof We use mathematical induction.

Basis Step The sum of the first (one) odd integer is 1, which is 1^2.
Induction Step Induction hypothesis: Assume $1 + 3 + 5 + \cdots + (2k-1) = k^2$ and prove

$$1 + 3 + 5 + \cdots + (2k-1) + (2k+1) = (k+1)^2$$

Proof Assume $1 + 3 + 5 + \cdots + (2k-1) = k^2$. Then

$$1 + 3 + 5 + \cdots + (2k-1) + (2k+1) = k^2 + (2k+1) = (k+1)^2$$

Hence, $1 + 3 + 5 + \cdots + (2k-1) + (2k+1) = (k+1)^2$.

Therefore, $1 + 3 + 5 + \cdots + (2n-1) = n^2$ for $n = 1, 2, 3, \ldots$ by the principle of mathematical induction. ∎

Usually the induction step is a direct proof, as in Example 5.4.2. Here is another example of a proof using mathematical induction.

EXAMPLE 5.4.3 Prove $2^{n-1} \leq 2^n - 1$ for $n = 1, 2, 3, \ldots$.

Proof We use mathematical induction again.

Basis Step $2^{1-1} = 2^0 = 1 \leq 2 - 1 = 2^1 - 1$
Induction Step The induction hypothesis is $2^{k-1} \leq 2^k - 1$. Hence, $2^k = 2(2^{k-1}) \leq 2(2^k - 1) \leq 2^{k+1} - 2 \leq 2^{k+1} - 1$

Therefore, $2^{n-1} \leq 2^n - 1$ for $n = 1, 2, 3, \ldots$ by the principle of mathematical induction. ∎

PRACTICE PROBLEM 1 Prove $1 + 2 + 2^2 + \cdots + 2^{n-1} = 2^n - 1$ for $n = 1, 2, \ldots$. [*Hint:* The induction hypothesis is $1 + 2 + 2^2 + \cdots + 2^{k-1} = 2^k - 1$. For the induction step, add 2^k to both sides of the equality.]

Example 5.4.4 gives the proof of a theorem that was used when we developed transitive closure in Chapter 2.

EXAMPLE 5.4.4 Prove: If R is a transitive binary relation on a set A, then $R^n \subseteq R$ for $n = 1, 2, 3, \ldots$.

Proof (By mathematical induction.) First recall that $R^1 = R$ and $R^{k+1} = R \circ R^k$.

Basis Step $(n = 1)$ $R^1 \subseteq R$ since $R^1 = R$.
Induction Step Induction hypothesis: Assume $R^k \subseteq R$. We will prove $R^{k+1} \subseteq R$ by the pick-a-point method. Assume $(u, v) \in R^{k+1}$. Then, $(u, v) \in R \circ R^k$ since $R^{k+1} = R \circ R^k$. Hence, $(u, z) \in R$ and $(z, v) \in R^k$ for some $z \in A$, by definition of $R \circ R^k$. By the induction hypothesis, $(z, v) \in R$. So, $(u, z) \in R$ and $(z, v) \in R$. Therefore, $(u, v) \in R$ since R is transitive. ∎

Intersection and Union of a Finite Number of Sets

We recall the definitions of the intersection and union of a finite number of sets given in Section 1.1. Let A_i for $i = 1, 2, 3, \ldots, n$ be a finite collection of sets. Then,

$A_1 \cap A_2 \cap \cdots \cap A_n = \{x : x \in A_i \text{ for all } i = 1, 2, 3, \ldots, n\}$
$A_1 \cup A_2 \cup \cdots \cup A_n = \{x : x \in A_i \text{ for at least one } i = 1, 2, 3, \ldots, n\}$

These definitions are natural generalizations of the definitions of the intersection and union of two sets. We can see this by noting that $x \in A_i$ for all $i = 1, 2$ is equivalent to $x \in A_1$ and $x \in A_2$; also, $x \in A_i$ for at least one $i = 1, 2$ is equivalent to $x \in A_1$ or $x \in A_2$.

Theorem 5.4.5 gives the generalized distributive law for sets. We give the proof for part a and leave the proof of part b as an exercise. (See Exercise 1, Exercise Set 5.4.) The proof is by mathematical induction.

5.4 INTRODUCTION TO MATHEMATICAL INDUCTION

THEOREM 5.4.5 For an arbitrary positive integer n,

a. $A \cup (B_1 \cap B_2 \cap \cdots \cap B_n) = (A \cup B_1) \cap (A \cup B_2) \cap \cdots \cap (A \cup B_n)$
b. $A \cap (B_1 \cup B_2 \cup \cdots \cup B_n) = (A \cap B_1) \cup (A \cap B_2) \cup \cdots \cup (A \cap B_n)$

Proof Basis Step ($n = 1$) $\quad A \cup (B_1) = A \cup B_1 = (A \cup B_1)$
Induction Step Induction hypothesis: Assume

$$A \cup (B_1 \cap B_2 \cap \cdots \cap B_k) = (A \cup B_1) \cap (A \cup B_2) \cap \cdots \cap (A \cup B_k)$$

Then

$$A \cup (B_1 \cap \cdots \cap B_k \cap B_{k+1}) = A \cup [(B_1 \cap \cdots \cap B_k) \cap B_{k+1}]$$
$$= [A \cup (B_1 \cap \cdots \cap B_k)] \cap [A \cup B_{k+1}]$$

by Boolean law 3a for set theory. By the induction hypothesis,

$$[A \cup (B_1 \cap \cdots \cap B_k)] \cap [A \cup B_{k+1}]$$
$$= [(A \cup B_1) \cap (A \cup B_2) \cap \cdots \cap (A \cup B_k)] \cap [A \cup B_{k+1}]$$

Therefore,

$$A \cup (B_1 \cap \cdots \cap B_k \cap B_{k+1})$$
$$= (A \cup B_1) \cap (A \cup B_2) \cap \cdots \cap (A \cup B_k) \cap (A \cup B_{k+1}) \quad \blacksquare$$

A note on the basis step: When the theorem requiring proof is of the form $p(n)$ for $n = b, b + 1, \ldots$, the basis step is to prove $p(b)$. In the examples discussed above, b was equal to 1, but b may be any integer. See Exercise Set 5.4 for other values of b. For example, Exercise 9 asks for a proof of the fact that $n^2 < 2^n$ for $n = 5, 6, 7, \ldots$. In this case the basis step is for $n = 5$.

By substituting for n we see that $2n + 1 \leq 2^n$ is false for $n = 1, 2$. However, $2n + 1 \leq 2^n$ is true for $n = 0, 3, 4, 5, \ldots$. To prove this, we need only substitute $n = 0$ to verify one case and then finish the following problem using mathematical induction with basis step $n = 3$.

PRACTICE PROBLEM 2 Prove $2n + 1 \leq 2^n$ for $n = 3, 4, \ldots$.

An example in which the basis step is for $n = 3$ is the following: A **convex** **n-gon** is an n-sided polygon with the property that a straight line segment between any two points of the polygon is entirely within the polygon. A triangle is a convex 3-gon. Since a polygon must have at least 3 sides, a triangle is the convex n-gon with the least number of sides.

EXAMPLE 5.4.6 Prove that the sum of the interior angles of a convex n-gon is $(n - 2)\pi$ radians.

Proof We know, as a fact of Euclidean geometry, that the interior angles of a triangle sum to π radians. So we have the basis step: The sum of the interior angles of a convex 3-gon is $(3 - 2)\pi$ radians. The induction hypothesis is:

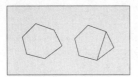

FIGURE 5.4.1
Convex $(k+1)$-gon =
Convex k-gon + Triangle

Assume that the sum of the interior angles of any convex k-gon is $(k-2)\pi$ radians.

An arbitrary convex $(k+1)$-gon can be decomposed into a convex k-gon and a triangle with a line drawn between two corners having a corner between them. By examining Figure 5.4.1, we see that the sum of the interior angles of the $(k+1)$-gon is equal to the sum of the interior angles of the convex k-gon plus the sum of the interior angles of the triangle. Hence, we get $(k-2)\pi + \pi = [(k+1) - 2]\pi$ as the sum of the interior angles of a convex $(k+1)$-gon. ■

PROGRAMMING NOTES

Program Verification and Loop Invariants Methods of logic are central to the study of algorithm verification. Here is an example illustrating how mathematical induction is used to prove an algorithm to be correct. Consider the following function in Pascal:

```
function Square(m : integer) : integer;
   var
      j, x : integer;
   begin
      j := 0;
      x := 0;
      while j < m do
         begin
            x := x + 2*j + 1;
            j := j + 1
         end;
      Square := x
   end;
```

The name we have given this function implies that the function computes the square of the integer m. After checking for a few values of m, we may believe that the function does square each integer m. How do we prove this fact?

We give a proof using induction and the idea of a loop invariant.

Proof Let j_n and x_n be the values of j and x, respectively, after the while loop is executed n times. After the loop has been executed m times, we have $j_m = m$ since j is initially 0 and is incremented by 1 each time the loop is executed. We will use induction to prove

$$x_n = (j_n)^2 \quad \text{for } n = 0, 1, 2, \ldots$$

We can then conclude that $x_m = m^2$.

Basis Step ($n = 0$) Before the loop is executed, both the variables x and j are equal to 0. Hence, $x_0 = (j_0)^2$.

Induction Step Induction hypothesis: Assume $x_k = (j_k)^2$. After one more time through the loop, we have

$$x_{k+1} = x_k + 2j_k + 1 \quad \text{and} \quad j_{k+1} = j_k + 1$$

5.4 INTRODUCTION TO MATHEMATICAL INDUCTION

Using the induction hypothesis, we obtain

$$x_{k+1} = (j_k)^2 + 2j_k + 1 = (j_k + 1)^2 = (j_{k+1})^2$$

Therefore, by mathematical induction, it follows that $x_n = (j_n)^2$ for $n = 0, 1, 2, \ldots$. ∎

The equation $x_n = (j_n)^2$ is a property of the variables x and j that remained true through each iteration of the loop. Such a property is an invariant property and is called a **loop invariant**. We write this loop invariant more simply as $x = j^2$. Loop invariants and mathematical induction are important tools in the area of algorithm verification.

EXERCISE SET 5.4

1. Prove Theorem 5.4.5b.

2. Use De Morgan's law for sets and mathematical induction to prove
$$(A_1 \cap A_2 \cap \cdots \cap A_n)' = A_1' \cup A_2' \cup \cdots \cup A_n'$$

3. Use De Morgan's law for sets and mathematical induction to prove
$$D - (A_1 \cup A_2 \cup \cdots \cup A_n) = (D - A_1) \cap (D - A_2) \cap \cdots \cap (D - A_n)$$

4. This problem involves Euclidean geometry. Use mathematical induction to prove: Given a line of unit length, it is possible to construct, with straightedge and compass, a line of length $\sqrt{n}$ for $n = 1, 2, 3, \ldots$.

Use mathematical induction to prove the statements in Exercises 5–13.

5. $1 + 2 + 3 + \cdots + n = \dfrac{n(n+1)}{2}$, for $n = 1, 2, 3, \ldots$

6. $1 + 8 + 27 + \cdots + n^3 = \left[\dfrac{n(n+1)}{2}\right]^2$, for $n = 1, 2, 3, \ldots$

7. $1 + \dfrac{1}{2} + \dfrac{1}{4} + \cdots + \dfrac{1}{2^n} = 2 - \dfrac{1}{2^n}$, for $n = 0, 1, 2, \ldots$

8. $1 + r + r^2 + \cdots + r^n = \dfrac{1 - r^{n+1}}{1 - r}$, for $n = 0, 1, 2, 3, \ldots$ and $r \neq 1$

9. $n^2 < 2^n$, for $n = 5, 6, 7, \ldots$ [*Hint:* You may use the result of Practice Problem 2.]

10. $3^n < n!$, for $n = 7, 8, \ldots$

11. $\dfrac{1}{1^2} + \dfrac{1}{2^2} + \cdots + \dfrac{1}{n^2} \leq \dfrac{2n}{n+1}$, for $n = 1, 2, 3, \ldots$

12. $1^3 + 3^3 + 5^3 + \cdots + (2n - 1)^3 = n^2(2n^2 - 1)$, for $n = 1, 2, 3, \ldots$

13. $1 \cdot 1! + 2 \cdot 2! + \cdots + n \cdot n! = (n + 1)! - 1$, for $n = 1, 2, 3, \ldots$

Exercises 14 and 15 contain two results that were used in the development of transitive closure in Chapter 2. Let R be a relation on a set A.

14. Assume R is transitive. Prove: If there is a path in R from $a \in A$ to $b \in A$, then $a \, R \, b$. [*Hint:* Use mathematical induction on the length of the path.]

15. Prove $M_{R^n} = (M_R)^n$ for $n = 1, 2, 3, \ldots$.

16. The following expression can be written as a function of n (sometimes called a **closed-form expression**). For example, the right side of the equation in Exercise 5 is the closed-form expression of the left side. Compute values of the left side of the following expression for several integers, and guess the closed-form expression; then use mathematical induction to prove your conjecture correct:

$$\left(1 - \frac{1}{4}\right)\left(1 - \frac{1}{9}\right) \cdots \left(1 - \frac{1}{n^2}\right) = f(n) \quad \text{for } n = 2, 3, 4, \ldots$$

17. The following expression can be written as a closed-form expression (see Exercise 16). Compute values of the left side for several integers, and guess the closed-form expression; then use mathematical induction to prove your conjecture correct:

$$\frac{1}{1 \cdot 2} + \frac{1}{2 \cdot 3} + \cdots + \frac{1}{n(n+1)} = g(n) \quad \text{for } n = 1, 2, 3, \ldots$$

Exercises 18 and 19 contain results useful in calculus. Let $D^n(f(x))$ stand for the nth derivative of the function f at x.

18. Prove $D^n(x^n) = n!$ for $n = 1, 2, \ldots$.

19. Prove $D^n(e^{bx}) = b^n e^{bx}$ for $n = 1, 2, \ldots$.

20. Prove $\dfrac{1}{\sqrt{1}} + \dfrac{1}{\sqrt{2}} + \cdots + \dfrac{1}{\sqrt{n}} > \sqrt{n}$ for $n = 2, 3, \ldots$.

21. Let A be a subset of $\mathbb{N}$. Suppose $0 \in A$ and, for every $k \in \mathbb{N}$, if $k \in A$, then $k + 1 \in A$. Prove $A = \mathbb{N}$.

22. Let A be a subset of $\mathbb{N}$. Suppose $0 \in A$ and, for every $k \in \mathbb{N}$, if $\{0, 1, \ldots, k\} \subseteq A$, then $k + 1 \in A$. Prove $A = \mathbb{N}$.

Find the errors in the incorrect proofs given in Exercises 23 and 24.

23. Prove that all the computers in any set of n computers are identical for $n = 1, 2, \ldots$.
Proof All computers in any set containing one computer are certainly identical. Assume that all computers in any set containing k computers are identical. Let $D = \{c_1, c_2, \ldots, c_k, c_{k+1}\}$ be an arbitrary set of $k + 1$ computers. Then, $A = \{c_1, c_2, \ldots, c_k\}$ and $B = \{c_2, \ldots, c_k, c_{k+1}\}$ are both sets containing k computers. So all the computers in A are identical, and all the computers in B are identical. But c_2 is in both sets. Hence, all the computers in D are identical to c_2. Therefore, all computers in D are identical. By the principle of mathematical induction, all computers in any set of n computers are identical for $n = 1, 2, \ldots$.

24. Prove that $7^n = 1$ for $n = 0, 1, 2, \ldots$.
Proof We have $7^0 = 1$ by definition. Assume $7^j = 1$ for all $0 \le j \le k$. We will prove $7^j = 1$ for all $0 \le j \le k + 1$. To do this we need only prove $7^{k+1} = 1$. But $7^{k+1} = (7^k \cdot 7^k)/7^{k-1} = (1 \cdot 1)/1 = 1$. By mathematical induction, $7^m = 1$ for all $0 \le m \le n$, where $n = 0, 1, 2, \ldots$. The result follows.

■ PROGRAMMING EXERCISES

25. Use mathematical induction to prove that the following program fragment generates $S = 2^n - 1$ for $n = 0, 1, 2, \ldots$ by proving that $S = 2^j - 1$ is a loop invariant.

```
S := 0;
j := 0;
while j < n do
    begin
        S := 2 * S + 1;
        j := j + 1
    end;
```

26. Use mathematical induction to prove that the following program fragment generates $C = m^3$ for $m = 0, 1, 2, \ldots$ by proving that $y = j^3$ is a loop invariant.

```
j := 0;
x := 0;
y := 0;
while j < m do
    begin
        y := y + 3*x + 3*j + 1;
        x := x + 2*j + 1;
        j := j + 1
    end;
C := y
```

27. Do a walk-through of the following program fragment, and guess the closed-form expression for S in terms of n. Prove your conjecture to be true. Note that $SQRT(m) = \sqrt{m}$.

```
S := 0;
for j := 1 to n do
    S := S + 2*SQRT(S) + 1
```

5.5
PREDICATES AND QUANTIFIERS

The elementary logic we have studied thus far is commonly called propositional logic. Propositional logic, which involves only statements, is inadequate for many purposes.

Statements are sentences that are either true or false but not both. The sentence "$-1 < 0$" is a statement that happens to be true in the context of the real numbers. Many other sentences used in mathematics are not statements. Examples include "$x < 5$" and "$\sin x = 1$." Each of these sentences has a variable. A **variable** is used to symbolize an arbitrary element from a given universal set U. Elements from U can be substituted for x in a sentence like "$x < 5$." A sentence of the form $p(x)$, in which a subsitution is possible and which becomes a statement after a substitution is made, is called a **predicate** (in one variable). If a variable occurs in several places in a predicate, as in "$x^5 + x^2 - 1 = 0$," then the same substitution must be made in each place where the variable occurs. Predicates frequently have more than one variable. The sentence "$x^2 + y^2 = 1$" has two variables, whereas "If $r < x$, then $1/x < t$" has three variables.

Predicate logic involves statements, predicates, and quantifiers. We will introduce quantifiers shortly. Just as with propositional calculus, we will learn many useful rules of manipulation. In addition, we will see many applications of predicate calculus.

We will always assume that the universal set U is nonempty. You may assume the universal set $U = \mathbb{R}$ unless otherwise specified.

EXAMPLE 5.5.1

PREDICATE	SUBSTITUTIONS THAT MAKE THE PREDICATE A:	
	TRUE STATEMENT	FALSE STATEMENT
a. $x \cdot 0 = 0$	Any	None
b. $x \cdot 5 = 0$	$x = 0$	$x \neq 0$
c. $\sin x = 2$	None	Any
d. If $x = 0$, then $x \cdot 5 = 0$.	Any	None
e. $x < 5$	$x = 3.1, 4$, etc.	$x = 5, 6.3$, etc.

Since part d might be puzzling, let us look at how substitutions work in that case. The substitution $x = 0$ yields the statement "$0 = 0$ implies $0 \cdot 5 = 0$," which is certainly true. The substitution $x = 3$ yields the statement "$3 = 0$ implies $3 \cdot 5 = 0$," which is also (vacuously) true since $3 = 0$ is false. ∎

All the substitutions from the universal set U that make a predicate true form a set called the **solution set** for that predicate. A solution set depends on both the predicate and the given universal set.

EXAMPLE 5.5.2

PREDICATE	UNIVERSAL SET	SOLUTION SET
a. $x < 5$	$\mathbb{R}$	$(-\infty, 5)$
b. $x < 5$	$\mathbb{N}$	$\{0, 1, 2, 3, 4\}$
c. $\sin x = 0$	$\mathbb{Z}$	$\{0\}$
d. $\sin x = 0$	$\mathbb{R}$	$\{n\pi : n \in \mathbb{Z}\}$
e. $\sin x = 2$	$\mathbb{R}$	$\emptyset$

∎

EXAMPLE 5.5.3 "The solution set for $x^2 - 3x + 2 = 0$ is $\{1, 2\}$" is a statement and not a predicate. The symbol x in this statement is not a variable for which a substitution from the reals can be made (try it yourself). Such a symbol is frequently called a dummy variable. ∎

EXAMPLE 5.5.4

PREDICATE	UNIVERSAL SET	SOLUTION SET
a. $x^2 + 3x + 2 = 0$	$\mathbb{N}$	$\emptyset$
b. $x^2 + 3^2 = 11$	$\mathbb{R}$	$\{-\sqrt{2}, \sqrt{2}\}$
c. $x^2 + 3^2 = 11$	$\mathbb{Q}$	$\emptyset$
d. $x \cdot 0 = 0$	$\mathbb{R}$	$\mathbb{R}$
e. $\sin^2 x + \cos^2 x = 1$	$\mathbb{R}$	$\mathbb{R}$

∎

5.5 PREDICATES AND QUANTIFIERS

The predicates in Examples 5.5.4d and 5.5.4e have the special property that their solution set is U. So, each is true for every possible substitution from U. Asserting that a predicate is true for all substitutions is so important we use a special symbol called the **universal quantifier**.

UNIVERSAL QUANTIFIER

The symbol $\forall x$ means "for all x" or "for any x."

EXAMPLE 5.5.5
  a. $\forall x \, (x \cdot 0 = 0)$ means "Any real number multiplied by 0 is equal to 0."
  b. $\forall x \, (\sin^2 x + \cos^2 x = 1)$ means "For any real number, the sum of the squares of the sine and cosine of that real number is equal to 1." It is this kind of assertion that is made in every trigonometric identity. ∎

Q1 PROOF WITH A UNIVERSAL QUANTIFIER

If $q(x)$ logically follows from the premises, where x is any element of the universal set, then $\forall x \, (q(x))$ logically follows from the premises.

Proofs using Rule Q1 frequently begin "Let x be arbitrary." It may also be understood and not mentioned that the element x is arbitrary. In any proof of a trigonometric identity, such as $\sin^2 x + \cos^2 x = 1$, it is understood that x is arbitrary.

Let us take another look at pick-a-point proofs. To prove $A \subseteq B$ we must take an arbitrary element $u \in U$ and prove that $u \in A$ implies $u \in B$. This method is a direct application of Rule Q1 to the definition of $A \subseteq B$. Recall that A is a subset of B if and only if $\forall u \, (u \in A$ implies $u \in B)$. Also, $A = B$ if and only if $\forall u \, (u \in A \leftrightarrow u \in B)$.

The assertion that the solution set for a predicate is nonempty is also important. The symbol used for this assertion is the **existential quantifier**.

EXISTENTIAL QUANTIFIER

The symbol $\exists x$ means "there exists an x" or "there is an x."

EXAMPLE 5.5.6
  a. $\exists x \, (x \cdot 5 = 7)$ means $x \cdot 5 = 7$ has a solution—in this case, only one.
  b. $\exists x \, (\sin x = 0)$ means $\sin x = 0$ has at least one solution. There are many solutions here. ∎

Q2 PROOF WITH AN EXISTENTIAL QUANTIFIER

If $q(x_0)$ logically follows from the premises, where x_0 is some (usually not arbitrary) element of the universal set, then $\exists x \, (q(x))$ follows logically from the premises.

EXAMPLE 5.5.7 **a.** To prove $\exists x \, (x \cdot 5 = 7)$, we need only exhibit the solution $x_0 = \frac{7}{5}$. Then $x_0 \cdot 5 = 7$ follows logically from the rules of arithmetic for $\mathbb{R}$.
b. To prove $\exists x \, (\sin x = 0)$, we need only give one of the solutions—for example, $x_0 = 2\pi$. ∎

The universal quantifier can be considered the generalized "and"; the existential quantifier can be considered the generalized "or." To see this, let U be the two-element set $\{a, b\}$. For $U = \{a, b\}$, $\forall x \, (p(x))$ is true means $p(a) \land p(b)$ is true, and $\exists x \, (p(x))$ is true means $p(a) \lor p(b)$ is true.

When the universal set U is given and $p(x)$ is a predicate in one variable, then both $\forall x \, (p(x))$ and $\exists x \, (p(x))$ are statements. Of course, these statements are not always true. By definition,

$\forall x \, (p(x))$ is true if and only if $p(u)$ is true for all substitutions u from U

$\exists x \, (p(x))$ is true if and only if there is at least one substitution u_0 from U for which $p(u_0)$ is true

EXAMPLE 5.5.8 Let $U = \mathbb{R}$.
a. $\forall x \, (x^2 + 2x + 5 > 0)$ is true. **b.** $\forall x \, (\sin x = 1)$ is false.
c. $\exists x \, (x^2 + 2x + 5 = 0)$ is false. **d.** $\exists x \, (\sin x \neq 1)$ is true. ∎

One reason we assume $U \neq \emptyset$ is that the implication "If $\forall x \, (q(x))$, then $\exists x \, (q(x))$" is a theorem for $U \neq \emptyset$ but is false when U is empty. To see this, let $U = \emptyset$ and $q(x)$ be $x \neq x$. Then, $\forall x \, (x \neq x)$ is (vacuously) true, and $\exists x \, (x \neq x)$ is false.

THEOREM 5.5.9 Let $U \neq \emptyset$ and $q(x)$ be any predicate. Then,

$$\forall x \, (q(x)) \text{ implies } \exists x \, (q(x))$$

Proof Assume $\forall x \, (q(x))$. Since U is nonempty and $q(u)$ is true for all substitutions u from U, there is an element $u_0 \in U$ such that $q(u_0)$ is true. Hence, $\exists x \, (q(x))$ by Rule Q2. ∎

The converse to Theorem 5.5.9 is not true. Here is one counterexample: Let $U = \mathbb{N}$ and $q(x)$ be "x is odd."

One of the important uses of quantified statements is in writing precise definitions. For many statements, more than one quantifier must be used.

EXAMPLE 5.5.10 **a.** u is an *upper bound* for a set A of real numbers if $\forall x \, (\text{If } x \in A, \text{ then } x \leq u)$.
b. A has an upper bound if $\exists u \, [\forall x \, (x \in A \text{ implies } x \leq u)]$.
c. u is a *greatest element* for a set A of real numbers if $u \in A$ and u is an upper bound for A. ∎

PRACTICE PROBLEM 1 Refer to Example 5.5.10. Use a quantified statement to define *greatest element*.

The statement on the right in Example 5.5.10b has two quantifiers and has the form $\exists u\, [\forall x\, (p(x, u))]$. We will usually write such a statement more efficiently as $\exists u\, \forall x\, (p(x, u))$. The order of quantifiers is important, as we will soon see. The statement $\exists u\, [\forall x\, (p(x, u))]$ is read "There exists a u such that, for all x, $p(x, u)$."

Properties of the integers often require the use of quantifiers, as shown in the next example.

EXAMPLE 5.5.11 The axioms for $\mathbb{Z}$ need more than one quantifier.

a. The sentence $x + y = y + x$ is a predicate in two variables. The sentence $\forall y\, (x + y = y + x)$ is a predicate in one variable, x. The *commutative law* is: $\forall x\, \forall y\, (x + y = y + x)$.
b. *Identity laws*: $\exists 0\, \forall x\, (x + 0 = x)$ and $\exists 1\, \forall x\, (x \cdot 1 = x)$.
c. *Additive inverse law*: $\forall x\, \exists (-x)\, (x + (-x) = 0)$. ∎

The order of the quantifiers is important when both kinds of quantifiers occur in the statement. For the commutative law it does not matter whether we write $\forall x\, \forall y$ or $\forall y\, \forall x$. However, the difference between statements like $\exists y\, \forall x\, (x + y = 0)$ and $\forall x\, \exists y\, (x + y = 0)$ is crucial. Indeed, $\forall x\, \exists y\, (x + y = 0)$ is true, but $\exists y\, \forall x\, (x + y = 0)$ is false in $\mathbb{Z}$.

PRACTICE PROBLEM 2 Convince yourself that $\forall x\, \exists y\, (x + y = 0)$ is true but $\exists y\, \forall x\, (x + y = 0)$ is false in $\mathbb{Z}$.

THEOREM 5.5.12 Let $p(x, y)$ be any predicate. Then,

$$\exists y\, \forall x\, (p(x, y)) \text{ implies } \forall x\, \exists y\, (p(x, y))$$

Proof Assume $\exists y\, \forall x\, (p(x, y))$. Then, since $U \neq \emptyset$, there is an element $y_0 \in U$ such that $\forall x\, (p(x, y_0))$ is true. Hence, $p(x, y_0)$ is true for all substitutions x from U. So, $\exists y\, (p(x, y))$ is true for all substitutions x from U. Therefore, $\forall x\, \exists y\, (p(x, y))$ is true. ∎

The discussion preceding Theorem 5.5.12 shows that the converse of this theorem is false.

Most definitions, axioms, and theorems are quantified statements. To decide whether or not a particular quantified statement is true, it is helpful to look at its negation.

EXAMPLE 5.5.13 To write "N has no upper bound" is easy: $\neg\, \exists u\, \forall x\, (x \in N \rightarrow x \leq u)$. But that is not too helpful. A better attempt is: $\forall u\, (u \text{ is not an upper bound for } N)$. We get $\forall u\, \neg\, \forall x\, (x \in N \rightarrow x \leq u)$. ∎

The following two rules for negating quantified statements are useful.

Q3

$\neg \forall x \, (p(x))$ if and only if $\exists x \, (\neg p(x))$

Q4

$\neg \exists x \, (p(x))$ if and only if $\forall x \, (\neg p(x))$

To see why the rules for negation make sense we give an example.

EXAMPLE 5.5.14 Let $p(x)$ stand for "x has a positive square" (x^2 is positive). Then, $\neg \forall x \, (p(x))$ means "It is not the case that all real numbers have positive squares." Also, $\exists x \, (\neg p(x))$ means "there is a real number that does not have a positive square." ∎

PRACTICE PROBLEM 3 Refer to Example 5.5.14. Read $\neg \exists x \, (p(x))$ and $\forall x \, (\neg p(x))$ aloud in English so you see why Rule Q4 makes sense.

Rules Q3 and Q4 are generalized De Morgan's laws. We can see this fact most easily by considering $U = \{a, b\}$. For this two-element universe, we write Rule Q3 as follows: $\neg \forall x \, (p(x))$ is true means $\neg(p(a) \wedge p(b))$ is true, whereas $\exists x \, (\neg p(x))$ is true means $(\neg p(a)) \vee (\neg p(b))$ is true. Hence, for a two-element universe, Rule Q3 is equivalent to De Morgan's law: $\neg(p(a) \wedge p(b))$ if and only if $(\neg p(a)) \vee (\neg p(b))$.

PRACTICE PROBLEM 4 Write Rule Q4 with a two-element universe, and convince yourself that Rule Q4 is a generalized De Morgan's law.

Let us consider again the definition for A is a subset of B.

EXAMPLE 5.5.15

DEFINITION	NEGATION
$A \subseteq B$	A is not a subset of B
$\forall u \, (u \in A \rightarrow u \in B)$	$\exists u \, (u \in A \wedge u \notin B)$

∎

PRACTICE PROBLEM 5 Refer to Example 5.5.15. Use Rule Q3 to derive the negation of $A \subseteq B$.

Here are some other statements and their negations.

EXAMPLE 5.5.16

STATEMENT	NEGATION
a. $\exists x \, \exists p \, (x + p = 0)$	$\forall x \, \forall p \, (x + p \neq 0)$
b. $\forall x \, \exists y \, (x + y = 0)$	$\exists x \, \forall y \, (x + y \neq 0)$
c. $\exists y \, \forall x \, (x + y = 0)$	$\forall y \, \exists x \, (x + y \neq 0)$
d. $\forall x \, \forall y \, [(x < y) \text{ implies } (x^2 < y^2)]$	$\exists x \, \exists y \, [(x < y) \wedge (y^2 \leq x^2)]$

5.5 PREDICATES AND QUANTIFIERS

Here is how Rules Q3 and Q4 are used to derive the negation in part b:

$$\neg \forall x \, \exists y \, (x + y = 0) \leftrightarrow \exists x \, \neg \, \exists y \, (x + y = 0) \leftrightarrow \exists x \, \forall y \, \neg \, (x + y = 0)$$
$$\leftrightarrow \exists x \, \forall y \, (x + y \neq 0)$$
∎

PRACTICE PROBLEM 6 There are eight statements (four statements and their negations) in Example 5.5.16. Derive the negations in parts a and c. Which four of the eight statements are true?

Disproving Statements

Which of the two statements in part d of Example 5.5.16 is true? Suppose we conjecture that $\forall x \, \forall y \, [(x < y) \text{ implies } (x^2 < y^2)]$. If it turns out that the negation is true, we have a counterexample to the conjecture. In this case the negation is true because $x = -3$ and $y = 2$ makes $(x < y) \land (y^2 \leq x^2)$ true. Hence, $x = -3$ and $y = 2$ is a counterexample to $\forall x \, \forall y \, (x < y \rightarrow x^2 < y^2)$. We have disproved the statement $\forall x \, \forall y \, [(x < y) \text{ implies } (x^2 < y^2)]$.

Suppose we make a conjecture of the form $\forall x \, (p(x))$. If its negation $\exists x \, \neg \, (p(x))$ is true, then we say an element x_0 such that $\neg p(x_0)$ is a *counterexample* to the conjecture. When faced with a conjecture, it is useful to write it in quantified form and then write its negation. By doing this we can frequently tell whether the conjecture is true and find a counterexample if it is false.

EXAMPLE 5.5.17 Let $U = \mathbb{N}$, let $p(x)$ stand for "x is a prime number," and let $o(x)$ stand for "x is an odd number."

STATEMENT	NEGATION
a. All prime numbers are odd:	Some prime numbers are not odd:
$\forall x \, [p(x) \rightarrow o(x)]$	$\exists x \, [p(x) \land \neg o(x)]$
b. Some prime numbers are odd:	No prime numbers are odd:
$\exists x \, [p(x) \land o(x)]$	$\forall x \, [p(x) \rightarrow \neg o(x)]$

Here is a derivation for the negation in part a:

$$\neg \forall x \, [p(x) \rightarrow o(x)] \leftrightarrow \neg \forall x \, [\neg p(x) \lor o(x)] \leftrightarrow \exists x \, \neg \, [\neg p(x) \lor o(x)]$$
$$\leftrightarrow \exists x \, [p(x) \land \neg o(x)]$$

Note also that $x = 2$ makes $\exists x \, [p(x) \land \neg o(x)]$ true. So, $x = 2$ is a counterexample to "All prime numbers are odd." ∎

PRACTICE PROBLEM 7 Derive the negation stated in part b of Example 5.5.17.

We discussed the four important properties of a relation in Chapters 2 and 3. Those definitions and their negations can be stated using quantifiers.

PRACTICE PROBLEM 8 Let R be a relation on a set A. Use quantifiers to state the definition of each of the following and its negation.

    a. R is reflexive.   **b.** R is symmetric.
    c. R is antisymmetric.   **d.** R is transitive.

It is sometimes useful to restrict the domain of a quantifier. This is usually accomplished by including a condition with the quantifier. Here is the definition for **restricted quantifiers**.

    a. $(\forall x \in B)(p(x))$ means $\forall x \, (x \in B \to p(x))$
    b. $(\exists x \in B)(p(x))$ means $\exists x \, (x \in B \land p(x))$

The next two examples illustrate the use of restricted quantifiers.

EXAMPLE 5.5.18
    a. $(\forall x \in \mathbb{N})(p(x))$ means $\forall x \, (x \in \mathbb{N} \to p(x))$
    b. $(\exists x > 0)(p(x))$ means $\exists x \, (x > 0 \land p(x))$ ∎

Notice that restriction of a universal quantifier is done using an "If...then" statement, whereas restriction of an existential quantifier is done using an "and" statement.

EXAMPLE 5.5.19
    a. $\mathbb{N}$ has no upper bound if and only if $\forall u \, (\exists x \in \mathbb{N})(u < x)$.
    b. Here is how we write "Every nonzero real number has a multiplicative inverse" with a restricted quantifier: $(\forall x \neq 0)(\exists y)(xy = 1)$. It is written $\forall x \, (x \neq 0 \to \exists y \, (xy = 1))$ with a standard quantifier. ∎

Intersection and Union of Any Collection of Sets

DEFINITION

> Let Ω be any collection of sets. The **generalized intersection** of Ω is
>
> $$\bigcap \Omega = \{x : (\forall A \in \Omega)(x \in A)\}$$
>
> The **generalized union** of Ω is
>
> $$\bigcup \Omega = \{x : (\exists A \in \Omega)(x \in A)\}$$

Since $\forall$ is the generalized "and" and $\exists$ is the generalized "or," these definitions are generalizations of the definitions of the intersection and union of a finite collection of sets given in Sections 1.1 and 5.4.

PRACTICE PROBLEM 9 Let $\Omega = \{A, B\}$. Convince yourself that $\bigcap \Omega = A \cap B$ and $\bigcup \Omega = A \cup B$.

EXAMPLE 5.5.20 State and prove a generalized distributive law for sets.

Solution The generalization of $D \cap (A_1 \cup A_2) = (D \cap A_1) \cup (D \cap A_2)$ is the generalized distributive law for sets: $D \cap \bigcup \Omega = \bigcup \{D \cap A : A \in \Omega\}$. To prove this we use a pick-a-point proof.

Let $x \in D \cap \bigcup \Omega$. Then, $x \in D \wedge x \in \bigcup \Omega$. Hence, $x \in D$ and there exists $A \in \Omega$ such that $x \in A$. So, $x \in D \cap A$ for some $A \in \Omega$. Therefore, $x \in \bigcup \{D \cap A : A \in \Omega\}$. This shows that $D \cap \bigcup \Omega \subseteq \bigcup \{D \cap A : A \in \Omega\}$. It is routine to prove the other set inclusion by reversing each of the steps above. ∎

PRACTICE PROBLEM 10 State and prove the other generalized distributive law.

PROGRAMMING NOTES

The Big-Oh Notation In Section 3.5 we introduced time-complexity functions. Time-complexity functions are used to estimate the run time of algorithms when they are implemented in computer programs. The main purpose of estimating time complexity is to compare the relative efficiency of different algorithms. The big-oh notation is very useful for comparing the growth rate of time-complexity functions.

We will define the big-oh notation and give several proofs of important facts related to the big-oh notation.

Asymptotic Behavior of Time-Complexity Functions We note that time-complexity functions depend on the number of items in a list and may be assumed to take on nonnegative real values since they measure the time (number of steps) taken to execute an algorithm. A time-complexity function has as domain the nonnegative integers and as codomain the nonnegative reals.

When comparing the relative efficiency of algorithms, we are most interested in the time complexity $T(n)$ for large n. For example, the function $T(n) = 5n^2$ increases as n increases and is proportional to n^2. We consider $5n^2$ and n^2 to have the same asymptotic behavior.

Since we are often interested in worst-case behavior, we introduce the idea of asymptotic dominance. The big-oh notation is very useful for describing asymptotic dominance. Donald Knuth wrote in Volume 1 of *The Art of Computer Programming*: "The **O**-notation is a big help in approximation work, since it briefly describes a concept which occurs frequently and it suppresses detailed information which is usually irrelevant."

Let f and g be time-complexity functions. When we say "$f(n)$ is big-oh of $g(n)$," we mean that $f(n)$ is less than or equal to a function proportional to $g(n)$ for all sufficiently large values of n. More precisely, $f(n)$ is $\mathbf{O}(g(n))$, read "$f(n)$ is **big-oh** of $g(n)$," if there is some positive constant t and some nonnegative integer m so that $f(n) \leq t \cdot g(n)$ for all $n > m$. Here is the definition using quantifiers:

DEFINITION

BIG-OH

$f(n)$ is $\mathbf{O}(g(n))$ if $(\exists m \in \mathbb{N})(\exists t \in \mathbb{R}^+)(\forall n > m)(f(n) \leq t \cdot g(n))$

For example, to show that $5n^2$ is $\mathbf{O}(n^2)$, we pick the constant $t = 6$ and note that $5n^2 \leq 6n^2$ for all $n > 1$.

How do we prove one function is big-oh of another function? According to the definition, we can choose the number t to be any positive real number. However, usually we are wise to pick a large number t. A careful reading of the definition also makes it clear that picking m large might be helpful, and it certainly cannot hurt.

EXAMPLE 5.5.21 We will show in Chapter 9 that the bubble-sort algorithm has time complexity $1 + 2 + 3 + \cdots + (n-1)$. We have previously showed that $1 + 2 + 3 + \cdots + (n-1) = n(n-1)/2$. Hence, the time-complexity function for the bubble sort is $f(n) = n(n-1)/2$. We will prove $f(n)$ is $\mathbf{O}(g(n))$, where $g(n) = n^2$.

Proof Note that $f(n) = n^2/2 - n/2$ and, hence, $f(n) = n^2/2 - n/2 \leq n^2$ for all n. So we choose $t = 1$ and $m = 0$. Then, $(\forall n > 0)(f(n) \leq g(n))$ is true. Therefore, $n(n-1)/2$ is $\mathbf{O}(n^2)$. ∎

EXAMPLE 5.5.22 Let $f(n) = 5n^3 + 12n$. We prove that $f(n)$ is $\mathbf{O}(g(n))$, where $g(n) = n^3$. In other words, we prove that $5n^3 + 12n$ is $\mathbf{O}(n^3)$.

Proof We need to find an integer m and a number t such that $5n^3 + 12n \leq tn^3$ for all $n > m$. Consider the quotient $(5n^3 + 12n)/n^3 = 5 + 12/n^2$. This quotient is less than or equal to 17 for $n > 0$. So, we can choose $t = 17$ and $m = 0$. We can also choose $m = 3$ and $t = 6$ because $5 + 12/n^2 \leq 6$ for $n > 3$. ∎

Some Properties of the Big-Oh Notation The big-oh notation defines a binary relation between time-complexity functions. We see that f is related to g whenever $f(n)$ is $\mathbf{O}(g(n))$. It is easy to prove that this relation is reflexive; it is also true but more difficult to prove that the relation is transitive. We record these facts here.

B1

$f(n)$ is $\mathbf{O}(f(n))$

B2

If $f(n)$ is $\mathbf{O}(g(n))$ and $g(n)$ is $\mathbf{O}(h(n))$, then $f(n)$ is $\mathbf{O}(h(n))$.

Does big-oh satisfy the symmetry property? No. It is an easy exercise to prove that n^2 is $\mathbf{O}(n^3)$. We will prove in Example 5.5.23 that n^3 is not $\mathbf{O}(n^2)$. Hence, big-oh is not a symmetric relation. In other words,

---

B3

It is possible that $f(n)$ is $\mathbf{O}(g(n))$ but $g(n)$ is not $\mathbf{O}(f(n))$.

To say $f(n)$ is $\mathbf{O}(g(n))$ is to say that $f(n)$ is dominated by a function proportional to $g(n)$ for all sufficiently large values of n. For example, when $f(n)$ is

5.5 PREDICATES AND QUANTIFIERS

$O(n^3)$, we can conclude that $f(n)$ does not grow faster than n^3; but we do not know whether $f(n)$ grows as fast as n^3.

Here are several more useful properties of the big-oh notation. You are asked to prove these properties in the exercise set.

B4

If $f(n)$ is $O(g(n))$, then $c \cdot f(n)$ is $O(g(n))$ for any positive constant c.

B5

If $f(n)$ and $g(n)$ are $O(h(n))$, then $a \cdot f(n) + b \cdot g(n)$ is $O(h(n))$ for any positive constants a and b.

B6

If $f(n)$ is $O(h(n))$ and $g(n)$ is $O(k(n))$, then $f(n) \cdot g(n)$ is $O(h(n) \cdot k(n))$.

Proving a Function Is Not Big-Oh of Another Function How do we prove that a given function $f(n)$ is not big-oh of another function $g(n)$? Consider the negation of $f(n)$ is $O(g(n))$. If we use quantifiers to write $f(n)$ is not $O(g(n))$, we obtain

$$f(n) \text{ is not } O(g(n)) \text{ if and only if } (\forall m \in \mathbb{N})(\forall t \in \mathbb{R}^+)(\exists n > m)(f(n) > t \cdot g(n))$$

Using this definition directly is unwieldy. A more efficient method is to do a proof by contradiction.

EXAMPLE 5.5.23 Prove that n^3 is not $O(n^2)$.

Proof Suppose n^3 is $O(n^2)$. Then, $n^3 \leq tn^2$ for some fixed t and m, and all $n > m$. Hence, $n = n^3/n^2 \leq t$. But this means there is some real number t bigger than all the integers. So, we have a contradiction to the fact that n increases without bound. ∎

EXERCISE SET 5.5

1. Find solution sets for each of the following predicates when $U = \mathbb{N}$ and when $U = \mathbb{R}$.
   a. $5x = 3$ **b.** $3x < 7$ **c.** $\sin(x + \pi/2) = \cos x$ **d.** $\sin x = 1$

2. Find solution sets for each of the following predicates when $U = \mathbb{N}$ and when $U = \mathbb{R}$.
   a. $|2 - x| < |x| + |2|$ **b.** $|2 - x| > |x| - |2|$ **c.** $\forall r \, (xr = r)$ **d.** $\exists r \, (xr = 1)$

3. Let $U = \mathbb{R}$. Determine whether each of the following statements is true or false.
   a. $\forall x \, (x^2 + 2x - 3 = 0)$ **b.** $\exists x \, (x^2 + 2x - 3 = 0)$
   c. $\forall x \, [\sin(x + \pi/2) = \cos x]$ **d.** $\exists x \, [\sin(x + \pi/2) = \cos x]$

4. Let $U = \mathbb{R}$. Determine whether each of the following statements is true or false.
   a. $\forall x \, (x^2 - 2x + 5 > 0)$ **b.** $\exists x \, (x^2 - 2x + 5 = 0)$
   c. $\forall x \, \exists r \, (xr = 1)$ **d.** $\forall x \, (\exists n \in \mathbb{N})(x < n)$

5. Write the negation for each statement in Exercise 3.

6. Write the negation for each statement in Exercise 4.

7. Use Rules Q3 and Q4 to derive the negation of the statement on the left to obtain the statement on the right in:
   a. Example 5.5.16d   b. Example 5.5.17b

8. Write the negation for each of the following. Determine whether the resulting statement is true or false. Let $U = \mathbb{R}$.
   a. $\forall x \, \exists m \, (x^2 < m)$ b. $\exists m \, \forall x \, (x^2 < m)$

9. Write the negation for each of the following. Determine whether the resulting statement is true or false.
   a. $\forall z \, (\exists w > 0)(0 \leq z < w \rightarrow z = 0)$ b. $\exists m \, \forall x \, [x/(|x| + 1) < m]$

10. Write the negation for each of the following. Determine whether the resulting statement is true or false.
    a. $\exists m \, (\forall n \in \mathbb{N})(m < n \vee n < m)$ b. $\forall m \, (\exists k > 0)[1/(m^2 + 1) < k]$

11. Let $f: D \rightarrow C$ be a function. Use quantified statements to write each of the following definitions.
    a. f is 1–1   b. f is onto

12. a. Use quantifiers to write the definition for a **lower bound** of a set B of real numbers.
    b. The **least element** of a set of real numbers is an element of the set that is also a lower bound of the set. Write the definition for the least element of a set B of real numbers using quantifiers and no words.
    c. Use quantifiers and no words to write "(0, 5] has no least element" as simply as possible.
    d. Prove there is no more than one least element of a set.

13. Let U be the set of all animals alive today, $g(x)$ be "x is a gibbon," and $c(x)$ be "x is carnivorous."
    a. Express "All gibbons are carnivorous" in symbolic form. Use Rule Q3 to derive the symbolic form for "Some gibbons are not carnivorous."
    b. Express "Some gibbons are carnivorous" in symbolic form. Use Rule Q4 to derive the symbolic form for "No gibbon is carnivorous."

14. Use a quantified statement to write the principle of mathematical induction for $\mathbb{N}$.

15. Let U be the set of all 2×2 Boolean matrices. Find a counterexample for each of the following conjectures. First write the negation of each statement in quantified form.
    a. If $M \otimes Q = 0$, then $M = 0$ or $Q = 0$.   b. $M \otimes Q = Q \otimes M$

16. State and prove the generalization of the following De Morgan's law:
    $(A \cup B)' = A' \cap B'$

17. State and prove the generalization of the following De Morgan's law:
    $D - (A \cap B) = (D - A) \cup (D - B)$

Exercises 18–28 involve the big-oh notation discussed in the Programming Notes.

18. a. Let $f(n) = n + 2000$. Prove $f(n)$ is $\mathbf{O}(n^2)$.
    b. Let $f(n) = an + b$, where a and b are arbitrary nonnegative real numbers. Prove $f(n)$ is $\mathbf{O}(n)$. Note that this proves $n + 2000$ is $\mathbf{O}(n)$.

19. Prove n^2 is $\mathbf{O}(n^3)$.

20. Prove $n^2 + n$ is $\mathbf{O}(n^2)$.

21. Prove n^2 is $O(2^n)$.
22. Prove n^2 is not $O(n + 2000)$.
23. Prove 2^n is not $O(n^2)$.
24. Show that big-oh is a reflexive relation. In other words, prove Property B1 from the Programming Notes.
25. Prove Property B2 from the Programming Notes.
26. Prove Property B4 from the Programming Notes.
27. Prove Property B5 from the Programming Notes.
28. Prove Property B6 from the Programming Notes.

5.6
APPLICATION: ARTIFICIAL INTELLIGENCE—AUTOMATED THEOREM PROVING

Artificial intelligence historically has been concerned with finding a mechanical analog of mental processes and, conversely, using the theory of computer processes to provide insight into neural functions. It is now primarily concerned with programming machines to accomplish tasks that appear to require intuition, creativity, and judgment. Automating logical and mathematical thought processes is a major part of artificial intelligence. Automated theorem proving is one example of this.

The idea of automated theorem proving is that, if computers can prove theorems, they can assist in the search for new mathematical knowledge. Proving theorems is harder for computers than for mathematicians. Human mathematicians rely on insight, intuition, ingenuity, and experience. They also use a wide variety of notational conventions and inference patterns. However, these complex factors must be simplified for a theorem-proving computer program. Simplification is accomplished by a reduction process. We will illustrate this reduction process within the context of propositional logic.

Conjunctive Normal Form

The first step in the reduction process is the transformation of statements in propositional logic to a normalized form. Any statement of propositional logic that contains no connectives is a **prime statement**. A **statement variable** is a symbol standing for a prime statement. We will use Greek letters for statement variables. For example, α is a statement variable symbolizing a prime statement α; but $\alpha \wedge \beta$ is not a prime statement. A **literal** is either a prime statement or the negation of a prime statement. For example, both δ and $\neg \delta$ are literals.

A **conjunction of statements** is a statement consisting of other statements with AND connectives between them. For example, $\alpha \wedge (\beta \rightarrow \delta) \wedge \neg \gamma$ is a conjunction of statements. Similarly, a **disjunction of statements** is a statement consisting of other statements with OR connectives between them.

A statement is in **conjunctive normal form (CNF)** whenever it is a conjunction of statements and each constituent statement is a literal or a disjunction of literals. For example, $\alpha \wedge (\beta \vee \delta) \wedge \neg \gamma$ is in conjunctive normal form. Conjunctive normal form is the normalized form we will use in the reduction process for automated theorem proving. Each constituent statement of a conjunctive normal form is called a **clause**. The clauses of $\alpha \wedge (\beta \vee \delta) \wedge \neg \gamma$ are α, $(\beta \vee \delta)$, and $\neg \gamma$.

First, we must show how to write every statement of propositional logic in conjunctive normal form.

THEOREM 5.6.1 Every statement of propositional logic can be written in conjunctive normal form.

Proof We give the steps of a process for transforming a given statement into CNF.

1. Remove all occurrences of $\leftrightarrow$ and $\to$ by applying the logical equivalences:

    $(p \leftrightarrow q) \equiv [(p \to q) \wedge (q \to p)]$

    $(p \to q) \equiv (\neg p \vee q)$

2. Remove negations or move them as far inside as possible using the rule of double negation and De Morgan's laws:

    $\neg(\neg p) \equiv p$

    $\neg(p \wedge q) \equiv \neg p \vee \neg q$

    $\neg(p \vee q) \equiv \neg p \wedge \neg q$

3. Distribute conjunctions over disjunctions using the equivalences:

    $(p \wedge q) \vee r \equiv (p \vee r) \wedge (q \vee r)$

    $p \vee (q \wedge r) \equiv (p \vee q) \wedge (p \vee r)$ ∎

EXAMPLE 5.6.2 Write $\neg(\alpha \to \beta) \to (\alpha \wedge \beta)$ in CNF.

Solution
$$\begin{aligned}
\neg(\alpha \to \beta) \to (\alpha \wedge \beta) &\equiv \neg\neg(\alpha \to \beta) \vee (\alpha \wedge \beta) \\
&\equiv (\alpha \to \beta) \vee (\alpha \wedge \beta) \\
&\equiv (\neg \alpha \vee \beta) \vee (\alpha \wedge \beta) \\
&\equiv [(\neg \alpha \vee \beta) \vee \alpha] \wedge [(\neg \alpha \vee \beta) \vee \beta] \\
&\equiv (\neg \alpha \vee \beta \vee \alpha) \wedge (\neg \alpha \vee \beta \vee \beta) \\
&\equiv [(\neg \alpha \vee \alpha) \vee \beta] \wedge [\neg \alpha \vee (\beta \vee \beta)] \\
&\equiv (\neg \alpha \vee \alpha) \wedge (\neg \alpha \vee \beta) \\
&\equiv (\neg \alpha \vee \beta)
\end{aligned}$$

Hence, there is one clause, $(\neg \alpha \vee \beta)$. ∎

The Resolution Principle

Historically, the biggest obstacle to automated theorem proving was how to program the machine to select correctly and efficiently a law of logic needed for the proof from a large set of alternatives, many of which were irrelevant. In 1965, the logician J. A. Robinson overcame this difficulty by proposing a *single* principle, called the **resolution principle**, which is both sufficient for the proofs of all theorems in propositional logic and easy to automate on a computer.

> **RESOLUTION PRINCIPLE**
>
> If p and q are clauses of a CNF and if p contains the negation of a literal contained in q, then delete the literal from q and the negation of the literal from p and form the disjunction of what remains. The resulting disjunction is called the **resolvent** of the two clauses. The CNF obtained by replacing the clauses p and q by their resolvent follows logically from the original CNF.

The resolution principle (RP) is a kind of cancellation law of propositional logic. For example, the resolvent of the clauses $\alpha \vee \beta$ and $\neg \alpha \vee \gamma$ is $\beta \vee \gamma$. Note that $(\alpha \vee \beta) \wedge (\neg \alpha \vee \gamma) \rightarrow (\beta \vee \gamma)$ is a tautology. In general, the resolution principle holds because the statement $p \vee q \rightarrow r$ is a tautology whenever r is the resolvent of clauses p and q.

Two special cases of the resolution principle require comment. First, note that the resolvent of $(\neg \alpha \vee \beta)$ and α is β. This follows from the resolution principle because $\alpha \equiv \alpha \vee O$. Hence, the resolvent of $(\neg \alpha \vee \beta)$ and α is the resolvent of $(\neg \alpha \vee \beta)$ and $(\alpha \vee O)$. And the resolvent of $(\neg \alpha \vee \beta)$ and $(\alpha \vee O)$ is $\beta \vee O \equiv \beta$. Since $\neg \alpha \vee \beta$ is the OR form of $\alpha \rightarrow \beta$, this instance of the resolution principle is modus ponens.

The second special case is that the resolvent of α and $\neg \alpha$ is O. This case follows from the tautology $p \wedge \neg p \rightarrow O$.

Automated theorem provers use the resolution principle along with the method of indirect proof. Suppose, for example, we wish to prove statement q using premises p_1, p_2, and p_3. The machine applies the resolution principle to the CNF of $\neg q \wedge p_1 \wedge p_2 \wedge p_3$ and attempts to derive a contradiction O.

Here is a proof given in Example 5.3.1 on page 196. We will prove this theorem using the resolution principle.

EXAMPLE 5.6.3   Premises $\neg(p \wedge n)$, $\neg q \rightarrow n$
Prove $p \rightarrow q$

Proof from Example 5.3.1
1. p Assumption
2. $\neg(p \wedge n)$ Premise

3. $\neg p \lor \neg n$ Step 2 and De Morgan's law
4. $p \rightarrow \neg n$ Step 3, OR form, and double negation
5. $\neg n$ Steps 1 and 4 and modus ponens
6. $\neg q \rightarrow n$ Premise
7. $\neg n \rightarrow q$ Step 6, contrapositive, and double negation
8. q Steps 5 and 7 and modus ponens

Notice that this proof required some thought and intuition for ordering the steps in a useful way and choosing the appropriate tautologies. In fact, there are several other ways in which a correct proof can be written. On the other hand, we show a more mechanical proof of the same theorem.

Resolution Proof
1. $\neg p \lor \neg n$ Premise written as a clause
2. $q \lor n$ Premise written as a clause
3. p Clause obtained by negating the conclusion $p \rightarrow q$
4. $\neg q$ Clause obtained by negating the conclusion $p \rightarrow q$
5. $\neg p \lor q$ Steps 1 and 2 and RP
6. q Steps 3 and 5 and RP
7. O Steps 4 and 6 and RP

The resolution proof we have given is still not quite mechanical since nothing has been said about the order in which the resolution principle is applied. For example, we could just as well have applied the resolution principle to steps 1 and 3 first, instead of 1 and 2. One method of mechanizing the process is to apply the resolution principle to the first literal and each of the other clauses in turn. Continue this strategy recursively with the remaining clauses and the new resolvents to complete the process.

Since most of the important theorems of mathematics are stated in terms of predicate logic, it is important to note that a more general version of the resolution principle applies to predicate logic. So, resolution theorem-proving methods are widely applicable in mathematics.

EXERCISE SET 5.6

In Exercises 1–5 supply a resolution proof for each of the given theorems.

1. Premises: $p \rightarrow m, m \rightarrow t, m$;   conclusion: $t \lor \neg p$
2. Premises: $p \rightarrow q, \neg p \land r \rightarrow s, \neg q$;   conclusion: $s \lor \neg r$
3. Premises: $p, p \land r \rightarrow s, \neg q$;   conclusion: $\neg s \rightarrow p \land \neg r$
4. Premises: $p \rightarrow \neg n, \neg p \rightarrow \neg s, \neg s \rightarrow q, r \rightarrow n$;   conclusion: $r \rightarrow \neg s$
5. Premises: $p \rightarrow \neg q, \neg p \rightarrow \neg s, s \lor q$;   conclusion: $p \rightarrow s$

KEY TERMS

Exclusive OR connective	NOR connective	Variable
Logical equivalence	NAND connective	Predicate
Rule of double negation	Boolean data type	Predicate logic
	Valid argument	Solution set
OR form of an implication	Follow logically	Universal quantifier
	Invalid argument	Existential quantifier
Contrapositive	Proof	Restricted quantifier
Negation	Direct proof	Generalized intersection
De Morgan's laws	Substitution principle	
Tautology	Indirect proof	Generalized union
Modus ponens	Proof by contradiction	Big-oh notation
Contradiction	Additive rule for inequalities	Prime statement
Boolean laws of logic		Statement variable
	Proof by contrapositive	Literal
Sufficient condition	Proof by cases	Conjunction of statements
Necessary condition	Pick-a-point method	
Necessary and sufficient condition	Principle of mathematical induction	Disjunction of statements
		Conjunctive normal form (CNF)
	Basis step	Clause
Counterexample	Induction step	Resolution principle (RP)
Functionally complete set of logical connectives	Induction hypothesis	
	Convex n-gon	Resolvent
	Loop invariant	

REVIEW EXERCISES

1. True or false? To show a statement is a contradiction, it is sufficient to find a counterexample for the statement. Explain your answer.

2. True or false? The negation of a contradiction is a tautology. Explain your answer.

3. Use mathematical induction to prove
$$\frac{1}{2^2 - 1} + \frac{1}{3^2 - 1} + \cdots + \frac{1}{(n+1)^2 - 1} = \frac{3}{4} - \frac{1}{2(n+1)} - \frac{1}{2(n+2)} \quad \text{for } n = 1, 2, \ldots$$

4. Use mathematical induction to prove $2^n < n!$ for $n = 4, 5, 6, \ldots$.

5. Let R be a relation on a set A. Prove the following.
   a. If R is transitive, then $R \circ R \subseteq R$. b. If $R \circ R \subseteq R$, then R is transitive.

6. Let R and S be relations on a set A, and let R^+ and S^+ denote the connectivity relations for R and S, respectively. Prove: If $R \subseteq S$, then $R^+ \subseteq S^+$.

7. Let R be a relation on a set A, and let R^+ denote the connectivity relation for R. Prove: If R is transitive, then $R^+ = R$.

8. Let R be a relation on a set A. Prove: The connectivity relation of R^+ is R^+ itself; that is, $(R^+)^+ = R^+$.

9. Assume that R, S, and T are relations on a set A. Prove: If $R \subseteq S$, then $R \circ T \subseteq S \circ T$.

10. Prove $A - (A - B) = A \cap B$.

11. Define the symmetric difference $A \Delta B$ of sets A and B by $A \Delta B = (A \cup B) - (A \cap B)$. Prove $A \Delta B = (A - B) \cup (B - A)$.

12. Prove $an^2 + b$ is $\mathbf{O}(n^2)$, where a and b are arbitrary nonnegative real numbers.

13. Prove $n(n-1)/2$ is not $\mathbf{O}(n)$.

14. Prove n is not $\mathbf{O}(\sqrt{n})$.

15. Prove $\sqrt{n}$ is $\mathbf{O}(n)$.

Recall the principle of mathematical induction with basis step at $n = 0$:

If $p(0)$ and, for all $k \in \mathbb{N}$, $p(k)$ implies $p(k+1)$, then $p(n)$ for all $n \in \mathbb{N}$.

The **principle of strong induction** *states:*

If $q(0)$ and, for all $k \in \mathbb{N}$, $q(j)$, $0 \leq j \leq k$, implies $q(k+1)$, then $q(n)$ for all $n \in \mathbb{N}$.

16. Use the principle of strong induction to prove the principle of mathematical induction. [*Hint:* To prove the principle of mathematical induction, you may assume $p(0)$ and, for all $k \in \mathbb{N}$, $p(k)$ implies $p(k+1)$. Then prove that, for any k, $p(j)$, $0 \leq j \leq k$, implies $p(k+1)$.]

17. Use the principle of mathematical induction to prove the principle of strong induction. [*Hint:* To prove the principle of strong induction, you may assume $q(0)$ and, for all $k \in \mathbb{N}$, if $q(j)$, $0 \leq j \leq k$, then $q(k+1)$. Let $p(m)$ be the statement $q(j)$, $0 \leq j \leq m$, where $m \in \mathbb{N}$. Then prove that, for any k, $p(k)$ implies $p(k+1)$.]

The **well-ordering principle** *for the natural numbers states that every nonempty subset of $\mathbb{N}$ has a least element.*

18. Use the well-ordering principle to prove the principle of mathematical induction. [*Hint:* To prove the principle of mathematical induction, you may assume $p(0)$ and, for all $k \in \mathbb{N}$, $p(k)$ implies $p(k+1)$. Let $A = \{n : p(n)\}$. Use the well-ordering principle to prove $A' = \emptyset$.]

19. Use the principle of strong induction to prove the well-ordering principle. [*Hint:* Use an indirect proof to prove the well-ordering principle. Start by supposing there is a nonempty subset A of $\mathbb{N}$ with no least element. Use the principle of strong induction to derive a contradiction.]

For Exercises 20–21 let $f : A \to B$, $g : B \to A$, and $h : B \to A$ be functions.

20. Prove: If f is onto and $g \circ f = h \circ f$, then $g = h$.

21. Prove: If f is 1–1 and $f \circ g = f \circ h$, then $g = h$.

REFERENCES

Aho, A. V., J. E. Hopcroft, and J. D. Ullman. *Data Structures and Algorithms*. Reading, MA: Addison-Wesley, 1983.

REFERENCES

Beckman, F. S. *Mathematical Foundations of Programming.* Reading, MA: Addison-Wesley, 1981.

De Long, H. *A Profile of Mathematical Logic.* Reading, MA: Addison-Wesley, 1971.

Enderton, H. *A Mathematical Introduction to Logic.* New York: Academic Press, 1972.

Knuth, D. D. *The Art of Computer Programming: Volume 1, Fundamental Algorithms,* 2nd ed. Reading, MA: Addison-Wesley, 1973.

Liu, C. *Elements of Discrete Mathematics,* 2nd ed. New York: McGraw-Hill, 1985.

Lucas, J. F. *Introduction to Abstract Mathematics.* Belmont, CA: Wadsworth, 1986.

Manna, Z. *Mathematical Theory of Computation.* New York: McGraw-Hill, 1974.

CHAPTER 6

INTEGERS AND BINARY NUMBERS

Digital computers store information and data as computer words, which are composed of strings of 0's and 1's called bit strings. Much of the time bit strings are manipulated as binary numbers. So computer scientists need a working knowledge of the arithmetic of binary numbers. The structure of the binary numbers with the usual arithmetic operations has much in common with the structure of the decimal integers.

In this chapter, we first study the integers, including divisibility and the Euclidean algorithm. We then introduce modular arithmetic for the integers. Finally, we develop binary arithmetic with application to digital computers.

APPLICATION

Cryptography (from the Greek *kryptos*, "hidden," and *graphein*, "to write") is the study of secret messages, called cryptograms. Cryptograms are messages intended to be incomprehensible to all except those who know the method to decipher the original text.

The earliest cryptographers were the Egyptians, the ancient Hebrews, and the Indians. The first recorded use of cryptography for correspondence was made by the Spartans, who used a cipher device for secret communications between the military commanders. The first systematic crypto-correspondence began around 1200 in the Papal territory of Italy and the Italian city-states.

Much progress has been made in cryptography since those early years. For example, cryptography had dramatic international effects during World War I and World War II. The recent development of powerful and inexpensive digital computers has freed cryptography from design and cost limitations. Furthermore, theoretical developments in computer science and information theory may lead to provably secure cryptosystems.

Number theory has played a central role in the development of cryptography. Most digital coding schemes make use of modular arithmetic and properties of primes. In Section 6.4, we will see how basic number theory is used to implement an important modern system called a public-key cryptosystem.

6.1
FACTORIZATION IN THE INTEGERS

We assume the standard rules of algebra for the set $\mathbb{Z}$ of all integers. In particular, if $xy = 0$, then $x = 0$ or $y = 0$. However, there is no multiplicative inverse for each integer. For example, there is no integer x such that $5x = 1$.

Because there is no multiplicative inverse for each integer, it is not always possible to find an integer x such that $xd = a$. That is, a/d is not always an integer for a given pair of integers a and d. For special cases, such as $a = 6$ and $d = 3$, we can certainly say that a is divisible by d, or d divides a. In this case, 6 is a multiple of 3, and 3 is a divisor of 6.

For the general case, let $a, d \in \mathbb{Z}$, with $d \neq 0$. The integer d **divides** a (written $d|a$) if there exists an integer m such that $a = md$.

The integer a is a **multiple** of d and d is a **divisor** of a when d divides a. A divisor is also called a **factor**.

The only divisors of 1 are 1 and -1.

THEOREM 6.1.1 Let a and d be integers, with $d \neq 0$. Then the following are true:

    a. $d|0$ **b.** $1|a$ **c.** $d|d$

Proof   **a.** Since $0 = 0 \cdot d$, we have $d|0$.

The proofs of parts b and c are left as exercises. (See Exercises 3 and 4, Exercise Set 6.1.) ∎

THEOREM 6.1.2 Let a, b, d, and e be integers, with $d \neq 0$ and $e \neq 0$. Then the following are true:

    a. If $d|a$ and $d|b$, then $d|(a + b)$.
    b. If $d|a$, then $d|ab$.
    c. If $d|e$ and $e|b$, then $d|b$.
    d. If $d|e$ and $e|d$, then $d = e$ or $d = -e$.

Proof The proofs of parts a and c are left as exercises. (See Exercises 5 and 6, Exercise Set 6.1.)

    b. Assume $d|a$. Then $a = md$ for some integer m. Hence, $ab = (md)b = (mb)d$. Therefore, $d|ab$ since mb is an integer.
    d. Assume $d|e$ and $e|d$. Then $e = md$ and $d = ne$ for some integers m and n. Hence, $e = mne$ and we have $mn = 1$. So $m|1$ and, therefore, $m = 1$ or $m = -1$. If $m = 1$, then $d = e$. If $m = -1$, then $d = -e$. ∎

Let p be a positive integer. By parts b and c of Theorem 6.1.1, we know 1 and p are factors of p. A positive integer p is a **prime** if p has exactly two positive factors. Since 1 has only one positive factor, 1 is not a prime.

EXAMPLE 6.1.3 Here is a list of the primes between 1 and 20, inclusive:

    2, 3, 5, 7, 11, 13, 17, 19   ∎

PRACTICE PROBLEM 1 Find all the primes between 20 and 50, inclusive.

Note that each of the numbers between 2 and 20 that is not a prime has prime factors. This is a special case of the following theorem.

THEOREM 6.1.4 Every integer $n \geq 2$ has a prime factor.

Proof The proof is by mathematical induction. We will prove that every integer in the set $\{2, 3, \ldots, n\}$ has a prime factor for $n = 2, 3, 4, \ldots$.

Basis Step ($n = 2$) The integer 2 clearly has a prime factor—namely, 2 itself.
Induction Step Induction hypothesis: Assume that every integer in the set $\{2, 3, \ldots, k\}$ has a prime factor. If $k + 1$ is a prime, then it has a prime factor, $k + 1$ itself. Assume $k + 1$ is not a prime. Then $k + 1$ has more than two positive factors. Hence, there exists a positive integer d such that $k + 1 = dm$ and $1 < d < k + 1$. By the induction hypothesis, d has a prime factor. Hence, $k + 1$ has a prime factor by Theorem 6.1.2b.

Therefore, every integer in the set $\{2, 3, \ldots, k, k + 1\}$ has a prime factor. ∎

Because of Theorem 6.1.4 and the Fundamental Theorem of Arithmetic (Theorem 6.1.10), to be proved shortly, the prime numbers are considered the "building blocks" of the integers. How many primes are there? The number of primes is infinite. We will prove in Theorem 6.1.5 that, no matter how big a positive integer one can think of, the cardinality of the set of primes is greater than that positive integer.

THEOREM 6.1.5 For every nonnegative integer n, there are at least $n + 1$ primes.

Proof The proof is by mathematical induction.

Basis Step ($n = 0$) We need only exhibit one prime. Pick your favorite.
Induction Step Induction hypothesis: Let k be a nonnegative integer, and assume that there are at least $k + 1$ primes. We list them: $p_0, p_1, \ldots, p_k$. We need only find one more prime. Let $m = 1 + p_0 \cdot p_1 \cdots \cdot p_k$. By Theorem 6.1.4, m has a prime factor q. If q is equal to any of $p_0, p_1, \ldots, p_k$, then q divides $p_0 \cdot p_1 \cdots \cdot p_k$. Since q divides m and $p_0 \cdot p_1 \cdots \cdot p_k$, then q divides $1 = m - p_0 \cdot p_1 \cdots \cdot p_k$. This is a contradiction. Hence, q is a prime different from $p_0, p_1, \ldots, p_k$. ∎

Theorem 6.1.5 is usually stated: There are an infinite number of primes. It was first proved by Euclid over 20 centuries ago. Euclid's proof did not use mathematical induction. It was a proof by contradiction.

PRACTICE PROBLEM 2 Supply a proof by contradiction for Theorem 6.1.5 using the integer m of the proof above.

■ 6.1 FACTORIZATION IN THE INTEGERS

Before going further into the arithmetic of the integers, we state without proof the **division algorithm** for the integers.

ALGORITHM 6.1.6 **THE DIVISION ALGORITHM FOR $\mathbb{Z}$** Let a and d be integers, with $d \neq 0$. There exist unique integers q and r such that

$$a = qd + r \quad \text{and} \quad 0 \leq r < |d|$$

As usual, $|d|$ denotes the absolute value of d. Note the divisor d cannot be zero. After applying the division algorithm to integers a and d, we say a is divided by d. The integers q and r are called the **quotient** and **remainder**, respectively.

The division algorithm is a formal statement of what we call long division. For example, if we use long division to divide 72 by 5 we get a quotient of 14 and a remainder of 2. That is, $72 = 14 \cdot 5 + 2$. Here we have $a = 72, d = 5, q = 14$, and $r = 2$.

PRACTICE PROBLEM 3 Use the division algorithm to find q and r such that $a = qd + r$ and $0 \leq r < |d|$ for each pair of integers a and d.

a. $a = 67, d = 4$ **b.** $a = 17, d = 23$
c. $a = -23, d = 5$ **d.** $a = 23, d = -3$

It is also clear, using the notation of the division algorithm, that $d|a$ if and only if the remainder $r = 0$.

We wish to prove the Unique Factorization Theorem for the integers, also called the Fundamental Theorem of Arithmetic (Theorem 6.1.10). In order to do that, we need a couple of more facts about prime factors.

In the proof of Theorem 6.1.7, we use an intuitively obvious fact about the nonnegative integers called the **well-ordering principle**. Before stating this property, we recall the definition of a least element for a set $B \subseteq \mathbb{Z}$. An integer u is a least element of B if $u \in B$ and $u \leq b$ for all $b \in B$.

Some sets do not contain least elements. For example, there is no least element for the set $\mathbb{Z}$ since, for every element $x \in \mathbb{Z}$, we have $x - 1 \in \mathbb{Z}$ and $x - 1 < x$.

WELL-ORDERING PRINCIPLE

If A is a nonempty subset of $\mathbb{N}$, then A has a least element.

The well-ordering principle is logically equivalent to the principle of mathematical induction. See Chapter 5 Review Exercises 18 and 19 for a proof of this fact.

THEOREM 6.1.7 Let a be a positive integer and p be a prime. If p is not a factor of a, then there exist integers m and n such that $1 = am + pn$.

Proof Consider the set of positive integers of the form $ax + py$. This set is nonempty since the integer $a \cdot 2 + p \cdot 3$, among others, is in the set. By the well-ordering principle, there exists a least positive integer v of the form $v = am + pn$.

We need only prove $v = 1$. To prove $v = 1$ it is sufficient to prove $v|p$ and $v|a$. (Why?) We will prove $v|p$ and leave the proof of $v|a$ as an exercise. (See Exercise 13, Exercise Set 6.1.)

By the division algorithm (Algorithm 6.1.6), there exist unique integers q and r such that $p = qv + r$ with $0 \leq r < v$. By substitution, $p = q(am + pn) + r$. Solving for r we obtain $r = (-mq)a + (1 - nq)p$. Hence, r is of the form $ax + py$ and $0 \leq r < v$. From this it follows that $r = 0$, since v is the least positive integer of the form $ax + py$. Therefore, $v|p$. ∎

PRACTICE PROBLEM 4 Let $a = 20$ and $p = 13$. Find integers m and n such that $1 = am + pn$.

THEOREM 6.1.8 Let p be a prime, and let a and b be positive integers. If p is a factor of ab, then p is a factor of a or p is a factor of b.

Proof Suppose $p|ab$ and p is not a factor of b. By Theorem 6.1.7, $1 = mb + np$ for some integers m, n. Multiplying both sides by a, we obtain $a = m(ab) + (na)p$. By parts b and a, respectively, of Theorem 6.1.2, $p|a$ since $p|ab$ and $p|p$. ∎

Note that Theorem 6.1.8 is not true if p is not a prime. For example, 6 is a factor of $24 = 3 \cdot 8$ but 6 is neither a factor of 3 nor a factor of 8.

COROLLARY 6.1.9 Let p be a prime, and let a_i be positive integers for $i = 1, 2, \ldots, n$. If p is a factor of the product $a_1 \cdot a_2 \cdot \cdots \cdot a_n$, then p is a factor of at least one of the a_i's for $i = 1, 2, \ldots, n$. ∎

The proof of Corollary 6.1.9 is by mathematical induction and is left as an exercise. (See Exercise 12, Exercise Set 6.1.)

We are now ready to prove the important **Fundamental Theorem of Arithmetic**. First, we note that prime factorizations will be considered identical if they differ only in the order of the prime factors.

THEOREM 6.1.10 **THE FUNDAMENTAL THEOREM OF ARITHMETIC** Every integer $n > 1$ has a unique factorization into primes. That is, there exist unique primes $p_1, p_2, \ldots, p_m$ such that $n = p_1 \cdot p_2 \cdot \cdots \cdot p_m$.

Proof The proof is by mathematical induction. We will prove the theorem for every positive integer j, $1 < j \leq n$ for $n = 2, 3, 4, \ldots$. The basis step is easy.

Induction Step Induction hypothesis: Assume the theorem is true for every positive integer j, $2 \leq j \leq k$. By Theorem 6.1.4, we know $k + 1$ has a prime factor. Hence, $k + 1 = pq$, where p is a prime and $1 \leq q \leq k$. If $q = 1$, we are through. If $q \neq 1$, then q has a unique factorization into primes by the induction hypothesis. Suppose $k + 1 = pq = r_0 \cdot r_1 \cdot r_2 \cdots r_t$ is a factorization into primes. Since p is a factor of $k + 1$, then $p = r_i$ for some $i = 0, 1, \ldots, t$ by Corollary 6.1.9. Assume that $p = r_0$. Then $q = r_1 \cdot r_2 \cdots r_t$. Since q has a unique factorization into primes, $k + 1 = pq = p \cdot r_1 \cdot r_2 \cdots r_t$ has a unique factorization into primes. ∎

Here is a nice application of the Fundamental Theorem of Arithmetic.

EXAMPLE 6.1.11 Prove that $\sqrt{3}$ is irrational.

Proof This is a proof by contradiction. Suppose $\sqrt{3}$ is a rational number. Then, $\sqrt{3} = m/n$, where m and n are positive integers with no common factors. Multiplying both sides by n and then squaring, we get $3n^2 = m^2$. By the Fundamental Theorem of Arithmetic, $3n^2$ and m^2 have the same prime factors. Hence, m^2 must have 3 as a prime factor. But, by Theorem 6.1.8, m has 3 as a prime factor also. So, m^2 has 9 as a factor. Hence, $3n^2 = 9k$ and $n^2 = 3k$. Therefore, n^2, and hence n, has 3 as a prime factor. This contradicts the fact that m and n have no common factors. Hence, $\sqrt{3}$ is an irrational number. ∎

PROGRAMMING NOTES

The programming language Pascal has two operators, div and mod, on the integers that are closely related to the division algorithm. First we note that the binary operator /, in Pascal, takes real numbers (or integers) x and y as inputs and outputs a real number x/y. The function trunc takes a real number w as input and outputs the integer part of w. So, for example, trunc(4.7) = 4 and trunc(-3.4) = -3.

In what follows, we study the relationship between a div d and q, where q is the quotient when a is divided by d using the division algorithm (Algorithm 6.1.6). Recall that in the division algorithm, the remainder is always nonnegative. For example, if $a = -9$, $d = 4$, then $-9 = (-3) \cdot 4 + 3$ where $q = -3$ and $r = 3$.

Assume a and d are integers and q is the quotient when a is divided by d.

If $d = 0$, an error occurs.

For $d \neq 0$, a div d is defined to be trunc(a/d).
   If a is positive, then a div $d = q$.
   If a is negative and d is positive, then (a div d) $- 1 = q$.
   If both a and d are negative, then (a div d) $+ 1 = q$.

EXAMPLE 6.1.12
a. 9 div 4 = 2 $9 = 2 \cdot 4 + 1$
b. 9 div (-4) = -2 $9 = (-2)(-4) + 1$
c. (-9) div 4 = -2 $-9 = (-3)4 + 3$
d. (-9) div (-4) = 2 $-9 = 3(-4) + 3$ ∎

For any integer a and positive integer d, $a \bmod d$ is defined to be the smallest nonnegative integer that can be subtracted from a to yield a multiple of d. Hence, $a \bmod d = r$, where r is the remainder when a is divided by d using the division algorithm. If $d < 0$, an error occurs. For example:

$$9 \bmod 4 = 1 \quad \text{since} \quad 9 = 4 \cdot 2 + 1$$
$$(-9) \bmod 4 = 3 \quad \text{since} \quad -9 = 4(-3) + 3$$

Notice that 4 div 9 = 0 and 4 mod 9 = 4.

PRACTICE PROBLEM 5 Find each of the following integers.

    a. 29 div 3   **b.** 29 mod 3   **c.** -21 div 4   **d.** -21 mod 4

EXERCISE SET 6.1

1. Write each of the following integers as a product of powers of primes.
   a. 210   **b.** 1024
2. Write each of the following integers as a product of powers of primes.
   a. 426   **b.** 8128
3. Prove Theorem 6.1.1b.
4. Prove Theorem 6.1.1c.
5. Prove Theorem 6.1.2a.
6. Prove Theorem 6.1.2c.
7. Use mathematical induction to prove $2^{2n} - 1$ is divisible by 3 for $n = 1, 2, 3, \ldots$.
8. Use mathematical induction to prove $n^2 + n$ is divisible by 2 for $n = 0, 1, 2, \ldots$.
9. Use mathematical induction to prove $n^3 + 2n$ is divisible by 3 for $n = 0, 1, 2, \ldots$.
10. Use mathematical induction to prove $n^4 + 3n^2$ is divisible by 4 for $n = 0, 1, 2, \ldots$.
11. Assume p is a prime. Prove that $1 + p^2$ has a prime factor different from p.
12. Prove Corollary 6.1.9.
13. Refer to the proof of Theorem 6.1.7. Prove $v | a$.
14. Prove that $\sqrt{2}$ is irrational.
15. Let p be any odd prime. Prove that $\sqrt{p}$ is irrational.
16. Prove that $\log_2 5$ is irrational, where $\log_2 5 = y$ if and only if $2^y = 5$.
17. Let p be any odd prime. Prove that $\log_2 p$ is irrational, where $\log_2 p = y$ if and only if $2^y = p$.
18. Prove: If a positive integer m has a positive factor greater than $\sqrt{m}$, then m also has a positive factor less than $\sqrt{m}$.
19. Assume n is a positive integer. Prove: If $2^n - 1$ is a prime, then n is a prime.
20. Prove $3 | (7^n - 4^n)$ for $n = 0, 1, 2, \ldots$.
21. Assume x and y are arbitrary unequal integers. Prove: $(x - y) | (x^n - y^n)$ for $n \geq 0$.

6.2 THE EUCLIDEAN ALGORITHM

Exercises 22 and 23 refer to the division algorithm (Algorithm 6.1.6).

22. Let d and k be positive integers. Suppose q is the quotient and r is the remainder when a is divided by d. Prove that q is the quotient and kr is the remainder when ka is divided by kd.

23. Let d and k be positive integers. Suppose q is the quotient when a is divided by d and s is the quotient when q is divided by k. Prove that s is the quotient when a is divided by kd.

■ **PROGRAMMING EXERCISES**

24. Find the integers 78 div 9 and 78 mod 9.
25. Find the integers -78 div -7 and -78 mod -7.
26. Find the integers -78 div 7 and -78 mod 7.
27. Find the integers 78 div -7 and 78 mod -7.
28. Explain why the following algorithm, written as a function in Pascal, is a valid test of whether a given positive integer is a prime. In particular, explain why the while loop tests all the necessary cases. It is assumed that positiveinteger is defined to be of type 1 .. maxint and n and divisor are variables of type positiveinteger. [*Hint:* See Exercise 18.]

```
function ISAPRIME(n: positiveinteger) : Boolean;
    if (n < 4) then
        ISAPRIME := (n > 1)
    else
        if n div 2 = 0 then
            ISAPRIME := false
        else
            begin
                divisor := 3;
                while (divisor <= SQRT(n)) and (n mod divisor <> 0) do
                    divisor := divisor + 2;
                ISAPRIME := (divisor > SQRT(n))
            end;
```

6.2

THE EUCLIDEAN ALGORITHM

In this section we will study the greatest common divisor and the Euclidean algorithm, which is useful for calculating the greatest common divisor. In addition we discuss a closely related concept, the least common multiple.

The Greatest Common Divisor

DEFINITION

Let a, b, and g be positive integers. The **greatest common divisor** g of a and b satisfies the properties:

a. $g|a$ and $g|b$ **b.** If $d|a$ and $d|b$, then $d|g$.

When g is the greatest common divisor of a and b, we write $g = \mathbf{gcd}(a, b)$. In the definition of the greatest common divisor g, Property a makes g a common divisor of a and b whereas Property b makes g the *greatest* common divisor.

EXAMPLE 6.2.1 **a.** $\gcd(18, 48) = 6$ **b.** $\gcd(18, 496) = 2$ ∎

There is a systematic method for finding the $\gcd(a, b)$. Let us find $\gcd(18, 496)$ by using the division algorithm repeatedly:

$$496 = 18 \cdot 27 + 10 \qquad b = aq_1 + r_1, \quad 0 \leq r_1 < a$$
$$18 = 10 \cdot 1 + 8 \qquad a = r_1 q_1 + r_2, \quad 0 \leq r_2 < r_1$$
$$10 = 8 \cdot 1 + 2 \qquad r_1 = r_2 q_3 + r_3, \quad 0 \leq r_3 < r_2$$
$$8 = 2 \cdot 4 + 0 \qquad r_2 = r_3 q_4 + 0$$

The last nonzero remainder is $\gcd(18, 496)$. Hence, $\gcd(18, 496) = 2$. While finding $\gcd(18, 496)$ above, we first divided 496 by 18. If we had divided 18 by 496 first, we would have obtained $18 = 496 \cdot 0 + 18$. So the next step would be to divide 496 by 18 and the remaining steps would be identical to those above. Hence, we need not be concerned whether $a \leq b$ or $b \leq a$.

PRACTICE PROBLEM 1   Find the $\gcd(48, 18)$ by using the division algorithm repeatedly.

Let us prove that the method illustrated above is an algorithm. First we prove a preliminary theorem.

THEOREM 6.2.2   Let a and d be positive integers. If $a = qd + r$ and $0 \leq r < d$, then $\gcd(a, d) = \gcd(d, r)$.

Proof   Assume $a = qd + r$ and $0 \leq r < d$. Let $v = \gcd(a, d)$ and $u = \gcd(d, r)$. Then $u \mid a$ since $u \mid d$, $u \mid r$, and $a = qd + r$. Hence, $u \mid v$ since $u \mid d$ and $v = \gcd(a, d)$. Next $v \mid r$ since $v \mid a$, $v \mid d$, and $r = a - qd$. So, $v \mid u$ since $v \mid d$ and $u = \gcd(d, r)$. Therefore, $v = u$ since $u \mid v$ and $v \mid u$. ∎

The **Euclidean algorithm** is as follows.

ALGORITHM 6.2.3   **EUCLIDEAN ALGORITHM**   To find $\gcd(a, b)$, where a and b are positive integers, apply the division algorithm repeatedly, as follows:

$$b = aq_1 + r_1, \qquad 0 \leq r_1 < a$$
$$a = r_1 q_2 + r_2, \qquad 0 \leq r_2 < r_1$$
$$\vdots$$
$$r_{m-2} = r_{m-1} q_m + r_m, \qquad 0 \leq r_m < r_{m-1}$$
$$r_{m-1} = r_m q_{m+1} + r_{m+1}, \qquad r_m > 0 \text{ and } r_{m+1} = 0$$

Then, $\gcd(a, b) = r_m$.

Proof Since $a > r_1 > r_2 > \cdots$, it is clear that the remainder must eventually be zero; say, $r_{m+1} = 0$. Since r_m is a factor of r_{m-1}, then $\gcd(r_{m-1}, r_m) = r_m$. By repeated use of Theorem 6.2.2, $\gcd(a, b) = \gcd(a, r_1) = \gcd(r_1, r_2) = \cdots = \gcd(r_{m-1}, r_m) = r_m$. ∎

The greatest common divisor of two positive integers $\gcd(a, b)$ can always be written in the form $am + bn$ for some $m, n \in \mathbb{Z}$. This fact is a corollary to Theorem 6.2.4.

THEOREM 6.2.4 Suppose a and b are positive integers. There is a common divisor d of a and b of the form $d = ax + by$, where x and y are integers. Furthermore, $d = \gcd(a, b)$.

Proof The proof is by mathematical induction on $n = a + b$. For the basis step, take $n = 2$. Then, $a = 1$ and $b = 1$. Choose $x = 1$ and $y = 0$. Hence, $d = 1$. Clearly, $d = \gcd(a, b)$.

Induction Step Induction hypothesis: Assume that the theorem is true for any positive integers c and f such that $c + f \leq k$. Suppose $a + b = k + 1$. If $a = b$, we choose $x = 1$ and $y = 0$. Then, $d = a = \gcd(a, b)$. Without loss of generality, we may assume $b < a$. Then, $a - b$ and b are positive integers, and $(a - b) + b \leq k$. By the induction hypothesis, there are integers x and y such that $d = (a - b)x + by$ is a common divisor of $a - b$ and b. Then, $d = ax + b(y - x)$. Since d is a common divisor of $a - b$ and b, then d divides $(a - b) + b = a$. Hence, d is a common divisor of a and b.

Any common divisor g of a and b must also divide $a - b$ and b. Hence, $g | d$, and it follows that $d = \gcd(a, b)$. ∎

COROLLARY 6.2.5 Let $g = \gcd(a, b)$. There exist integers x and y such that $w = ax + by$ if and only if $g | w$.

Proof ($\rightarrow$) This part follows easily from Theorem 6.1.2.

($\leftarrow$) This follows from Theorem 6.2.4 and is left as an exercise. (See Exercise 11, Exercise Set 6.2.) ∎

EXAMPLE 6.2.6 Let $a = 46$ and $b = 64$.

a. Apply the Euclidean algorithm (Algorithm 6.2.3) to find $\gcd(46, 64)$.
b. Use the work in part a to find integers m and n such that $2 = 46m + 64n$.

Solution a. $64 = 46 \cdot 1 + 18$
$46 = 18 \cdot 2 + 10$
$18 = 10 \cdot 1 + 8$
$10 = 8 \cdot 1 + 2$
$8 = 2 \cdot 4$

Hence, $\gcd(46, 64) = 2$.

b. By part a we see that

$$18 = 64 - 46 \cdot 1$$
$$10 = 46 - 18 \cdot 2$$
$$8 = 18 - 10 \cdot 1$$
$$2 = 10 - 8 \cdot 1$$

Working backwards and substituting, we obtain

$$2 = 10 - (18 - 10 \cdot 1)1 = 2 \cdot 10 - 18 = 2(46 - 18 \cdot 2) - 18$$
$$= 2 \cdot 46 - 5 \cdot 18 = 2 \cdot 46 - 5(64 - 46 \cdot 1) = 7 \cdot 46 + (-5)64$$

Hence, $m = 7$ and $n = -5$ are two possible integers. ∎

PRACTICE PROBLEM 2 Let $a = 155$ and $b = 20$. Find $\gcd(a, b)$ and $m, n \in \mathbb{Z}$ such that $\gcd(155, 20) = 155m + 20n$.

Least Common Multiple

Let a and b be positive integers. If we write the sum of two fractions $1/a + 1/b$ as a single fraction, so that the numerator and denominator have no factor in common, the denominator will be the least common multiple of a and b.

EXAMPLE 6.2.7 Let $a = 46$ and $b = 64$.

$$\frac{1}{a} + \frac{1}{b} = \frac{1}{2 \cdot 23} + \frac{1}{2 \cdot 32} = \frac{32 + 23}{2 \cdot 23 \cdot 32}$$

∎

DEFINITION

Let a, b, and v be positive integers. The **least common multiple** v of a and b satisfies the properties:

a. $a|v$ and $b|v$ **b.** If $a|c$ and $b|c$, then $v|c$.

When v is the least common multiple of a and b, we write $v = \mathbf{lcm}(a, b)$. Property a of the definition says v is a common multiple of a and b. Property b says v is the *least* common multiple of a and b.

As we saw in Example 6.2.7, $\mathrm{lcm}(46, 64) = 2 \cdot 23 \cdot 32$. We have already shown $\gcd(46, 64) = 2$. Note that $46 \cdot 64 = (2 \cdot 23)(2 \cdot 32) = 2(2 \cdot 23 \cdot 32) = \gcd(46, 64) \cdot \mathrm{lcm}(46, 64)$. This is no accident, as we see in Theorem 6.2.8.

Before proving Theorem 6.2.8, let us use the Fundamental Theorem of Arithmetic to find $\gcd(60, 100)$ and $\mathrm{lcm}(60, 100)$. The prime factorizations are: $60 = 2^2 \cdot 3 \cdot 5$ and $100 = 2^2 \cdot 5^2$. We take the minimum power of each prime factor to obtain $\gcd(60, 100) = 2^2 \cdot 5$. It is clear that $2^2 \cdot 5$ is a common divisor and that it is the greatest common divisor of 60 and 100. To find $\mathrm{lcm}(60, 100)$, we take the maximum power of each prime factor to obtain

6.2 THE EUCLIDEAN ALGORITHM

$\text{lcm}(60, 100) = 2^2 \cdot 3 \cdot 5^2$. We can see that $2^2 \cdot 3 \cdot 5^2$ is a common multiple and that it is the least common multiple of 60 and 100.

THEOREM 6.2.8 Let a and b be positive integers. Then, $ab = \gcd(a, b) \cdot \text{lcm}(a, b)$.

Proof Using the Fundamental Theorem of Arithmetic, we write each of a and b as a product of powers of primes. So, $a = p_1^{k_1} \cdot p_2^{k_2} \cdot \cdots \cdot p_m^{k_m}$ and $b = p_1^{j_1} \cdot p_2^{j_2} \cdot \cdots \cdot p_m^{j_m}$, where some of the exponents may be zero. By the definition of the greatest common divisor of a and b, we see that

$$\gcd(a, b) = p_1^{v_1} \cdot p_2^{v_2} \cdot \cdots \cdot p_m^{v_m}, \quad \text{where } v_i = \min(k_i, j_i)$$

Similarly,

$$\text{lcm}(a, b) = p_1^{u_1} \cdot p_2^{u_2} \cdot \cdots \cdot p_m^{u_m}, \quad \text{where } u_i = \max(k_i, j_i)$$

But,

$$k_i + j_i = v_i + u_i$$

Therefore,

$$ab = p_1^{k_1+j_1} \cdot p_2^{k_2+j_2} \cdot \cdots \cdot p_m^{k_m+j_m} = p_1^{v_1+u_1} \cdot p_2^{v_2+u_2} \cdot \cdots \cdot p_m^{v_m+u_m}$$
$$= \gcd(a, b) \cdot \text{lcm}(a, b) \qquad \blacksquare$$

Theorem 6.2.8 provides a nice method for calculating $\text{lcm}(a, b)$ from the product ab and the greatest common divisor $\gcd(a, b)$ because

$$\text{lcm}(a, b) = \frac{ab}{\gcd(a, b)}$$

EXAMPLE 6.2.9 Refer to Example 6.2.1.

a. Let $a = 18$ and $b = 496$. We know $\gcd(18, 496) = 2$. So,

$$\text{lcm}(18, 496) = \frac{18 \cdot 496}{2} = 9 \cdot 496 = 4464$$

b. Let $a = 18$ and $b = 48$. We know $\gcd(18, 48) = 6$. So,

$$\text{lcm}(18, 48) = \frac{18 \cdot 48}{6} = 3 \cdot 48 = 144 \qquad \blacksquare$$

PRACTICE PROBLEM 3 Let $a = 155$ and $b = 20$. Find $\text{lcm}(a, b)$ by using the Fundamental Theorem of Arithmetic. Verify that $155 \cdot 20 = \gcd(155, 20) \cdot \text{lcm}(155, 20)$.

■ **PROGRAMMING NOTES**

The following function written in Pascal takes as input positive integers a and b and outputs the greatest common divisor of a and b. We assume that a and b are previously defined as positive integers and that Temp has not been previously defined.

```
function gcd(a, b: integer): integer;
   var
      Temp: integer;
   begin
      while (a > 0) do
         begin
            Temp := a;
            a := b mod a;
            b := Temp
         end;
      gcd := Temp
   end;
```

It is left as an exercise (see Exercise 28, Exercise Set 6.2) to verify that this function implements the Euclidean algorithm and, hence, computes the greatest common divisor of two positive integers.

EXERCISE SET 6.2

1. Find gcd(a, b) and m and n such that $ma + nb = $ gcd(a, b).
   a. $a = 29, b = 11$ **b.** $a = 32, b = 76$

2. Find gcd(a, b) and m and n such that $ma + nb = $ gcd(a, b).
   a. $a = 93, b = 496$ **b.** $a = 45, b = 105$

3. Find lcm(a, b) for each of the following pairs a and b.
   a. $a = 40, b = 68$ **b.** $a = 175, b = 450$

4. Find lcm(a, b) for each of the following pairs a and b.
   a. $a = 196, b = 302$ **b.** $a = 47, b = 90$

5. **a.** Find lcm(21, 96) using the Fundamental Theorem of Arithmetic (Theorem 6.1.10).
   b. Find gcd(21, 96) using the Euclidean algorithm (Algorithm 6.2.3).
   c. Verify that $21 \cdot 96 = $ gcd(21, 96) $\cdot$ lcm(21, 96).

6. **a.** Find lcm(45, 225) using the Fundamental Theorem of Arithmetic.
   b. Find gcd(45, 225) using the Euclidean algorithm.
   c. Verify that $45 \cdot 225 = $ gcd(45, 225) $\cdot$ lcm(45, 225).

7. Use a theorem from this section to justify your answer to each of the following questions.
   a. Can you find integers x and y such that $35x + 49y = 3$?
   b. Can you find integers x and y such that $35x + 49y = 21$?

8. Use a theorem from this section to justify your answer to each of the following questions.
   a. Can you find integers x and y such that $42x + 66y = 3$?
   b. Can you find integers x and y such that $47x + 11y = 1$?

9. **a.** Find gcd(10, 33) and m and n such that $ma + nb = $ gcd(10, 33).
   b. Suppose you have two unmarked beakers that will hold exactly 10 cc and 33 cc, respectively. Describe how you would measure out exactly 1 cc of liquid using only the two beakers.

6.3 MODULAR ARITHMETIC AND BINARY NUMBERS

10. Suppose you have two unmarked beakers that will hold exactly 10 cc and 48 cc, respectively. Can you measure out exactly 1 cc of liquid using only the two beakers? Use a theorem from this section to justify your answer.
11. Finish the proof of Corollary 6.2.5.
12. Use Corollary 6.2.5 to prove Theorem 6.1.7.
13. Suppose p is a prime. Prove that either $\gcd(a, p) = 1$ or $\gcd(a, p) = p$.
14. Assume that all the common divisors of a and b are common divisors of c and d and vice versa. Prove $\gcd(a, b) = \gcd(c, d)$.
15. Prove that $\gcd(a, b) = 1$ if and only if there exist integers m and n such that $1 = am + bn$.
16. Prove: If $\gcd(ab, c) = 1$, then $\gcd(a, c) = 1$ and $\gcd(b, c) = 1$.
17. Suppose $\gcd(a, b) = c$, and write $a = cd$ and $b = cf$. Prove $\gcd(d, f) = 1$.
18. Prove: If $\gcd(a, b) = 1$ and $b|ac$, then $b|c$.
19. Prove: If $\gcd(a, b) = 1$, $ad|c$, and $bd|c$, then $abd|c$.
20. Suppose $\gcd(a, b) = 1$ and $a > b$. Determine the possible values of $\gcd(a + b, a - b)$. Justify your answer.
21. Suppose a is odd and b is even, or vice versa, and $a > b$. Prove $\gcd(a, b) = \gcd(a + b, a - b)$.
22. Prove that $\text{lcm}(b, c) = bc$ if and only if $\gcd(b, c) = 1$.
23. Prove $\text{lcm}(a, ab) = ab$.

    Exercises 24–26 use the following information: Both the $\gcd$ *and the* lcm *can be considered as binary operations on* $\mathbb{Z}$. *Let* $\cdot$ *and* $+$ *be defined by* $a \cdot b = \gcd(a, b)$ *and* $a + b = \text{lcm}(a, b)$.

24. Prove that the operations $\cdot$ and $+$ are commutative and associative.
25. **a.** Find an integer e such that $a + e = a$ for all a.
    b. Either find an integer f such that $a \cdot f = a$ for all a, or explain why no such element exists.
26. Prove a distributive law: $a + (b \cdot c) = (a + b) \cdot (a + c)$.

PROGRAMMING EXERCISES

27. Write a computer program to implement the function gcd given in the Programming Notes.
28. Verify that the function gcd given in the Programming Notes implements the Euclidean algorithm and, hence, computes the greatest common divisor of two positive integers. [*Hint:* Let t_n = Temp after n times through the loop. Prove that $t_{n-1} = t_n q_n + t_{n+1}$ for $n = 1, 2, \ldots$.]

6.3
MODULAR ARITHMETIC AND BINARY NUMBERS

DEFINITION

Let m and n be integers, and let d be some positive integer. Then,

$n \equiv m \ (\textbf{mod } \textbf{\textit{d}})$ if $n - m$ is divisible by d

The statement $n \equiv m \pmod{d}$ is read "n is **congruent** to m **modulo** d."

EXAMPLE 6.3.1 Let $d = 3$.

a. Then $17 \equiv 2 \pmod{3}$ since $17 - 2$ is a multiple of 3.
b. Also, $-5 \equiv 1 \pmod{3}$ since $-5 - 1$ is a multiple of 3. ∎

Note that $k \equiv 0 \pmod{3}$ whenever k is divisible by 3. In this case we say "k is 0 mod 3."

For more complex examples, the division algorithm (Algorithm 6.1.6) is useful.

EXAMPLE 6.3.2 Find r such that $517 \equiv r \pmod{7}$ and $0 \leq r < 7$.

Solution Use long division to get $517/7 = 73 + 6/7$. Hence, $517 \equiv 6 \pmod{7}$. ∎

PRACTICE PROBLEM 1 Use long division to find the remainder r such that

a. $2043 \equiv r \pmod{3}$ and $0 \leq r < 3$ **b.** $759 \equiv r \pmod{8}$ and $0 \leq r < 8$

We proved that the relation $\equiv$ is an equivalence relation on the integers $\mathbb{Z}$ in Section 5.3. Hence, $\equiv$ partitions the integers into equivalence classes $[x]$. We use the division algorithm to characterize these equivalence classes.

It follows from the division algorithm that, for each positive integer d and any integer n, there is a unique r with properties: $n \equiv r \pmod{d}$ and $0 \leq r < d$. So, for a given integer d, every integer n satisfies one of the following: $n \equiv 0 \pmod{d}, n \equiv 1 \pmod{d}, \ldots, n \equiv d - 1 \pmod{d}$. Hence, the equivalence classes are $[0], [1], [2], \ldots, [d-1]$.

EXAMPLE 6.3.3
a. For $d = 2$, the equivalence class $[0]$ contains all the even integers and $[1]$ contains all the odd integers. In other words, $n \equiv 0 \pmod{2}$ means n is even; $n \equiv 1 \pmod{2}$ means n is odd.
b. For any integer n, either $n \equiv 0 \pmod{3}$ or $n \equiv 1 \pmod{3}$ or $n \equiv 2 \pmod{3}$. Hence, the equivalence classes for $d = 3$ are $[0], [1]$, and $[2]$. ∎

In general, when $n \equiv r \pmod{d}$ and $0 \leq r < d$, we say "n is r mod d." So every integer is either 0 mod 3, 1 mod 3, or 2 mod 3. In Example 6.3.2, we found that 517 is 6 mod 7.

Modular Arithmetic

We let $\mathbf{Z}_d = \{[0], [1], \ldots, [d-1]\}$ be the set of equivalence classes. For simplicity, we write $[x]$ as x. So, $\mathbf{Z}_d = \{0, 1, 2, \ldots, d-1\}$.

Let $d = 12$ and consider the clock face shown in Figure 6.3.1. We have designated noon (or midnight) as 0 since 12 is 0 mod 12. Imagine advancing the hand seven hours from 0 to 7. Then advance it nine more hours from 7.

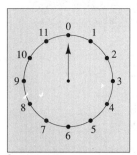

FIGURE 6.3.1

6.3 MODULAR ARITHMETIC AND BINARY NUMBERS

The hand is now pointing to 4. This is an example of modular addition. We have $7 + 9 = 4 \mod 12$. We define **addition mod d**, written $+_d$, on Z_d in the following way:

$$[x] +_d [y] = [x + y]$$

For the current example, $[7] +_d [9] = [7 + 9] = [16] = [4]$.

In Theorem 6.3.4 we will prove that addition mod d is well-defined (in other words, is a binary operation).

PRACTICE PROBLEM 2 Find $[x] +_d [y]$ for

a. $d = 12$, $x = 8$, $y = 7$ b. $d = 5$, $x = 1$, $y = 3$ c. $d = 5$, $x = 3$, $y = 4$

TABLE 6.3.1

$+_5$	0	1	2	3	4
0	0	1	2	3	4
1	1	2	3	4	0
2	2	3	4	0	1
3	3	4	0	1	2
4	4	0	1	2	3

We will write $[x] +_d [y]$ more simply as $x +_d y$, and the result is written as an integer mod d. So, for example, $3 +_5 3 = 1$.

Table 6.3.1 defines addition for Z_5. By the table we see that $4 +_5 2 = 1$, $3 +_5 1 = 4$, etc. We can verify that $+_5$ is a binary operation using Table 6.3.1. Moreover, we can prove that $+_d$ is a binary operation on Z_d for any positive integer d.

THEOREM 6.3.4 Addition mod d, $+_d$, is a binary operation on Z_d.

Proof It is easy to see that $[x] +_d [y] \in Z_d$ if $[x], [y] \in Z_d$. What we need to prove is that the evaluation of $[x] +_d [y]$ does not depend on the choice of element in the equivalence classes $[x]$ or $[y]$.

Assume $[x] = [u]$ and $[y] = [v]$. Then, $x - u$ is a multiple of d, and $y - v$ is also a multiple of d. Hence, $x = u + kd$ and $y = v + jd$ for some integers k and j. Adding we obtain $x + y = u + v + d(k + j)$. Hence, $(x + y) - (u + v)$ is a multiple of d. So, $[x + y] = [u + v]$ and, therefore, $[x] +_d [y] = [u] +_d [v]$. ∎

Multiplication mod d is defined by

$$[x] *_d [y] = [xy]$$

TABLE 6.3.2

$*_5$	0	1	2	3	4
0	0	0	0	0	0
1	0	1	2	3	4
2	0	2	4	1	3
3	0	3	1	4	2
4	0	4	3	2	1

Notational simplifications similar to those for addition apply to multiplication also. For example, $[2] *_5 [4] = [2 \cdot 4] = [8] = [3]$, and this is also written $2 *_5 4 = 3$.

Table 6.3.2 defines multiplication for Z_5. As with addition, we can verify that $*_5$ is a binary operation on Z_5 using Table 6.3.2. We can also prove that $*_d$ is a binary operation on Z_d for any positive integer d.

THEOREM 6.3.5 Multiplication mod d, $*_d$, is a binary operation on Z_d.

Proof It is easy to see that $[x] *_d [y] \in Z_d$ if $[x], [y] \in Z_d$ since $[xy]$ is an element of Z_d by definition. It is left as an exercise to prove that $*_d$ is well-defined. In other words, to prove: If $[x] = [u]$ and $[y] = [v]$, then $[x] *_d [y] = [u] *_d [v]$. (See Exercise 5.) ∎

Binary Numbers

When discussing the set $\mathbb{Z}$ of integers and the set Z_d of integers mod d, all numbers have been written in **base 10** using **decimal** notation. For example, when we write $m = 529$ we mean $m = 5 \cdot 10^2 + 2 \cdot 10 + 9$. In general, $m = d_0 d_1 \ldots d_j$ means $m = d_0 10^j + d_1 10^{j-1} + \cdots + d_j$, where $d_i \in \{0, 1, \ldots, 9\}$.

Decimal numbers are written in terms of powers of 10. When a number is written in terms of the powers of a positive integer b, then b is the **base** of the representation for that number.

For many purposes it is useful to write integers in a base other than 10. The base 2 for binary numbers has become significant since the advent of digital computers. When an integer $n = (b_0 b_1 \ldots b_j)_2$ is written in **base 2** or **binary** notation, it means that

$$n = b_0 2^j + b_1 2^{j-1} + \cdots + b_j, \quad \text{where } b_i \in \{0, 1\}$$

For example, $35 = (100011)_2$ since $35 = 1 \cdot 2^5 + 0 \cdot 2^4 + 0 \cdot 2^3 + 0 \cdot 2^2 + 1 \cdot 2^1 + 1$. In addition, $27 = (11011)_2$ and $43 = (101011)_2$.

As we have seen, the integers 0 and 1 can be considered as Boolean variables. In that role they are binary digits, usually abbreviated bits. The value of binary notation for digital machines is that any integer can be directly encoded in its binary representation as a bit string. Hence, the integer 35 can be represented as the bit string 100011.

EXAMPLE 6.3.6 How many numbers can be represented by 7 bits.

Solution From our previous work we know that there are $2^7 = 128$ bit strings of length 7. Hence, we can represent 128 integers using 7 bits. For example, we can represent $(0000000)_2 = 0, (0000001)_2 = 1, (0000010)_2 = 2, \ldots, (1111111)_2 = 127$. ∎

It is very helpful to have a systematic way of converting from binary to decimal and vice versa. It is easier going from binary to decimal.

EXAMPLE 6.3.7 Write $(11000101)_2$ in decimal notation.

Solution Read the bit string 11000101 *right to left*. Starting with $1 = 2^0$, increase the exponent as each digit is read. Each time the digit is a 1, write the corresponding power of 2 as a summand. For $(11000101)_2$ we obtain $2^0 + 2^2 + 2^6 + 2^7 = 1 + 4 + 64 + 128 = 197$. ∎

The method used in Example 6.3.7 can easily be generalized to an algorithm for converting any nonnegative integer in binary form to a decimal number. It is left as an exercise to write an algorithm for converting the

EXAMPLE 6.3.8 Write 29 in binary notation.

Solution Take the highest power of 2 that is less than or equal to 29; this is 16. Write $29 = 16 + 13$. Take the highest power of 2 that is less than or equal to 13; this is 8. Write $29 = 16 + 8 + 5$. Repeat, until we obtain 29 as a sum of powers of 2. Hence, $29 = 16 + 8 + 4 + 1 = 2^4 + 2^3 + 2^2 + 2^0$. Therefore, $29 = (11101)_2$. ∎

We discuss an algorithm for converting nonnegative integer decimal numbers to the corresponding binary number shortly. (See Algorithm 6.3.10.) In the meantime the method illustrated in Example 6.3.8 can be used for relatively small decimal numbers.

PRACTICE PROBLEM 3
a. Write $(10011010)_2$ in decimal notation.
b. Write 169 in binary notation.

Addition of Binary Numbers

When we add two decimal numbers we place them so that the corresponding digits are in the same column. Then we add column by column right to left while keeping track of the carry from the previous column. Following are two examples of the addition of pairs of decimal integers. In both cases there are carries of 0 or 1 from a column to the adjacent column on the left.

$$
\begin{array}{r}
6735 \\
+\ 569 \\
\hline
7304
\end{array}
\qquad
\begin{array}{r}
1\ 1\ 1 \\
6\ 7\ 3\ 5 \\
0\ 5\ 6\ 9 \\
\hline
7\ 3\ 0\ 4
\end{array}
$$

$$
\begin{array}{r}
9673 \\
+\ \ 56 \\
\hline
9729
\end{array}
\qquad
\begin{array}{r}
0\ 1\ 0 \\
9\ 6\ 7\ 3 \\
0\ 0\ 5\ 6 \\
\hline
9\ 7\ 2\ 9
\end{array}
$$

Addition of binary numbers works in a similar way. An example follows:

$$
\begin{array}{r}
10110 \\
+\ \ 101 \\
\hline
11011
\end{array}
\qquad
\begin{array}{r}
0\ 1\ 0\ 0 \\
1\ 0\ 1\ 1\ 0 \\
0\ 0\ 1\ 0\ 1 \\
\hline
1\ 1\ 0\ 1\ 1
\end{array}
$$

In binary notation, $(10110)_2 + (101)_2 = (11011)_2$. The corresponding addition in decimal notation is $22 + 5 = 27$, which checks.

EXAMPLE 6.3.9 Find the sum $(101101)_2 + (11001)_2$ using binary addition. Then check the result with the corresponding decimal numbers.

Solution

$$\begin{array}{rr} 101101 & 45 \\ +\ 11001 & +25 \\ \hline 1000110 & 70 \end{array}$$

■

PRACTICE PROBLEM 4 Verify the solution to Example 6.3.9.

Converting Decimal Numbers to Binary Numbers

In Example 6.3.8 we illustrated a method for converting decimal numbers to binary numbers by converting 29 to its base 2 equivalent. Here is a different method involving repeated division, which can be implemented as an algorithm more easily:

Divide 29 by 2.	$29 = 2 \cdot 14 + \mathbf{1}$
Divide the previous quotient (14) by 2.	$14 = 2 \cdot 7 + \mathbf{0}$
Divide the previous quotient (7) by 2.	$7 = 2 \cdot 3 + \mathbf{1}$
Divide the previous quotient (3) by 2.	$3 = 2 \cdot \mathbf{1} + \mathbf{1}$
Stop when the quotient is less than 2.	

Obtain the digits for the binary number by reading, from the bottom up, the last quotient followed by all the remainders to obtain 11101. This result, written *left to right*, is the desired binary number.

Let us see why this method works. Recall that $29 = 16 + 8 + 4 + 1$. We reverse the steps taken above, substituting for each quotient in turn, starting with the next to last quotient:

$7 = 2 \cdot 3 + 1 = 2(2 \cdot 1 + 1) + 1$ Since $3 = 2 \cdot 1 + 1$
$14 = 2 \cdot 7 + 0 = 2[2(2 \cdot 1 + 1) + 1] + 0$ By substitution from the last step
$29 = 2 \cdot 14 + 1 = 2\{2[2(2 \cdot 1 + 1) + 1] + 0\} + 1 = 16 + 8 + 4 + 1$

Recall that n mod 2 and n div 2 are the remainder and quotient, respectively, after dividing n by 2. Here is the algorithm written in pseudocode.

ALGORITHM 6.3.10 This algorithm converts a nonnegative integer N written in base 10 to a base 2 number.

$Q := N$ and $J := 0$.
Repeat until $Q < 2$
   $J := J + 1$ and
   $R := Q$ mod 2 and
   $Q := Q$ div 2 and
   $R[J] := R$.
Return the binary number $(QR[J]R[J-1]\ldots R[1])_2$.

PRACTICE PROBLEM 5 Use Algorithm 6.3.10 to convert the decimal number 784 to a binary number.

Storage of Numbers in a Digital Computer

Storage of numbers in a digital computer is accomplished by placing them in registers. We illustrate this idea with a 4-bit register. See Figure 6.3.2. A clock face is used to represent the register to make clear that the register has a modulus. The 4-bit register in Figure 6.3.2 is a mod-16 register. A 4-bit register has 16 positions and can be used to store any number from 0 to 15.

As with the clock in Figure 6.3.1, addition of numbers in the register can be accomplished by successively advancing the pointer. For example, to find $3 + 5 = (0011)_2 + (0101)_2$, we advance the pointer to the position 0011, then advance it further five places to 1000, thus obtaining $(1000)_2 = 8$ as the answer. When adding two numbers using this method, the result will be correct only when that result falls in the range of 0 to 15. Otherwise, the answer is correct mod 16. If we add 13 to 9 on this register, we obtain $13 + 9 = 22 \equiv 6 \pmod{16}$.

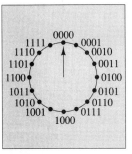

FIGURE 6.3.2

EXERCISE SET 6.3

1. Find the integer mod 7 for each of the following.
   a. 738   **b.** 93

2. Find the integer mod 6 for each of the following.
   a. 731   **b.** 9304

3. Find the integer mod d for 517 when
   a. $d = 3$ **b.** $d = 6$

4. Find the integer mod d for 843 when
   a. $d = 7$ **b.** $d = 8$

5. Prove Theorem 6.3.5. That is, prove: If $x \equiv u \pmod{d}$ and $y \equiv v \pmod{d}$, then $xy \equiv uv \pmod{d}$.

6. Without using Theorem 6.3.5 or any equivalent statement, prove: If $x \equiv y \pmod{d}$, then $wx \equiv wy \pmod{d}$ for any integer w.

For Exercises 7–10 write each of the given binary numbers as a decimal number.

7. **a.** $(1011)_2$ **b.** $(11001)_2$
8. **a.** $(1101)_2$ **b.** $(10110)_2$
9. **a.** $(1001110)_2$ **b.** $(110011101)_2$
10. **a.** $(1000111)_2$ **b.** $(1010011101)_2$

In Exercises 11–14 compute the indicated sum of two binary numbers using binary addition. Convert the numbers to decimal notation. Check your results.

11. $(10010)_2 + (1101)_2$
12. $(10011)_2 + (11010)_2$
13. $(100101)_2 + (11011)_2$
14. $(110101)_2 + (1001)_2$

For Exercises 15–18 write each of the given decimal numbers as a binary number. Use the method of Example 6.3.8 and then use Algorithm 6.3.10 for each number.

15. 42
16. 60
17. 105
18. 123

19. Find a proof or a counterexample for the conjecture: If $ab \equiv 0 \pmod{n}$, then $a \equiv 0 \pmod{n}$ or $b \equiv 0 \pmod{n}$.
20. Find a proof or a counterexample for the conjecture: If p is a prime and $ab \equiv 0 \pmod{p}$, then $a \equiv 0 \pmod{p}$ or $b \equiv 0 \pmod{p}$.
21. Prove: If p is a prime and $a^2 \equiv b^2 \pmod{p}$, then $a \equiv \pm b \pmod{p}$.
22. Prove: If $ac \equiv bc \pmod{n}$ and $\gcd(c, n) = 1$, then $a \equiv b \pmod{n}$.
23. Prove: If m is a factor of n and $a \equiv b \pmod{n}$, then $a \equiv b \pmod{m}$.

■ PROGRAMMING EXERCISES

24. Write a procedure in Pascal to implement the method used in Example 6.3.7 for converting a binary number to the corresponding decimal number.
25. Write a computer program to implement Algorithm 6.3.10.

6.4
APPLICATION: CRYPTOGRAPHY AND FACTORING

A **public-key cryptosystem** allows communication of secret messages over public channels in such a way that eavesdroppers cannot decipher the content. The **RSA scheme**, named after its inventors Rivest, Shamir, and Adleman, is an implementation of such a system.

In a public-key system, a receiver (of messages) places an encryption key in the public file, so it is available to everyone, including the sender of messages and any potential eavesdroppers. Only the receiver has the decryption keys, which allow him or her to decipher the messages rapidly. Eavesdroppers need to perform an exponential amount of computation to crack the code.

First we describe how the RSA scheme works, then we illustrate it by a simple example. The RSA scheme has three components: the sender of the message, the receiver of the message, and the public file. The receiver chooses two distinct large prime numbers p and q. These are unknown to everyone else including the sender.

The receiver places two positive integers n and r in the public file, where

a. $n = pq$, and
b. r is a large integer chosen so that the greatest common divisor of r and $(p-1)(q-1)$ is 1.

To prepare for deciphering messages from the sender, the receiver calculates the **multiplicative inverse** $s \bmod (p-1)(q-1)$ of the integer r. In other words, the receiver finds an integer s such that

$$rs \equiv 1 \pmod{(p-1)(q-1)}$$

We note that $rs = 1 + t(p-1)(q-1)$ for some integer t.

The calculation for s is fast for the receiver, who knows p and q. The eavesdropper, for whom p and q are unknown, cannot find s.

The integers n and r are the **encryption keys** placed in the public file. The integers p, q, and s are the **decryption keys**, known only to the receiver of messages. Everyone can be a receiver of messages by placing his or her own encryption keys in the public file, which is a directory of the encryption keys of each user. Anyone wishing to send a message to a particular user needs to look up the encryption keys of that user.

The sender can represent the message as a string of bits. The string of bits, when considered as a binary number, has an integer value M. We can assume $0 \leq M \leq n - 1$, since, if the integer M is too large, the sender can break a long message into smaller blocks. At this point, the summary of information is:

The sender knows M.
Everyone knows n and r.
The receiver knows p, q, and s.

See Figure 6.4.1.

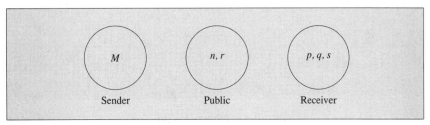

FIGURE 6.4.1

After looking up the encryption keys n and r of the receiver in the public file, the sender disguises the message M by transmitting C over public channels, where

$C \equiv M^r \pmod{n}$

Note that C is the remainder of M^r after division by n. After receiving C, the receiver uses modular arithmetic to calculate C^s to recover the original message M as follows:

$C^s \equiv (M^r)^s$ Since $C \equiv M^r \pmod{n}$
$\equiv M^{rs}$
$\equiv M^{1+t(p-1)(q-1)}$ Since $rs = 1 + t(p-1)(q-1)$ for some integer t
$\equiv M \pmod{n}$

To justify the final step of the calculation, we need the following important theorem, which was stated in 1640 by the mathematician Pierre de Fermat (1601–1665). **Fermat's Theorem** is as follows.

THEOREM 6.4.1   **FERMAT'S THEOREM**   If a is an integer and p is a prime such that $\gcd(a, p) = 1$, then

$a^{p-1} \equiv 1 \pmod{p}$

In our application, $a = M$. By Fermat's Theorem, we have

$$M^{p-1} \equiv 1 \pmod{p}$$

for all M such that $\gcd(M, p) = 1$. In other words, p does not divide M. Since $p - 1$ is a factor of $(p - 1)(q - 1)$, we have

$$M^{(p-1)(q-1)} \equiv 1 \pmod{p} \quad \text{and} \quad M^{1+t(p-1)(q-1)} \equiv M \pmod{p}$$

for all M such that p does not divide M.

If p divides M, then $M^{1+t(p-1)(q-1)} \equiv M \pmod{p}$ holds trivially. It follows that

$$M^{1+t(p-1)(q-1)} \equiv M \pmod{p}$$

for all M, where $0 \leq M \leq n - 1$.

By a similar argument for q, we have

$$M^{1+t(p-1)(q-1)} \equiv M \pmod{q}$$

for all M, where $0 \leq M \leq n - 1$.

Since p and q are primes and $n = pq$, it follows from elementary number theory (see Exercise 7, Exercise Set 6.4) that

$$M^{1+t(p-1)(q-1)} \equiv M \pmod{n}$$

for all M, where $0 \leq M \leq n - 1$.

We note that the eavesdropper is unable to calculate C^s without knowing s. Although the eavesdropper knows n and r, he or she cannot calculate s without knowing p and q. Furthermore, the sender uses n and r from the public file; the sender does not need to know p and q. Therefore, the receiver does not have to risk communicating p and q to the sender. Of course, the receiver should be aware that the adversary might discover p and q and keep that fact a secret. Therefore, the receiver should periodically change the primes for security reasons.

We now illustrate the RSA scheme by a very simple example with small primes.

EXAMPLE 6.4.2 Suppose the receiver chooses two primes, $p = 3$ and $q = 5$. Next, the receiver chooses an integer r such that $\gcd(r, (p - 1)(q - 1)) = 1$; say, $r = 11$. Note that $(p - 1)(q - 1) = 8$ and $\gcd(11, 8) = 1$. The receiver calculates the multiplicative inverse $s \pmod{(p - 1)(q - 1)}$ of r:

$$rs \equiv 1 \pmod{(p - 1)(q - 1)} \quad \text{or} \quad 11s \equiv 1 \pmod{8}$$

One method of finding the multiplicative inverse of 11 is to divide each integer of the form $8k + 1$, where $k = 1, 2, 3, \ldots$, by 11 until we find one that is divisible by 11. The resulting quotient is the multiplicative inverse we need. In this case, 11 divides $8 \cdot 4 + 1 = 33$ and, hence, $s = 3$.

6.4 APPLICATION: CRYPTOGRAPHY AND FACTORING

Now, the public file has the encryption keys $n = 15$ and $r = 11$. The receiver has the decryption keys $p = 3$, $q = 5$, and $s = 3$.

Suppose the sender wants to send a message $M = 12$. Note that $0 \leq 12 \leq 14 = 15 - 1$. The sender looks at the public encryption keys $n = 15$ and $r = 11$, then calculates $C \equiv 12^{11} \pmod{15}$ and transmits C.

The receiver calculates C^s to find

$$C^s \equiv (M^r)^s \equiv (12^{11})^3 \equiv 12^{33} \equiv 12^{1+4 \cdot 8} \equiv 12 \pmod{15}$$

The last step of this equation follows from Fermat's Theorem (Theorem 6.4.1). Now, the receiver has recovered the message $M = 12$. ∎

Since $n = pq$ is publicly known and it is crucial that the adversary not discover the primes p and q, the receiver must create and use large primes p and q. In attempting to find the primes p and q and, hence, s, the adversary needs to factor a large integer n into primes to find p and q. Factoring large integers into primes is presumed, though never proved, to be a computationally intractable problem. A problem is *computationally intractable* if no fast algorithm will solve all instances of it. Thus, the adversary is faced with a very difficult problem of factoring large integers into primes. It is this apparent computational difficulty that provides the security of the RSA scheme.

If the receiver suspects that the adversary has discovered p and q, then p and q can be changed without having to send any secret keys to the sender. This is a significant advantage of public-key systems in general.

In summary, we have shown how a presumably computationally intractable problem in number theory—namely, factoring large integers into primes—can be used to create a relatively secure system for transmitting messages over public communication channels. Since the security of the RSA public-key cryptosystem depends on the difficulty of factoring large integers into primes, the absolute guarantee of security can be achieved only when the factoring problem is proved to be computationally intractable.

EXERCISE SET 6.4

1. Find a number s such that $3s \equiv 1 \pmod 4$.
2. Find an integer s such that $17s \equiv 1 \pmod{72}$.
3. Suppose the receiver picks $p = 5$, $q = 2$, and $r = 7$. Describe how the RSA system is used to encrypt and decrypt the message $M = 2$.
4. Repeat Exercise 3 for $p = 7$, $q = 11$, $r = 7$, and $M = 15$.
5. Factor the number 9699690 into primes.
6. Find the greatest common divisor of 1729 and 15,210.
7. Suppose p and q are nonidentical primes and $n = pq$. Prove that, if $a \equiv b \pmod p$ and $a \equiv b \pmod q$, then $a \equiv b \pmod n$.

KEY TERMS

Divides
Multiple
Divisor
Factor
Prime
Division algorithm
Quotient
Remainder
Well-ordering principle
Fundamental Theorem of Arithmetic
Greatest common divisor (gcd)
Euclidean algorithm
Least common multiple (lcd)
Mod d

Congruent (modulo d)
Z_d
Addition mod d ($+_d$)
Multiplication mod d ($*_d$)
Base 10
Decimal
Base 2
Binary
Public-key cryptosystem
RSA scheme
Multiplicative inverse s mod n
Encryption keys
Decryption keys
Fermat's Theorem

REVIEW EXERCISES

1. **a.** Find lcm(46, 120) using the Fundamental Theorem of Arithmetic (Theorem 6.1.10).
    b. Find gcd(46, 120) using the Euclidean algorithm (Algorithm 6.2.3).
    c. Verify that $46 \cdot 120 = \gcd(46, 120) \cdot \text{lcm}(46, 120)$.

2. **a.** Find lcm(82, 205) using the Fundamental Theorem of Arithmetic.
    b. Find gcd(82, 205) using the Euclidean algorithm.
    c. Verify that $82 \cdot 205 = \gcd(82, 205) \cdot \text{lcm}(82, 205)$.

In Exercises 3 and 4 compute the indicated sum of two binary numbers using binary addition. Convert the numbers to decimal notation and check your results.

3. $(100101)_2 + (11101)_2$

4. $(1001)_2 + (110110)_2$

For Exercises 5 and 6 use both the method of Example 6.3.8 and Algorithm 6.3.10 to write each of the given decimal numbers as a binary number.

5. 63

6. 130

7. Use mathematical induction to prove $5 | (8^n - 3^n)$ for $n = 0, 1, 2, \ldots$.

8. Prove that $4^n \equiv 1 \pmod{3}$ for $n = 0, 1, 2, \ldots$.

9. Assume $\gcd(a, b) = 1$. Prove: If $a | c$ and $b | c$, then $ab | c$.

10. Prove that $6 | (n^3 - n)$ for $n = 0, 1, 2, \ldots$.

11. Prove that $n^5 \equiv n \pmod{30}$ for $n = 0, 1, 2, \ldots$. [*Hint:* Use Exercises 9 and 10.]

12. Find a proof or a counterexample for the conjecture: If $\gcd(a, b) = d$ and ad is a factor of cb, then a is a factor of c.

13. Prove: If p is a prime and p is a factor of a^n, then p is a factor of a for $n = 1, 2, \ldots$.

14. Prove that $\gcd(ac, bc) = c \cdot \gcd(a, b)$.

15. Prove that $\text{lcm}(ac, bc) = c \cdot \text{lcm}(a, b)$.

16. Define $f: \mathbb{N} \times \mathbb{N} \to \mathbb{N}$ by $f(m, n) = 2^m 3^n$. Prove that f is a 1–1 function.

REFERENCES

Apostol, T. M. *Introduction to Analytic Number Theory.* New York: Springer-Verlag, 1976.

Diffie, W., and M. E. Hellman. "New Directions in Cryptography." *IEEE Transactions on Information Theory*, 22, 6 (1976): 644–654.

Hardy, G. H., and E. M. Wright. *An Introduction to the Theory of Numbers*, 5th ed. NewYork: Oxford University Press, 1979.

Hellman, M. E. "The Mathematics of Public-Key Cryptography." *Scientific American*, 241, 2 (1979): 146–157.

Niven, I., and H. S. Zuckerman. *Introduction to the Theory of Numbers.* New York: Wiley, 1980.

Rivest, R. L., A. Shamir, and L. M. Adleman. "A Method for Obtaining Digital and Public-Key Cryptosystems." *Communication of the ACM*, 21, 2 (1978): 120–126.

Wilf, H. S. *Algorithms and Complexity.* Englewood Cliffs, NJ: Prentice-Hall, 1986.

CHAPTER 7

BOOLEAN ALGEBRA

Logic design underlies data processing systems development, both hardware and software. It is applied to determine how basic logic units, called logic gates, are combined to form a logic network that will perform a specific task.

In this chapter, we will study Boolean algebras, Boolean expressions, Boolean functions, and their applications to logic networks. Taken together, these topics constitute the field of logic design. Furthermore, we will learn to simplify Boolean expressions and thereby enhance the efficiency of logic network design.

APPLICATION

The seven-segment display on most calculators and digital clocks is a familiar sight. Each digit from 0 through 9 is displayed by lighting the appropriate line segments in a seven-segment configuration, as shown in the figure below.

Seven-segment display for the digits 0, 1, ..., 9

Without the seven-segment display or something similar, calculators would be virtually useless. The application in Section 7.5 shows how Boolean algebra and logic networks are used to design electronic circuits that activate appropriate segments in response to specific input.

7.1
BOOLEAN ALGEBRA

The mathematical structure now called Boolean algebra was formulated by George Boole (1815–1864) in his 1854 publication "An Investigation of the Laws of Thought." In 1938, C. E. Shannon showed that Boolean algebra was essential for the analysis and synthesis of logic networks. Since then, the application of Boolean algebra has been central to logic design.

EXAMPLE 7.1.1 Suppose a single light bulb in the middle of a long hallway is controlled by switches at each end of the hallway. The light bulb should be on when either, but not both, of the switches is on. Otherwise it should be off. How might we design a logic network to control the light bulb? This network has two inputs and one output. The inputs are the states of the switches, and the output is the state of the light bulb. The switches can either be on or off, and the light bulb can either be on or off. We can picture the switches as inputs to some unspecified logic network, represented by the black box, as shown in Figure 7.1.1. We will use Boolean algebra to analyze this problem more fully in Example 7.1.3. ∎

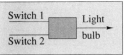

FIGURE 7.1.1

Boolean algebra is an essential tool for designing networks to solve input and output problems like this simple example and others far more complex. It is the basic mathematics required for the study of the logic design of digital systems.

Digital computers use two-state devices that produce two distinct output signals. The two states can be symbolized as 0 and 1. We will show that the set $\{0, 1\}$, with appropriate operations, forms a Boolean algebra (see Example 7.1.2). This is the two-element Boolean algebra that is most useful to logic design.

Although applications of Boolean algebra to logic design motivate its study, one of the most familiar examples of a Boolean algebra is from set theory. Consider the set of all subsets of U, denoted $P(U)$. The set union ($\cup$) and set intersection ($\cap$) are binary operations on $P(U)$, and the set complement ($'$) is a unary operation on $P(U)$. Recall that the following ten laws hold for all X, Y, and Z in $P(U)$.

BOOLEAN LAWS FOR SET THEORY

1. a. $X \cup Y = Y \cup X$ b. $X \cap Y = Y \cap X$
2. a. $(X \cup Y) \cup Z = X \cup (Y \cup Z)$ b. $(X \cap Y) \cap Z = X \cap (Y \cap Z)$
3. a. $X \cup (Y \cap Z) = (X \cup Y) \cap (X \cup Z)$ b. $X \cap (Y \cup Z) = (X \cap Y) \cup (X \cap Z)$
4. a. $X \cup \emptyset = X$ b. $X \cap U = X$
5. a. $X \cup X' = U$ b. $X \cap X' = \emptyset$

The system $(P(U), \cup, \cap, ', \emptyset, U)$ is an example of a Boolean algebra. With this system as a model, we state the definition of a Boolean algebra.

DEFINITION

A **Boolean algebra** $(B, +, \cdot, ', 0, 1)$ consists of a nonempty set B, binary operations $+$ and $\cdot$ on B, a unary operation $'$ on B, and specified elements 0 and 1 in B such that the following **axioms for Boolean algebra** hold for all x, y, and z in B.

BOOLEAN ALGEBRA AXIOMS

1. a. $x + y = y + x$ b. $x \cdot y = y \cdot x$
2. a. $(x + y) + z = x + (y + z)$ b. $(x \cdot y) \cdot z = x \cdot (y \cdot z)$
3. a. $x + (y \cdot z) = (x + y) \cdot (x + z)$ b. $x \cdot (y + z) = x \cdot y + x \cdot z$
4. a. $x + 0 = x$ b. $x \cdot 1 = x$
5. a. $x + x' = 1$ b. $x \cdot x' = 0$

Since $+$ and $\cdot$ are binary operations on B, $x + y$ and $x \cdot y$ are in B if x and y are in B. Similarly, because $'$ is a unary operation on B, $x' \in B$ if $x \in B$.

Each of these axioms has a descriptive name. Axiom 1 is called the *commutative law*; Axiom 2, the *associative law*; Axiom 3, the *distributive law*; Axiom 4, the *identity law*; and Axiom 5, the *complement law*. Axioms 4a and 4b define the elements 0 and 1; Axioms 5a and 5b define x' for $x \in B$. We call x' the *complement* of x.

These axioms are given in pairs. Each statement in a pair (such as Axioms 3a and 3b) is the dual of the other statement in the pair. The **dual** of a statement is obtained by simultaneously replacing $+$ by $\cdot$, $\cdot$ by $+$, 0 by 1, and 1 by 0. In a Boolean algebra, if a statement is true, then its dual is also true. This is called the **principle of duality**.

The 0 and 1 in the Boolean algebra $(B, +, \cdot, ', 0, 1)$ are called the **zero element** and the **one element**, respectively. The operations $+$, $\cdot$, and $'$ are called the *sum*, *product*, and *complement*, respectively.

When the context is clear and there is no danger of confusion, we write B, instead of $(B, +, \cdot, ', 0, 1)$, to denote a Boolean algebra.

The operations for a Boolean algebra follow a precedence hierarchy. Parenthesized expressions are performed first, as in any algebraic procedure. Among the operations $+$, $\cdot$, and $'$, the unary operation $'$ has the highest precedence. Next is the binary operation $\cdot$, followed by $+$. Finally, the equality sign in an equation has the lowest precedence. Parentheses are used for clarity but can be omitted in many cases by using the precedence hierarchy. So, $(x \cdot y) + z$ can be written as $x \cdot y + z$, and $z \cdot (y') + x$ as $z \cdot y' + x$. For simplicity, we will usually write $x \cdot y$ as xy. Hence, $x \cdot (y' + x \cdot z) = x(y' + xz)$.

EXAMPLE 7.1.2   **TWO-ELEMENT BOOLEAN ALGEBRA**   Let $(B, +, \cdot, ', 0, 1)$ be a system with $B = \{0, 1\}$, where the operations $+$, $\cdot$, and $'$ are defined as follows:

$0 + 0 = 0$ $0 \cdot 0 = 0$ $0' = 1$
$0 + 1 = 1$ $0 \cdot 1 = 0$ $1' = 0$
$1 + 0 = 1$ $1 \cdot 0 = 0$
$1 + 1 = 1$ $1 \cdot 1 = 1$

7.1 BOOLEAN ALGEBRA

With these operations, the system $(B, +, \cdot, ', 0, 1)$ is a Boolean algebra. The zero element is 0, and the one element is 1. It is possible, though tedious, to verify the ten Boolean algebra axioms for all different choices of elements in $B = \{0, 1\}$. We omit the details here. ∎

PRACTICE PROBLEM 1 Refer to Example 7.1.2. Verify Boolean algebra axioms 4a and 5b with $x = 1$.

EXAMPLE 7.1.3 **LIGHT BULB PROBLEM** Consider the light bulb problem posed in Example 7.1.1. The light in the middle of a long hallway is controlled by switches on each end. Let S_1 and S_2 denote the switches and L denote the light bulb. We will use 0 for the off state, and 1 for the on state. We want L to be 1 when either of the switches is 1, and L to be 0 when both of the switches are 0 or when both of the switches are 1. The relationship between the light bulb and the switches can be described by Table 7.1.1. The rightmost column shows the states of L, which depend on the states of S_1 and S_2. ∎

TABLE 7.1.1

S_1	S_2	L
0	0	0
0	1	1
1	0	1
1	1	0

PRACTICE PROBLEM 2 Modify the light bulb problem as follows. Suppose we want the light to go on only when both switches are on. As in Table 7.1.1, describe the state of the light bulb (L) in terms of the states of the two switches (S_1 and S_2).

In the Boolean algebra $(P(U), \cup, \cap, ', \emptyset, U)$, note that $B = P(U)$; the operations are $\cup$, $\cap$, and $'$; and the zero and one elements are the empty set $\emptyset$ and the universal set U, respectively. For any set U, the system $(P(U), \cup, \cap, ', \emptyset, U)$ is a Boolean algebra. If U is a set of n elements, then $P(U)$ has 2^n elements, and the system $(P(U), \cup, \cap, ', \emptyset, U)$ is a Boolean algebra with 2^n elements.

PRACTICE PROBLEM 3 Let $U = \{a, b\}$. Verify Boolean algebra axioms 5a and 5b for all sets X in the power set $P(U)$.

The 0 and 1 can be considered the generic zero element and one element of a Boolean algebra. Each example of a Boolean algebra will have a specific zero and a specific one, which will usually be designated with symbols other than 0 and 1. As we have noted, the zero and one in the Boolean algebra $(P(U), \cup, \cap, ', \emptyset, U)$ are the empty set $\emptyset$ and the set U, respectively.

There are many important but elementary theorems in Boolean algebra, which can be proved from the Boolean algebra axioms. Once we have proved them, we will have useful facts that are true for every example of a Boolean algebra. Many of these facts will be necessary for the applications we discuss later in this chapter.

Each Boolean algebra theorem has parts a and b. Part b is the dual statement of part a, and vice versa. Since the dual of a theorem is a theorem by the principle of duality, we need only prove either part a or part b in a theorem. We usually refer to the axioms and theorems in Boolean algebra as **Boolean algebra laws**. We prove several of the most important theorems in this section. The proofs of many Boolean algebra laws are left for Exercise Set 7.1.

Consider a Boolean algebra $(B, +, \cdot, ', 0, 1)$. We assume that x and y are arbitrary elements of B in what follows.

Following is a simplification theorem that will be particularly useful as a basis for Karnaugh maps, discussed in Section 7.3.

THEOREM 7.1.4 **a.** $xy + xy' = x$ **b.** $(x + y)(x + y') = x$

Proof of a $xy + xy' = x(y + y') = x \cdot 1 = x$ ∎

The next theorem will be very useful for proving subsequent theorems of Boolean algebra.

THEOREM 7.1.5 **UNIQUE COMPLEMENT LAW** For each element $x \in B$, x' is the unique element in B satisfying the properties $x + x' = 1$ and $xx' = 0$.

Proof   We need to prove: If $x + y = 1$ and $xy = 0$, then $y = x'$. Assume $x + y = 1$ and $xy = 0$. Then

$$y = y \cdot 1 = y(x + x')$$
$$= yx + yx' = xy + x'y$$
$$= 0 + x'y = xx' + x'y$$
$$= x'x + x'y = x'(x + y)$$
$$= x' \cdot 1 = x'$$ ∎

PRACTICE PROBLEM 4 Supply a reason for each step in the proof of Theorem 7.1.5.

THEOREM 7.1.6 **INVOLUTION LAWS**

a. $(x')' = x$ **b.** $1' = 0$ and $0' = 1$

Proof   We give a proof of the involution law $1' = 0$. This proof uses the unique complement law proved above.

$1 + 0 = 1$ and $1 \cdot 0 = 0 \cdot 1 = 0$ Axioms 4a, 1b, and 4b

But

$1 + 1' = 1$ and $1 \cdot 1' = 0$ Axioms 5a and 5b

Hence,

$1' = 0$ Unique complement law   ∎

It is left as an exercise to prove the other involution laws. (See Exercise 17, Exercise Set 7.1.)

THEOREM 7.1.7 **DE MORGAN'S LAWS**

a. $(x + y)' = x'y'$ **b.** $(xy)' = x' + y'$

Proof   We will prove part b using the unique complement law.

$$xy + (x' + y') = (xy + x') + y'$$
$$= (x' + xy) + y' = (x' + x)(x' + y) + y'$$
$$= 1(x' + y) + y' = (x' + y) + y'$$
$$= x' + (y + y') = x' + 1 = 1$$

The justification of the last equality is left as an exercise. (See Exercise 14a.)

$$xy(x' + y') = xyx' + xyy' = yxx' + xyy' = y \cdot 0 + y \cdot 0 = 0 + 0 = 0$$

Since $xy + (x' + y') = 1$ and $xy(x' + y') = 0$, we have $x' + y' = (xy)'$ by the unique complement law. ∎

The proof of part a is left as an exercise. (See Exercise 18, Exercise Set 7.1.)

After proving several theorems, we should reflect on what has been accomplished. We now know that all the theorems proved are true of any Boolean algebra. Hence, it is not necessary to verify any of the preceding theorems for any specific Boolean algebra.

For our purpose, the most important Boolean algebra is the two-element Boolean algebra $(B, +, \cdot, ', 0, 1)$ with $B = \{0, 1\}$, described in Example 7.1.2. Suppose, for example, we wanted to verify Theorem 7.1.4a for this Boolean algebra. We would need to show $xy + xy' = x$ for all possible choices of x and y in $B = \{0, 1\}$. Even this simple chore would involve the verification of four cases—namely, $x = 0$, $y = 0$; $x = 0$, $y = 1$; $x = 1$, $y = 0$; and $x = 1$, $y = 1$.

There are many other interesting Boolean algebras. One based on the greatest common divisor (gcd) and the least common multiple (lcm) of two positive integers follows. For a given positive integer n, we denote the set of all positive divisors of n by D_n.

EXAMPLE 7.1.8   Let $B = D_{14} = \{1, 2, 7, 14\}$. Define $x' = 14/x$, where / is ordinary division, $x + y = \text{lcm}(x, y)$, and $x \cdot y = \gcd(x, y)$. It is not difficult to verify that ' defines a unary operation on D_{14} and that both $+$ and $\cdot$ define binary operations on D_{14}. With patience and a little elementary number theory, we can verify that the system $(D_{14}, +, \cdot, ', 1, 14)$ is a Boolean algebra. Note that for this Boolean algebra the zero element is 1 and the one element is 14. ∎

PRACTICE PROBLEM 5   Write the ten axioms of Boolean algebra using the notation of Example 7.1.8.

Example 7.1.9 gives two examples very similar to Example 7.1.8. However, neither of them is a Boolean algebra.

EXAMPLE 7.1.9 Define $x' = 12/x$, $x + y = \text{lcm}(x, y)$, and $x \cdot y = \gcd(x, y)$.

a. Let $B = \{1, 3, 12\}$. Then the system $(B, +, \cdot, ', 1, 12)$ is not a Boolean algebra because $'$ is not a unary operation on B. To see this, we note that $3' = 12/3 = 4$. But $4 \notin B$. Hence, the operation $'$ is not a unary operation on B.

b. Let $D_{12} = \{1, 2, 3, 4, 6, 12\}$. Consider the system $(D_{12}, +, \cdot, ', 1, 12)$. Convince yourself that $'$ is a unary operation on D_{12} and that $+$ and $\cdot$ are both binary operations on D_{12}. The element 2 is not the zero element of the system since $3 + 2 \neq 3$. However, $2 \cdot 2' = 2 \cdot 6 = \gcd(2, 6) = 2$. This violates Axiom 5b. Hence, the system $(D_{12}, +, \cdot, ', 1, 12)$ is not a Boolean algebra. ∎

PRACTICE PROBLEM 6 Refer to Example 7.1.9b. Verify that the system $(D_{12}, +, \cdot, ', 1, 12)$ does not satisfy Boolean algebra axiom 5a.

The **idempotent laws**, $x + x = x$ and $x \cdot x = x$, and the **identity laws**, $x + 1 = 1$ and $x \cdot 0 = 0$, are useful theorems. (See Exercises 13 and 14.)

EXERCISE SET 7.1

1. Determine whether $^\#$ is a unary operation on the given set. If not, give a counterexample.
   a. $B = \mathbb{R}$; $x^\# = x$
   b. $B = $ the set of all finite subsets of $\mathbb{N}$; $X^\# = \mathbb{N} - X$, where $X \in B$
   c. $B = \mathbb{N}$; $x^\# = 1/x$

2. Determine whether $^\#$ is a unary operation on the given set. If not, give a counterexample.
   a. $B = \{1, 2, 3, 4, 12\}$; $x^\# = 12/x$
   b. $B = \mathbb{Z}$; $x^\# = -x$
   c. $B = $ the set of all finite subsets of $\mathbb{N}$; $X^\# = X - \{0, 1\}$, where $X \in B$

3. Determine whether $*$ is a binary operation on the given set B. Give a reason if not.
   a. $B = \{1, 2, 3, 4, 12\}$; $x * y = \text{lcm}(x, y)$
   b. $B = \{1, 2, 3, 4, 12\}$; $x * y = \gcd(x, y)$
   c. $B = $ the set of all infinite subsets of $\mathbb{N}$; $X * Y = X \cap Y$

4. Determine whether $*$ is a binary operation on the given set B. Give a reason if not.
   a. $B = $ the set of all infinite subsets of $\mathbb{N}$; $X * Y = X \cup Y$
   b. $B = \mathbb{N}$; $x * y = 2x + 1$
   c. $B = \mathbb{Z}$; $x * y = x/y$

5. Write the dual of each Boolean equation or statement:
   a. $xy' + x' = x' + y'$
   b. $(x + 0) \cdot 1 + x' = 1$
   c. If $x \cdot y = 1$, then $x = 1$ and $y = 1$.

6. Write the dual of each Boolean statement:
   a. If $x + y = y + z$ and $x' + y = x' + z$, then $y = z$.
   b. If $x + y = y$, then $x \cdot y' = 1$.

7. Let $x' = 24/x$, $x + y = \text{lcm}(x, y)$, and $x \cdot y = \gcd(x, y)$. For each of the given sets B, determine whether $(B, +, \cdot, ', 1, 24)$ is a Boolean algebra.
   a. $B = \{1, 2, 3, 12, 24\}$ b. $B = D_{24} = \{1, 2, 3, 4, 6, 8, 12, 24\}$

8. Let $x' = 24/x$, $x + y = \text{lcm}(x, y)$, and $x \cdot y = \gcd(x, y)$. For each of the given sets B, determine whether $(B, +, \cdot, ', 1, 24)$ is a Boolean algebra.
   a. $B = \{1, 2, 3, 8, 12, 24\}$ b. $B = \{1, 3, 8, 24\}$

9. For a given positive integer n, let $B = D_n$, $x' = n/x$, $x + y = \text{lcm}(x, y)$, and $x \cdot y = \gcd(x, y)$. For each of the given values of n, determine whether $(D_n, +, \cdot, ', 1, n)$ is a Boolean algebra.
   a. $n = 1$ b. $n = 2$ c. $n = 4$

10. For a given positive integer n, let $B = D_n$, $x' = n/x$, $x + y = \text{lcm}(x, y)$, and $x \cdot y = \gcd(x, y)$. For each of the given values of n, determine whether $(D_n, +, \cdot, ', 1, n)$ is a Boolean algebra.
    a. $n = 12$ b. $n = 15$ c. $n = 30$

11. Prove the following simplification laws in a Boolean algebra.
    a. $x + xy = x$ b. $(x + y)(x + y') = x$

12. Prove the following simplification laws in a Boolean algebra.
    a. $x(x + y) = x$ b. $x + yx' = x + y$

13. Prove the following **idempotent laws** in a Boolean algebra.
    a. $x + x = x$ b. $x \cdot x = x$

14. Prove the following **identity laws** in a Boolean algebra.
    a. $x + 1 = 1$ b. $x \cdot 0 = 0$

If we consider Boolean algebra axioms 4a and 4b carefully, we see they guarantee the existence of a zero element and a one element. But these axioms say nothing about uniqueness. In other words, are 0 and 1 the only elements satisfying Axioms 4a and 4b? Furthermore, they say nothing about whether 0 and 1 are distinct. Exercises 15 and 16 clarify these issues.

15. Consider a Boolean algebra $(B, +, \cdot, ', 0, 1)$.
    a. Prove that the element 0 is unique. In other words, prove that 0 is the only element satisfying Axiom 4a. [*Hint:* Suppose y is also an element such that $x + y = x$ for all x.]
    b. Prove that the element 1 is unique.

16. Consider a Boolean algebra $(B, +, \cdot, ', 0, 1)$. Prove that the elements 0 and 1 are distinct if B has more than one element. [*Hint:* Assume B has more than one element and suppose $0 = 1$. Do a proof by contradiction.]

17. a. Prove $(x')' = x$. [*Hint:* Use the unique complement law.]
    b. Prove $0' = 1$. [*Hint:* You may use $0 = 1'$ and part a.]

18. Prove De Morgan's law $(x + y)' = x'y'$ in two ways:
    a. As in the proof of $(xy)' = x' + y'$. [See Theorem 7.17.]
    b. By using $(ab)' = a' + b'$ and $(c')' = c$.

19. Use Boolean algebra laws to simplify each of the following.
    a. $(x + y)(x' + y)$ b. $(xy)(x' + y)$

20. Use Boolean algebra laws to simplify each of the following.
    a. $(x + 0)(y + 1)$ b. $[(xy)z + (xy)z'] + xy'$

By the associative law, $(x + y) + z = x + (y + z)$, we can write $x + y + z$ unambiguously to mean either $(x + y) + z$ or $x + (y + z)$. Similarly, $xyz = (xy)z = x(yz)$. In general, the terms in the sum and product of any number of variables can be associated unambiguously in any way. For example, $xyzw = x(yzw) = (xy)(zw) = (xyz)w$. In Exercises 21–29 the generalized associative laws are assumed.

21. **a.** Prove an extended De Morgan's law: $(x + y + z)' = x'y'z'$.
    b. State the dual of this version of De Morgan's law.
22. **a.** Prove an extended distributive law: $w + xyz = (w + x)(w + y)(w + z)$.
    b. State the dual of this version of the distributive law.
23. Use Boolean algebra laws to simplify $xy + xy' + x'y + x'y'$.
24. Use Boolean algebra laws to simplify $xy + (xy)'z + z'$.
25. Use Boolean algebra laws to simplify $xyz + xy'z + xy'z' + xyz'$.
26. Use Boolean algebra laws to simplify $x(x' + y) + z + y$.
27. Use Boolean algebra laws to simplify $(x + xy' + z')' + xz$.
28. Use Boolean algebra laws to prove $x'w + x'y' + yz' + x'z = x' + yz'$.
29. Use Boolean algebra laws to prove $(x'yw + x'z'w + x'y'w' + yz' + x'z)' = x(y' + z)$.
    [*Hint:* You may wish to use the result of Exercise 28.]

7.2
BOOLEAN FUNCTIONS AND BOOLEAN EXPRESSIONS

Boolean functions are important because they precisely describe the relation between the inputs and outputs of logic networks. When we discussed the light bulb problem in Example 7.1.3, we used Table 7.1.1 to describe the relation between the states of the switches and that of the light bulb. As we will see, that table represents a Boolean function. We will also see that the basic Boolean operations, symbolized by x', $x + y$, and $x \cdot y$, correspond to Boolean functions.

In Example 7.1.2, we studied the two-element Boolean algebra ($\{0, 1\}$, $+$, $\cdot$, $'$, 0, 1). We restate the binary operations and unary operation on $\{0, 1\}$ in Tables 7.2.1–7.2.3.

TABLE 7.2.1 ADDITION

x	y	$x + y$
0	0	0
0	1	1
1	0	1
1	1	1

TABLE 7.2.2 MULTIPLICATION

x	y	$x \cdot y$
0	0	0
0	1	0
1	0	0
1	1	1

TABLE 7.2.3 COMPLEMENT

x	x'
0	1
1	0

For the rest of this chapter, we will be primarily concerned with the two-element Boolean algebra ($\{0, 1\}$, $+$, $\cdot$, $'$, 0, 1). We will frequently use the set $\{0, 1\}$ to denote the Boolean algebra ($\{0, 1\}$, $+$, $\cdot$, $'$, 0, 1).

7.2 BOOLEAN FUNCTIONS AND BOOLEAN EXPRESSIONS

DEFINITION

> **BOOLEAN FUNCTION**
>
> Let $\{0, 1\}^n$ denote the set of all n-tuples of 0's and 1's. A **Boolean function** is a function of n variables
>
> $$f: \{0, 1\}^n \to \{0, 1\}, \quad \text{where } n \geq 1$$

For simplicity, $f((x_1, \ldots, x_n))$ will be written as $f(x_1, \ldots, x_n)$. We note that $f(x_1, \ldots, x_n) = 0$ or $f(x_1, \ldots, x_n) = 1$ for any Boolean function f.

Recall that we discussed Boolean-valued functions in Chapter 3. A Boolean function is a specific type of Boolean-valued function.

EXAMPLE 7.2.1 Consider the Boolean function $f: \{0, 1\} \to \{0, 1\}$ defined by $f(0) = 1$ and $f(1) = 0$. We can represent the function f by Table 7.2.4.

TABLE 7.2.4

x	$f(x)$
0	1
1	0

Note that the Boolean function of Example 7.2.1 is described by the same table as the complement operation on $\{0, 1\}$ (Table 7.2.3). Thus, the complement operation on $\{0, 1\}$ corresponds to the Boolean function of Example 7.2.1. Similarly, the two binary operations on $\{0, 1\}$ correspond to Boolean functions.

By our work in Chapter 4, we know there are four different Boolean functions with domain $\{0, 1\}$ and codomain $\{0, 1\}$.

PRACTICE PROBLEM 1 Find all four Boolean functions from $\{0, 1\}$ to $\{0, 1\}$. One of these functions is given in Example 7.2.1.

Boolean functions are usually represented by tables called truth tables. A **truth table** specifies the values of a Boolean function for every possible combination of values in $\{0, 1\}$ of the variables.

EXAMPLE 7.2.2 Construct the truth table for $g: \{0, 1\}^3 \to \{0, 1\}$ defined by $g(0, 0, 0) = 0$, $g(0, 0, 1) = 0$, $g(0, 1, 0) = 0$, $g(0, 1, 1) = 0$, $g(1, 0, 0) = 0$, $g(1, 0, 1) = 1$, $g(1, 1, 0) = 0$, and $g(1, 1, 1) = 1$.

Solution See Table 7.2.5.

TABLE 7.2.5

x	y	z	$g(x, y, z)$
0	0	0	0
0	0	1	0
0	1	0	0
0	1	1	0
1	0	0	0
1	0	1	1
1	1	0	0
1	1	1	1

PRACTICE PROBLEM 2 Let $B = \{0, 1\}$.

a. Find $\text{Card}(B)$ and $\text{Card}(B^3)$.
b. Find the number of Boolean functions from B^3 to B.

To relate a Boolean function to an algebraic formula, we first need the concept of a Boolean expression.

DEFINITION

> A **Boolean expression** in n variables $x_1, \ldots, x_n$ is a finite string of symbols involving $x_1, \ldots, x_n$ together with the Boolean operations $+, \cdot,$ and $'$ and parentheses such that the string is formed by the following rules:
>
> **a.** The symbols 0 and 1 and $x_1, \ldots, x_n$ are Boolean expressions.
> **b.** If E and F are Boolean expressions, then $E', F', (E + F),$ and $(E \cdot F)$ are also Boolean expressions.

EXAMPLE 7.2.3 The expressions $x_1, (x_2 + x_3), x_2', ((x_1 \cdot x_2) + x_3), (x_1 \cdot x_1),$ and $(x_2' + x_3)$ are Boolean expressions in variables $x_1, x_2,$ and x_3. The expressions 0 and 1 can be considered to be Boolean expressions of n variables for any $n \geq 1$. ∎

When working with Boolean expressions, we usually omit the outside parentheses and write $x_1 x_2$ for $x_1 \cdot x_2$. Furthermore, we follow the usual precedence; that is, evaluate expressions in parentheses first, then evaluate complements, then evaluate products, and finally evaluate sums.

In addition, we will freely use the associative laws and write $x_1 x_2 x_3$ for $(x_1 x_2) x_3$ or $x_1 (x_2 x_3)$, and $x_1 + x_2 + x_3$ for $(x_1 + x_2) + x_3$ or $x_1 + (x_2 + x_3)$. For Boolean expressions involving only a few variables, we will use $x, y,$ and z instead of $x_1, x_2,$ and x_3.

Boolean Expressions Define Boolean Functions

If we substitute 0 or 1 for each occurrence of a variable in a Boolean expression, we obtain a new expression involving $+, \cdot, ', 0, 1,$ and parentheses. This new expression has a value in the two-element Boolean algebra $\{0, 1\}$. For example, if we replace x by 0 and y by 1 in the Boolean expression $x + y$, we obtain the expression $0 + 1$, which is equal to 1 in the Boolean algebra $\{0, 1\}$. So, a Boolean expression, along with all possible combinations of 0 and 1 substitutions and resulting outputs, represents a Boolean function. Hence, we can use a Boolean expression to define a Boolean function.

EXAMPLE 7.2.4 Define a Boolean function of three variables corresponding to the Boolean expression $xy + z'x$.

7.2 BOOLEAN FUNCTIONS AND BOOLEAN EXPRESSIONS

Solution   We define $f:\{0,1\}^3 \to \{0,1\}$ by the formula $f(x, y, z) = xy + z'x$. For instance, $f(1, 1, 0) = 1 \cdot 1 + 0' \cdot 1 = 1 + 1 \cdot 1 = 1$. See Table 7.2.6 for the truth table representing f.

TABLE 7.2.6

x	y	z	$f(x, y, z) = xy + z'x$
0	0	0	0
0	0	1	0
0	1	0	0
0	1	1	0
1	0	0	1
1	0	1	0
1	1	0	1
1	1	1	1

By the truth table method illustrated in Example 7.2.4, we see that every Boolean expression defines a Boolean function.

When a Boolean expression E defines a Boolean function f, we say E **corresponds** to the function f, or vice versa. We have already observed that x', $x + y$, and $x \cdot y$ correspond to Boolean functions. The respective truth tables for x', $x + y$, and $x \cdot y$ (Tables 7.2.1–7.2.3) define those Boolean functions.

PRACTICE PROBLEM 3   Consider the Boolean expression $yx + xz' + yy'$. Use a truth table to find the corresponding Boolean function $g:\{0,1\}^3 \to \{0,1\}$. Observe that the function g is the same as the function f in Example 7.2.4.

PRACTICE PROBLEM 4   Use a truth table to define a Boolean function $h:\{0,1\}^2 \to \{0,1\}$ corresponding to the Boolean expression xx'.

The Boolean expression 0 defines the constant Boolean function from $\{0,1\}^n$ to $\{0,1\}$ by the formula $f(x_1, \ldots, x_n) = 0$ for all $(x_1, \ldots, x_n) \in \{0,1\}^n$. Similarly, the Boolean expression 1 defines the constant Boolean function from $\{0,1\}^n$ to $\{0,1\}$ by the formula $g(x_1, \ldots, x_n) = 1$ for all $(x_1, \ldots, x_n) \in \{0,1\}^n$.

The Boolean expression $x_1 x_1'$ can be used to define a Boolean function $h:\{0,1\}^n \to \{0,1\}$. Note that $h(x_1, \ldots, x_n) = 0$ for any $(x_1, \ldots, x_n)$ in $\{0,1\}^n$. (See Practice Problem 4.)

In general, a Boolean expression in m variables can define a Boolean function $f:\{0,1\}^n \to \{0,1\}$ for $n \geq m$. Whether we regard $x + yx$ as a Boolean expression of two or of more variables depends on the context. If the context is not clear, we must clearly state how $x + yx$ is regarded.

EXAMPLE 7.2.5

a. Define a Boolean function $f:\{0,1\}^2 \to \{0,1\}$ by the Boolean expression $x + yx$.
b. Define a Boolean function $g:\{0,1\}^3 \to \{0,1\}$ by the Boolean expression $x + yx$.

Solution

a. Clearly, $x + yx$ is to be regarded as a Boolean expression of two variables because the domain of f is the set of 2-tuples of 0's and 1's. We use the formula $f(x, y) = x + yx$, and present f in a truth table, as shown in Table 7.2.7.
b. Since the domain of g is $\{0,1\}^3$, we must regard $x + yx$ as a Boolean expression of three variables. The Boolean function g is given in the truth table shown in Table 7.2.8.

TABLE 7.2.7

x	y	$f(x, y) = x + yx$
0	0	0
0	1	0
1	0	1
1	1	1

TABLE 7.2.8

x	y	z	$g(x, y, z) = x + yx$
0	0	0	0
0	0	1	0
0	1	0	0
0	1	1	0
1	0	0	1
1	0	1	1
1	1	0	1
1	1	1	1

∎

The use of Boolean expressions to define Boolean functions is similar to our experience with algebraic expressions and algebraic functions on the set of real numbers $\mathbb{R}$.

Constructing a Boolean expression as a string of symbols is a syntactic issue, whereas giving meaning to a Boolean expression is a semantic issue. A Boolean function gives *meaning* to the Boolean expression corresponding to it. So we consider Boolean expressions as the syntax and Boolean functions as the semantics of logic design.

It is easy to verify that the Boolean expressions $x + y$ and $y + x$ define the same Boolean function. Similarly, we can show that the Boolean expressions $x(y + z)$, $xy + xz$, and $xy + xz + xx'$ define the same Boolean function. We leave the verification to Practice Problem 5.

PRACTICE PROBLEM 5 Use a truth table to show that the Boolean expressions $x(y + z)$, $xy + xz$, and $xy + xz + xx'$ define the same Boolean function from $\{0, 1\}^3$ to $\{0, 1\}$.

Two Boolean expressions of n variables are considered to be **equivalent** if they define the same Boolean function from $\{0, 1\}^n$ to $\{0, 1\}$. Though the Boolean expressions $x(y + z)$, $xy + xz$, and $xy + xz + xx'$ are not formally equal (symbol for symbol), they are equivalent. We can write $E = F$ without causing confusion when E and F are equivalent Boolean expressions.

Besides using a truth table, we can apply Boolean algebra laws to establish the equivalence of two Boolean expressions. For example, $x(y + z) = xy + xz$ by Boolean algebra axiom 3b. Also, $xy + xz + xx' = xy + xz + 0 = xy + xz$.

EXAMPLE 7.2.6 Show that $(x + y'z)(yz)'$ is equivalent to $xy' + xz' + y'z$ by Boolean algebra laws.

Solution
$$\begin{aligned}
(x + y'z)(yz)' &= (x + y'z)(y' + z') && \text{De Morgan's law} \\
&= x(y' + z') + y'z(y' + z') && \text{Axioms 1b and 3b} \\
&= xy' + xz' + y'zy' + y'zz' && \text{Axiom 3b} \\
&= xy' + xz' + y'z + y' \cdot 0 && \text{Axioms 1b, 5b and the idempotent law} \\
&= xy' + xz' + y'z && \text{Identity law and Axiom 4a}
\end{aligned}$$ ∎

PRACTICE PROBLEM 6 Use Boolean algebra laws to show that $(x + y)'(y' + x) + x'$ is equivalent to x'.

The notion of equivalence on the set S of Boolean expressions in n variables is actually an equivalence relation on S. We leave the verification of this to Exercise 23, Exercise Set 7.2.

Standard Sum of Products

Every Boolean expression in n variables is equivalent to a unique Boolean expression called the standard sum of products, to be defined shortly.

A Boolean expression consisting of a single variable or its complement is called a **literal**. For example, x_1, x_2', x_3', and x_1' are literals. A **minterm** in n variables is a product of n literals such that each variable x_i occurs exactly once either as x_i or x_i', but not both. For example, $x_1 x_2' x_3$, $x_1' x_2 x_3$, and $x_1' x_2' x_3$ are some minterms in three variables. The expressions $x_1 x_3$ and $x_1 x_2$ are not minterms in three variables. Also, the expression $x_1 x_2 x_1' x_3$ is not a minterm in three variables since both x_1 and x_1' occur in the product.

There are eight minterms of three variables x, y, z: $x'y'z'$, $x'y'z$, $x'yz'$, $x'yz$, $xy'z'$, $xy'z$, xyz', and xyz. Each of these eight minterms defines a Boolean function of three variables that has the value 1 on exactly one of the eight 3-tuples in $\{0, 1\}^3$ and has the value 0 elsewhere. For example, the minterm $x'y'z$ has the value 1 on $(0, 0, 1)$ and has the value 0 elsewhere. Note that x and y are complemented since they correspond to the 0 entries in $(0, 0, 1)$; z is not complemented since it corresponds to the 1 entry in $(0, 0, 1)$. See Table 7.2.9, page 270.

TABLE 7.2.9

x	y	z	x'y'z'	x'y'z	x'yz'	x'yz	xy'z'	xy'z	xyz'	xyz
0	0	0	1	0	0	0	0	0	0	0
0	0	1	0	1	0	0	0	0	0	0
0	1	0	0	0	1	0	0	0	0	0
0	1	1	0	0	0	1	0	0	0	0
1	0	0	0	0	0	0	1	0	0	0
1	0	1	0	0	0	0	0	1	0	0
1	1	0	0	0	0	0	0	0	1	0
1	1	1	0	0	0	0	0	0	0	1

PRACTICE PROBLEM 7 Find the number of minterms of n variables.

DEFINITION

Given a Boolean expression E in n variables, if E is equivalent to a sum of distinct minterms, then this sum of distinct minterms is called the **standard sum of products** (or **disjunctive normal form**) for E.

We can find the standard sum of products for any Boolean expression.

EXAMPLE 7.2.7 Given the Boolean expression xz as an expression of three variables, find the standard sum of products for xz.

Solution First we define a Boolean function $f(x, y, z) = xz$ from $\{0, 1\}^3$ to $\{0, 1\}$ in a truth table as shown in Table 7.2.10.

Now we form a minterm for each 1 in the rightmost column (the values of f). In the sixth row, $f(x, y, z) = 1$ for $(x, y, z) = (1, 0, 1)$. The minterm of this row is $xy'z$, where only y is complemented since y has value 0, whereas x has the value 1 and z has the value 1. In the eighth row, $f(x, y, z) = 1$ for $(x, y, z) = (1, 1, 1)$. The necessary minterm is xyz. Since the minterm $xy'z$ has the value 1 on $(1, 0, 1)$ and has the value 0 elsewhere, and since the minterm xyz has the value 1 on $(1, 1, 1)$ and has the value 0 elsewhere, we conclude that the expression $xy'z + xyz$ defines the function f. Hence, the standard sum of products for xz is $xy'z + xyz$. ∎

TABLE 7.2.10

x	y	z	$f(x, y, z) = xz$
0	0	0	0
0	0	1	0
0	1	0	0
0	1	1	0
1	0	0	0
1	0	1	1
1	1	0	0
1	1	1	1

The standard sum of products for a Boolean expression is unique except for the order in which the literals appear in the minterms and the order in which the minterms appear in the sum.

The method of using truth tables to find the standard sum of products for a given Boolean expression is called the **truth table method**. The truth table method can be used to find a standard sum of products for any Boolean function.

PRACTICE PROBLEM 8 Use the truth table method to find the standard sum of products for the Boolean expression $(x + y)(x' + y')$ in two variables.

7.2 BOOLEAN FUNCTIONS AND BOOLEAN EXPRESSIONS

PRACTICE PROBLEM 9 Consider the Boolean function $g:\{0, 1\}^3 \to \{0, 1\}$ defined by the truth table below. Find the standard sum of products that corresponds to g.

x	y	z	g(x, y, z)
0	0	0	0
0	0	1	1
0	1	0	0
0	1	1	0
1	0	0	0
1	0	1	1
1	1	0	1
1	1	1	0

There is a one-to-one correspondence between the Boolean functions of n variables and the Boolean expressions of n variables in the standard sum of products form. This fact will be applied in the next section, where we discuss the application of Boolean functions and Boolean expressions to the design of logic networks.

Functionally Complete Operations

All Boolean expressions may be written using only the three basic operations $\cdot$, $+$, and $'$. It is possible and frequently useful to write Boolean expressions using only one or two operations. For example, all statements can be written in terms of the operations $+$ and $'$ since $x \cdot y = (x' + y')'$ by one of De Morgan's laws (Theorem 7.1.7) and the involution law (Theorem 7.1.6).

A set of Boolean operations is **functionally complete** if, for every Boolean function, there is a corresponding Boolean expression written in terms of the operations in the set. The set of the three operations $\cdot$, $+$, and $'$ is functionally complete. We can show that a set A of operations is functionally complete if we show that the three basic operations can all be written in terms of the operations in A. We have just shown that the set $\{', +\}$ is functionally complete using that method.

PRACTICE PROBLEM 10 Use one of De Morgan's laws and the involution law to show that the set of operations $\{', \cdot\}$ is functionally complete.

Functionally complete sets have important applications to logic networks, which are studied in Section 7.3. Each Boolean operation has an analogous logic gate. All logic networks can be designed using only the three basic AND ($\cdot$), OR ($+$), and NOT ($'$) gates. Because $\{', +\}$ is functionally complete, it is possible to design every logic network using only NOT gates and OR gates. Similarly, it is possible to design every logic network using only NOT gates and AND gates.

Is it possible to design a single logic gate so that all logic networks can be written in terms of that logic gate? In other words, is there a Boolean operation such that all Boolean expressions can be written in terms of that operation? You will see in the following example and in Exercise 24 (Exercise Set 7.2) that there are two such Boolean operations.

If a functionally complete set consists of one operation, we say that the **operation is functionally complete**.

EXAMPLE 7.2.8

A FUNCTIONALLY COMPLETE BOOLEAN OPERATION Define $x \downarrow y$ by the truth table given in Table 7.2.11. Clearly $x \downarrow y = (x + y)'$. The Boolean operation $\downarrow$ is called a **NOR** (not or) **operation**. To show that the operation $\downarrow$ is functionally complete, we will write each of the three basic operations in terms of $\downarrow$. Since $\{', \cdot\}$ is functionally complete, it is sufficient to write $'$ and $\cdot$ in terms of $\downarrow$.

a. $x' = (x + x)' = x \downarrow x$
b. $x \cdot y = (x')'(y')' = (x' + y')' = x' \downarrow y' = (x \downarrow x) \downarrow (y \downarrow y)$ ∎

TABLE 7.2.11

x	y	$x \downarrow y$
1	1	0
1	0	0
0	1	0
0	0	1

The logic gate for $x \downarrow y$ is called a **NOR gate**. So any logic network can be written entirely in terms of NOR gates. There is also a **NAND** (not and) **operation** written with the symbol $\uparrow$. You are asked to show that the operation $\uparrow$ is functionally complete in Exercise 24, Exercise Set 7.2. The logic gate for $x \uparrow y$ is the **NAND gate**, and any logic network can be designed with only NAND gates.

EXERCISE SET 7.2

1. Find the value in the Boolean algebra $\{0, 1\}$ of the Boolean expression $x(y + z)' + yx + (xz)'$ by replacing x by 0, y by 1, and z by 1.

2. Find the value in the Boolean algebra $\{0, 1\}$ of the Boolean expression $(x + y + z)'(xzy) + zz'$ by replacing x by 1, y by 0, and z by 1.

3. Find the standard sum of products for the Boolean function defined by the truth table in Table 7.2.12.

TABLE 7.2.12

x	y	$f(x, y)$
0	0	1
0	1	0
1	0	1
1	1	1

TABLE 7.2.13

x	y	z	$g(x, y, z)$
0	0	0	1
0	0	1	0
0	1	0	0
0	1	1	0
1	0	0	1
1	0	1	0
1	1	0	0
1	1	1	1

4. Find the standard sum of products for the Boolean function defined by the truth table in Table 7.2.13.

5. Define a Boolean function $f:\{0, 1\}^2 \to \{0, 1\}$ by the Boolean expression $y(x' + y)$. Use a truth table.

6. Define a Boolean function $g:\{0, 1\}^3 \to \{0, 1\}$ by the Boolean expression $x(z + yy') + z'$. Use a truth table.

7. Verify that the Boolean expressions $(x + y)'z$ and $x'y'z$ define the same Boolean function of three variables.
 a. Use the truth table method. **b.** Use Boolean algebra laws.

8. Repeat Exercise 7 for the Boolean expressions $xx' + y(z + z')$ and $y + zz'$.

9. Find the standard sum of products for each of the following Boolean expressions of two variables by the truth table method.
 a. $(x + y')x$ **b.** $(x + y')(x' + y)$

10. Find the standard sum of products for each of the following Boolean expressions of three variables by the truth table method.
 a. $f(x, y, z) = (xy')'(x + z)$ **b.** $g(x, y, z) = xy' + x'z$
 c. $h(x, y, z) = x + z(x + y')$

11. Consider the Boolean expression $E = (xy')'(x + z)$.
 a. Find the standard sum of products for E.
 b. Use Boolean algebra laws to verify the solution to part a.

12. Consider the Boolean expression $E = xy' + x'z$.
 a. Find the standard sum of products for E.
 b. Use Boolean algebra laws to verify the solution to part a.

13. Consider the Boolean expression $E = x + z(x + y')$.
 a. Find the standard sum of products for E.
 b. Use Boolean algebra laws to verify the solution to part a.

14. Repeat Exercise 13 for each of the following Boolean expressions.
 a. $(x + y')x$ **b.** $(x + y')(x' + y)$

15. Find the standard sum of products for the following Boolean function of two variables:

$$f(x, y) = \begin{cases} 1 & \text{if } x = 0 \text{ and } y = 1 \\ 0 & \text{otherwise} \end{cases}$$

[*Hint:* Make a truth table for the function f.]

16. Find the standard sum of products for the following Boolean function of three variables:

$$g(x, y, z) = \begin{cases} 1 & \text{if } x = 0 \text{ and } (y = 1 \text{ or } z = 1) \\ 0 & \text{otherwise} \end{cases}$$

17. Find the standard sum of products for the following Boolean function of three variables:

$$g(x, y, z) = \begin{cases} 1 & \text{if } x = 0 \text{ or } (y = 0 \text{ and } z = 1) \\ 0 & \text{otherwise} \end{cases}$$

*The duals of minterm and standard sum of products are called **maxterm** and **standard product of sums**, respectively. Specifically, a **maxterm** in n variables is a sum of n literals such that each variable x_i occurs exactly once either as x_i or x'_i, but not both. By De*

Morgan's law it is not hard to see that the complement of a minterm is a maxterm. For example, $(x'yz)' = x + y' + z'$. A product of distinct minterms is called the **standard product of sums** (or **conjunctive normal form**). Exercises 18 and 19 illustrate that the standard product of sums form is sometimes useful for writing a Boolean expression more efficiently.

18.
  a. Define a Boolean function $f:\{0, 1\}^3 \to \{0, 1\}$ by $f(x, y, z) = 0$ if and only if $x = 0$ and $y = 0$ and $z = 1$. Let E be a Boolean expression corresponding to f. Find the standard sum of products for E'.
  b. Take the complement of E' and thus write E as a maxterm.

19. Let $E = x'y'z' + x'y'z + x'yz' + x'yz + xy'z' + xy'z$.
  a. Find the standard sum of products for E'.
  b. Take the complement of E' and thus write E as a standard product of sums.

In Exercises 20–22, find the standard product of sums for a Boolean expression E as follows: First find the standard sum of products F for E'. Taking complements of both sides of $E' = F$ and applying De Morgan's law, we obtain $E = F'$, where F' is the standard product of sums for E.

20. Find the standard product of sums for each of the following Boolean expressions.
  a. $xy + x'y'$ **b.** $xy'z + x'yz + x'yz' + x'y'z + x'y'z'$

21. Find the standard product of sums for each of the following Boolean expressions.
  a. $xy + x'y + x'y'$ **b.** $x'z' + x'y + xy' + xz$

22. Find the standard product of sums for each of the following Boolean functions.
  a. $f:\{0, 1\}^3 \to \{0, 1\}$, where $f(0, 0, 0) = 1$, $f(1, 0, 0) = 1$, $f(0, 0, 1) = 0$, $f(0, 1, 0) = 0$, $f(0, 1, 1) = 1$, $f(1, 0, 1) = 1$, $f(1, 1, 0) = 1$, and $f(1, 1, 1) = 1$
  b. $g:\{0, 1\}^3 \to \{0, 1\}$, where $g(x, y, z) = xz' + yz$

23. Let S be the set of all Boolean expressions in n variables. Define a binary relation $\approx$ on S as follows: For all E and F in S, $E \approx F$ if and only if E and F define the same Boolean function from $\{0, 1\}^n$ to $\{0, 1\}$. Prove that $\approx$ is an equivalence relation on S.

24. Define $x \uparrow y$ by the truth table in Table 7.2.14. It is clear that $x \uparrow y = (x \cdot y)'$. The logic gate for the operation $\uparrow$ is called a **NAND** (not and) **gate**. We call $\uparrow$ the **Sheffer stroke**. Prove the Sheffer stroke is functionally complete by showing:
  a. $x \uparrow x = x'$ **b.** $(x \uparrow x) \uparrow (y \uparrow y) = x + y$ **c.** $(x' + y')' = x \cdot y$

TABLE 7.2.14

x	y	$x \uparrow y$
1	1	0
1	0	1
0	1	1
0	0	1

TABLE 7.2.15

x	y	$x \veebar y$
1	1	0
1	0	1
0	1	1
0	0	0

25. Define $x \veebar y$ by the truth table in Table 7.2.15. It is clear from Table 7.2.15 that $x \veebar y = x \cdot y' + x' \cdot y$. The logic gate for this operation is called the **XOR** (exclusive or) **gate**. Prove each of following.
  a. $x \veebar x = 0$ **b.** $x \veebar x' = 1$ **c.** $(x \veebar y)' = x' \veebar y = x \veebar y'$

7.3
LOGIC NETWORKS AND KARNAUGH MAPS

In this section, we will first discuss the relationship between Boolean expressions and logic networks, then introduce a minimization process for Boolean expressions.

Logic Networks

A **logic gate** is a binary electronic device with one or more input signals to produce one output signal, where the input signals and output signals can be of two possible values. The two values will be symbolized by the elements in the Boolean algebra $\{0, 1\}$. There are three basic logic gates, called the **OR gate**, the **AND gate**, and the **NOT gate** (**inverter**). Figure 7.3.1 shows these gates in the standard ANSI/IEEE (American National Standards Institute, Institute of Electronics and Electrical Engineers) symbols.

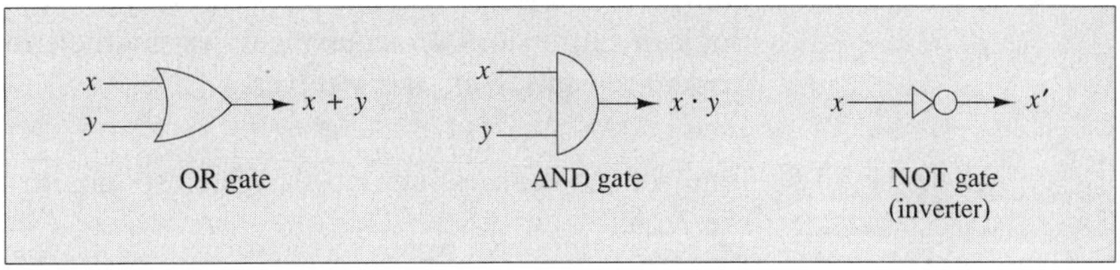

FIGURE 7.3.1

The OR gate represents Boolean addition in $\{0, 1\}$. The output of an OR gate is 1 if one or both input values are 1; it is 0 otherwise. The AND gate represents Boolean multiplication in $\{0, 1\}$. The output of an AND gate is 1 only when both input values are 1; it is 0 otherwise. The NOT gate (inverter) represents Boolean complement in $\{0, 1\}$. The output of a NOT gate is 0 when the input value is 1, and the output is 1 when the input value is 0. Refer to Tables 7.2.1–7.2.3 (page 264), which define addition, multiplication, and complement for the two-element Boolean algebra.

A **logic network** (also called switching network, gating network, combinational network, logic circuit, etc.) is some combination of OR, AND, and NOT gates. When we use single variables for inputs, the output of a logic network is represented by a Boolean expression of the input variables.

EXAMPLE 7.3.1 Figure 7.3.2 shows two logic networks with x and y as input variables and their corresponding Boolean expressions $(x + y)(x' + y')$ and $x' \cdot y + x \cdot y'$.

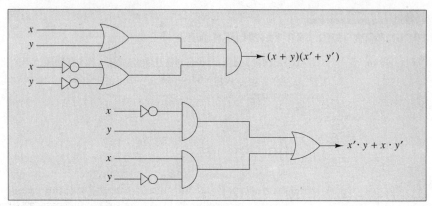

FIGURE 7.3.2

PRACTICE PROBLEM 1 Draw a logic network with x, y, and z as inputs and $xy' + z'y$ as its output Boolean expression.

The two logic networks in Example 7.3.1 define the same Boolean function $f:\{0, 1\}^2 \to \{0, 1\}$ since $(x + y)(x' + y') = x'y + xy'$. When two output Boolean expressions are equivalent, we say the two corresponding logic networks are **equivalent**, and vice versa.

EXAMPLE 7.3.2 Find the Boolean expression that corresponds to the logic network in Figure 7.3.3.

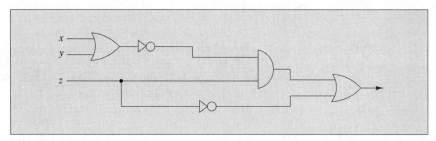

FIGURE 7.3.3

Solution The leftmost OR gate yields $x + y$; applying the inverter to $(x + y)$ gives $(x + y)'$; the AND gate combines $(x + y)'$ with z to yield $(x + y)'z$. The variable z is input to an inverter giving z'; and, finally, the rightmost OR gate combines $(x + y)'z$ with z' to yield $(x + y)'z + z'$ as the output Boolean expression. ∎

PRACTICE PROBLEM 2 Find the corresponding Boolean expression for the logic network given in Figure 7.3.4.

7.3 LOGIC NETWORKS AND KARNAUGH MAPS 277

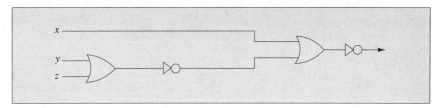

FIGURE 7.3.4

Since every Boolean expression defines a Boolean function, every logic network also defines a Boolean function. For example, the logic network in Example 7.3.2 defines the Boolean function $f:\{0, 1\}^n \to \{0, 1\}$ corresponding to $(x + y)'z + z'$.

PRACTICE PROBLEM 3 Find the Boolean function that corresponds to the logic network shown in Figure 7.3.5.

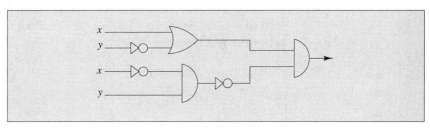

FIGURE 7.3.5

Although many equivalent Boolean expressions (hence, logic networks) can correspond to the same Boolean function g, there is only one standard sum of products form that corresponds to g.

Logic design problems are often stated in ordinary English. We must translate the English into a Boolean function that describes the desired behavior of the logic network.

EXAMPLE 7.3.3 Construct a logic network to instruct a robot to answer the telephone according to the following rules:

The robot answers the telephone

if the switch is on or

if the switch is not on and the master does not answer the telephone before the fifth ring

Solution Let

$x = 1$ if the master answers the telephone before the fifth ring
$x = 0$ if the master does not answer the telephone before the fifth ring
$y = 1$ if the switch is on
$y = 0$ if the switch is not on

So, we want the robot to answer if and only if $y = 1$ or ($y = 0$ and $x = 0$). Hence, the Boolean function $f(x, y) = y + y'x'$ describes the desired behavior of our logic network. The logic network is shown in Figure 7.3.6.

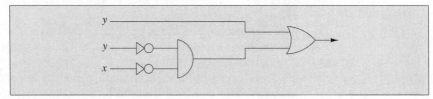

FIGURE 7.3.6

TABLE 7.3.1

x	y	g(x, y)
0	0	0
0	1	1
1	0	1
1	1	0

The Boolean function g that described the light bulb problem in Example 7.1.3 can be presented in Table 7.3.1. The function g corresponds to the standard sum of products form $xy' + x'y$. A logic network for the function g corresponding to that standard sum of products form is given in Figure 7.3.7.

The logic network in Figure 7.3.7 is a solution for the light bulb problem we discussed in Example 7.1.3. It is also part of the logic network called a half-adder in Example 7.3.4.

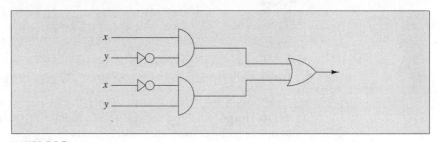

FIGURE 7.3.7

EXAMPLE 7.3.4 The **half-adder network** has two outputs, s and c, as described in Table 7.3.2. The output s is the binary sum of the two inputs. The output c is the carry

TABLE 7.3.2

x	y	s	c
0	0	0	0
0	1	1	0
1	0	1	0
1	1	0	1

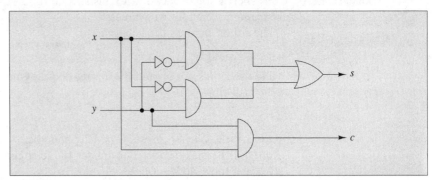

FIGURE 7.3.8 Half-adder network

7.3 ◾ LOGIC NETWORKS AND KARNAUGH MAPS

resulting when the two input digits are added. See Table 7.3.2. We see that $s = xy' + x'y$ and $c = xy$. Figure 7.3.8 shows a logic network for the half-adder. ◾

Minimizing Expressions

The design of logic networks can be viewed as a three-step process:

1. Specify the input and output task with a Boolean function f.
2. Find the Boolean expression E, in the standard sum of products form, corresponding to f.
3. Minimize E.

We have discussed steps 1 and 2. Now we address step 3. We will restrict our attention to a specific type of minimized expression called a minimal sum of products.

Recall that a single variable or its complement is called a literal. Each product in a sum of products is a product of literals in which no variable appears more than once.

DEFINITION

A **minimal sum of products** F for a Boolean expression E has the following properties:

1. $F = E$
2. F is a sum of products.
3. If G is any other sum of products such that $G = E$, then G contains at least as many products as F.
4. If H is any other sum of products such that $H = E$ and H has the same number of products as F, then H contains at least as many literals as F.

Unlike a standard sum of products, there is not necessarily a unique minimal sum of products equivalent to a given Boolean expression. As we will see in Example 7.3.10, both $xyu + yz'u + x'y'z'$ and $xyu + x'z'u + x'y'z'$ are minimal sums of products for $xyzu + xyz'u + x'yz'u + x'y'z'u + x'y'z'u'$.

Furthermore, a minimal sum of products form may not require the fewest possible logic gates of all equivalent Boolean expressions. For example, the Boolean expression $xy' + xz$ is a minimal sum of products for $xyz + xy'z + xy'z'$, and the logic network for $xy' + xz$ requires four logic gates. But $xy' + xz = x(y' + z)$, and the logic network for $x(y' + z)$ requires only three logic gates.

If a minimal sum of products may not be "minimal" in the sense of having the fewest gates in its logic network, why then is it useful? In general, there is no known algorithm for reducing a given Boolean expression to an equivalent expression that requires a minimal number of logic gates in its logic network. Furthermore, finding another equivalent expression requiring fewer gates than a minimal sum of products usually requires guesswork and ingenuity. However, there is a method known as the Karnaugh map method for reducing a Boolean expression in its standard sum of products form to a minimal sum of products. Before introducing the Karnaugh map, we first examine the theoretical basis for this method.

Consider how the Boolean expression $xyz + xyz'$ might be simplified. The Boolean algebra law $ab + ab' = a$ (see Theorem 7.1.4) can be applied to the preceding Boolean expression to yield the expression xy as follows:

$$xyz + xyz' = (xy)z + (xy)z' = xy$$

We say two minterms **differ in one literal** y if y is present in one minterm, y' is present in the other, and the minterms contain the same literals otherwise. We saw above that xyz and xyz' differ in the literal z and that $xyz + xyz'$ can be reduced to xy. As another example, note that $xy'zu'$ and $xyzu'$ differ in one literal y, and $xy'zu' + xyzu'$ can be reduced to xzu'. Thus, we have the following basic **minimization rule**.

MINIMIZATION RULE

If a Boolean expression is the sum of two minterms differing in one literal, that literal can be removed and the sum written as the product of the remaining literals.

In order to make the above minimization process easy to apply, we use the Karnaugh map method. The **Karnaugh map** is a representation of a Boolean expression written in standard sum of products form. It is a rectangular array of boxes with entries of 1's and blanks such that each entry 1 represents a minterm in the given standard sum of products. The rows and columns of boxes in the Karnaugh map are labeled so that products in the Boolean expression differing in one literal are represented by 1's in adjacent boxes. For example, the Karnaugh map for a Boolean expression in two variables x and y is a 2×2 array of boxes, where the rows are labeled x and x' and the columns are labeled y and y', as shown in Figure 7.3.9(a). Figure 7.3.9(b) shows the representation of the Boolean expression $x'y + x'y'$. Note that the minterms $x'y$ and $x'y'$ differ in the literal y and are represented by 1's in adjacent boxes.

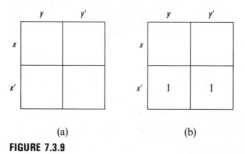

FIGURE 7.3.9

7.3 LOGIC NETWORKS AND KARNAUGH MAPS

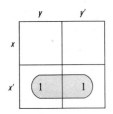

FIGURE 7.3.10

Using the minimization rule, we can write $x'y + x'y'$ as x'. The y and y' are removed from $x'y + x'y'$, which is written in the minimized form x'. This is pictured in the Karnaugh map where the 1's are looped as shown in Figure 7.3.10. The minimization process will be explained further below.

Karnaugh maps are most easily applied to Boolean expressions in two, three, or four variables. The next few examples illustrate the minimization process for expressions in three or four variables.

EXAMPLE 7.3.5 Use a Karnaugh map to minimize $xyz + xyz'$.

Solution The Karnaugh map for $xyz + xyz'$ is a 2×4 array of eight boxes for a standard sum of products E in three variables. See Figure 7.3.11. Note that products differing in one literal are in adjacent boxes.

The adjacent 1's in the Karnaugh map in Figure 7.3.11 represent the two products xyz and xyz' differing in one literal—namely, z. To use the Karnaugh map method for minimization, the adjacent 1's are looped, as shown in Figure 7.3.12.

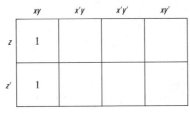

FIGURE 7.3.11

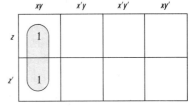

FIGURE 7.3.12

Now write the product with nondiffering literals—namely, xy—to replace the original sum $xyz + xyz'$. Hence, we have reduced $xyz + xyz'$ to xy. ∎

Before listing the steps for minimizing a standard sum of products, we give some typical three-variable Karnaugh maps.

EXAMPLE 7.3.6
a. The Karnaugh map for $xyz + xy'z + x'yz' + x'y'z'$ is shown in Figure 7.3.13. The Karnaugh map method reduced the Boolean expression $xyz + xy'z + x'yz' + x'y'z'$ to $xz + x'z'$.
b. The Karnaugh map for $x'yz + x'yz' + x'y'z'$ is shown in Figure 7.3.14.

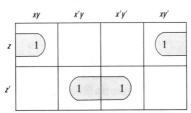

FIGURE 7.3.13

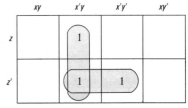

FIGURE 7.3.14

This Karnaugh map method reduced the expression $x'yz + x'yz' + x'y'z'$ to $x'y + x'z'$. The idempotent law $a = a + a$ is used when a 1 is in two loops (see Exercise 13, Exercise Set 7.1). In this case,

$$x'yz + x'yz' + x'y'z' = x'yz + (x'yz' + x'yz') + x'y'z'$$
$$= (x'yz + x'yz') + (x'yz' + x'y'z')$$
$$= x'y + x'z'$$

c. Figure 7.3.15 shows the Karnaugh map for $x'yz + x'y'z + x'yz' + x'y'z'$. The square of four 1's gives the minimization, $x'yz + x'y'z + x'yz' + x'y'z' = x'z + x'z' = x'$.

d. The Karnaugh map for $xyz + x'yz + x'y'z + xy'z$ is shown in Figure 7.3.16. The rectangle of four 1's yields $xyz + x'yz + x'y'z + xy'z = z$.

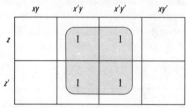

FIGURE 7.3.15   **FIGURE 7.3.16**   ∎

ALGORITHM 7.3.7   **KARNAUGH MAP ALGORITHM**   To find a minimal sum of products form representing a given Boolean expression of two, three, or four variables:

1. Write the Boolean expression as a standard sum of products.
2. For each box of the Karnaugh map corresponding to a product in the standard sum of products, mark that box with a 1. A box with a 1 is called a marked box.
3. Write a product for any marked box that is not adjacent to any other marked box.
4. For each marked box that can be combined with an adjacent marked box in exactly one way to form a pair, draw a loop around that pair.
5. For each marked box that is not affected by steps 3 or 4 and that can be combined in a square or rectangle of four marked boxes in exactly one way, draw a loop around the block of four 1's.
6. For each marked box that is not affected by steps 3, 4, or 5 and that can be combined in exactly one way into a rectangle of eight marked boxes, draw a loop around the block of eight 1's.
7. For each marked box not affected by steps 3, 4, 5, or 6, arbitrarily loop it with the largest block that includes it.
8. Write a product of nondiffering terms for each looped block.

EXAMPLE 7.3.8   Use the Karnaugh map algorithm to minimize a four-variable Boolean function.

7.3 LOGIC NETWORKS AND KARNAUGH MAPS

Solution The Karnaugh map in Figure 7.3.17 represents a Boolean expression for which steps 1 and 2 of the algorithm have already been done.

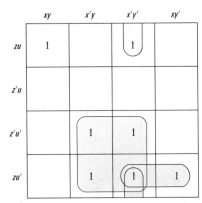

By step 3: Write $xyzu$.
By step 4: Write $x'y'z$ and $y'zu'$.
By step 5: Write $x'u'$.

FIGURE 7.3.17

So, the minimal sum of products is $xyzu + x'y'z + y'zu' + x'u'$.

Example 7.3.9 shows why the algorithm does not call for looping the bigger blocks of 1's first.

EXAMPLE 7.3.9 Let $E = xyz'u' + x'yzu + x'yz'u + x'yz'u' + x'y'z'u + x'y'z'u' + x'y'zu' + xy'z'u$. By looping the four-square block first, we reduce E to $x'z' + x'yu + yz'u' + x'y'u' + y'z'u$. See Figure 7.3.18.

The Karnaugh map for E is shown in Figure 7.3.19. By following the Karnaugh map algorithm, we reduce E to $x'yu + yz'u' + x'y'u' + y'z'u$.

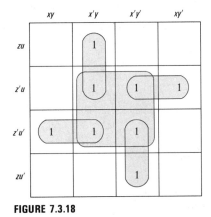

FIGURE 7.3.18   **FIGURE 7.3.19**

EXAMPLE 7.3.10 Show that a minimal sum of products for a given Boolean expression may not be unique.

Solution We will find two minimal sums of products for $E = xyzu + xyz'u + x'yz'u + x'y'z'u + x'y'z'u'$. First we loop the uniquely formed pairs of marked boxes for $xyzu + xyz'u$ and $x'y'z'u + x'y'z'u'$. This leaves a marked box that may be looped in two different ways, as shown in the following Karnaugh map. See Figure 7.3.20. So, either $xyu + yz'u + x'y'z'$ or $xyu + x'z'u + x'y'z'$ is a valid minimal sum of products for E.

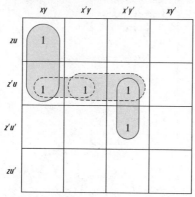

FIGURE 7.3.20

It may be necessary to design a logic network with, say, twelve different inputs. These inputs will be encoded as bit strings. More than three variables (bits) are required, since three variables can represent at most eight different inputs. Four variables allow sixteen different inputs. In this situation, four of the inputs will never be used. Inputs in a logic network that are not used are called **don't cares**. Don't cares can be helpful in the minimization process because they can be considered as either 1 or blank. The don't cares are labeled with a # in the Karnaugh map.

EXAMPLE 7.3.11 **AN EXAMPLE WITH FOUR DON'T CARES** The Karnaugh map for $xyzu + xyz'u + x'yzu + x'y'z'u$ is shown in Figure 7.3.21. This Karnaugh map yields $yu + z'u$.

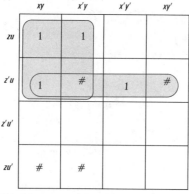

FIGURE 7.3.21

7.3 LOGIC NETWORKS AND KARNAUGH MAPS

Before giving the final example in this section, we introduce logic gates with more than two inputs. The AND and OR gates we have been using have two inputs x and y; they are binary gates. We know, by the associative laws of Boolean algebra, that both xyz and $x + y + z$ are unambiguous expressions. We can use this fact to construct ternary (three-input) AND and OR gates. See Figure 7.3.22.

[Figure showing AND gate with inputs x, y, z producing xyz, where $xyz = (xy)z = x(yz)$, and OR gate with inputs x, y, z producing $x + y + z$, where $x + y + z = (x + y) + z = x + (y + z)$]

FIGURE 7.3.22 Ternary gates

Of course the associative laws also allow us to construct n-ary AND and OR gates for any positive integer n. We will use both ternary (3-ary) and 4-ary gates in the next example.

EXAMPLE 7.3.12 Design a logic network to add (ignoring the carry) three binary digits.

Solution Define a Boolean function f by $f(x, y, z) = 1$ if an odd number of the variables x, y, and z have value 1. The truth table is shown in Table 7.3.3. The Karnaugh map for this function is represented in Figure 7.3.23. This Karnaugh map provides no minimization. So, the minimal sum of products is the same as the standard sum of products. Therefore,

$$f(x, y, z) = xyz + x'yz' + x'y'z + xy'z'$$

The logic network for this Boolean expression is shown in Figure 7.3.24.

TABLE 7.3.3

x	y	z	f
1	1	1	1
1	1	0	0
1	0	1	0
1	0	0	1
0	1	1	0
0	1	0	1
0	0	1	1
0	0	0	0

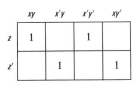

	xy	$x'y$	$x'y'$	xy'
z	1		1	
z'		1		1

FIGURE 7.3.23

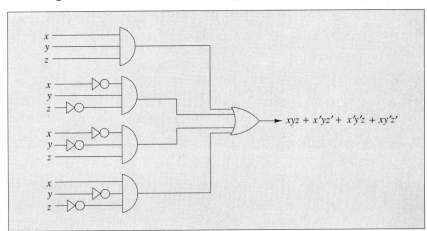

FIGURE 7.3.24

EXERCISE SET 7.3

1. Write each of the following as a minimal sum of products.
   a. $x'y'z + xy'z' + x'y'z' + xy'z$
   b. $xyz'u + xyzu' + xyz'u' + x'yz'u + x'yzu'$
   c. $xyz'u + xyz'u' + x'yz'u' + x'y'z'u' + x'y'zu'$
   d. $xyzu + xy'zu + xy'zu'$

2. Write each of the following as a minimal sum of products.
   a. $x + y + z$ b. $xz' + xy' + xzu'$
   c. $x'y'z + x'yu' + x'y'u + yz'u'$ d. $xyz + (x' + y')u$

3. Recall De Morgan's laws: $(a + b)' = a'b'$ and $(ab)' = a' + b'$. Use De Morgan's laws to write the complement of each given sum of products as a product of sums.
   a. $xy' + x'y$ b. $xyz + x'yz'$
   c. $xyz + xyz' + x'yz$ d. $xyzu + xy'zu + xy'zu'$

4. a. Find two distinct minimal sums of products for $xyz + x'yz + x'y'z + xyz' + x'y'z' + xy'z'$.
   b. Write the following as a minimal sum of (two) products: $x'yz'u + x'y'z'u + xy'zu + xy'z'u + xy'z'u' + xy'zu'$.

5. Write each of the following as a minimal sum of products.
   a. $f(x, y) = \begin{cases} 1 & \text{if } x = 0 \text{ or } y = 1 \\ 0 & \text{otherwise} \end{cases}$
   b. $f(x, y, z) = \begin{cases} 1 & \text{if } x = 0 \text{ or } (y = 1 \text{ or } z = 1) \\ 0 & \text{otherwise} \end{cases}$

6. Write each of the following as a minimal sum of products.
   a. $f(x, y, z, u) = \begin{cases} 1 & \text{if } (x = 1 \text{ or } y = 1) \text{ and } (xy = 0) \text{ and } z = 1 \\ 0 & \text{otherwise} \end{cases}$
   b. $f(x, y, z, u) = \begin{cases} 1 & \text{if } x + y + z + u = 1 \\ 0 & \text{otherwise} \end{cases}$

7. Refer to Exercise Set 5.2 for the definition of "If x then y else z."
   a. Write "If x then y else z" as a minimal sum of products.
   b. Draw the logic network for that minimal sum of products.

8. Recall that # stands for "don't care." Write each of the following as a minimal sum of products.
   a. $f(x, y, z) = \begin{cases} 1 & \text{if } xyz = 1 \\ \# & \text{if } xy' = 1 \\ 0 & \text{otherwise} \end{cases}$
   b. $g(x, y, z, u) = \begin{cases} 1 & \text{if } x = 0 \text{ and } (zu = 1 \text{ or } yu = 1) \\ \# & \text{if } x = 1 \\ 0 & \text{otherwise} \end{cases}$

9. Write each of the following as a minimal sum of products.
   a. $g(x, y, z) = \begin{cases} 1 & \text{if } x + y = 0 \\ \# & \text{if } xyz = 1 \text{ or } xy' = 1 \\ 0 & \text{otherwise} \end{cases}$
   b. $f(x, y, z, u) = \begin{cases} 1 & \text{if } (x + y = 0) \text{ or } (x'zu = 1) \\ \# & \text{if } xz = 1 \\ 0 & \text{otherwise} \end{cases}$

10. **a.** Use Boolean algebra laws to show that $xy' + x'y = (x + y)(xy)'$.
    b. The logic network we gave for the minimal sum of products $xy' + x'y$ in Example 7.3.1 had five logic gates. How many logic gates does the network for $(x + y)(xy)'$ have?

11. **a.** Give the Boolean expression corresponding to the logic network in Figure 7.3.25.
    b. Use Boolean algebra laws to simplify the Boolean expression obtained in part a.
    c. Draw the logic network for the simplified Boolean expression obtained in part b.

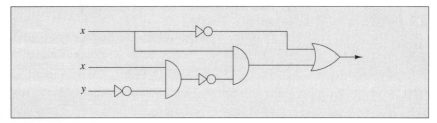

FIGURE 7.3.25

12. **a.** Give the Boolean expression corresponding to the logic network in Figure 7.3.26.
    b. Find the standard sum of products form of the Boolean expression obtained in part a.
    c. Find the minimal sum of products form of the Boolean expression obtained in part a.
    d. Draw the logic network for the Boolean expression obtained in part c.

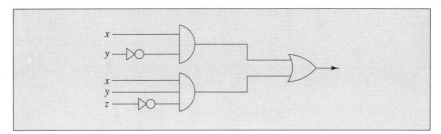

FIGURE 7.3.26

13. **a.** Give the Boolean expression corresponding to the logic network in Figure 7.3.27.

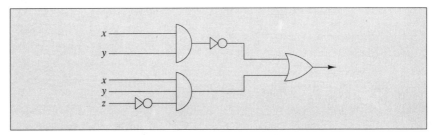

FIGURE 7.3.27

b. Find the standard sum of products form of the Boolean expression obtained in part a.
c. Find the minimal sum of products form of the Boolean expression obtained in part a.
d. Draw the logic network for the Boolean expression obtained in part c.

14. **a.** Give the Boolean expression corresponding to the logic network in Figure 7.3.28.
    b. Find the standard sum of products form of the Boolean expression obtained in part a.
    c. Find the minimal sum of products form of the Boolean expression obtained in part a.
    d. Draw the logic network for the Boolean expression obtained in part c.

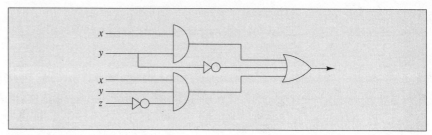

FIGURE 7.3.28

15. A **full-adder network** takes three bits x, y, and z as input and outputs the binary sum s and the carry c resulting from adding the three bits.
    a. Give a truth table defining the Boolean functions s and c.
    b. Find minimal sums of products for the functions s and c.
    c. Draw a full-adder network based on your solution to part b with the two outputs s and c.

Exercises 16–19 contain results concerning NOR and NAND gates. Refer to Section 7.2 for the definition and relevant facts about the NAND and NOR operations. Figure 7.3.29 shows the symbols used in logic networks for two-input NOR and NAND gates.

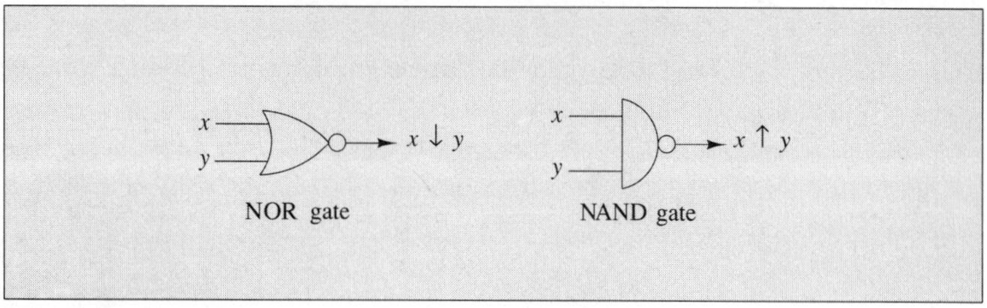

FIGURE 7.3.29

16. Design each of the following logic networks using only NAND gates.
    a. An inverter   **b.** An OR network   **c.** An AND network

17. Design each of the following logic networks using only NOR gates.
    a. An inverter   b. An AND network   c. An OR network
18. a. Prove that the NAND operation is commutative.
    b. Show that the NAND operation is not associative by finding a counterexample to $x \uparrow (y \uparrow z) = (x \uparrow y) \uparrow z$.
19. a. Prove that the NOR operation is commutative.
    b. Show that the NOR operation is not associative by finding a counterexample to $x \downarrow (y \downarrow z) = (x \downarrow y) \downarrow z$.

7.4
THE STRUCTURE OF BOOLEAN ALGEBRA (Optional)

In Section 7.1 we stated and proved the most important theorems pertaining to the algebraic structure of a Boolean algebra. In Section 7.3 we applied laws of Boolean algebra to the design of logic networks. But we are still not able to answer easily the question "Is this a Boolean algebra?" when faced with an arbitrary algebraic structure. To answer this question for finite Boolean algebras, we need to prove a structure theorem. First we will consider a partial ordering on a Boolean algebra.

Partial Ordering for a Boolean Algebra

We define a **partial ordering**, $\preceq$, on any Boolean algebra as follows:

$$x \preceq y \quad \text{if and only if} \quad x \cdot y = x$$

To illustrate the above ordering in a specific Boolean algebra, we examine the Boolean algebra $(P(U), \cup, \cap, ', \emptyset, U)$, in which the condition $x \cdot y = x$ is $X \cap Y = X$. This is equivalent to $X \subseteq Y$. So the analogous ordering for $(P(U), \cup, \cap, ', \emptyset, U)$ is the subset relation.

In any Boolean algebra, we have $x \preceq 1$ for any x because $x \cdot 1 = x$ for any x.

PRACTICE PROBLEM 1 Prove $0 \preceq x$ for any x.

THEOREM 7.4.1 The three conditions $xy = x$, $x + y = y$, and $xy' = 0$ are logically equivalent. Hence, these conditions are all equivalent to $x \preceq y$.

Proof
1. Prove $(xy' = 0) \rightarrow (xy = x)$. We assume $xy' = 0$. By a simplification theorem, $x = xy + xy'$. Hence, $x = xy$.
2. Prove $(xy = x) \rightarrow (y = x + y)$. Assume $xy = x$. By simplifying we obtain $y = yx + yx'$. So, $y = xy + yx' = x + yx' = (x + y)(x + x') = (x + y)1 = x + y$. Therefore, $y = x + y$.
3. Prove $(x + y = y) \rightarrow (xy' = 0)$. Assume $x + y = y$. By De Morgan's law, $y' = (x + y)' = x'y'$. Hence, $xy' = x(x'y') = (xx')y = 0 \cdot y' = 0$.

By the transitivity of the implication, the three conditions $xy = x$, $x + y = y$, and $xy' = 0$ are logically equivalent. Therefore, $x \preceq y$ is logically equivalent to any one of the above three conditions. ∎

The implication is transitive because $[(p \rightarrow q) \wedge (q \rightarrow r)] \rightarrow (p \rightarrow r)$ is a tautology. (See Exercise 16, Exercise Set 5.1.)

THEOREM 7.4.2 $xy \preceq y \preceq x + y$

Proof   Since $(xy)y = xy$, then $xy \preceq y$ follows immediately from the definition of $\preceq$. Since $y + (x + y) = x + y$, we have $y \preceq x + y$ by Theorem 7.4.1. ∎

Here is another example of the ordering $\preceq$ in a Boolean algebra. Consider the Boolean algebra $(D_{14}, +, \cdot, ', 1, 14)$, introduced in Section 7.1, with $D_{14} = \{1, 2, 7, 14\}$; $x + y = \text{lcm}(x, y)$; $x \cdot y = \gcd(x, y)$, and $x' = 14/x$ for arbitrary x and y in D_{14}. In this Boolean algebra, the condition $x \cdot y = x$ is $\gcd(x, y) = x$. This holds if and only if x divides y. So, in this Boolean algebra, $x \preceq y$ means x divides y. Remember this example and the subset ordering as you consider the following theorems.

The next three theorems prove $\preceq$ is a partial order.

THEOREM 7.4.3   **REFLEXIVE LAW FOR $\preceq$**

$x \preceq x$

Proof   By the idempotent laws, $xx = x$ (see Exercise 13, Exercise Set 7.1). Hence, $x \preceq x$. ∎

THEOREM 7.4.4   **ANTISYMMETRY LAW FOR $\preceq$**   If $x \preceq y$ and $y \preceq x$, then $x = y$.

Proof   Assume $x \preceq y$ and $y \preceq x$. Hence, $xy = x$ and $yx = y$. By Boolean algebra axiom 1b, $xy = yx$. Therefore, $x = y$. ∎

THEOREM 7.4.5   **TRANSITIVITY LAW FOR $\preceq$**   If $x \preceq y$ and $y \preceq z$, then $x \preceq z$.

Proof   Assume $x \preceq y$ and $y \preceq z$. Hence, $xy = x$ and $yz = y$. So, $xz = (xy)z = x(yz) = xy = x$. Therefore, $x \preceq z$. ∎

By the last three theorems, the ordering $\preceq$ is reflexive, antisymmetric, and transitive. Hence, $\preceq$ is a partial ordering. When we wish to emphasize the partial ordering $\preceq$, we write $(B, +, \cdot, ', 0, 1, \preceq)$.

THEOREM 7.4.6   Assume $x \preceq y$ and $w \preceq z$.

     a. $x + w \preceq y + z$ **b.** $xw \preceq yz$

Proof of a

$$(x+w)(y+z) = (x+w)y + (x+w)z$$
$$= y(x+w) + z(x+w)$$
$$= yx + yw + zx + zw$$
$$= xy + xz + wy + wz$$
$$= x + xz + wy + w$$
$$= x(1+z) + w(y+1)$$
$$= x \cdot 1 + w \cdot 1 = x + w$$

Hence, $x + w \leq y + z$. ∎

The proof of part b is left as an exercise. (See Exercise 4, Exercise Set 7.4.)

Now consider the Boolean algebra $(D_{30}, +, \cdot, ', 1, 30)$ with $D_{30} = \{1, 2, 3, 5, 6, 10, 15, 30\}$; $x + y = \text{lcm}(x, y)$; $x \cdot y = \gcd(x, y)$, and $x' = 30/x$ for arbitrary x and y in D_{30}. Each element $x \in D_{30}$ can be written as $x = 2^i 3^j 5^k$, where $i, j, k = 0$ or 1. In other words, all the elements of this Boolean algebra are built from the elements 2, 3, and 5. Similarly for the Boolean algebra $(P(U), \cup, \cap, ', \emptyset, U)$, all elements (the subsets of U) can be written as a union of some collection of singleton sets. For example, with $U = \{a, b, c, d\}$, the subset $\{a, c, d\} = \{a\} \cup \{c\} \cup \{d\}$. Of course, the empty set is the union of *no* singleton sets. This is analogous to $1 = 2^0 3^0 5^0$ in D_{30}.

The primes 2, 3, and 5 in D_{30} and the singletons in U act as building blocks, called the atoms of the respective Boolean algebra.

Let $(B, +, \cdot, ', 0, 1, \leq)$ be a Boolean algebra. An element $a \in B$ is an **atom** if $a \neq 0$, and, for all $x \in B$, if $x \leq a$ and $x \neq a$, then $x = 0$.

Whenever $x \leq y$ and $x \neq y$ we say x is **strictly less than** y and write $x < y$. An atom has no nonzero element strictly less than itself.

PRACTICE PROBLEM 2 Find the atoms in the Boolean algebra $(D_{231}, +, \cdot, 1, 231)$.

PRACTICE PROBLEM 3 Let a_1 and a_2 be atoms. Prove: If $a_1 \leq a_2$, then $a_1 = a_2$.

THEOREM 7.4.7 An element a is an atom if and only if, for all x,

$$ax = a \quad \text{or} \quad ax = 0$$

Proof ($\rightarrow$) Assume a is an atom. Suppose $ax \neq a$. Since $ax \leq a$, it follows immediately that $ax = 0$.

($\leftarrow$) Assume $ax = a$ or $ax = 0$ for all x. To prove a is an atom, we may assume x is arbitrary with $x \leq a$ and $x \neq a$. But $x \leq a$ implies $xa = x$. So, by our first assumption, $x = a$ or $x = 0$. Therefore, $x = 0$ since $x \neq a$. ∎

THEOREM 7.4.8 Let a be an atom, and let x and y be arbitrary elements of B.

a. $a \leq x + y$ if and only if $a \leq x$ or $a \leq y$
b. $a \leq xy$ if and only if $a \leq x$ and $a \leq y$
c. $a \leq x$ if and only if $\neg(a \leq x')$

Proof **a.** ($\leftarrow$) Assume $a \leq x$ or $a \leq y$. By Theorem 7.4.2, $x \leq x + y$ and $y \leq x + y$. Hence, in either case, $a \leq x + y$ by the transitive law for $\leq$.

($\rightarrow$) Assume $a \leq x + y$. By definition, $a(x + y) = a$. Hence, $ax + ay = a$. We consider two cases.
  Case 1: Assume $ax = 0$. If so, then $ay = a$. Hence, $a \leq y$.
  Case 2: Assume $ax \neq 0$. By Theorem 7.4.7, $ax = a$. Hence, $a \leq x$.

b. ($\rightarrow$) Assume $a \leq xy$. Then $a \leq x$ and $a \leq y$ follows from the transitive law and the fact that $xy \leq x$ and $xy \leq y$.

($\leftarrow$) Assume $a \leq x$ and $a \leq y$. Then $aa \leq xy$ by Theorem 7.4.6b. But $aa = a$, so $a \leq xy$.

c. ($\rightarrow$) Assume $a \leq x$ and suppose $a \leq x'$. Then $a = aa \leq xx' = 0$. Contradiction.

($\leftarrow$) Since $a \leq 1$, then $a \leq x + x'$. Hence, by part a of this theorem, $a \leq x$ or $a \leq x'$. Therefore, using the OR form of an implication, we obtain $\neg(a \leq x') \rightarrow a \leq x$. ∎

DEFINITION

For $x \in B$, define **Atom**$[x] = \{a : a$ is an atom of B and $a \leq x\}$. We denote the set of all atoms in B by W.

We see that Atom$[x]$ is composed of all the atoms less than or equal to x.

EXAMPLE 7.4.9 Consider the Boolean algebra $(D_{30}, +, \cdot, 1, 30)$. We see that Atom$[15] = \{3, 5\}$ and Atom$[30] = \{2, 3, 5\} = W$. Note that $15 = \text{lcm}(3, 5) = 3 + 5$ and $30 = \text{lcm}(2, \text{lcm}(3, 5)) = 2 + 3 + 5$ in this Boolean algebra. ∎

PRACTICE PROBLEM 4 Let $U = \{a, b, c\}$ in the Boolean algebra $(P(U), \cup, \cap, ', \emptyset, U)$. Find Atom$[\{a, c\}]$.

Here are several basic facts you are asked to prove in Exercise 14, Exercise Set 7.4.

a. Atom$[0] = \emptyset$ **b.** Atom$[1] = W$
c. If a is an atom, then Atom$[a] = \{a\}$.

THEOREM 7.4.10 **a.** Atom$[x + y] = $ Atom$[x] \cup$ Atom$[y]$
b. Atom$[xy] = $ Atom$[x] \cap$ Atom$[y]$
c. Atom$[x'] = W - $ Atom$[x]$

The proof of Theorem 7.4.10 is equivalent to Theorem 7.4.8, and its proof is left as an exercise. (See Exercise 16, Exercise Set 7.4.)

7.4 THE STRUCTURE OF BOOLEAN ALGEBRA (OPTIONAL)

We say that $(B, +, \cdot, ', 0, 1, \preceq)$ is a **finite Boolean algebra** whenever B is a finite set. We write Card(B) for the number of elements in a finite Boolean algebra.

THEOREM 7.4.11 Let B be a finite Boolean algebra. For any nonzero element x in B, there is an atom a such that $a \preceq x$.

Proof Pick any nonzero element x. If x is an atom, we are done. If x is not an atom, then there exists an element $x_1 \neq 0$ such that $x_1 \prec x$. Similarly, if x_1 is not an atom, there exists an element $x_2 \neq 0$ such that $x_2 \prec x_1$. Continuing in this way we have $x \succ x_1 \succ x_2 \succ \cdots$. But B is finite, so this process must terminate. We have $x \succ x_1 \succ x_2 \succ \cdots \succ x_k \succ 0$. We have an element $x_k \preceq x$ with no nonzero element strictly less than x_k. Therefore, x_k is the atom we are seeking. ∎

If, for every nonzero element x in a Boolean algebra, there exists an atom a such that $a \preceq x$, the Boolean algebra is said to be **atomistic**. In Theorem 7.4.11 we proved that every finite Boolean algebra is atomistic. See Exercise 19 of Exercise Set 7.4 for an atomistic Boolean algebra that is not finite.

Every atomistic Boolean algebra satisfies two properties that will make it possible for us to describe the structure of such a Boolean algebra.

THEOREM 7.4.12 Let B be an atomistic Boolean algebra.

a. Atom$[x] = \emptyset$ if and only if $x = 0$.
b. Atom$[x] = $ Atom$[y]$ if and only if $x = y$.

Proof **a.** ($\leftarrow$) It is left as an exercise to prove that Atom$[0] = \emptyset$. (See Exercise 14a, Exercise Set 7.4.)

($\rightarrow$) This follows from the definition of atomistic because, for every $x \neq 0$, Atom$[x] \neq \emptyset$ in an atomistic Boolean algebra.

b. ($\leftarrow$) This follows from the definition of Atom$[x]$.

($\rightarrow$) Assume Atom$[x] = $ Atom$[y]$. Then, Atom$[x] \cap (W - $ Atom$[y]) = \emptyset$. Hence, Atom$[xy'] = $ Atom$[x] \cap $ Atom$[y'] = \emptyset$. Therefore, $xy' = 0$ by part a. Similarly, $x'y = 0$. Therefore, $x \preceq y$ and $y \preceq x$. So we have $x = y$. ∎

Recall that W is the set of all atoms in a Boolean algebra B. So $P(W)$ is the collection of all sets of atoms in B.

COROLLARY 7.4.13 Let B be an atomistic Boolean algebra. There is a 1–1 function $f: B \to P(W)$.

Proof Define $f: B \to P(W)$ by $f(x) = $ Atom$[x]$. This function is well-defined and 1–1 by Theorem 7.4.12b. ∎

Corollary 7.4.13 tells us that every atomistic Boolean algebra can be mapped by a 1–1 function into a collection of sets. So an atomistic Boolean algebra must "look like" a collection of sets. As we will see in Corollary 7.4.15, the situation is even more completely described for finite Boolean algebras. The following example illustrates these ideas.

EXAMPLE 7.4.14 Consider the Boolean algebra $B = (D_{30}, +, \cdot, ', 1, 30)$. Then, $W = \{2, 3, 5\}$. The function $f: B \to P(W)$ defined by $f(x) = \text{Atom}[x]$ is shown in Table 7.4.1.

TABLE 7.4.1

x	$f(x) = \text{Atom}[x]$
1	$\emptyset$
2	$\{2\}$
3	$\{3\}$
5	$\{5\}$
6	$\{2, 3\}$
10	$\{2, 5\}$
15	$\{3, 5\}$
30	$\{2, 3, 5\}$

COROLLARY 7.4.15 Let B be a finite Boolean algebra. Then there is a bijective function $f: B \to P(W)$.

Proof As in Table 7.4.1, define $f: B \to P(W)$ by $f(x) = \text{Atom}[x]$. By Theorems 7.4.11 and 7.4.13, we only need prove that f is onto. Let $Y \subseteq W$ be a finite collection of atoms, and let y be the sum of the atoms in Y. We leave it as an exercise to show that $\text{Atom}[y] = Y$. (See Exercise 15, Exercise Set 7.4.) Since W is finite it follows that, for every set $Y \in P(W)$, $Y = \text{Atom}[y]$ for some $y \in B$. Therefore, f is onto. ∎

We state, without proof, a fundamental fact about finite sets and bijective functions.

CARDINALITY PRINCIPLE

Let $\text{Card}(A) = m$ and $\text{Card}(B) = n$.

If there exists a bijective function $f: A \to B$, then $m = n$.

COROLLARY 7.4.16 If the finite Boolean algebra B contains k atoms, then $\text{Card}(B) = 2^k$.

Proof By Corollary 7.4.15 there is a bijective function mapping the elements of a finite Boolean algebra onto the power set of W, that is, a bijective function

7.4 THE STRUCTURE OF BOOLEAN ALGEBRA (OPTIONAL)

$f: B \to P(W)$. But $\text{Card}(P(W)) = 2^k$. Therefore, $\text{Card}(B) = 2^k$ by the cardinality principle. ∎

As an example of how to apply Corollary 7.4.16, consider $D_{36} = \{1, 2, 3, 4, 6, 9, 12, 18, 36\}$. Is it possible to form a Boolean algebra using $B = D_{36}$? The answer is no, since the cardinality of D_{36} is 9 and $9 \neq 2^n$ for any n.

EXERCISE SET 7.4

1. Prove: If $z \leq y$ and $z \leq y'$, then $z = 0$.
2. Prove: If $z = 1$ and $z \leq y$, then $y = 1$.
3. Prove: If $x \leq y$, then $y' \leq x'$.
4. Assume $x \leq y$ and $w \leq z$. Prove part b of Theorem 7.4.6: $xw \leq yz$.
5. Prove $x \leq y$ if and only if $x' + y = 1$.
6. For each of the following sets B, decide whether a Boolean algebra $(B, +, \cdot, ', 0, 1)$ can be formed. In each case justify your answer.
   a. $B = D_{25}$ b. $B = D_{26}$
7. For each of the following sets B, decide whether a Boolean algebra $(B, +, \cdot, ', 0, 1)$ can be formed. In each case justify your answer.
   a. $B = D_{32}$ b. $B = D_{31}$
8. Consider the Boolean algebra $(D_{42}, +, \cdot, ', 1, 42)$.
   a. Find $P(W)$.
   b. Specify the bijective function of Corollary 7.4.15.
9. Consider the Boolean algebra $(D_{105}, +, \cdot, ', 1, 105)$.
   a. Find $P(W)$.
   b. Specify the bijective function of Corollary 7.4.15.
10. Consider the Boolean algebra $(P(U), \cup, \cap, ', \emptyset, U)$.
    a. Verify that $W = \{\{x\} : x \in U\}$.
    b. Specify the 1–1 function of Corollary 7.4.13.
11. As we have seen, for certain positive integers n we can form a Boolean algebra $(D_n, +, \cdot, ', 1, n)$, and for others (for example, $n = 36$) we cannot. Here, as usual, we are assuming $x + y$ is $\text{lcm}(x, y)$, $x \cdot y$ is $\gcd(x, y)$, and $x' = n/x$. Prove: $(D_n, +, \cdot, ', 1, n)$ is a Boolean algebra if and only if there is no integer $m > 1$ such that m^2 divides n.
12. Here is another proof that x' is unique. Fill in the missing steps of the proof. Assume $yx = 0$ and $y + x = 1$.
    a. $y = y \cdot 1 = y(x + x') = \cdots = yx'$. Hence, $y = yx'$ and, therefore, $y \leq x'$.
    b. $y = y + 0 = y + xx' = \cdots = y + x'$. Hence, $y = y + x'$ and, therefore, $x' \leq y$. Therefore, $y = x'$.
13. Let a_1 and a_2 be atoms. Prove: $a_1 \neq a_2$ if and only if $a_1 a_2 = 0$.
14. Prove:
    a. $\text{Atom}[0] = \emptyset$ b. $\text{Atom}[1] = W$
    c. If a is an atom, then $\text{Atom}[a] = \{a\}$.

15. Let $Y \subseteq W$ be a finite collection of atoms, and let y be the sum of the atoms in Y. Prove that $\text{Atom}[y] = Y$.
16. Prove Theorem 7.4.10.
17. For $x, y \in B$, define $x \, \theta \, y = x'y + xy'$. Recall that $A \, \Delta \, B = (A - B) \cup (B - A)$. Prove: $\text{Atom}[x \, \theta \, y] = \text{Atom}[x] \, \Delta \, \text{Atom}[y]$.
18. Let $W = \{a_1, a_2, \ldots, a_k\}$ be all the atoms of the finite Boolean algebra B. Prove: Every element x in B is the sum of atoms in $\text{Atom}[x]$.
19. This exercise provides an atomistic Boolean algebra that is not finite. Let S be the set of all subsets A of $\mathbb{N}$, where A is finite or A' is finite.
    a. Prove that $(S, \cup, \cap, ', \emptyset, \mathbb{N})$ is a Boolean algebra. [*Hint:* You need only verify that $\cup$ and $\cap$ are binary operations on S and that $'$ is a unary operation on S.]
    b. Prove that this Boolean algebra is atomistic.
20. Refer to the atomistic Boolean algebra defined in Exercise 19.
    a. Find $P(W)$ and specify the 1-1 function of Corollary 7.4.13.
    b. Show that the 1-1 function specified in part a is not bijective.
21. For this exercise do not use Corollary 7.4.16 or any equivalent statement. Let B be a finite Boolean algebra.
    a. Prove: If $0 = 1$, then $\text{Card}(B) = 1$.
    b. Prove: If $x = x'$ for some $x \in B$, then $\text{Card}(B) = 1$.
    c. Prove: If $\text{Card}(B) > 1$, then $\text{Card}(B)$ is even.

7.5

APPLICATION: SEVEN-SEGMENT DISPLAY

The **seven-segment display** on most calculators and digital clocks is a familiar sight. Each of the digits from 0 to 9 is displayed by lighting the appropriate line segments in the seven-segment configuration. The seven segments are labeled a, b, c, d, e, f, and g. See Figure 7.5.1. For example, 0 is displayed by lighting the six outside segments a, b, c, d, e, and f; and 1 is displayed by lighting the two segments b and c.

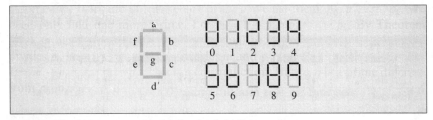

FIGURE 7.5.1 Seven-segment display for the digits $0, 1, \ldots, 9$

There are ten different inputs, one for each decimal digit $0, 1, \ldots, 9$. We assume the input for each digit is its corresponding 4-bit binary number; in other words, **binary coded decimal** (**BCD**) input. The BCD inputs correspond to $0 = 0000$, $1 = 0001$, $2 = 0010, \ldots, 9 = 1001$.

7.5 APPLICATION: SEVEN-SEGMENT DISPLAY

We need to design logic networks that activate the appropriate segments in response to specific BCD input. Figure 7.5.2 illustrates the design problem.

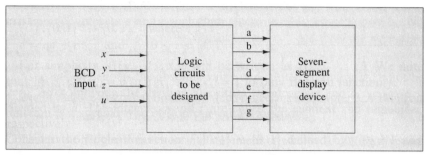

FIGURE 7.5.2

Since 4 bits will allow 16 different outputs and we need only 10 inputs, there will be 6 don't cares in this example. We construct a Boolean function for each segment. The output for each Boolean function will signify on (1) or off (0) for the corresponding segment. Table 7.5.1 shows the truth tables (in one table) defining the seven Boolean functions a, b, c, d, e, f, and g. Although the truth tables have 16 lines, we will omit the last 6 lines since they are the don't care lines.

TABLE 7.5.1

DECIMAL DISPLAYED	BCD INPUTS				OUTPUTS (1 = on, 0 = off)						
	x	y	z	u	a	b	c	d	e	f	g
0	0	0	0	0	1	1	1	1	1	1	0
1	0	0	0	1	0	1	1	0	0	0	0
2	0	0	1	0	1	1	0	1	1	0	1
3	0	0	1	1	1	1	1	1	0	0	1
4	0	1	0	0	0	1	1	0	0	1	1
5	0	1	0	1	1	0	1	1	0	1	1
6	0	1	1	0	0	0	1	1	1	1	1
7	0	1	1	1	1	1	1	0	0	0	0
8	1	0	0	0	1	1	1	1	1	1	1
9	1	0	0	1	1	1	1	0	0	1	1

We will design a logic circuit for segment d. The other segments are left as exercises.

EXAMPLE 7.5.1 Find the minimal sum of products for segment d.

Solution From Table 7.5.1 we obtain the Karnaugh map for the Boolean function d. The Karnaugh map in Figure 7.5.3 yields the minimal sum of products $yz'u + y'z + y'u' + zu'$.

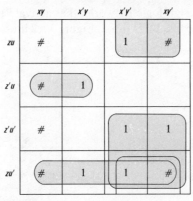

FIGURE 7.5.3

EXAMPLE 7.5.2 Draw the logic network for $yz'u + y'z + y'u' + zu'$.

Solution Figure 7.5.4 shows the required logic network.

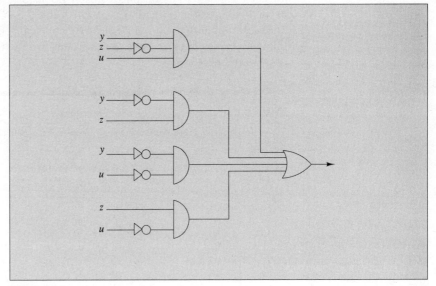

FIGURE 7.5.4

EXERCISE SET 7.5
1. Draw the logic network for segment c in the seven-segment display corresponding to the standard product of sums form.
2. Draw the logic network for segment c in the seven-segment display corresponding to the minimal sum of products form.

3. Draw the logic network for segment e in the seven-segment display corresponding to the minimal sum of products form.
4. Draw the logic network for segment a in the seven-segment display corresponding to the standard product of sums form.

KEY TERMS

Boolean algebra
Axioms for Boolean algebra
Dual
Principle of duality
Zero element
One element
Boolean algebra laws
Unique complement law
Involution laws
De Morgan's laws
Boolean function
Truth table
Boolean expression
Correspondence between a Boolean expression and a Boolean function
Equivalent Boolean expressions
Literal

Minterm
Standard sum of products
Disjunctive normal form
Truth table method
Functionally complete set of operations
Functionally complete operation
NOR operation
NOR gate
NAND operation
NAND gate
Logic gate
OR gate
AND gate
NOT gate (inverter)
Logic network
Equivalent logic networks

Half-adder network
Minimal sum of products
Differ in one literal
Minimization rule
Karnaugh map
Don't care inputs
Partial ordering for a Boolean algebra
Atom
Strictly less than
Atom[x]
Finite Boolean algebra
Atomistic Boolean algebra
Seven-segment display
Binary coded decimal (BCD)

REVIEW EXERCISES

1. Let $B = \{0, 1\}$. Define binary operations $+$ and $\cdot$ by $x + y = \max(x, y)$ and $x \cdot y = \min(x, y)$, respectively. Define the unary operation $'$ by $0' = 1$ and $1' = 0$. It can be shown that $(B, +, \cdot, ', 0, 1)$ is a Boolean algebra. Verify Boolean algebra axioms 3a and 4b by checking all possible cases. [$\max(x, y)$ is the maximum value of the elements in the set $\{x, y\}$; for example, $\max(0, 1) = 1$, $\min(0, 1) = 0$.]

2. Give an example of a Boolean algebra $(B, +, \cdot, ', 0, 1)$ where B has exactly eight elements. Specify B, $+$, $\cdot$, $'$, 0, and 1 for this Boolean algebra.

3. Can you give an example of a Boolean algebra $(B, +, \cdot, ', 0, 1)$ where B has exactly ten elements? Explain. [*Hint:* This is most easily done using a theorem of Section 7.4.]

4. Use the Karnaugh map method to find a minimal sum of products form for the Boolean function given in the table on page 300.

x	y	z	u	g(x, y, z, u)
1	1	1	1	1
1	1	1	0	1
1	1	0	1	1
1	1	0	0	1
1	0	1	1	0
1	0	1	0	1
1	0	0	1	0
1	0	0	0	1
0	1	1	1	1
0	1	1	0	1
0	1	0	1	1
0	1	0	0	1
0	0	1	1	0
0	0	1	0	0
0	0	0	1	0
0	0	0	0	0

5. Find the standard product of sums form for the Boolean function in Exercise 4.

6. Use the Karnaugh map method to find a minimal sum of products form for each of the following Boolean expressions.
  a. $xy + yz + x'z$ (The result you obtain here, $xy + yz + x'z = xy + x'z$, is called the **Consensus Theorem**.)
  b. $xy'z'u' + x'y'z'u' + x'y'z'u + xyz'u' + x'yzu + xyz'u$

7. Give an example of a Boolean algebra $(B, +, \cdot, ', 0, 1)$ such that:
  a. There exist three elements $x, y, z \in B$ satisfying $x + y = x + z$ but $x \neq 1$ and $y \neq z$
  b. There exist three elements $u, v, w \in B$ satisfying $u \cdot v = u \cdot w$ but $u \neq 0$ and $v \neq w$

8. **a.** Give the Boolean expression corresponding to the logic network shown in Figure (A).
  b. Use Boolean algebra laws to simplify the Boolean expression obtained in part a.
  c. Draw the logic network for the simplified Boolean expression obtained in part b.

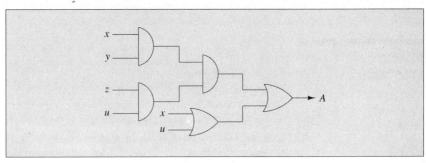

FIGURE (A)

9. **a.** Give the Boolean expression corresponding to the logic network shown in Figure (B).
  b. Use Boolean algebra laws to simplify the Boolean expression obtained in part a.

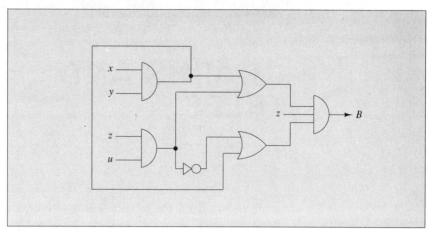

FIGURE (B)

    c. Draw the logic network for the simplified Boolean expression obtained in part b.

10. Use the Karnaugh map method to verify that $xy + x'z + y'z' = x'y' + xz' + yz$.

11. Let $E = x'(y' + u) + xzu'$.
    a. Write E as a standard sum of products.
    b. Find E' as a standard sum of products.
    c. Use part b to write E as a standard product of sums.

12. How many minterms with seven variables are there?

13. How many minterms with seven variables have the variables x and y complemented?

14. How many minterms with seven variables have exactly two variables complemented?

15. How many four-variable standard sums of products are there?

16. How many four-variable standard sums of products are there that include the minterms $xy'zu$ and $xyz'u'$ as part of the sum?

17. Define a relation R on a finite Boolean algebra as follows: $x\ R\ y$ if $\text{Card}(\text{Atom}[x]) = \text{Card}(\text{Atom}[y])$.
    a. Prove R is an equivalence relation.
    b. Specify the equivalence classes with respect to R for the Boolean algebra $(D_{30}, \text{lcm}, \text{gcd}, ', 1, 30)$, where $x' = 30/x$.

REFERENCES

D'Angelo, H. *Microcomputer Structures*. New York: BYTE Publications, 1981.

Karnaugh, M. "The Map Method for Synthesis of Combinatorial Logic Circuits." *Transactions of the AIEE, Part I*. 72 (1953), 593–599.

Mott, J. L., A. Kandel, and T. P. Baker. *Discrete Mathematics for Computer Scientists*. Reston, VA: Reston, 1983.

Shannon, C. E. "A Symbolic Analysis of Relay and Switching Circuits." *Transactions of the AIEE*, 57 (1938): 713–723

Shannon, C. E. "The Synthesis of Two-Terminal Switching Circuits." *Bell System Technical Journal*, 28 (1949): 59–98.

Roth, C. H. *Fundamentals of Logic Design*, 2nd ed. St. Paul, MN: West, 1979.

Taub, H. *Digital Circuits and Microprocessors*. New York: McGraw-Hill, 1982.

CHAPTER 8

GRAPHS AND TREES

Graph theory is an old subject that has received a tremendous renewal of interest in recent years. It was invented by Leonhard Euler (1707–1783) in 1736 when he published his solution and generalizations of the famous Königsberg bridge problem. In the 19th century, Gustav Robert Kirchhoff (1824–1887) developed a graph theory approach to the analysis of electrical networks. During the 20th century, graph theory has been widely applied in electrical engineering, transportation, information processing, computer science, and genetics, among others.

Trees are special kinds of graphs. Arthur Cayley (1821–1895) discovered trees in his study of chemical compounds of the form C_nH_{2n+2}. Trees are also used to represent algebraic expressions, codes, and information chains. There are many applications of trees to searching and sorting algorithms. Parse trees are also used in the study of formal languages and compiler design. In addition, trees are helpful for analyzing the complexity of recursive algorithms.

In Chapter 2, we studied directed graphs and reachability. In this chapter, we give an introduction to graphs and trees with many examples and applications. We will review some familiar terms as we study new concepts.

APPLICATION

Sorting census data in the early part of the 20th century was the first major application of digital computers to a social science problem. Mathematicians and computer scientists have been analyzing sorting techniques ever since the power of automatic data processing was demonstrated in the 1920 U.S. Census. Many algorithms for sorting have been proposed and studied over the years. A key problem emerges: Given a list of n objects, which sorting algorithm will arrange these objects in ascending (or descending) order most efficiently?

The heapsort is among the most efficient of all sorting algorithms. Heapsort is designed using a tree structure called a heap. We will see how tree structures are used to implement a heapsort in Section 8.5.

8.1
GRAPHS

As a first example of a graph, Figure 8.1.1 shows a simplified airline map with routes between some major U.S. cities. Vertices represent cities, and edges represent routes.

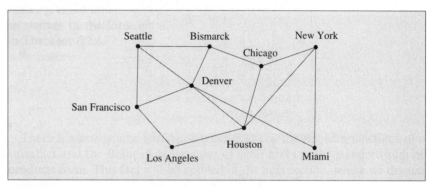

FIGURE 8.1.1

The idea of a map of routes between locations generalizes to the concept of a graph. The map shown in Figure 8.1.1 is the picture of a graph. Each dot in the picture represents a vertex, and a line between two vertices represents an edge.

DEFINITION

> A **graph**, denoted $G = (V, E)$, consists of a nonempty set V of **vertices** (or **nodes**) and a set E of **edges** (or **arcs**) such that each edge corresponds to a unique unordered pair of distinct vertices $\{u, v\}$ and no more than one edge corresponds to $\{u, v\}$. The sets V and E are assumed to be finite.

When an edge e corresponds to $\{u, v\}$, we say e *joins* the vertices u and v; the vertices u and v are *incident* with the edge e; the edge e is *incident* with each of its vertices u and v; and u and v are called *endpoints* of the edge e. An edge e is *incident* with an edge f if e and f have a vertex in common. For simplicity, we will often identify $\{u, v\}$ as the edge joining u to v.

When two or more edges correspond to $\{u, v\}$, we say there are **multiple edges** joining the vertices u and v. By definition, a graph cannot contain multiple edges incident with any pair of vertices. A **multigraph** is defined like a graph except that multiple edges are allowed. Every graph is a multigraph but not conversely.

A **loop** corresponds to a singleton set containing a vertex. A **loopgraph** is defined like a graph except that it contains loops.

We emphasize that a *graph* has a finite number of vertices and edges but includes neither loops nor multiple edges.

Now we consider some examples of graphs, multigraphs, and loopgraphs.

EXAMPLE 8.1.1 Figure 8.1.2 is a picture of three graphs. The structures pictured in Figure 8.1.3 are not graphs. Figure 8.1.3(a) shows a loopgraph, and the structure in Figure 8.1.3(b) is a multigraph.

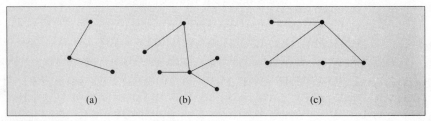

FIGURE 8.1.2

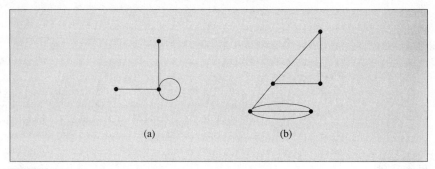

FIGURE 8.1.3

Graphs can be useful for scheduling and work assignments, as illustrated by the next two examples.

EXAMPLE 8.1.2 **COMMITTEE SCHEDULING** Suppose we have seven university committees on Space, Public Relations, Course Review, Travel, Grading Policy, Scholarship, and Parking. These committees must meet weekly from 10:00 A.M. to 12:00 noon on either Mondays, Wednesdays, or Fridays. It is imperative not to schedule simultaneous meetings of two committees that have a member in common.

In order to keep track of the committees that are not to be scheduled at the same time, we can use a graph with seven vertices representing the committees and an edge between any two committees with a member in common. See Figure 8.1.4.

By studying the graph in Figure 8.1.4, we see, for example, that the Parking committee and the Scholarship committee must not meet at the same time. After some deliberation, we see that one way to schedule the meetings is:

Monday: Grading Policy, Travel, and Scholarship committees
Wednesday: Course Review, Parking, and Public Relations committees
Friday: Space committee

8.1 GRAPHS

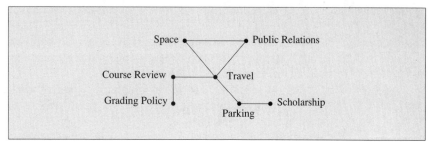

FIGURE 8.1.4

PRACTICE PROBLEM 1 Refer to Example 8.1.2. Find a different grouping of committees that allows a meeting schedule without time conflicts.

In Example 8.1.2, suppose that, instead of assigning meeting times to the committees, we assign a color to each vertex of the graph G in Figure 8.1.4. To avoid time conflicts, we must assign colors in such a way that two vertices joined by an edge will get different colors. We would normally want to find a schedule by using the least number of meeting times. In other words, we want to color the vertices of G using the least number of colors.

In general, if the vertices of a graph can be colored by n colors so that two vertices joined by an edge get different colors, then the graph is said to be **n-colorable**. The smallest number n such that the graph is n-colorable is called the **chromatic number** of the graph.

The graph in Figure 8.1.4 has chromatic number three. (Why?) See Exercise Set 8.1 for some graph-coloring problems.

EXAMPLE 8.1.3 **WORK ASSIGNMENT** Suppose a group of people {a, b, c, d, e, f, g} are working on a project. In order to spread the work evenly and generate new ideas, the supervisor assigns people to work in pairs on certain tasks. The supervisor wants a record of which pairs have worked together so that different pairs can be assigned to the next task. Suppose that

a has worked with c, d, f, and g
b has worked with c, e, and g
c has worked with a and b
d has worked with a and g
e has worked with b
f has worked with a
g has worked with a, b, and d

A graph is a natural way to record this information. We let the vertex set consist of the letters representing the people. Two vertices are joined by an edge if and only if the two corresponding people have worked together. Figure 8.1.5 shows the graph $G = (V, E)$ with $V = $ {a, b, c, d, e, f, g}.

It is not difficult to determine which people have not yet been paired from the graph in Figure 8.1.5.

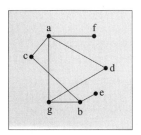

FIGURE 8.1.5

PRACTICE PROBLEM 2 Determine from Figure 8.1.5 which workers have not been paired yet.

Refer to Example 8.1.3. If every person had been paired with every other person, there would be an edge joining every pair of vertices of the graph in Figure 8.1.5.

A graph $K = (V, E)$ in which every pair of vertices is joined by an edge is called a **complete graph**. We will frequently use K_n to specify a complete graph with n vertices.

PRACTICE PROBLEM 3 For any positive integer n, how many edges does a complete graph K_n have?

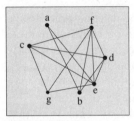

FIGURE 8.1.6

To decide which people in Example 8.1.3 have not been paired yet, we look for edges in the complete graph K with vertex set $V = \{a, b, c, d, e, f, g\}$ but not in the graph $G = (V, E)$. Figure 8.1.6 shows a picture of the graph with edges in K but not in G.

When $G = (V, E)$ is a graph (or multigraph), we will write $V(G)$ for the set of vertices of G and $E(G)$ for the set of edges of G. The **complement of a graph** G is the graph $\bar{G}$ with $V(\bar{G}) = V(G)$ and $E(\bar{G}) = E(K) - E(G)$, where K is the complete graph with $V(K) = V(G)$. In other words, G and $\bar{G}$ have the same vertex set; and, for all distinct vertices v and w in $V(G)$, the edge joining v and w is in $E(\bar{G})$ if and only if the edge joining v and w is not in $E(G)$.

The graphs in Figure 8.1.5 and Figure 8.1.6 are complements of each other.

THEOREM 8.1.4 Let G be a graph with n vertices and m edges. Then

$$n \geq 1 \quad \text{and} \quad 0 \leq m \leq C(n, 2), \quad \text{where } C(n, 2) = \frac{n(n-1)}{2}$$

Proof Since the vertex set is nonempty, it follows that $n \geq 1$. There can be at most $C(n, 2)$ edges in a graph because there are $C(n, 2)$ two-element subsets of V and a graph has no more than one edge corresponding to any set of vertices (v, w). Hence, $m \leq C(n, 2)$. Since $m \geq 0$, we have $0 \leq m \leq C(n, 2)$. ∎

Note that, if K_n is a complete graph with m edges, then $m = C(n, 2)$, which is the result of Practice Problem 3.

Subgraphs

Sometimes it is useful to study graphs that are parts of other graphs.

DEFINITION A graph H is a **subgraph** of a graph G, denoted by $H \subseteq G$, if $V(H) \subseteq V(G)$ and $E(H) \subseteq E(G)$. If H is a subgraph of G, then H is a **proper subgraph** of G whenever $V(H) \neq V(G)$ or $E(H) \neq E(G)$. A submultigraph and proper submultigraph can be similarly defined.

8.1 GRAPHS

Assume K is the complete graph with vertex set $V(K)$. Let G be any graph such that $V(G) = V(K)$. Then both G and its complement $\bar{G}$ are subgraphs of K.

Among the most important subgraphs are those that arise from a given graph when a vertex or an edge is deleted. When a vertex v is deleted from G, we write the resulting graph as $G - v$. To obtain the graph $G - v$, we delete from G the vertex v and all edges that are incident with v. So, $V(G - v) = V(G) - \{v\}$ and $E(G - v)$ equals the set of edges in G not incident with v.

When an edge e is deleted from a graph G, we write $G - e$ for the resulting graph. The deletion of an edge does not remove vertices incident with that edge. Hence, we have $V(G - e) = V(G)$ and $E(G - e) = E(G) - \{e\}$.

See Figure 8.1.7 for graphs G, $G - v$, and $G - e$.

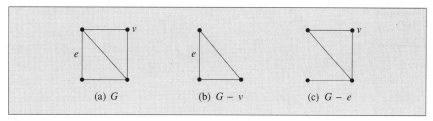

FIGURE 8.1.7

Components

Recall that every graph is a multigraph. So any definition or theorem that concerns multigraphs also concerns graphs. In other words, the term *multigraph* is equivalent to *graph or multigraph*.

DEFINITION

A **path** from vertex v to vertex w in a multigraph is an alternating sequence of vertices and edges $(v_0, e_1, v_1, e_2, v_2, \ldots, e_n, v_n)$, where $v_0 = v$, $v_n = w$, and each edge e_i joins vertices v_{i-1} and v_i for $i = 1, \ldots, n$.

The vertex v is called the **initial vertex** and w the **terminal vertex** of the path; we also say this path is *from v to w*. A path with n edges is said to have **length** n. A path is said to *include* a vertex v or an edge e if v or e appears in the alternating sequence.

Let v and w be vertices in a multigraph. Then v is **connected to** w if there is a path from v to w or if $v = w$. So, "connected to" defines a relation on the set of vertices of a multigraph. It is not hard to verify that this is an equivalence relation.

PRACTICE PROBLEM 4 Verify that "connected to" defines an equivalence relation on the set of vertices of a multigraph.

The relation "connected to" partitions the vertex set V in a multigraph G into a collection of equivalence classes. Each equivalence class V_i of vertices, along with all the edges incident with any of the vertices in V_i, forms a submultigraph H_i of G. These submultigraphs are called the **components** of the multigraph G. In particular, if V is the vertex set in a graph, each equivalence class V_i gives rise to a subgraph of the graph. These subgraphs are called the components of the graph.

If every vertex in a multigraph is connected to every other vertex, then the multigraph is said to be **connected**. Therefore, a multigraph is connected if and only if it has only one component. The graph in Figure 8.1.8 is not connected since it has three components. Note that each of the three components in Figure 8.1.8 is itself a connected graph. Furthermore, observe that any connected subgraph of the graph of Figure 8.1.8 must be contained in one of the three components.

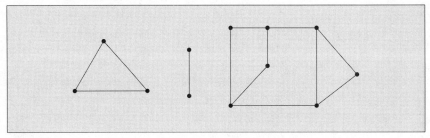

FIGURE 8.1.8 A graph with 12 vertices, 13 edges, and 3 components

A submultigraph H of a multigraph G is a **maximal connected submultigraph** if H is connected and, for any other connected submultigraph K of G, if $H \subseteq K$, then $H = K$.

For a graph G, a subgraph H is a **maximal connected subgraph** if H is connected and, for any other connected subgraph K of G, if $H \subseteq K$, then $H = K$. The proof that each component of a graph is a maximal connected subgraph is left as an exercise. (See Exercise 16, Exercise Set 8.1.)

A connected subgraph H of a connected graph G is said to be a **spanning subgraph** of G if $V(H) = V(G)$. So, a spanning subgraph of a connected graph G is the result of deleting zero or more edges from G.

EXAMPLE 8.1.5   **A HIGHWAY SYSTEM**   Consider the connected graph in Figure 8.1.9 representing a highway system. The vertices represent towns and the edges represent roads.

Most highway systems have redundant roads so that the system still functions even when some roads are closed due to catastrophic conditions. For example, the system represented by the graph of Figure 8.1.9 will function even if the roads between Sunset Beach and Green Valley and between Desert Springs and Lost Hills are closed. The remaining roads and all the towns form a spanning subgraph of the graph in Figure 8.1.9.

8.1 GRAPHS

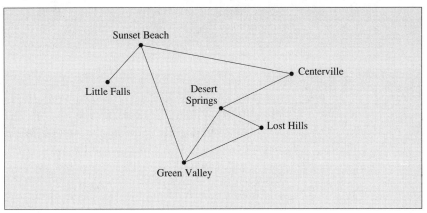

FIGURE 8.1.9

PRACTICE PROBLEM 5 Draw a spanning subgraph, other than the one mentioned in Example 8.1.5, for the graph in Figure 8.1.9.

Vertex Degree

The **degree** of a vertex v, degree(v), of a multigraph G is the number of edges in G incident with v. For example, in Figure 8.1.9, degree (Sunset Beach) = 3 because there are three roads leaving Sunset Beach. Figure 8.1.10 shows two graphs where the numeral at each vertex indicates the degree of the vertex.

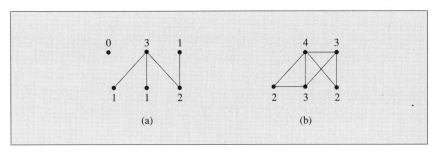

FIGURE 8.1.10

When the degrees of all vertices in the graph of Figure 8.1.10(a) are added, the resulting sum is 8; for the graph in Figure 8.1.10(b), the sum of the degrees is 14. In both cases the sum is equal to twice the number of edges in the respective graphs. That this is no accident is recorded as Theorem 8.1.6.

THEOREM 8.1.6 Let G be a multigraph with m edges and n vertices $v_1, v_2, \ldots, v_{n-1}, v_n$. Then,

$$\sum_{i=1}^{n} \text{degree}(v_i) = 2m$$

Proof Every edge is incident with exactly two vertices. So, when the degrees of the two vertices incident with an edge are added, that edge is counted twice. Hence, the sum of the degrees of all vertices must equal $2m$. ∎

COROLLARY 8.1.7 The number of vertices of odd degree in any multigraph is even.

Proof Let G be a multigraph with a set of vertices V. Suppose W is the subset of all vertices with odd degree. Let $U = V - W$. Then, U is the subset of vertices with even degree. Let w be the sum of the degrees of the vertices in W and u be the sum of the degrees of the vertices in U. By Theorem 8.1.6, the sum $w + u$ is even. But u is also even since every vertex in U has even degree. Hence, w is even. Since the sum of an odd number of odd numbers is odd, it follows that the number of vertices in W must be even. ∎

Representation of Graphs

If we wish to represent a graph for use in a computer application, an adjacency matrix is one useful method. Let us assume that $G = (V, E)$ is a graph, that $V = \{v_1, v_2, v_3, \ldots, v_n\}$, and that e_{ij} is the edge joining v_i and v_j. The **adjacency matrix** $M_G = [b_{ij}]$ for G relative to the ordered set of vertices V is defined by

$$b_{ij} = \begin{cases} 1 & \text{if } e_{ij} \in E \\ 0 & \text{otherwise} \end{cases}$$

Since e_{ij} and e_{ji} denote an identical edge, we have $e_{ij} = e_{ji}$ for $i = 1, \ldots, n$ and $j = 1, \ldots, n$. Hence, an adjacency matrix for a graph is always symmetric about the main diagonal.

Refer to Example 8.1.3 and the graph shown in Figure 8.1.5. The adjacency matrix M for that graph relative to the ordered set of vertices $\{a, b, c, d, e, f, g\}$ is

$$M = \begin{array}{c} \\ a \\ b \\ c \\ d \\ e \\ f \\ g \end{array} \begin{array}{c} \begin{matrix} a & b & c & d & e & f & g \end{matrix} \\ \begin{bmatrix} 0 & 0 & 1 & 1 & 0 & 1 & 1 \\ 0 & 0 & 1 & 0 & 1 & 0 & 1 \\ 1 & 1 & 0 & 0 & 0 & 0 & 0 \\ 1 & 0 & 0 & 0 & 0 & 0 & 1 \\ 0 & 1 & 0 & 0 & 0 & 0 & 0 \\ 1 & 0 & 0 & 0 & 0 & 0 & 0 \\ 1 & 1 & 0 & 1 & 0 & 0 & 0 \end{bmatrix} \end{array}$$

This definition of an adjacency matrix is analogous to that given in Chapter 2 for directed graphs. One difference is that the adjacency matrix for a graph is always symmetric about the main diagonal, as noted above. Another difference is that the adjacency matrix for a graph has only 0's on the main diagonal since a graph contains no loops.

EXERCISE SET 8.1

Exercises 1–4 refer to the graph G pictured in Figure 8.1.11.

1. Find the adjacency matrix for G.
2. Draw the complement $\bar{G}$. How many components of $\bar{G}$ are there?
3. a. How many connected subgraphs of G are there?
   b. How many spanning subgraphs of G are there?
4. Find a proper spanning subgraph of G. How many proper spanning subgraphs of G are there?

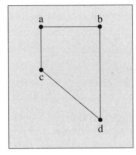

FIGURE 8.1.11   **FIGURE 8.1.12**

Exercises 5–8 refer to the graph G pictured in Figure 8.1.12.

5. a. How many connected subgraphs of G are there?
   b. How many spanning subgraphs of G are there?
6. Find a proper connected spanning subgraph of G. How many proper connected spanning subgraphs of G are there?
7. Find the adjacency matrix for G.
8. Draw the complement $\bar{G}$. How many components of $\bar{G}$ are there?
9. The Civic Club has five committees on Program, Outreach, Finance, Research, and Acquisition. Each of these committees must meet once a week on either Friday evenings, Saturday mornings, or Sunday afternoons. Each of the following pairs of committees has a member in common: the Program and Acquisition committees, the Research and Outreach committees, the Program and Outreach committees, the Acquisition and Outreach committees, and the Research and Acquisition committees.
   a. Draw a graph to show the relationship of the committees and common members, as in Example 8.1.2.
   b. Find two different ways to schedule the meeting times to avoid simultaneous meetings of any two committees that have a common member.
10. Refer to Exercise 9.
    a. What is the chromatic number of the graph in Exercise 9?
    b. Can we schedule the meetings, without time conflict, if the committees can only meet on either Friday evenings or Saturday mornings?

11. The following six councils need to have meetings scheduled. Find the least number of meeting times to avoid simultaneous meetings for committees with a common member.

$C_1 = \{\text{Smith, Lee, Jones, Takata, James}\}$
$C_2 = \{\text{Lee, Madison, Valentine, Morales}\}$
$C_3 = \{\text{Valentine, Robinson, Lucas, Wong}\}$
$C_4 = \{\text{Lucas, Albert, Lee, Davies}\}$
$C_5 = \{\text{Lopez, Vuong, Bergman}\}$
$C_6 = \{\text{Vuong, Green, Jones}\}$

12. **a.** Find the chromatic number of the graph in Figure 8.1.9.
    b. Find the chromatic number of the graph in Figure 8.1.1.

13. Find the chromatic number of the graph in Figure 8.1.13.

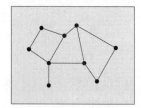

FIGURE 8.1.13

14. Draw a complete graph K_4. How many edges does it have?

15. Draw a complete graph K_5. How many edges does it have?

16. Let H be a component of a graph G. Prove that H is a maximal connected subgraph of G. That is, prove:
    a. H is a connected subgraph.
    b. If K is any other connected subgraph such that $H \subseteq K$, then $H = K$.

17. Prove that, if H is a connected subgraph of G, then H is contained in a component of G. Deduce that a maximal connected subgraph of G is a component of G.

18. **a.** Give an example of a connected graph $G = (V, E)$ such that $G - e$ is not connected for any $e \in E$.
    b. Give an example of a connected graph $G = (V, E)$ such that $G - e$ is connected for any $e \in E$.

Let $G = (V, E)$ be a connected graph. A vertex $v \in V$ is an **articulation point** of G if $G - v$ is not connected. Exercises 19 and 20 are concerned with articulation points.

19. Draw a graph with five vertices and no articulation points.

20. Draw a graph with five vertices and exactly one articulation point.

21. Suppose a group of eleven persons is working on a project. Is it possible that every person in the group has worked with exactly three other persons in the group? If it is possible, draw a graph illustrating the situation. If it is not possible, explain why not.

22. Suppose a group of eight persons is working on a project. Is it possible that every person in the group has worked with exactly three other persons in the group? If it is possible, draw a graph illustrating the situation. If it is not possible, explain why not.

*Exercises 23–26 concern bipartite graphs. A graph $G = (V, E)$ is a **bipartite graph** if V is a union of two nonempty disjoint sets A and B in such a way that every edge in E joins a vertex in A to a vertex in B. In this case, we write the bipartite graph as $G = (A \cup B, E)$. When a bipartite graph has the property that every vertex in A is joined to every vertex in B by an edge in E, the graph is a **complete bipartite graph**. We write the complete bipartite graph $(A \cup B, E)$ as $K_{m,n}$ when $\mathrm{Card}(A) = m$ and $\mathrm{Card}(B) = n$.*

23. Draw each of the following complete bipartite graphs.
    a. $K_{1,3}$ **b.** $K_{2,3}$

24. Draw each of the following complete bipartite graphs.
    a. $K_{3,3}$ **b.** $K_{2,4}$

25. Find a formula for the number of edges in $K_{m,n}$. Note that the number of edges in $K_{m,n}$ is an upper bound on the numbers of edges in all bipartite graphs $(A \cup B, E)$ when $\mathrm{Card}(A) = m$ and $\mathrm{Card}(B) = n$.

26. Determine whether the graph $G = (V, E)$ in Figure 8.1.14 is a bipartite graph. If so, find the sets A and B such that $G = (A \cup B, E)$.

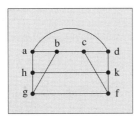

FIGURE 8.1.14

8.2
PATHS, CIRCUITS, AND CYCLES

A multigraph can be considered as a map of routes between locations, such as an airline route map or a road map. We think of the vertices as the locations, and the edges as the routes or roads. In order to make a trip from location A to location B, we must find a path from the vertex corresponding to A to the vertex corresponding to B. Since there may be multiple roads between locations on a map, we will be concerned with multigraphs.

To determine whether a multigraph G is connected, we can delete any multiple edges between vertices, thus obtaining a graph H with exactly one edge between every pair of vertices that are joined in the original multigraph. It is easy to see that the multigraph G is connected if and only if the graph H is connected.

How can we determine whether there is a path between given vertices in a graph H? Each graph can be represented by an adjacency matrix. Hence, a very good systematic method of determining whether there is a path from one vertex to another is to find the reachability matrix corresponding to the graph. As we know from Chapter 2, this can be done using Warshall's algorithm or by matrix multiplication. So the multigraph G will be connected if and only if every entry in the reachability matrix of H is 1.

A path from a vertex v to the same vertex v in a multigraph is called a **closed path**. A path with no repeated edges is a **simple path**. A closed simple path is called a **circuit**. A circuit that includes the initial vertex exactly twice and includes the other vertices exactly once is a **cycle**. In Figure 8.2.1(a), we can find examples of closed paths that are circuits but not cycles. Figure 8.2.1(b) shows a graph in which every circuit is a cycle.

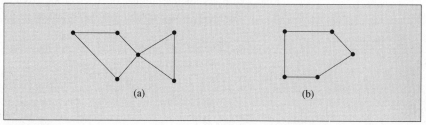

FIGURE 8.2.1

We will consider two basic questions concerning the existence of circuits and cycles.

Question 1: Does there exist a circuit in a connected multigraph G that includes each edge of the multigraph? Such a circuit is called an **Eulerian tour**, named after the mathematician Leonhard Euler. An Eulerian tour begins and ends at the same vertex and includes each edge exactly once and each vertex at least once.

Question 2: Does there exist a cycle in a connected graph G that includes each vertex of the graph? Such a cycle is called a **Hamiltonian cycle**, named after the mathematician Sir William Rowan Hamilton (1805–1865). A Hamiltonian cycle begins and ends at the same vertex and includes each vertex, other than the initial vertex, exactly once.

First, let us consider the following specific example of Question 1. It is this example and its solution that started graph theory as a field of mathematics. The following problem was solved by Euler, and its solution was published in the first paper on graph theory, in 1736.

8.2 PATHS, CIRCUITS, AND CYCLES

EXAMPLE 8.2.1 **KÖNIGSBERG BRIDGE PROBLEM** Euler's paper began by considering a puzzle called the Königsberg bridge problem. The old Prussian city of Königsberg was situated on both sides and on two islands of the Pregel River. There were seven bridges connecting the various sections of the town. See Figure 8.2.2 for a schematic of the city of Königsberg.

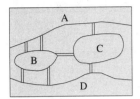

FIGURE 8.2.2

A question arose among the people who walked the city. Was it possible to plan a walking tour so that it began and ended at the same place but crossed each bridge exactly once? As we will see, the question is whether an Eulerian tour can be found. By considering each section of town as a vertex and each bridge as an edge, we can draw a multigraph representing this map. See Figure 8.2.3.

Euler proved that, starting at any vertex, it is impossible to find a circuit that includes each edge. In other words, there is no Eulerian tour for the multigraph in Figure 8.2.3. In his paper, Euler proved Theorem 8.2.3 (on the following page). It gives a necessary and sufficient condition for a connected multigraph to contain an Eulerian tour. ∎

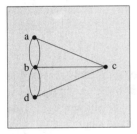

FIGURE 8.2.3

It is clear that a multigraph with an Eulerian tour must be connected. Furthermore, each vertex of the multigraph must have an even degree since each time the Eulerian tour includes a vertex it uses two edges. Euler proved that a connected multigraph G has an Eulerian tour if and only if each vertex of G has an even degree. This gives an easy way to determine whether a multigraph has an Eulerian tour. Before giving the proof of Euler's Theorem, we first consider an example. The proof of Euler's Theorem is a generalization of the construction used in Example 8.2.2.

EXAMPLE 8.2.2 Build an Eulerian tour for the multigraph G in Figure 8.2.4.

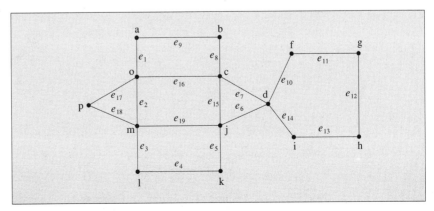

FIGURE 8.2.4

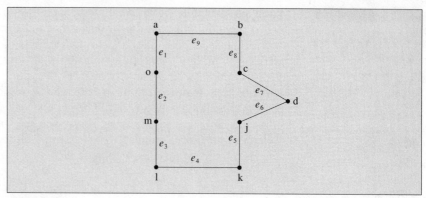

FIGURE 8.2.5

Solution   We note that the multigraph G is connected and that each vertex has even degree. Suppose we begin at vertex a and build a circuit $C = $ (a, e_1, o, e_2, m, e_3, l, e_4, k, e_5, j, e_6, d, e_7, c, e_8, b, e_9, a) as shown in Figure 8.2.5. Since each vertex has even degree, we can always depart from each vertex, other than a, after entering it. Now we consider the multigraph of remaining edges, shown in Figure 8.2.6. Although the multigraph is no longer connected, each vertex is still of even degree. Each component has an Eulerian tour—namely, (c, e_{16}, o, e_{17}, p, e_{18}, m, e_{19}, j, e_{15}, c) and (d, e_{10}, f, e_{11}, g, e_{12}, h, e_{13}, i, e_{14}, d). These two Eulerian tours can now be inserted in the circuit C at vertices d and c to obtain the following Eulerian tour of the multigraph G: (a, e_1, o, e_2, m, e_3, l, e_4, k, e_5, j, e_6, d, e_{10}, f, e_{11}, g, e_{12}, h, e_{13}, i, e_{14}, d, e_7, c, e_{16}, o, e_{17}, p, e_{18}, m, e_{19}, j, e_{15}, c, e_8, b, e_9, a).

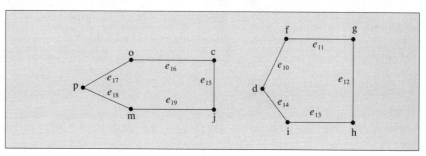

FIGURE 8.2.6

THEOREM 8.2.3   **EULER'S THEOREM**   Let G be a connected multigraph. G contains an Eulerian tour if and only if the degree of every vertex in G is even.

Proof   ($\rightarrow$) Assume G contains an Eulerian tour. Let the circuit go from vertex v back to vertex v. Then degree(v) must be even since the circuit must leave and return to v the same number of times. For any other vertex w, the

8.2 PATHS, CIRCUITS, AND CYCLES

circuit must arrive at w and leave w the same number of times. Hence, degree(w) must be even.

($\leftarrow$) Assume the degree of every vertex is even. We will build an Eulerian tour starting at some vertex v. The path begins with an edge incident to v and continues to include unused edges until we enter a vertex from which we cannot depart because all the edges incident with it have been included in the path. This must eventually occur because there are finitely many edges.

The vertex from which we cannot depart must be vertex v by the following three facts. First, if we start a path at v leaving on some edge, then the number of unused edges incident with v is odd until we return to v. Second, a path entering a vertex of even degree may leave on an edge not yet included in the path. Third, entering and leaving a vertex u each time decreases the number of unused edges incident with u by two.

Let C denote the circuit that we have built beginning and ending at vertex v. If there are edges still left to include, pick a vertex w that has been included in C but is incident with edges not yet included. Build a new circuit beginning and ending at w as we did for the vertex v, not including any edges in C.

Now we can enlarge the circuit C by going from v to w in C, then going from w back to w in the new circuit, and finally returning to v in the unused portion of C.

Since the number of edges is finite, we can repeat this enlargement process until all edges have been included. Hence, we have constructed an Eulerian tour. ∎

An **Eulerian path** in a multigraph G is a path from vertex v to vertex w that includes each edge of G exactly once.

PRACTICE PROBLEM 1   Find necessary and sufficient conditions for a connected multigraph to contain an Eulerian path from vertex v to a different vertex w.

Question 1 is often referred to as the highway inspector problem since a route that goes through each road exactly once is ideal for someone inspecting the road surface or painting a center line. There is another, related question called the mail carrier problem.

> **MAIL CARRIER PROBLEM**
> Given a connected multigraph G, is it possible to find a closed path in G such that each edge of G is included exactly twice?

This problem is so named because of the need for mail carriers to travel both sides of each street on a given route. It is left as an exercise to solve the mail carrier problem. (See Exercise 11, Exercise Set 8.2.)

Question 2, concerning Hamiltonian cycles, is often called the **traveling salesperson problem**. That is because a businessperson is concerned about including each town on a given trip exactly once but is not concerned about including each road exactly once.

In 1857, Sir William Rowan Hamilton invented a puzzle involving what we now call Hamiltonian cycles. The object of the puzzle was to include the edges of a dodecahedron and each corner point (vertex) exactly once. A dodecahedron is a polyhedron that has regular pentagons for each of its 12 faces and has 3 edges meeting at each of its 20 vertices. See Figure 8.2.7 for a graph representing the surface of a dodecahedron in the plane. The cycle drawn on the dodecahedron in Figure 8.2.7 is a Hamiltonian cycle and, hence, a solution to Hamilton's puzzle.

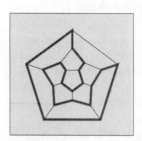

FIGURE 8.2.7

Euler had earlier considered problems involving Hamiltonian cycles. For example, Euler solved the problem of the knight's tour of the chessboard described in the next example.

EXAMPLE 8.2.4   **KNIGHT'S TOUR**   A chessboard has 64 squares that can be considered as the vertices of a graph. Each move for a knight consists of one square in a vertical direction followed by two squares in a horizontal direction, or two squares in a vertical direction and one square in a horizontal direction. A knight's tour of the chessboard involves beginning and ending at a given square, occupying each square exactly once. Figure 8.2.8 shows one solution to this problem. Each vertex represents a square on a chessboard. ∎

FIGURE 8.2.8

The traveling salesperson problem looks very much like the highway inspector problem. Since Euler found an efficient algorithm (Theorem 8.2.3 and its proof) for solving the highway inspector problem, it might seem that we can obtain an efficient algorithm for solving the traveling salesperson problem. Unfortunately, the traveling salesperson problem is apparently very difficult. No efficient algorithm for solving this problem has yet been found.

We will discuss the lack of an efficient algorithm for solving the traveling salesperson problem after we consider a more general version of the problem. A generalization of the traveling salesperson problem can be discussed after introducing weighted graphs.

Weighted Graphs

DEFINITION

A **weighted graph** (V, E, w) is a graph $G = (V, E)$ with a **weight function** $w: E \to \mathbb{R}^+$. For each $e \in E$, $w(e)$ is the **weight** for edge e.

If we think of a graph as a map of routes or roads, then we can think of the weight on each edge as the cost of traveling that route or as the length of the road represented by the edge.

8.2 PATHS, CIRCUITS, AND CYCLES

EXAMPLE 8.2.5 Table 8.2.1 is a reference table of distances in miles between the given cities. Using Table 8.2.1, we can easily draw a weighted graph representing the distances between the given cities. See Figure 8.2.9.

TABLE 8.2.1

				Atlanta
			Boston	1116
		Chicago	1009	722
	Cleveland	355	654	734
Dallas	1198	933	1795	800

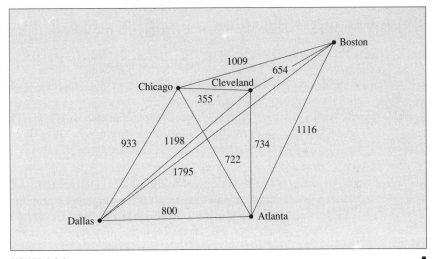

FIGURE 8.2.9

PRACTICE PROBLEM 2 Find two spanning subgraphs for the graph in Figure 8.2.9.

Now the general traveling salesperson problem can be stated in terms of weighted graphs.

TRAVELING SALESPERSON PROBLEM

Given a connected weighted graph G, find a Hamiltonian cycle $(v_0, e_1, v_1, e_2, v_2, \ldots, v_{n-1}, e_n, v_0)$ so that the sum of the weights $w(e_1) + w(e_2) + \cdots + w(e_n)$ is minimized.

This problem arises because of a salesperson's need to minimize the distance traveled while still visiting each city on a given trip exactly once. Let us consider the problem for the weighted graph in Figure 8.2.9. For specificity, we will begin and end in Boston. Since this is a complete graph, we can get from any vertex to any other vertex along a single edge. Note that there are 4!, or 24, possible Hamiltonian cycles starting at Boston. This is so because we can choose any of the other 4 cities as the first destination, then any of the remaining 3 cities as the second destination, and so on, finally returning to Boston.

Of the 24 possible Hamiltonian cycles, we need only be concerned with 12 since for each cycle there is a cycle with the cities in reverse order, and including the cities in reverse order has no effect on the sum of the weights on the edges of the cycle. Hence, we can solve this problem by finding the 12 Hamiltonian cycles and calculating the sum of the weights on the edges of each cycle.

PRACTICE PROBLEM 3 Solve the traveling salesperson problem for Example 8.2.5.

Note that, if the graph G is complete, there is always a Hamiltonian cycle in G.

PRACTICE PROBLEM 4 Prove that, if G is a complete graph with $m + 1$ vertices, there are $(m + 1)!$ Hamiltonian cycles in G, where $m \geq 2$. Note that there are $m!$ Hamiltonian cycles starting from a specific vertex.

The technique of finding half the Hamiltonian cycles starting from a specific vertex in a given complete weighted graph and then calculating the sum of the weights for each cycle is not practical for graphs with a large number of vertices. For instance, if the number of vertices is 26, then the number of cycles to be found and checked is $7.76 \cdot 10^{24}$.

Even though numerous mathematicians and computer scientists have spent many years of research on the problem of finding an efficient algorithm for solving the traveling salesperson problem, no such algorithm has been found. The best-known algorithms require time proportional to $n \cdot 2^n$ for a complete graph with n vertices. For $n = 26$, $n \cdot 2^n$ is about $1.74 \cdot 10^9$, which is considerably smaller than $7.76 \cdot 10^{24}$. However, for large n, the number $n \cdot 2^n$ is much too large to make such an algorithm practical.

Most of the algorithms used to solve the traveling salesperson problem for large values of n are designed to find a Hamiltonian cycle that has a weight close to the minimum weight. One such algorithm is called the **closest-neighbor algorithm**. It constructs the Hamiltonian cycle by traveling from a given vertex v to the vertex u, where u has not already been visited and the edge $\{v, u\}$ has the least weight of all such edges incident with v. In case of a tie, pick any of the possible edges.

ALGORITHM 8.2.6 **CLOSEST-NEIGHBOR ALGORITHM** Let $G = (V, E, w)$ be a complete weighted graph with n vertices. A closest-neighbor cycle from v_1 to v_1 is $(v_1, e_2, v_2, \ldots, v_{n-1}, e_n, v_1)$, where the vertices v_i are found by the following algorithm. The set $\{u, v\}$ denotes the unique edge with endpoints u and v.

$V_1 := V - \{v_1\}$.
For k = 2 to n do
  Begin
    $v_k := v$, where $w(\{v_{k-1}, v\})$ is the minimum of
      $\{w(\{v_{k-1}, u\}) : u \in V_{k-1}\}$.
    $V_k := V_{k-1} - \{v_k\}$
  End.

PRACTICE PROBLEM 5 Find a closest-neighbor cycle, starting in Boston, for the graph of Example 8.2.5. Is the closest-neighbor cycle the cycle of minimum weight?

The closest-neighbor algorithm is an example of a **greedy algorithm**. This algorithm is greedy in the sense that the best choice is made at each iteration without regard to previous or future choices. We will encounter other greedy algorithms later in this chapter.

EXERCISE SET 8.2

1. Draw a graph that does not contain a Hamiltonian cycle.
2. Let G be a weighted graph such that the weight function w is a constant function—e.g., $w(e) = 1$ for every edge e. Prove that the solution to the traveling salesperson problem for G will be any Hamiltonian cycle in G. (Of course, G may have no Hamiltonian cycle.)
3. Consider the multigraph in Figure 8.2.10.

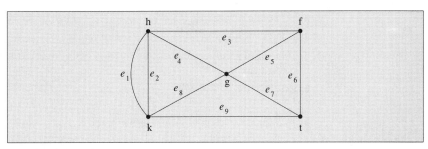

FIGURE 8.2.10

a. Either find an Eulerian tour or explain why one cannot be found.
b. Find an Eulerian path. Specify the initial and terminal vertices.
c. Either find a Hamiltonian cycle or explain why one cannot be found.

4. Consider the multigraph in Figure 8.2.11.

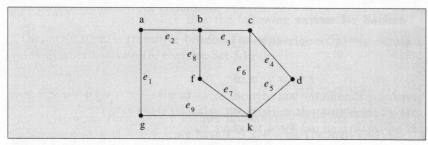

FIGURE 8.2.11

   a. Either find an Eulerian tour or explain why one cannot be found.
   b. Find an Eulerian path. Specify the initial and terminal vertices.
   c. Either find a Hamiltonian cycle or explain why one cannot be found.

5. Use the techniques in Example 8.2.2 and in the proof of Theorem 8.2.3 to find an Eulerian tour in the multigraph shown in Figure 8.2.12.

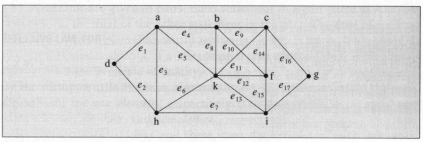

FIGURE 8.2.12

6. Draw a graph that contains an Eulerian tour but not a Hamiltonian cycle.
7. Draw a graph that contains a Hamiltonian cycle but no Eulerian tour.
8. Draw a graph containing an Eulerian tour and a Hamiltonian cycle that are not identical.
9. When does the complete graph K_n have an Eulerian tour? Prove that your answer is correct.
10. When does the complete bipartite graph $K_{m,n}$ have an Eulerian tour? Prove that your answer is correct.
11. Solve the mail carrier problem.
12. When does the complete bipartite graph $K_{m,n}$ have a Hamiltonian cycle? Prove that your answer is correct.
13. Let $G = (V, E)$ be a connected graph. Prove: If $e \in E$ is an edge in some circuit of G, then $G - e$ is connected.

14. Show that the graph in Figure 8.2.13 has no Hamiltonian cycle.

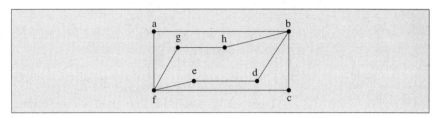

FIGURE 8.2.13

15. Give an example of a connected seven-edge graph G such that G contains an Eulerian path and G has an edge e such that $G - e$ is not connected.

16. Let G be a graph, and let v and w be distinct vertices. Prove: If G has a path from v to w, then G has a path from v to w with no repeated vertices.

17. Prove: A multigraph G has an Eulerian path from vertex v to a different vertex w if and only if G is connected and v and w are the only two vertices of odd degree.

18. Prove: If a graph G contains a closed path beginning and ending at a vertex v, then G contains a circuit beginning and ending at v.

For Exercises 19 and 20, consider the six-city traveling salesperson problem with the cost matrix given in Table 8.2.2.

TABLE 8.2.2

FROM \ TO	v_1	v_2	v_3	v_4	v_5	v_6
v_1	—	4	4	3	8	4
v_2	4	—	4	5	6	12
v_3	4	4	—	3	7	2
v_4	3	5	3	—	4	3
v_5	8	6	7	4	—	5
v_6	4	12	2	3	5	—

19. Refer to Table 8.2.2. Use the closest-neighbor algorithm (Algorithm 8.2.6) to find a closest-neighbor cycle starting with v_1. Is this closest-neighbor cycle the cycle of minimum cost?

20. Use the closest-neighbor algorithm to find a closest-neighbor cycle starting with v_3. Is this closest-neighbor cycle the cycle of minimum cost?

■ PROGRAMMING EXERCISES

21. Refer to page 85 for Floyd's shortest-path algorithm for digraphs. Write a computer program to implement Floyd's algorithm for finding the length of the shortest path between vertices in a graph.

22. Write a computer program to implement the closest-neighbor algorithm.

8.3
TREES

DEFINITION

A **tree** is a graph such that there is a unique simple path between each pair of vertices.

Note that a tree is connected.

Trees are very useful in computer science applications. For example, trees represent hierarchical data structures and are useful for analyzing algorithms.

The graphs in Figures 8.3.1(a) and 8.3.1(b) are trees, whereas the graphs in Figures 8.3.1(c) and 8.3.1(d) are not trees.

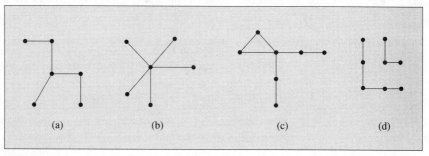

FIGURE 8.3.1

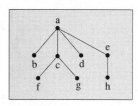

FIGURE 8.3.2

A **rooted tree** is a tree in which a specific vertex is designated as the **root**. Figure 8.3.2 shows a rooted tree with root a.

In a rooted tree with root a, the **level** of a vertex v is the length of the unique simple path from a to v. When the level of a vertex is n, we will frequently say the vertex is at level n. Figure 8.3.2 shows how a rooted tree is drawn. The root is placed at the top. Below the root, we place any vertices at level 1. Below each of these vertices, we place vertices at level 2. We continue in this way until the entire tree is drawn. In the rooted tree in Figure 8.3.2, vertices f, g, and h are at level 2, whereas vertices b, c, d, and e are at level 1.

EXAMPLE 8.3.1 Consider the rooted tree T in Figure 8.3.3 (page 325). T has nine vertices and eight edges. The vertices d and g are at level 3, whereas vertex f is at level 5. ∎

PRACTICE PROBLEM 1 Determine the level of vertices e and k in the rooted tree of Figure 8.3.3. How many vertices are there at level 5?

If vertices v and w are joined by an edge and the level of v is equal to the level of w plus 1, then w is the **parent** of v and v is the **child** of w. The root

8.3 TREES

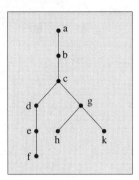

FIGURE 8.3.3

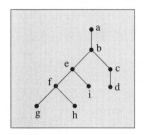

FIGURE 8.3.4

has no parent. Each vertex v other than the root has exactly one parent since there is a unique simple path from the root to v. Vertices with no children are called **leaves**; vertices with one or more children are called **parents** or **internal vertices**. If two vertices have the same parent, they are called **siblings**.

All the children of v together with their children and their children's children, etc., are **descendants** of v. All the parents of v together with their parents and their parents' parents, etc., are **ancestors** of v.

In Figure 8.3.4, the vertex e has b as its parent, f and i as its children, and c as its sibling. Also, the vertices g, h, i, and d are leaves; the vertices a, b, c, e, and f are internal vertices. The descendants of e are f, i, g, and h. The ancestors of e are a and b.

PRACTICE PROBLEM 2 Consider the rooted tree T in Figure 8.3.2.

a. Find all the leaves of T.
b. Find all the internal vertices of T.
c. Name the siblings of the vertex c.

DEFINITION

A rooted tree in which every internal vertex has at most m children is called an **m-ary tree**. A 2-ary tree is usually called a **binary tree**. A 3-ary tree is also called a **ternary tree**.

A binary tree in which every parent has exactly two children is called a **full binary tree**. Similarly, a **full m-ary tree** is an m-ary tree in which every parent has exactly m children.

Figure 8.3.5(a) on page 326 shows a binary tree that is not a full binary tree, whereas Figure 8.3.5(b) shows a full ternary tree.

The next two theorems give fundamental relationships among the numbers of vertices, internal vertices, and leaves and the number m for full m-ary trees.

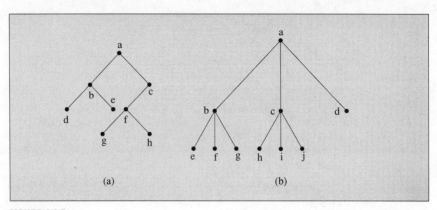

FIGURE 8.3.5

THEOREM 8.3.2 If a full m-ary tree T has t internal vertices, then the total number of vertices in T is $mt + 1$.

Proof Since each internal vertex has m children, there are mt vertices not including the root. Hence, there are $mt + 1$ vertices. ∎

THEOREM 8.3.3 Let T be a full m-ary tree.

a. If T has t internal vertices, then it has $(m - 1)t + 1$ leaves.
b. If T has k leaves, then it has $(k - 1)/(m - 1)$ internal vertices and $(mk - 1)/(m - 1)$ vertices.

Proof Let n be the total number of vertices, t be the number of internal vertices, and k be the number of leaves in T.

a. We have $k = n - t$. By Theorem 8.3.2, $n = mt + 1$. Hence, $k = n - t = (mt + 1) - t = (m - 1)t + 1$.
b. Since we have $k = (m - 1)t + 1$ from part a, we have $k - 1 = (m - 1)t$. Hence, $t = (k - 1)/(m - 1)$. Therefore, $n = k + t = k + (k - 1)/(m - 1) = (mk - 1)/(m - 1)$. ∎

PRACTICE PROBLEM 3 Let T be a full m-ary tree with n vertices.

a. Show that T has $(n - 1)/m$ internal vertices.
b. Show that T has $[(m - 1)n + 1]/m$ leaves.

Applications

EXAMPLE 8.3.4 **AN ORGANIZATION CHART** A rooted tree is frequently used to describe the hierarchical relationships in the structure of a company. The highest office usually corresponds to the root of the tree, whereas lower offices are represented by the remaining vertices. Figure 8.3.6 shows the organization of a small company.

8.3 TREES

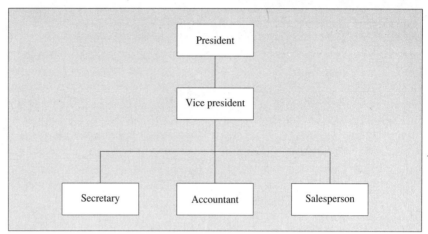

FIGURE 8.3.6

EXAMPLE 8.3.5 **A TELEPHONE TREE** Suppose a cultural group has 61 members. The president of this group wishes to set up a full 5-ary "telephone" tree such that, when she wants to send a telephone message to all members, she need call only five people. Each of these five people calls five other different people. This process is repeated until all members of the group have received the message.

    a. Is it possible to construct a full 5-ary tree with 61 vertices (members)?
    b. Assuming no repetitions, how many members will not have to make any telephone calls?

Solution **a.** By Theorem 8.3.2, $n = mt + 1$. Since $61 = 5 \cdot 12 + 1$, we have $n = 61$, $m = 5$, and $t = 12$. Therefore, we can construct a full 5-ary tree with 61 vertices and 12 internal vertices.
    b. Since $61 - 12 = 49$, there are 49 leaves. Hence, there are 49 members who will not have to make a telephone call.

PRACTICE PROBLEM 4 **a.** Solve the problem in Example 8.3.5 by using the formula derived in Practice Problem 3b.
    b. Solve the problem in Example 8.3.5 under the assumption that a 7-ary tree is used. Note that a full 7-ary tree with 61 vertices is not possible here. (Why?)

EXAMPLE 8.3.6 **A SINGLE-ELIMINATION TOURNAMENT** We use a full binary tree to set up a single-elimination tournament. For example, to design a tree for a single-elimination tournament with five people entered, we use a full binary tree with eight leaves. Five of the leaves are labeled with the name of an entrant, and three ($3 = 8 - 5$) of those leaves have a sibling labeled "bye," which indicates that the entrant does not have to play a first-round match. Matches at the same level are said to be in the same round. There are three rounds in this

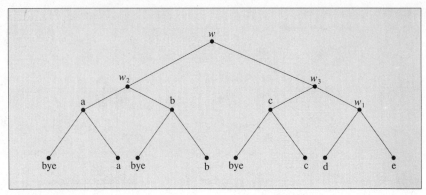

FIGURE 8.3.7

tournament. In general, if the level of the leaves is q, there will be q rounds. See Figure 8.3.7.

The entrants in this five-person tournament are a, b, c, d, and e. The label w_1 stands for the winner of the match between d and e; w_2 is the winner of the match between a and b; w_3 is the winner of the match between c and w_1; and w is the winner of the tournament. ∎

Expression Trees and Tree Traversal

It is useful, especially in computer science, to represent algebraic expressions with full binary trees. We will use full binary trees to investigate two methods of writing algebraic expressions.

It is customary to call variables, such as a, b, c, and d, **operands**; the operations $+$, $-$, $\cdot$, and $\div$ are called **operators**, which operate on pairs of operands or expressions. Operands are represented by leaves of a full binary tree, and operators are represented by internal vertices of a full binary tree.

EXAMPLE 8.3.7

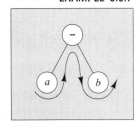

FIGURE 8.3.8

The algebraic expression $a - b$ can be represented by the full binary tree in Figure 8.3.8.

When we visit and record each vertex of a tree, we say we **traverse** the tree. For example, we obtain the algebraic expression $a - b$ when we traverse the binary tree in the left–root–right manner indicated by the curved, directed line shown in Figure 8.3.8. Explicitly, we visit and record the vertex a first, visit and record the vertex $-$ next, and visit and record the vertex b last. The traversal shown in Figure 8.3.8 is called an **inorder traversal**. The expression $a - b$ so obtained is said to be written using **infix notation**.

If we traverse the binary tree in Figure 8.3.9 in the left–right–root manner by visiting and recording the vertex a first, visiting and recording the vertex b next, and visiting and recording the vertex $-$ last, the resulting algebraic expression is $a\ b\ -$. The symbols $a\ b\ -$ indicate that b is to be subtracted

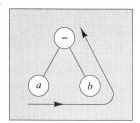

FIGURE 8.3.9

from a. This is an example of **postfix notation** (or **reverse Polish notation**), and the traversal shown is called a **postorder traversal**.

Before comparing infix and postfix notation with a more complex example, we introduce some useful terminology for binary trees. Let T be a binary tree. In T, each child of an internal vertex is designated as either a **left child** or a **right child**. In a diagram, the left child is drawn to the left of its parent and the right child is drawn to the right of its parent. The **left subtree** of an internal vertex v of T is the tree with the left child of v as the root along with all the descendants of that left child. The **right subtree** of v is similarly defined.

We now give the algorithms for inorder and postorder traversal. These algorithms are recursive in the sense that they call themselves. Example 8.3.10 following will help clarify how the algorithms work.

ALGORITHM 8.3.8   **INORDER TRAVERSAL ALGORITHM**   This algorithm traverses a binary tree in the left–root–right manner. A binary tree T with root R is the input.

> If R is empty, then return.
> Call Algorithm 8.3.8 to traverse the left subtree of R.
> Traverse R.
> Call Algorithm 8.3.8 to traverse the right subtree of R.

ALGORITHM 8.3.9   **POSTORDER TRAVERSAL ALGORITHM**   This algorithm traverses a binary tree in the left–right–root manner. A binary tree T with root R is the input.

> If R is empty, then return.
> Call Algorithm 8.3.9 to traverse the left subtree of R.
> Call Algorithm 8.3.9 to traverse the right subtree of R.
> Traverse R.

A binary tree T is said to *represent* an expression E, written using infix notation, if an inorder traversal of T produces E.

EXAMPLE 8.3.10   **a.** Draw a binary tree T that represents the expression

$$(a - 7) \div [(2 \cdot b + c) \cdot (d - 4)]$$

In other words, construct a binary tree T such that an inorder traversal of T will produce the algebraic expression $(a - 7) \div [(2 \cdot b + c) \cdot (d - 4)]$.
b. Find the postfix expression for the same algebraic expression by doing a postorder traversal of T.

Solution   **a.** We first examine the given expression for the order in which the operators are applied. Since $\div$ is applied last, it is the root of the tree. The expression $a - 7$ will be represented by the left subtree of $\div$, and $(2 \cdot b + c) \cdot (d - 4)$ will be represented by the right subtree of $\div$. The root of the left subtree will be $-$. The root of the right subtree will be $\cdot$ since it is the last operator applied in $(2 \cdot b + c) \cdot (d - 4)$. The left subtree of $\cdot$ will represent $2 \cdot b + c$, and the right subtree will represent $d - 4$. See Figure 8.3.10 for the partially completed binary tree. Continuing in this way, we obtain the binary tree T shown in Figure 8.3.11.

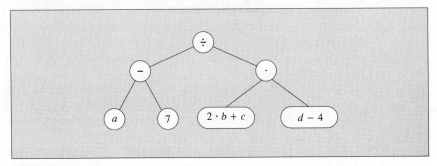

FIGURE 8.3.10

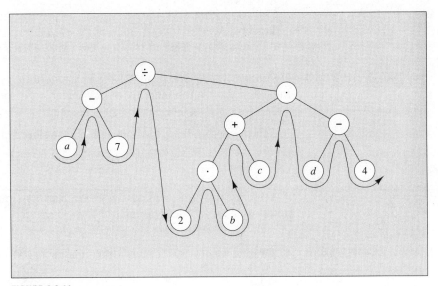

FIGURE 8.3.11

Now we apply the inorder traversal algorithm (Algorithm 8.3.8) to the tree we have constructed. (See Figure 8.3.11.) Since the root is $\div$ and therefore not empty, Algorithm 8.3.8 is called to traverse the left subtree of the root $\div$. Since the root of this subtree is $-$ and therefore not empty, Algorithm 8.3.8 is called to traverse the left subtree of $-$, which is a.

The subtree a has root a with empty left and right subtrees. When Algorithm 8.3.8 is called to traverse the left subtree of a, it returns. The algorithm now traverses the root, which is a. Now Algorithm 8.3.8 is called to traverse the right subtree of a and again it returns.

Next the vertex $-$ is traversed. Then Algorithm 8.3.8 traverses the right subtree of $-$ to obtain 7. Now we have the infix expression $a - 7$. This completes the traversal of the left subtree of the root $\div$.

Next Algorithm 8.3.8 traverses the root $\div$. Then Algorithm 8.3.8 traverses the right subtree of $\div$. Since the root of this subtree is $\cdot$, Algorithm 8.3.8 traverses the left subtree of the vertex $\cdot$. Continuing in this way, we obtain the inorder traversal for $(a - 7) \div [(2 \cdot b + c) \cdot (d - 4)]$. We read this expression, in infix notation, by following the curved, directed line shown in Figure 8.3.11.

During an inorder traversal, the vertices representing the symbols of an expression are traversed in the order: left operand, operator, right operand.

b. For a postorder traversal, the order is: left operand, right operand, operator. Figure 8.3.12 shows the order of the postorder traversal for the same algebraic expression. By using Algorithm 8.3.9, we obtain the postfix notation

$$a\ 7 - 2\ b \cdot c + d\ 4 - \cdot \div$$

for the algebraic expression $(a - 7) \div [(2 \cdot b + c) \cdot (d - 4)]$.

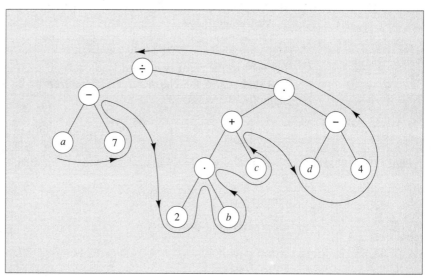

FIGURE 8.3.12

Postfix expressions may look peculiar to students more familiar with standard infix expressions. However, postfix notation has certain advantages over infix notation. Postfix expressions require no parentheses for unambiguous evaluation. Many compilers translate infix expressions to postfix notation. In addition, some hand calculators require algebraic expressions to be entered in postfix notation.

Coding

Full binary trees are also useful for coding.

EXAMPLE 8.3.11 **HUFFMAN CODES** Huffman codes represent textual characters by variable-length bit strings. Shorter bit strings are used to represent the more frequently used characters such as the letters e and t. Longer bit strings are used to represent less frequently used characters such as the colon (:) and the letter z. Huffman codes generally use much less space than ASCII codes, which represent characters by fixed-length bit strings (usually eight bits for a character). A Huffman code is readily defined by a rooted full binary tree, such as the one in Figure 8.3.13. Each edge is labeled 1 or 0; characters are represented by the leaves.

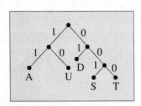

FIGURE 8.3.13

To decode a bit string, we begin at the root and move through the edges toward higher levels until we reach a character. We move to the right at the bit 0 and to the left at the bit 1 as we read the bit string from left to right.

As an example, we will decode the bit string 0110001000 using the Huffman code in Figure 8.3.13. Beginning at the root in Figure 8.3.13, we move right since the first bit is 0. Next, we move left since the second bit is 1. Now we reach the character D. To decode the next character, we begin at the root again. The bit 1 causes us to move left, the bit 0 causes us to move right. We reach the character U. The next character we reach is S by moving from the root to the right twice and to the left once. Finally, the last group of bits 000 leads us to the character T. Therefore, the bit string 0110001000 represents the word DUST.

We can also encode English words into bit strings with a Huffman code.

PRACTICE PROBLEM 5 Encode the words SAD and ADA by a bit string using the tree in Figure 8.3.13.

Characterizing Trees

We conclude this section by proving several standard theorems that characterize trees.

THEOREM 8.3.12 Let T be a graph. T is a tree if and only if T is connected and contains no cycles.

Proof Assume T is a tree. By definition there is a unique simple path between every pair of distinct vertices in T. Hence, T is connected because there is a path between each pair of vertices. Suppose T has a cycle $(v_0, e_1, v_1, e_2, \ldots,$

v_{n-1}, e_n, v_0). It follows that there are at least two different paths from v_0 to v_1—namely, the path (v_0, e_1, v_1) and the path $(v_0, e_n, v_{n-1}, \ldots, e_2, v_1)$. This contradicts the definition of a tree.

Conversely, assume T is connected and has no cycles. Assume a and b are distinct vertices in T. There is a path from a to b since T is connected. By Exercise 16, Exercise Set 8.2, for any path from a to b, we can find a corresponding path with no repeated edges or vertices. If there are two different paths with no repeated vertices from a to b, then we can find a cycle in T. This contradicts the assumption that T has no cycles. Hence, there is a unique path with no repeated edges joining a and b. ∎

Before proving that every tree with n vertices has exactly $n - 1$ edges, we need the following theorem.

THEOREM 8.3.13 Any tree with two or more vertices has at least one vertex of degree 1.

Proof First, we designate one vertex as the root. Since a tree has finitely many vertices, the root can have only finitely many descendants. By Exercise 24, Exercise Set 8.3, not every vertex can be a parent. In other words, there must exist at least one leaf. This leaf has degree 1. ∎

THEOREM 8.3.14 A tree T with n vertices has exactly $n - 1$ edges.

Proof We use mathematical induction on the number of vertices in the tree T. If T has one vertex, then there is no edge. Hence, our theorem is true for $n = 1$.

Assume that every tree with k vertices has exactly $k - 1$ edges, where $k \geq 1$. Now, consider a tree T with $k + 1$ vertices. By Theorem 8.3.13, a tree with two or more vertices must have a vertex of degree 1. Let b be a vertex of degree 1. Remove vertex b and the edge incident with b to obtain $T - b$. The graph $T - b$ is a tree since it is connected and contains no cycles. The tree $T - b$ has k vertices. By the induction hypothesis, $T - b$ has exactly $k - 1$ edges. Therefore, T has $(k - 1) + 1 = k$ edges since T has one more edge than $T - b$ does. This completes the proof. ∎

PRACTICE PROBLEM 6 Refer to Example 8.3.5. Use Theorem 8.3.14 to find the total number of telephone calls that must be made.

Theorem 8.3.15 gives another characterization of a tree.

THEOREM 8.3.15 Let $T = (V, E)$ be a graph. Then T is a tree if and only if $\text{Card}(E) = \text{Card}(V) - 1$ and T has no cycles.

Proof Assume that T is a tree and that $\text{Card}(V) = n$. By Theorem 8.3.12, T has no cycles. By Theorem 8.3.14, G has $n - 1$ edges, and, therefore, $\text{Card}(E) = n - 1 = \text{Card}(V) - 1$.

For the converse, assume that $\text{Card}(E) = \text{Card}(V) - 1$ and that T has no cycles. We wish to prove that T is connected. Suppose T has m components $T_1, \ldots, T_m$. Let $k_1, \ldots, k_m$ be the number of vertices of $T_1, \ldots, T_m$, respectively. Since each T_i is connected and contains no cycles, it is a tree. Hence, T_i has $k_i - 1$ edges by Theorem 8.3.14. It is left as an exercise (see Review Exercise 20) to prove that every vertex and every edge is in some component. Therefore, the number of edges in T is $(k_1 - 1) + (k_2 - 1) + \cdots + (k_m - 1) = k_1 + \cdots + k_m - m = \text{Card}(V) - m$. In other words, $\text{Card}(E) = \text{Card}(V) - m$. Since $\text{Card}(E) = \text{Card}(V) - 1$, then $\text{Card}(V) - m = \text{Card}(V) - 1$ and $m = 1$. Hence, T has one component and is connected. Therefore, T is a tree. ∎

THEOREM 8.3.16 Let $T = (V, E)$ be a graph. Then T is a tree if and only if $\text{Card}(E) = \text{Card}(V) - 1$ and T is connected.

The proof of Theorem 8.3.16 is left as an exercise. (See Exercise 25.)

In summary, we note that a graph $T = (V, E)$ is a tree if and only if it satisfies any two of the following conditions:

i. T is connected;
ii. $\text{Card}(E) = \text{Card}(V) - 1$;
iii. T contains no cycles.

EXERCISE SET 8.3

Exercises 1 and 2 refer to the tree $T = (V, E)$ pictured in Figure 8.3.14.

1. a. Verify that T is a tree.   **b.** Verify that $\text{Card}(E) = \text{Card}(V) - 1$.
2. Find the degree of each vertex in T.

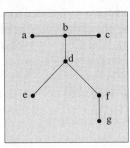

FIGURE 8.3.14

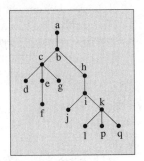

FIGURE 8.3.15

Exercises 3 and 4 refer to the rooted tree pictured in Figure 8.3.15.

3. a. Find the ancestors of k.   **b.** Find the descendants of c.
   c. Find the siblings of e.   **d.** Find the children of i.
4. a. Find the ancestors of f.   **b.** Find the descendants of h.
   c. Find the siblings of g.   **d.** Find the children of b.

8.3 TREES

5. Suppose an organization with 101 members sets up a full 4-ary telephone tree as described in Example 8.3.5. How many members will not have to make any telephone calls?

6. Repeat Exercise 5 for a 101-member organization with a 6-ary tree.

7. Let T be a full binary tree, and suppose the level of each leaf is q.
   a. How many leaves are there in T?   **b.** How many vertices are there in T?

8. Let T be a full m-ary tree, and suppose the level of each leaf is q.
   a. How many leaves are there in T?   **b.** How many vertices are there in T?

9. Suppose 29 people enter a single-elimination tennis tournament.
   a. How many rounds must be played?   **b.** How many matches must be played?

10. Suppose 100 people enter a single-elimination tennis tournament.
    a. How many rounds must be played?   **b.** How many matches must be played?

11. Draw a binary tree representing each algebraic expression.
    a. $a - (b \cdot c)$ **b.** $(3 \cdot b) + (5 \cdot c - d)$

12. Draw a binary tree representing each algebraic expression.
    a. $(a \cdot b) + c$ **b.** $(2 \cdot a) - (8 \cdot c + d)$

13. Write each algebraic expression in postfix notation.
    a. $a - (b \cdot c)$ **b.** $(3 \cdot b) + (5 \cdot c - d)$

14. Write each algebraic expression in postfix notation.
    a. $(a \cdot b) + c$ **b.** $(2 \cdot a) - (8 \cdot c + d)$

15. Consider the binary tree T shown in Figure 8.3.16.

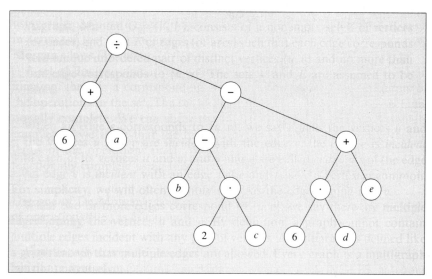

FIGURE 8.3.16

a. Do an inorder traversal of T to obtain the expression written in infix notation (use parentheses whenever necessary).

b. Do a postorder traversal of T to obtain the expression written in postfix notation.

16. Consider the algebraic expression $(4 - a) \cdot [(b + 3 \cdot c) \div (4 \cdot d - e)]$.
    a. Draw a binary tree representing the algebraic expression.
    b. Do a postorder traversal of the binary tree, and write the expression in postfix notation.

17. Consider the algebraic expression $[(b + a) \div (b \cdot (3 + c))] - c \cdot d$.
    a. Draw a binary tree representing the algebraic expression.
    b. Do a postorder traversal of the binary tree, and write the expression in postfix notation.

Consider the binary tree pictured in Figure 8.3.17. The traversal indicated by the curved, directed line is called **preorder traversal**. *The expression obtained from the preorder traversal shown is:* $+ \cdot - a \, b \, c \div d \, e$. *An expression obtained from a preorder traversal is said to be in* **prefix notation**. *Prefix notation is also called* **Polish notation**. *For example, the prefix notation for the infix expression* $a - b$ *is* $- a \, b$. *A preorder traversal is done by visiting the symbols in the order: operator, operand, operand.*

As with postfix notation, no parentheses are required and no conventions are needed regarding precedence of operators. To understand better how the expression $+ \cdot - a \, b \, c \div d \, e$ *is evaluated, we can use parentheses to obtain* $+ [\cdot (- a \, b) c] (\div d \, e)$.

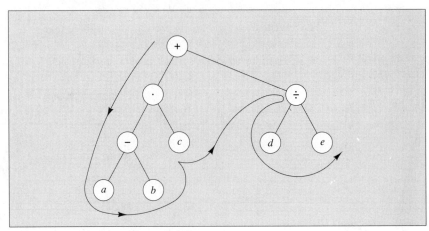

FIGURE 8.3.17

18. Consider the binary tree T shown in Figure 8.3.17.
    a. Do an inorder traversal of T to obtain the expression written in infix notation.
    b. Do a postorder traversal of T to obtain the expression written in postfix notation.

19. Do a preorder traversal of the tree in Figure 8.3.16 to obtain the expression in prefix notation.

20. Consider the algebraic expression $(a \div b) \cdot [(c - d) + e]$ written in infix notation.
    a. Draw a binary tree T to represent the expression.
    b. Write the expression in prefix notation by doing a preorder traversal of T.

21. Use the Huffman code defined by the binary tree in Figure 8.3.18 to decode each of the following bit strings.
    a. 00011001   b. 1001111001   c. 00011011001100000

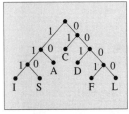

FIGURE 8.3.18

22. Use the Huffman code in Figure 8.3.18 to encode each of the following words:
  a. CAL   **b.** ACIDIC   **c.** ADA
23. Let $T = (V, E)$ be a tree. Prove that the sum of degrees is $2 \cdot \text{Card}(V) - 2$.
24. Prove that any tree with two or more vertices has a vertex that is not a parent.
25. Prove Theorem 8.3.16.
26. Prove that any tree with more than one vertex has at least two vertices of degree 1.

8.4 SPANNING TREES

Suppose the regional cable television company wants to string cable so that several distribution centers are connected to one another. Furthermore, assume the company wants to do this in the most economical way possible. We imagine each distribution center as a vertex and each cable connecting two centers as an edge of a graph. Since economy is the goal, there is no reason to have more than one simple path between any two centers. Hence, what we need is a tree that includes all centers.

DEFINITION

> A spanning subgraph T of a connected graph G is called a **spanning tree** of G whenever T is a tree.

The next example illustrates this concept.

EXAMPLE 8.4.1 Figure 8.4.1(a) shows a graph. Figures 8.4.1(b) and 8.4.1(c) each show a spanning tree of the graph in Figure 8.4.1(a).

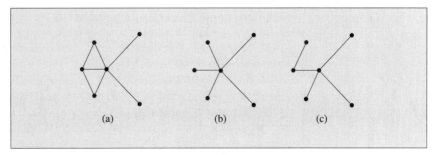

FIGURE 8.4.1

PRACTICE PROBLEM 1 Find all the spanning trees for the graph pictured in Figure 8.4.1(a).

THEOREM 8.4.2 Every connected graph has a spanning tree.

Proof Assume that G is a connected graph. We use mathematical induction on the number m of edges in G. If $m = 0$, then G has no cycles. Hence, G is a tree by Theorem 8.3.12.

Assume that any connected graph with k edges has a spanning tree. Now suppose that G is a connected graph with $k + 1$ edges. If G contains no cycles, then G is a spanning tree. Suppose G contains a cycle. Remove an edge e from that cycle in G. By Exercise 13, Exercise Set 8.2, the remaining graph $G - e$ is connected. Clearly, $G - e$ has k or fewer edges. By the induction hypothesis, $G - e$ has a spanning tree T. And T is also a spanning tree of G since $G - e$ has all of the vertices of G. ∎

Theorem 8.4.2 has theoretically established the existence of a spanning tree of a connected graph. If a connected graph has one vertex, then the spanning tree of the graph is the graph itself. We now give an algorithm for constructing a spanning tree in a given connected graph with two or more vertices. Recall that $V(G)$ is the set of vertices in a graph G.

ALGORITHM 8.4.3 **SPANNING TREE ALGORITHM** Let $G = (V, F)$ be a connected graph with n vertices, where $n \geq 2$. This algorithm finds a spanning tree $T = (V, E)$ for G. Initially E is empty and $V = V(G)$.

> While $F \neq \emptyset$ and the number of edges in E is less than $n - 1$, do
>   Pick any edge $e \in F$. (G has at least one edge.)
> $F := F - \{e\}$.
>   If e does not form a cycle with the edges in E then
> $E := E \cup \{e\}$.

The resultant subgraph $T = (V, E)$ is a spanning tree of G.

Why does Algorithm 8.4.3 produce a spanning tree of G? Note that, after each iteration of the while loop, the set $E \cup F$ contains all the edges originally in F except those that form a cycle with the edges in E. Hence, the graph $(V, E \cup F)$ is connected after each iteration of the while loop, by Exercise 13, Exercise Set 8.2.

When the loop terminates, there are two possible cases: either $F = \emptyset$ or $F \neq \emptyset$. If $F = \emptyset$, then $T = (V, E) = (V, E \cup F)$ is connected and contains no cycles. Hence, T is a spanning tree. If $F \neq \emptyset$, then $T = (V, E)$ has $n - 1$ edges and no cycles. Then T must be connected since V contains n vertices, E contains $n - 1$ edges, each edge joins two vertices, and T has no cycles. By Theorem 8.3.15, T is a tree. Therefore T is a spanning tree.

8.4 SPANNING TREES

Minimal Spanning Trees

A **minimal spanning tree** of a connected weighted graph G is a spanning tree of G with the least total weight among all spanning trees of G.

There are many applications of minimal spanning trees. For example, if we want to build a telephone network connecting a number of cities and if we know the construction cost of telephone lines between each pair of cities, then we can use a minimal spanning tree connecting all the cities in order to minimize cost.

EXAMPLE 8.4.4 Let G be the weighted graph pictured in Figure 8.4.2. With case-by-case elimination, we see that the tree T shown in Figure 8.4.3 is a minimal spanning tree of G.

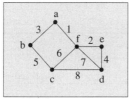

FIGURE 8.4.2   **FIGURE 8.4.3** ∎

The task of finding a minimal spanning tree of a graph is called the minimal connector problem. We describe a procedure for solving the minimal connector problem in the following algorithm, named after the mathematician J. B. Kruskal, Jr.

ALGORITHM 8.4.5 **KRUSKAL'S ALGORITHM** Let G be a connected weighted graph with n vertices, where $n \geq 2$. This algorithm constructs a minimal spanning tree $T = (V, E)$ of G. Initially E is empty and $V = V(G)$.

Order and label the edges of G as $e_1, \ldots, e_m$ so that their weights are in nondecreasing order. That is, $w(e_1) \leq w(e_2) \leq \cdots \leq w(e_m)$.

Pick e_1 as the first edge in E.

While the number of edges in E is less than n − 1, do
  Adjoin to E any remaining edge in G with least subscript that does not form a cycle with the edges already in E.

The resultant subgraph $T = (V, E)$ is a minimal spanning tree of G.

From the explanation following the spanning tree algorithm (Algorithm 8.4.3), we know that T is a spanning tree of G. The proof that T is minimal is somewhat intricate and is therefore omitted.

EXAMPLE 8.4.6 Use Kruskal's algorithm (Algorithm 8.4.5) to find a minimal spanning tree of the connected weighted graph G in Figure 8.4.4 (page 340).

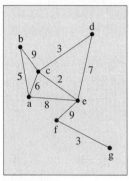
FIGURE 8.4.4

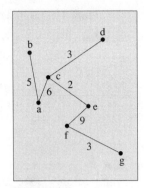
FIGURE 8.4.5

Solution Following Kruskal's algorithm, we label and list the edges as e_i with weights $w(e_i)$ in nondecreasing order:

$e_1 = \{c, e\},\quad w(e_1) = 2 \qquad e_2 = \{c, d\},\quad w(e_2) = 3 \qquad e_3 = \{f, g\},\quad w(e_3) = 3$

$e_4 = \{a, b\},\quad w(e_4) = 5 \qquad e_5 = \{a, c\},\quad w(e_5) = 6 \qquad e_6 = \{d, e\},\quad w(e_6) = 7$

$e_7 = \{a, e\},\quad w(e_7) = 8 \qquad e_8 = \{b, c\},\quad w(e_8) = 9 \qquad e_9 = \{e, f\},\quad w(e_9) = 9$

At each stage we let E be the set of edges chosen, $V = V(G)$, and $T = (V, E)$.

We choose e_1 as the first edge in E. Next we add e_2 and e_3 to E. Then we add e_4 and e_5 to E. There are no cycles in T with each addition. We consider adding e_6 since Card$(E) = 5 < 7 - 1$. However, we will not add e_6 to E since doing so will create a cycle in T. For the same reason, we will not add e_7 and e_8 to E. Finally, we add e_9 to E. Now the resultant graph $T = (V, E)$ is a minimal spanning tree of G. The total weight of T is $w(e_1) + w(e_2) + w(e_3) + w(e_4) + w(e_5) + w(e_9) = 2 + 3 + 3 + 5 + 6 + 9 = 28$.

The minimal spanning tree T is pictured in Figure 8.4.5. ∎

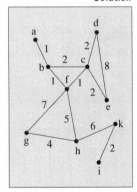
FIGURE 8.4.6

PRACTICE PROBLEM 2 For the connected weighted graph G in Figure 8.4.6, use Kruskal's algorithm to find a minimal spanning tree of G.

Another algorithm for finding a minimal spanning tree of a connected weighted graph is named after the mathematician R. C. Prim.

ALGORITHM 8.4.7 **PRIM'S ALGORITHM** Let G be a connected weighted graph with n vertices, where $n \geq 1$. This algorithm finds a minimal spanning tree of G. Let V be a set of vertices and E be a set of edges. Both V and E are initially empty.

Pick any vertex v of G as the first vertex in V.

While the number of edges in E is less than n − 1, do
  Adjoin to E any remaining edge e in G that is incident to a vertex in V and a vertex not in V and has minimal weight among all such edges. Then adjoin the vertex not in V incident with e to V.

The resultant subgraph $T = (V, E)$ is a minimal spanning tree of G.

8.4 SPANNING TREES

At each stage of Prim's algorithm, an edge not in E is adjoined to E and a vertex not in V is adjoined to V in a way to assure that T remains connected and that no cycle is formed in T. Hence, the subgraph T is a tree at each stage of the algorithm. As with Kruskal's algorithm (Algorithm 8.4.5), we see that T is a spanning tree of G. We omit the proof that Prim's algorithm results in a minimal tree.

Let us apply Prim's algorithm to solve the minimal connector problem of Example 8.4.6. Refer to Figure 8.4.4.

Choose the vertex c of G and let $V = \{c\}$. Now the edge $\{c, e\}$ is incident to c, and, of all edges incident to c, it is of minimal weight. Add $\{c, e\}$ to E and e to V. So, $E = \{\{c, e\}\}$ and $V = \{c, e\}$. Consider all the edges not in E incident to c or e, and choose one of minimal weight that is also incident with a vertex not in V. Hence, we add $\{c, d\}$ to E and d to V. Now we have $E = \{\{c, e\}, \{c, d\}\}$ and $V = \{c, d, e\}$. Next, among all the edges not in E incident to c, d, or e, the edge $\{a, c\}$ is of minimal weight and is also incident with a vertex not in V. Add $\{a, c\}$ to E and a to V. Next we add $\{a, b\}$ to E and b to V. At this point we have $E = \{\{c, e\}, \{c, d\}, \{a, c\}, \{a, b\}\}$ and $V = \{a, b, c, d, e\}$.

Now, among the edges not in E incident to a, b, c, d, or e, the edge $\{d, e\}$ has minimal weight. We will not add $\{d, e\}$ to E since $\{d, e\}$ is not incident with a vertex not in V. We cannot add $\{a, e\}$ or $\{b, c\}$ for the same reason. Add $\{e, f\}$ and $\{f, g\}$ to E and f and g to V. Now, we have the tree T, pictured in Figure 8.4.7. Since the number of vertices in G is seven and T has six edges, the tree T is a minimal spanning tree of G.

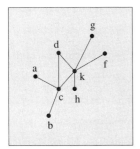

FIGURE 8.4.7

PRACTICE PROBLEM 3 Use Prim's algorithm (Algorithm 8.4.7) to find a minimal spanning tree for the connected weighted graph G in Figure 8.4.6.

EXERCISE SET 8.4
1. Find all the spanning trees of the graph G in Figure 8.4.8.
2. Find all the spanning trees of the graph G in Figure 8.4.9.

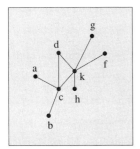

FIGURE 8.4.8

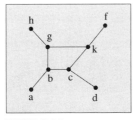

FIGURE 8.4.9

3. Prove that, if $G = (V, E)$ is a connected graph of n vertices, then $\text{Card}(E) \geq n - 1$.
4. Give an example of a connected graph $G = (V, E)$, where $\text{Card}(V) = 8$ and $\text{Card}(E) = 10$.

5. Suppose G is a connected graph. Prove: G is a tree if the number of vertices is one more than the number of edges in G.

6. Prove that a connected graph G is a tree if $G - e$ is not connected, where e is any edge in G.

7. Apply Kruskal's algorithm (Algorithm 8.4.5) to find a minimal spanning tree for each of the graphs in Figure 8.4.10.

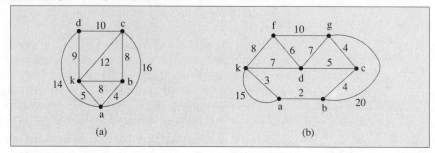

FIGURE 8.4.10

8. Apply Prim's algorithm (Algorithm 8.4.7) to find a minimal spanning tree for each of the graphs in Figure 8.4.10.

9. Apply Prim's algorithm (Algorithm 8.4.7) to find a minimal spanning tree for each of the graphs in Figure 8.4.11.

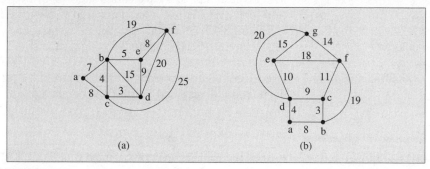

FIGURE 8.4.11

10. Apply Kruskal's algorithm (Algorithm 8.4.5) to find a minimal spanning tree for each of the graphs in Figure 8.4.11.

11. Find a minimal spanning tree for the graph G in Figure 8.4.11(b) by the following method: While possible, continue removing an edge from G with maximum weight such that removal of that edge leaves G connected.

12. Use the method in Exercise 11 to find a minimal spanning tree for the graph in Figure 8.4.11(a).

13. Prove that any edge of a connected graph G is an edge of some spanning tree of G.

14. Prove that, for any two distinct edges of a connected graph G, there exists a spanning tree containing the two edges.

15. Prove: If the weight of an edge e in a graph G is less than the weight of every other edge in G, then every minimal spanning tree of G contains the edge e.

16. Prove: A graph that has $n - 1$ edges, no cycles, and at most n vertices is connected.

17. Prove: If the edges in a connected graph G have distinct weights, then G has a unique minimal spanning tree.

18. Prove that, if G is connected and an edge e is included in a cycle of G, then $G - e$ is connected.

■ **PROGRAMMING EXERCISES**

19. Write a computer program to implement Kruskal's algorithm (Algorithm 8.4.5). Test your program by entering the graph of Example 8.4.6. Compare the output with the graph pictured in Figure 8.4.5.

20. Write a computer program to implement Prim's algorithm (Algorithm 8.4.7).

8.5
APPLICATION: HEAPSORT

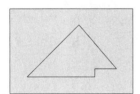

FIGURE 8.5.1

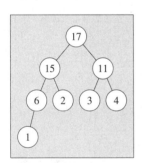

FIGURE 8.5.2

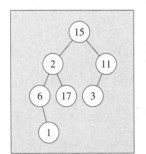

FIGURE 8.5.3

Heapsort is an efficient sorting algorithm that uses the tree structure called a heap. If we want to use the heapsort to arrange a given set of unordered data into an ascending list, first we put the data in a heap. What is a heap? How does a heapsort work?

A **heap** is a rooted tree with a value in each vertex such that

a. Every parent has exactly two children except for possibly one parent at the second highest level, which could have only one child.

b. All leaves are on the highest and the second-highest levels, and they are as far left as possible.

c. The value in the root is greater than or equal to the values in its children (if any), and the left and right subtrees (if any) are again heaps.

We note that condition b above gives the **shape** property of the heap, while condition c gives the **order** property of the heap. The picture in Figure 8.5.1 depicts the shape of a heap. The tree in Figure 8.5.2 is a heap, whereas that in Figure 8.5.3 is not.

In what follows, we use tree structures to illustrate the building of a heap and the process of the heapsort algorithm.

Building the Heap

What is the procedure for building a heap from a set of numbers S? The heap is built level by level beginning with the root. First we note that a binary tree with only one vertex is a heap. Therefore, we insert the first number of S in a vertex as the root. Assume that a heap with the first k numbers has been built. Insert the $(k + 1)$st number in a new leaf in one of the following ways:

1. Insert left: Insert a number in the left child of a leftmost former leaf or

2. Insert right: Insert a number in the right child of a one-child parent

Next, we apply

3. Interchange: Compare the new number with the value in its parent and interchange the two numbers if the value in the parent is smaller.

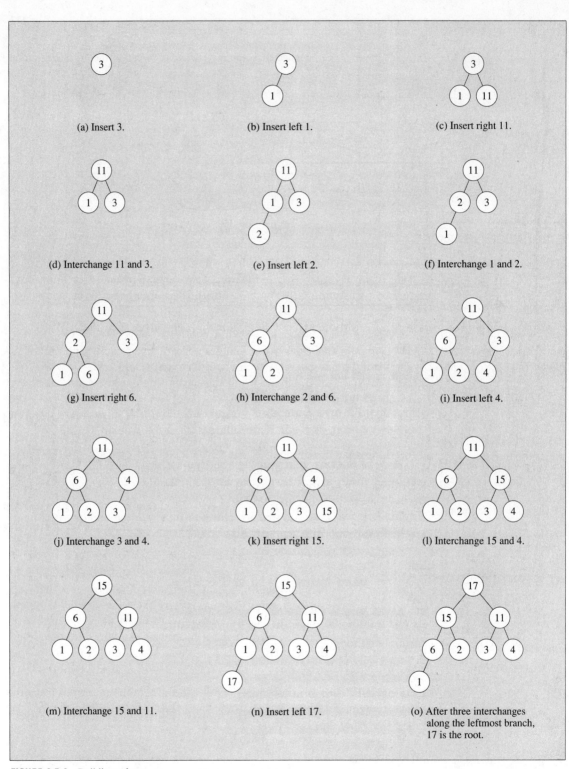

FIGURE 8.5.4 Building a heap

8.5 APPLICATION: HEAPSORT

Continue the interchange until the value in each parent is at least as big as the values in its children. Several interchanges may be required. Repeat the process until the set S is exhausted.

Figures 8.5.4(a)–8.5.4(o) show the various stages of building a heap from the set $S = \{3, 1, 11, 2, 6, 4, 15, 17\}$. In (a), 3 is inserted in the root; in (b), 1 is inserted in the left child; in (c), 11 is inserted in the right child of a one-child parent; in (d), 3 and 11 are interchanged. In Figure 8.5.4(g), 6 is inserted in the new leaf as the right child of the one-child parent containing 2; in Figure 8.5.4(i), 4 is inserted in the new leaf as the left child of the leftmost former leaf containing 3. Figure 8.5.4(o) shows the heap constructed from the set S.

In a computer program, the data in any heap may be represented in an array. Figure 8.5.5(a) illustrates the order in which the data in the heap in our example are stored in an array. We have array$[1] = 17$, array$[2] = 15$, array$[3] = 11$, array$[4] = 6$, array$[5] = 2$, array$[6] = 3$, array$[7] = 4$, and array$[8] = 1$. Figure 8.5.5(b) shows how the data in the heap are stored in the array. The index of each vertex is shown in Figure 8.5.5(a).

Consider the vertex with index 3. The indices of its children are 6 and 7, as shown in Figure 8.5.5(a). Note that $3 = 6$ div 2 and $3 = 7$ div 2. For a heap, if j is the index of a vertex and i is the index of either child, then $j = i$ div 2.

In general, a heap with n elements stored in an array satisfies the following conditions:

$$\text{Array}\,[i] \leq \text{Array}[j] \quad \text{for } 2 \leq i \leq n \quad \text{and} \quad j = i \text{ div } 2$$

The given set S can be made into a heap by storing the elements from S in an array satisfying the above conditions.

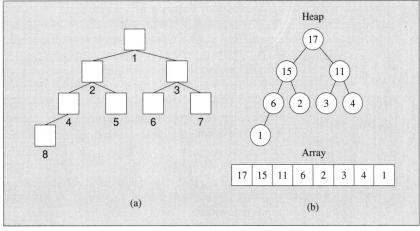

FIGURE 8.5.5

The Heapsort Algorithm

Now, we discuss how the heapsort algorithm sorts the elements of $S = \{3, 1, 11, 2, 6, 4, 15, 17\}$ into an ascending list. We note that the value in the root is greater than or equal to all the other values in the heap. First, we remove the value from the root and place it at the end of the ascending list we are building. To make room for the value from the root, we place the value in the last leaf (rightmost leaf at the highest level) of the heap in a temporary variable called Temp. In our example, we remove 17 from the root, place it at the end of the array, and place 1 in Temp. Figure 8.5.6(a) shows the resulting tree with 17 removed and the value in the rightmost leaf placed in Temp. Next, we reestablish the heap by promoting the largest of the values in Temp or in the children (15 and 11) of the vacant root to that root.

We use the above procedure recursively to pick a value to insert in the vacant root for the subtree whose root's value was promoted to a smaller level. In our example, the value (15) in the root of the left subtree has been promoted and the root is vacant. See Figure 8.5.6(b). So, we promote 6 to the vertex formerly occupied by 15. Now, the tree is no longer a heap. In order to maintain the shape property of the heap, we reinsert 1, the value in the temporary variable, in the leaf as the left child of the subtree with the root containing 6. As before, we place the value (4) in the rightmost leaf at the largest level in Temp. See Figure 8.5.4(c). We now have a new heap.

Next, we repeat this procedure until the heap is empty and the sorted list complete. This final list is: 1, 2, 3, 4, 6, 11, 15, 17. Therefore, the heapsort has arranged the data in the heap into an ascending list. See Figures 8.5.6(a)–8.5.6(g) for the various stages of the heapsort algorithm. The arrow indicates the separation between the sorted data and the unsorted ones. The unsorted data lie to the left of the arrow, whereas the sorted data lie to the right.

In summary, the heapsort involves two phases. First, an unordered set of data is put in a heap, which is a binary tree with the shape and order properties. Second, the heapsort algorithm sorts the data in the heap into an ascending list. If there are repeated pieces of data in the data set, the sorted list will be in a nondecreasing, instead of ascending, order.

The heapsort is important for sorting large sets of data because it is of order $\mathbf{O}(n(\log n))$, where n is the number of inputs; it requires minimal space; and it can be written as a recursive procedure. A sorting algorithm is said to be of order $\mathbf{O}(f(n))$, where f is a function of the number of inputs n, if the number of comparisons and movements of items in the sorting algorithm is at most a constant times $f(n)$. Any sorting procedure of order $\mathbf{O}(n(\log n))$ is considered to be relatively efficient. Few sorting procedures are of order $\mathbf{O}(n(\log n))$. Although quicksort is of order $\mathbf{O}(n(\log n))$ on the average, it is of order $\mathbf{O}(n^2)$ in the worst case. Insertion sort, bubble sort, and selection sort are all $\mathbf{O}(n^2)$. A sorting procedure of order $\mathbf{O}(n(\log n))$ is more efficient than one of order $\mathbf{O}(n^2)$.

For a discussion of various sorting algorithms, see the references at the end of this chapter.

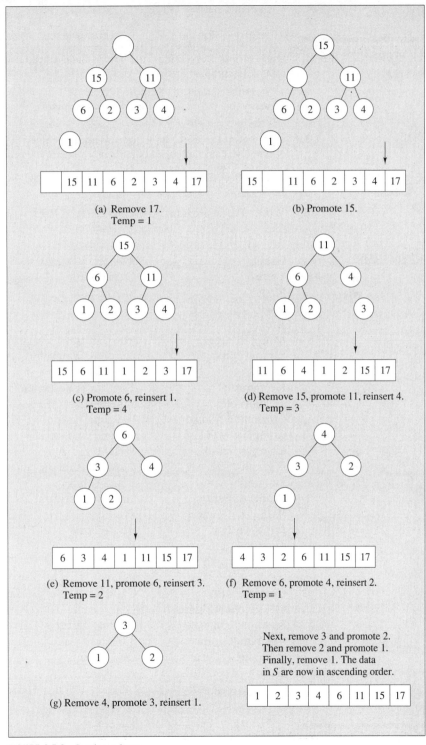

FIGURE 8.5.6 Sorting a heap

EXERCISE SET 8.5

1. Draw various trees to illustrate how to build a heap from the set $A = \{g, c, d, l, e, s\}$ using lexicographic order.
2. **a.** Store the data in the heap from Exercise 1 in an array.
   b. Verify that the indices of the array in part a satisfy the relation $\text{Array}[i] \leq \text{Array}[j]$ for $2 \leq i \leq n$ and $j = i$ div 2.
3. Draw various trees to show how the heapsort algorithm sorts the data in the set A of Exercise 1 into an ascending list in lexicographic order.

PROGRAMMING EXERCISES

4. Write a Pascal procedure to build a heap from the data set $S = \{9, 78, 61, 7, 9, 15, 21, 9, 71, 34, 17, 31, 55, 29, 15, 39\}$.
5. Write a Pascal program to sort the data set S in Exercise 4 into nondecreasing order.

KEY TERMS

Graph	Adjacency matrix	M-ary tree
Vertex	Closed path	Binary tree
Node	Simple path	Ternary tree
Edge	Circuit	Full binary tree
Arc	Cycle	Full m-ary tree
Multiple edges	Eulerian tour	Operands
Multigraph	Hamiltonian cycle	Operators
Loop	Eulerian path	Traverse (a tree)
Loopgraph	Traveling salesperson problem	Inorder traversal
n-Colorable		Infix notation
Chromatic number	Weighted graph	Postfix notation
Complete graph	Weight function	Reverse Polish notation
Complement of a graph	Weight	
	Closest-neighbor algorithm	Postorder traversal
Subgraph		Left child
Proper subgraph	Greedy algorithm	Right child
Path	Tree	Left subtree
Initial vertex	Rooted tree	Right subtree
Terminal vertex	Root	Spanning tree
Length of path	Level of a vertex	Minimal spanning tree
Connected vertices	Parent	Kruskal's algorithm
Components	Child	Prim's algorithm
Connected multigraph	Leaf	Heapsort
Maximal connected submultigraph	Internal vertex	Heap
	Sibling	Shape property
Spanning subgraph	Descendant	Order property
Degree of a vertex	Ancestor	

REVIEW EXERCISES

1. From the following adjacency matrix M, draw the corresponding graph and determine the number of edges and the number of vertices:

REVIEW EXERCISES

$$M = \begin{array}{c} \\ a \\ b \\ c \\ d \\ e \\ f \\ g \end{array} \begin{array}{c} a\ b\ c\ d\ e\ f\ g \\ \begin{bmatrix} 0 & 0 & 1 & 1 & 0 & 1 & 1 \\ 0 & 0 & 0 & 1 & 1 & 1 & 1 \\ 1 & 0 & 0 & 0 & 0 & 0 & 0 \\ 1 & 1 & 0 & 0 & 0 & 1 & 1 \\ 0 & 1 & 0 & 0 & 0 & 0 & 0 \\ 1 & 1 & 0 & 1 & 0 & 0 & 1 \\ 1 & 1 & 0 & 1 & 0 & 1 & 0 \end{bmatrix} \end{array}$$

2. Draw a graph with six vertices and no articulation point. (See page 312 for a definition of articulation point.)

3. Draw a different graph with at least six vertices to illustrate each of the following.
   a. A graph containing a circuit that is not a cycle
   b. A graph containing a closed path that is neither a circuit nor a cycle

4. Students {a, b, c, d, e, f, g, h} in a geometry class need to solve major exercises in pairs throughout the semester. The teacher wants the students to work with different partners on each major exercise. Suppose that so far this semester

   a has worked with b, g, h.
   b has worked with a, e, f.
   c has worked with d, e.
   d has worked with c, e, g, h.
   e has worked with b, c, d, f.
   f has worked with b, e, g.
   g has worked with a, d, f, h.
   h has worked with a, d, g.

   a. Draw the graph G with students as vertices. Two vertices are joined by an edge if the two students have worked together.
   b. Draw the complete graph K_8 with a, b, ..., h as vertices.
   c. From the complement of the graph G, find which students have not worked together.

5. Given the connected weighted graph in Figure (A), find a closest-neighbor cycle beginning with vertex a.

6. Find a minimal spanning tree of the graph in Figure (A).

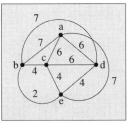

FIGURE (A)

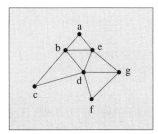

FIGURE (B)

7. Determine the chromatic number of the graph in Figure (B).

8. Find three different connected spanning graphs of the graph in Figure (B).
9. Find three different spanning trees of the graph in Figure (B).
10. Suppose an organization with 84 members sets up a 5-ary telephone tree, as described in Example 8.3.5 (page 327). How many members will not have to make any phone calls? Is the actual telephone tree a full 5-ary tree?
11. Determine the chromatic number of a complete graph K_n, where n is a positive integer.
12. Find the chromatic number of the graphs in Figure (C).

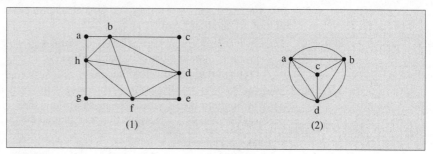

FIGURE (C)

13. Refer to the rooted tree shown in Figure (D).
    a. Is this tree a 3-ary tree?
    b. What is the largest level of this tree?
    c. Find the siblings of vertex e.
    d. Find the descendants of vertex i.

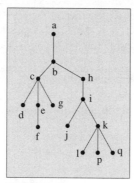

FIGURE (D)

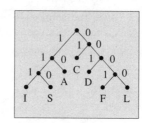

FIGURE (E)

14. Use the Huffman code in Figure (E) to decode each of the following bit strings.
    a. 110100011110001   b. 100011000100110
15. Use the Huffman code in Figure (E) to encode each of the following words.
    a. FAIL   b. AID   c. CAL
16. Draw a binary tree to represent each algebraic expression.
    a. $(3 \cdot f) + [7 - (9 - g)] - (5 \cdot h)$ b. $(7 \cdot a) - [(b + 4) \cdot (8 + a)]$

17. Give the expression in postfix notation corresponding to each of the following expressions in infix notation.
  a. $(3 \cdot f) + [(7 - (d - g)) - (5 \cdot h)]$ **b.** $(7 \cdot a) - [(b + 4) \cdot (8 + a)]$

18. Determine the number of graphs with n vertices.

19. Determine the number of graphs with n vertices and k edges.

20. **a.** Prove that any vertex of a graph G is contained in some component of G.
  b. Prove that any edge of a graph G is contained in some component of G.

21. Prove that, if a particular vertex of a graph G is connected to every other vertex of G, then G is connected.

22. Prove that a complete graph is connected.

23. **a.** Find a Hamiltonian cycle in the graph G in Figure (F).
  b. Prove that the graph G in Figure (F) does not contain an Eulerian tour.

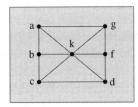

FIGURE (F)

24. Draw a graph with seven edges that has exactly two articulation points.

REFERENCES

Deo, N. *Graph Theory with Applications in Engineering and Computer Science.* Englewood Cliffs, NJ: Prentice-Hall, 1974.

Floyd, R. W. "Algorithm 97: Shortest Path." *Communications of the ACM*, 5:6 (1962): 345.

Gonzales, R. C., and P. Wintz. *Digital Image Processing.* Reading, MA: Addison-Wesley, 1977.

Harary, F. *Graph Theory.* Reading, MA: Addison-Wesley, 1969.

Knuth, D. E. *The Art of Computer Programming, Volume 3, Sorting and Searching*, 2nd ed. Reading, MA: Addison-Wesley, 1973.

Kruskal, Jr., J. B. "On the Shortest Spanning Tree of a Graph and the Traveling Salesman Problem." *Proceedings of the American Mathematical Society*, 7 (1956): 48–50.

Ore, O. *Graphs and Their Uses.* New York: Random House, 1963.

Phillips, D., and A. Garcia-Diaz. *Fundamentals of Network Analysis.* Englewood Cliffs, NJ: Prentice-Hall, 1981.

Prim, R. C. "Shortest Connection Networks and Some Generalizations," *Bell System Technical Journal*, 36 (1957): 1389–1401.

Singh, B., and T. Naps. *Introduction to Data Structures.* St. Paul, MN: West, 1985.

Stein, S. *Mathematics: The Man-Made Universe*, 2nd ed. San Francisco, CA: Freeman, 1969.

Tucker, A. *Applied Combinatorics*, 2nd ed. New York: Wiley, 1984.

CHAPTER 9

RECURRENCE RELATIONS

Many problems of mathematics and computer science have sequences for their solutions. In order to evaluate sequence values or, better yet, to find an explicit formula for the sequence, we look for patterns. Recurrence relations are equations giving the nth sequence value in terms of previous values; that is, they provide a pattern. The structure of a recurrence relation is like that of mathematical induction.

Recurrence relations are encountered frequently in computer science applications of mathematics. They are useful in three ways: for analyzing problems that have sequence solutions, for calculating sequence values when an explicit formula is unavailable or not easy to calculate, and for finding an explicit formula. We will use recurrence relations to analyze and solve many problems in this chapter. A computer program with either a loop structure or recursion is the best way to calculate sequence values using recurrence relations. For some sequences this is the best method of calculating sequence values.

To find an explicit formula for a sequence defined by a recurrence relation, we need to solve the recurrence relation. In this chapter we will investigate two methods of solving recurrence relations: the characteristic equation method and the generating function method. Generating functions are very useful in combinatorics, and we will see some applications besides that of solving recurrence relations.

APPLICATION

Many sorting and searching algorithms are classified as divide and conquer algorithms because they accomplish the task by dividing a given problem into smaller but similar problems. For example, to sort a list of names alphabetically, we might first divde the list into sublists corresponding to the first letter of each name. Some of the most efficient algorithms known, including mergesort and quicksort, are divide and conquer algorithms.

Recurrence relations are especially useful tools for predicting the time complexity of divide and conquer algorithms. We will discuss how to use recurrence relations for analyzing the time complexity of divide and conquer algorithms in Section 9.4.

9.1
RECURSION AND RECURRENCE RELATIONS

Recurrence relations are used to define sequences. This is frequently called definition by recursion.

DEFINITION

> A **recurrence relation** for a sequence $(a_n)_{n=0,1,2,\ldots}$ is an equation that expresses the nth term a_n in terms of its predecessors $a_0, a_1, \ldots, a_{n-1}$. One or more of the a_i's may be given specific values; the given values are called **initial conditions**.

For simplicity and when no confusion can arise, we will frequently write (a_n) for $(a_n)_{n=0,1,2,\ldots}$. So, a_n stands for the nth term of the sequence and (a_n) stands for the sequence itself. For example, $(1/2^n)$ stands for the sequence $(1, 1/2, 1/4, \ldots)$, and the nth term of the sequence is $1/2^n$.

Let us look again at the factorial function. Recall that

$1! = 1$

$2! = 2 \cdot 1$

$3! = 3 \cdot 2 \cdot 1 = 3 \cdot 2!$

$4! = 4 \cdot 3!$

In general, $n! = n(n-1)!$.

The rule $n! = n(n-1)!$ is a recurrence relation. If we let $t_n = n!$, then we have $t_n = nt_{n-1}$. So, $t_n = nt_{n-1}$ for $n = 1, 2, 3, \ldots$ is the recurrence relation defining $n!$. However, $t_n = nt_{n-1}$ does not define a unique sequence since $\pi, \pi, 2\pi, 3 \cdot 2\pi, 4 \cdot 3 \cdot 2\pi, \ldots$ also satisfies this recurrence relation. If we require that $t_0 = 1$ and $t_n = nt_{n-1}$ for $n = 1, 2, 3, \ldots$, we obtain $n!$ uniquely. This illustrates the need for initial conditions if we want a recurrence relation to define a sequence uniquely. The initial condition $t_0 = 1$ for the factorial sequence is $0! = 1$.

Recurrence relations arise naturally in the area of finance.

EXAMPLE 9.1.1 **MONEY MARKET ACCOUNT** We invest $500 in a money market account that earns 7% interest compounded annually. Let A_n be the amount of money in the account at the end of n years. We can determine a recurrence relation for the sequence $(A_0, A_1, A_2, \ldots, A_n, \ldots)$ as follows: The initial condition is $A_0 = 500$, $A_1 = 500 + (0.07)(500) = (1.07)(500) = (1.07)A_0$, $A_2 = A_1 + (0.07)A_1 = (1.07)A_1$. In general, $A_n = (1.07)A_{n-1}$. Therefore, we have the recurrence relation $A_n = (1.07)A_{n-1}$ with initial condition $A_0 = 500$. ∎

EXAMPLE 9.1.2 **ANNUITY ACCOUNT** We invest $2000 at the end of each year in an annuity that earns 9% interest compounded annually. Let A_n be the amount of money in the account at the end of n years. We can determine a recurrence relation for the sequence (A_n) as follows: $A_1 = 2000$, $A_2 = 2000 + (1.09)(2000)$, $A_3 = 2000 + (1.09)A_2$. In general, $A_n = 2000 + (1.09)A_{n-1}$ with initial condition $A_1 = 2000$. ∎

If you are considering an annuity such as the one in Example 9.1.2 for part of your retirement plan, you would like to know how much money the account will contain after a certain number of years. Suppose you want to know how much is in the account at the end of 20 years.

There are two methods of solving this problem. One method is to find an algebraic formula for A_n and then calculate A_{20} using the formula. We will find such a formula later in this section. Another method is to write a computer program that uses the recurrence relation to compute A_{20}. Since recurrence relations are mathematical cousins of recursion in Pascal, it is natural and elegant to write a function using recursion. See the Programming Notes at the end of this section for a function in Pascal to calculate the value of an annuity.

EXAMPLE 9.1.3 Find a recurrence relation for the number of subsets of a finite set.

Solution Let a_n be the number of subsets of a set B with n elements. We will find a recurrence relation for the sequence (a_n). The initial condition is $a_0 = 1$, since the empty set is the only subset of a set with 0 elements. To find the recurrence relation, we let $B = \{b_1, b_2, \ldots, b_n\}$ and $D = B - \{b_n\}$. So, a_n is the number of subsets of B, and a_{n-1} is the number of subsets of D. Every subset of B either contains b_n or does not contain b_n. Moreover, for each subset A of D, there are two distinct subsets of B: A and $A \cup \{b_n\}$. Hence, B has exactly twice as many subsets as D. Thus, the recurrence relation is $a_n = 2a_{n-1}$. ∎

To calculate the number of subsets of a set with n elements, using the recurrence relation $a_n = 2a_{n-1}$, requires that we apply the relation n times. That is, we must use the multiplication operation n times. So, as n gets larger, the evaluation of a_n requires an increasing number of operations. We would like to find an explicit formula for a_n that requires, at most, a fixed number of operations for its evaluation.

Given a sequence (t_n) and a formula $f(n)$ equal to t_n, the formula $f(n)$ is a **closed-form expression** for the sequence (t_n) if there exists a fixed upper bound on the number of arithmetic operations required to evaluate $f(n)$. A recurrence relation defining a sequence is not a closed-form expression for the sequence.

Here is an example to make the concept of a closed-form expression clear. Consider the sequence (b_n), where b_n is the sum of the first n positive integers. We can write $b_n = 1 + 2 + 3 + \cdots + n$. This is not a closed-form expression for b_n since this expression requires $n - 1$ addition operations to evaluate b_n and $n - 1$ is not bounded above. We proved in Chapter 3 that $b_n = [n(n + 1)]/2$. The formula $[n(n + 1)]/2$ is a closed-form expression for

b_n since it requires only one addition, one multiplication, and one division no matter how large n is.

To **solve** a recurrence relation for a sequence $(a_0, a_1, \ldots, a_n, \ldots)$ is to find a closed-form expression for the general term a_n.

Now we solve the recurrence relation of Example 9.1.3 noting that

$$a_1 = 2a_0 = 2 \cdot 1 = 2$$
$$a_2 = 2a_1 = 2 \cdot 2 = 2^2$$
$$a_3 = 2a_2 = 2 \cdot 2^2 = 2^3$$

We conclude from the pattern that $a_n = 2^n$. This is called a **solution by iteration** and is often the easiest method for solving a recurrence relation. We could prove $a_n = 2^n$ for $n = 0, 1, 2, \ldots$ using mathematical induction. This result follows from Theorem 9.1.4 by letting $r = 2$ and $a_0 = b = 1$.

The recurrence relation of Theorem 9.1.4 includes Examples 9.1.1 and 9.1.3 as special cases. We will give the explicit formula solution and prove it is correct using mathematical induction.

THEOREM 9.1.4 Let b and r be fixed positive numbers. Define the recurrence relation by $a_n = ra_{n-1}$ for $n = 1, 2, 3, \ldots$ and the initial condition by $a_0 = b$. The explicit formula solution is $a_n = br^n$ for $n = 0, 1, 2, \ldots$.

Proof We use mathematical induction.

Basis Step ($n = 0$) $a_0 = b = b \cdot 1 = br^0$
Induction Step Induction hypothesis: Assume $a_k = br^k$. By the recurrence relation, $a_{k+1} = ra_k$. Hence, $a_{k+1} = r(br^k)$ by the induction hypothesis. Therefore, $a_{k+1} = br^{k+1}$. ∎

The sequence defined in Theorem 9.1.4 is called a geometric progression with ratio r. A **geometric progression** is a sequence with the property that each term is a constant multiple of the previous term. An **arithmetic progression** is a sequence with the property that each term is the sum of the previous term and a fixed constant.

Here is an example of an arithmetic progression.

EXAMPLE 9.1.5 Given an arithmetic progression $(a_0, a_1, a_2, \ldots, a_n, \ldots)$, where the common difference of a term and its predecessor is 5, and initial condition $a_0 = 9$, the recurrence relation is

$$a_n = a_{n-1} + 5$$ ∎

PRACTICE PROBLEM 1 Find the explicit formula solution for the arithmetic progression in Example 9.1.5. Prove that your formula is correct using mathematical induction.

Let us try to solve the recurrence relation of Example 9.1.2. In other words, let us find a formula telling how much money is in the annuity after n years.

Recall that $A_1 = 2000$ and $A_n = 2000 + (1.09)A_{n-1}$ for $n = 2, 3, \ldots$. To solve this by iteration requires that we find a pattern:

$$A_2 = 2000 + 1.09 \cdot 2000 = A_1 + 1.09 A_1$$
$$A_3 = A_1 + 1.09 A_2 = A_1 + 1.09(A_1 + 1.09 A_1) = A_1(1 + 1.09 + 1.09^2)$$

In general, $A_n = A_1(1 + 1.09 + 1.09^2 + \cdots + 1.09^{n-1})$. But, $1 + 1.09 + 1.09^2 + \cdots + 1.09^{n-1}$ is a geometric sum with ratio $r = 1.09$. In Exercise 8 on page 209, we found that a geometric sum satisfies the following formula:

$$1 + r + r^2 + \cdots + r^{n-1} = \frac{1 - r^n}{1 - r} = \frac{r^n - 1}{r - 1} \quad \text{for } r \neq 1 \text{ and } n = 1, 2, \ldots$$

Hence, $1 + 1.09 + 1.09^2 + \cdots + 1.09^{n-1} = (1.09^n - 1)/0.09$. Therefore, we have $A_n = (2000/0.09)(1.09^n - 1)$. Using a calculator it is easy to find $A_{20} = 102{,}320$. So, after 20 years, the annuity has over \$102,000 although only \$40,000 was deposited. This is a good example of the power of compound interest.

EXAMPLE 9.1.6   **BUBBLE SORT**   The bubble sort is a well-known method for sorting items, such as numbers or records, into a specified order. Suppose we have a list $A(1), A(2), \ldots, A(n)$ of n numbers and wish to sort them in ascending order. You may think of $A(1)$ as the top of the list. The idea behind the bubble sort is to compare consecutive numbers in the list and bubble the smallest number to the top of the list. First, $A(n)$ is compared to $A(n-1)$. If $A(n)$ is smaller than $A(n-1)$, then the numbers are interchanged. Otherwise the two numbers are left where they are. Next, $A(n-1)$, which is either the original $A(n-1)$ or the original $A(n)$ depending on the result of the first comparison, is compared to $A(n-2)$. Again, the smaller of the two becomes the new $A(n-2)$. These steps are repeated until the smallest number is assigned to $A(1)$. Then this process is repeated for the list $A(2), A(3), \ldots, A(n)$.

Let b_n be the number of comparisons it takes for the bubble sort to sort a list of n numbers into ascending order. The number of comparisons done on the original list of n numbers to bubble the smallest number to the top [that is, to assign it to $A(1)$], is $n - 1$. Then it remains to sort the list $A(2), A(3), \ldots, A(n)$ of $n - 1$ numbers. Hence, $b_n = b_{n-1} + (n - 1)$.

Using the terminology of Section 3.5, b_n is the time-complexity function for the bubble sort. Therefore, the time-complexity function b_n for the bubble sort satisfies the recurrence relation $b_n = b_{n-1} + (n - 1)$. It is clear that $b_0 = 0$ and $b_1 = 0$. We will find b_n by the method of iteration:

$$b_2 = b_1 + 1 = 1$$
$$b_3 = b_2 + 2 = 1 + 2 = 3$$
$$b_4 = b_3 + 3 = 1 + 2 + 3 = 6$$
$$\vdots$$
$$b_n = 1 + 2 + \cdots + (n - 1)$$

■ 9.1 RECURSION AND RECURRENCE RELATIONS 357

Using the equation $1 + 2 + \cdots + m = [m(m+1)]/2$, we obtain $b_n = [n(n-1)]/2$. ■

PRACTICE PROBLEM 2 Refer to Example 9.1.6. Use mathematical induction to prove $b_n = 1 + 2 + \cdots + (n-1)$ for $n = 1, 2, 3, \ldots$.

EXAMPLE 9.1.7 **BINARY SEARCH** The binary search algorithm is a method for searching a linearly ordered list. Suppose we are given a list $A(1), A(2), \ldots, A(n)$ of n numbers that are in ascending order. Given a number c, the goal is to determine whether c is in the list. The first step in the binary search is to compare the number c to the number in the middle of the list. We take $A(m)$ to be the middle number of the list with $m = \lfloor n/2 \rfloor$, where $\lfloor x \rfloor$ is the greatest integer less than or equal to x. Hence, $A(\lfloor n/2 \rfloor)$ is the middle number of the list. For example, when $n = 9$, $\lfloor n/2 \rfloor = \lfloor 4.5 \rfloor = 4$. So, $A(4)$ is the middle number of a list of nine numbers.

After comparing c with $A(\lfloor n/2 \rfloor)$, we have three cases:

1. $c = A(\lfloor n/2 \rfloor)$ and we are done.
2. $c < A(\lfloor n/2 \rfloor)$ and we compare c to the middle number in the first half, $A(1), \ldots, A(\lfloor n/2 \rfloor - 1)$, of the list.
3. $c > A(\lfloor n/2 \rfloor)$ and we compare c to the middle number in the last half, $A(\lfloor n/2 \rfloor + 1), \ldots, A(n)$, of the list.

For $n = 32$, say, the middle number is $A(16)$, the first half of the list has 15 elements, and the second half has 16 elements. The binary search algorithm continues comparing c to the middle numbers in successively smaller lists until either c is found or the list is exhausted.

Let $T(n)$ be the greatest number of comparisons needed to ascertain whether c is in a list of n numbers. For simplicity, we assume $n = 2^r$. We will analyze this case and leave the general solution as an exercise. (See Exercise 27, Exercise Set 9.1.)

Assume $T(0) = 0$. Then, $T(1) = 1 = 1 + T(0)$. In general, $T(2^m) = 1 + T(2^{m-1})$ since the second half of a list with 2^m elements has 2^{m-1} elements. By iteration we see that

$$T(1) = 1 + T(0) = 1$$
$$T(2) = 1 + T(1) = 2$$
$$T(4) = 1 + T(2) = 3$$
$$\vdots$$
$$T(2^m) = 1 + T(2^{m-1}) = 1 + m$$

So, doubling the length of the list adds only one step to the search. ■

We will give solutions for the following two examples in the next section.

EXAMPLE 9.1.8 **FIBONACCI SEQUENCE** In his book *Liber Abaci* (1202), Leonardo of Pisa (Fibonacci) posed the following problem. Suppose we have one pair of rabbits at the beginning of the first month. After one month of life, a pair of rabbits reproduces. For the second month and each month thereafter of a rabbit's life, every pair of rabbits produces another pair. See Figure 9.1.1. Assuming no rabbits die, how many rabbits will there be at the beginning of the second year?

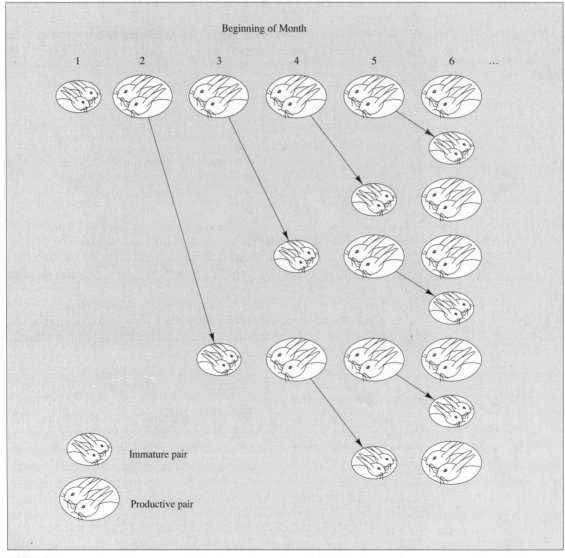

FIGURE 9.1.1

9.1 RECURSION AND RECURRENCE RELATIONS

Let F_n be the number of pairs of rabbits alive at the beginning of the nth month. For convenience we let $F_0 = 0$. We are given that $F_1 = 1$. Because of the rules for reproduction, we get $F_2 = 1$ and $F_3 = 2$. Of the two pairs alive at the beginning of month 3, one pair can reproduce at the beginning of month 4 and one pair first reproduces in month 5. So, $F_4 = 3 = 2 + 1 = F_3 + F_2$. Similarly, $F_5 = 3 + 2 = F_4 + F_3$. Indeed, the recurrence relation for the Fibonacci sequence is $F_n = F_{n-1} + F_{n-2}$. We have seen that the initial conditions are $F_0 = 0$ and $F_1 = 1$. ∎

PRACTICE PROBLEM 3 Refer to Example 9.1.8. How many rabbits will be alive at the beginning of the second year? In other words, find F_{13}.

Recall that a binary representation of an integer is a string consisting of the two digits 0 and 1. Similarly, an octal representation of an integer consists of the eight digits $0, 1, 2, \ldots, 7$.

EXAMPLE 9.1.9 Find a recurrence relation for the number of n-digit octal representations of integers with no consecutive 0's.

Solution Let t_n be the number of n-digit octal numbers with no consecutive 0's. There are eight 1-digit octal numbers and sixty-four 2-digit octal numbers. One of the 2-digit octal numbers is 00. So, $t_1 = 8$ and $t_2 = 63$.

Every n-digit octal number either begins with 0 or it does not. If an n-digit octal number with no consecutive 0's does not begin with 0, the $(n-1)$-digit octal number to the right of the first digit can be any $(n-1)$-digit octal number with no consecutive 0's. Since the first digit of an n-digit octal number not beginning with 0 can be any of the remaining 7 digits, the number of n-digit octal numbers with no consecutive 0's that do not begin with 0 is $7t_{n-1}$.

If an n-digit octal number with no consecutive 0's does begin with 0, the digit adjacent can only be one of the remaining 7 digits. Since the second digit is not 0, the $(n-2)$-digit octal number to the right of the second digit can be any $(n-2)$-digit octal number with no consecutive 0's. Hence, the number of n-digit octal numbers with no consecutive 0's that do begin with 0 is $7t_{n-2}$. Therefore, $t_n = 7t_{n-1} + 7t_{n-2}$.

If we let $t_0 = 1$, the initial conditions are $t_0 = 1$ and $t_1 = 8$. Note that $t_2 = 7 \cdot 8 + 7 \cdot 1 = 63$, which agrees with our original calculation of t_2. There are two reasons for making $t_0 = 1$ one of the initial conditions. In the following two sections, we will solve the recurrence relation $t_n = 7t_{n-1} + 7t_{n-2}$ and others like it. The calculations for these solutions are easier if one of the initial conditions is at the $n = 0$ level. Of course, the choice of initial condition must be consistent with any other initial condition and the recurrence relation. The other reason is that we can think of a 0-digit octal number as the empty string, and there is exactly one of those. We will discuss the empty string in Chapter 10. ∎

EXAMPLE 9.1.10 **BINOMIAL COEFFICIENTS** Prove that $C(n, k) = C(n - 1, k - 1) + C(n - 1, k)$ for $0 < k < n$ and $n = 2, 3, \ldots$ is a recurrence relation for the binomial coefficients.

Solution   To prove this, we use a combinatorial argument. If we are to select k objects from n objects, there are $C(n - 1, k)$ ways to select k objects such that a certain object is always left out, and there are $C(n - 1, k - 1)$ ways such that this certain object is always included. Therefore,

$$C(n, k) = C(n - 1, k - 1) + C(n - 1, k) \quad \text{for } 0 < k < n$$

The initial conditions are $C(n, 0) = C(n, n) = 1$ for $n = 0, 1, \ldots$. ∎

In Chapter 4 we used algebra to establish the recurrence relation in Example 9.1.10. The recurrence relation in this example is a **double recurrence**, since it is defined on a pair of integers n and k.

Here is another double recurrence relation. It defines a function notorious for fast growth.

EXAMPLE 9.1.11 **ACKERMANN'S FUNCTION** Ackermann's function A is defined as follows:

1. $A(n, 0) = A(n - 1, 1)$ for $n = 1, 2, \ldots$
2. $A(n, k) = A(n - 1, A(n, k - 1))$ for $n, k = 1, 2, \ldots$
3. Initial conditions: $A(0, k) = k + 1$ for $k = 0, 1, \ldots$ ∎

Ackermann's function is an important example for theoretical computer science. The definition of Ackermann's function provides an algorithm to compute it effectively, but there is no practical method of carrying out the computation. The difficulty arises because the function grows so fast that no computer can have enough memory to carry out the computation of $A(n, k)$ for even as small an integer as $n = 4 = k$. As you will see below, $A(4, 4)$ is a stupendously huge number.

Here are some sample computations:

$A(0, 0) = 1$ By the initial conditions
$A(1, 0) = A(0, 1) = 2$ By steps 1 and 3
$A(1, 1) = A(0, A(1, 0)) = A(0, 2) = 2 + 1 = 3$
$A(2, 0) = A(1, 1) = 3$

PRACTICE PROBLEM 4   Refer to Example 9.1.11. Calculate $A(2, 2)$.

You will be asked to prove each of the following facts about Ackermann's function in Exercise Set 9.1 (see Exercises 24–26).

$A(1, m) = m + 2 = 2 + (m + 3) - 3$ for $m = 0, 1, \ldots$
$A(2, m) = 2m + 3 = 2(m + 3) - 3$ for $m = 0, 1, \ldots$
$A(3, m) = 2^{m+3} - 3$ for $m = 0, 1, \ldots$

9.1 RECURSION AND RECURRENCE RELATIONS

Examination of the equations above reveals a pattern that goes from addition to multiplication to exponentiation. What is next? How big will $A(4, m)$ be? It can be shown that:

$$A(4, m) = \underbrace{2^{2^{2^{\cdot^{\cdot^{\cdot^{2}}}}}}}_{m+3} - 3, \quad \text{so } A(4, 4) = 2^{2^{2^{2^{2^{2^{2}}}}}} - 3$$

The number $A(4, 4)$ is incredibly large. Its decimal representation has more than $10^{19,199}$ digits. Since the total number of elementary particles in the known universe is estimated to be 10^{80}, it is impossible to comprehend just how large $A(4, 4)$ is.

■ **PROGRAMMING NOTES**

Recurrence relations are closely related to recursion available in some high-level languages such as Pascal. Just as recurrence relations define each sequence value in terms of previous values, recursion is a method for defining values or actions by calling the previous values or actions. We give two examples below.

A Pascal Function to Compute $n!$ The function is written in Pascal using recursion. Note how the "else" line is used to define the function value for NFactorial in terms of the previous function value; that is, NFactorial calls itself.

As usual we assume "nonnegativeinteger" and "positiveinteger" are previously defined types.

```
function NFactorial (n : nonnegativeinteger) : positiveinteger;
    begin
        if n = 0 then
            NFactorial := 1
        else
            NFactorial := n*NFactorial(n - 1)
    end;
```

PRACTICE PROBLEM 5 Refer to the function above. Use mathematical induction to prove NFactorial$(m) = m!$ for $m = 0, 1, 2, \ldots$.

A Pascal Function Using Recursion to Compute the Value of an Annuity As usual, we assume "positiveinteger" is a previously defined type.

```
function Annuity(n : positiveinteger) : real;
    begin
        if n = 1 then
            Annuity := 2000
        else
            Annuity := 2000 + 1.09*Annuity(n - 1)
    end;
```

PRACTICE PROBLEM 6 Refer to Example 9.1.2. Use the function Annuity as part of a computer program to calculate the value of the annuity after 20 years.

EXERCISE SET 9.1

1. Suppose you are depositing $2000 in an account paying 8% interest compounded annually. Let A_n be the amount in the account at the end of the nth year.
   a. Find a recurrence relation and an initial condition for the sequence (A_n), $n = 0, 1, \ldots$.
   b. Solve for A_n.

2. Suppose you are depositing $3000 in an account paying 8% interest compounded monthly. Let A_n be the amount in the account at the end of the nth year.
   a. Find a recurrence relation and an initial condition for the sequence (A_n), $n = 0, 1, \ldots$.
   b. Solve for A_n.

3. Suppose you are depositing $3000 each year in an annuity account paying 8% interest compounded annually. Let A_n be the amount in the account at the end of the nth year.
   a. Find a recurrence relation and an initial condition for the sequence (A_n), $n = 1, 2, \ldots$.
   b. Solve for A_n.

4. Suppose you are depositing $1500 each year in an annuity account paying 10% interest compounded annually. Let A_n be the amount in the account at the end of the nth year.
   a. Find a recurrence relation and an initial condition for the sequence (A_n), $n = 1, 2, \ldots$.
   b. Solve for A_n.

5. Refer to the Fibonacci sequence in Example 9.1.8. Calculate F_{15}.

6. Refer to the Fibonacci sequence in Example 9.1.8. Calculate F_{18}.

7. Refer to the binomial coefficients in Example 9.1.10. Given that $C(7, k)$ are, in order from $k = 0$ to $k = 7$, equal to 1, 7, 21, 35, 35, 21, 7, 1, use the recurrence relation for the binomial coefficients to find all values $C(8, k)$ for $k = 0, 1, \ldots, 8$.

8. Refer to the binomial coefficients in Example 9.1.10. Use the solution to Exercise 7 and the recurrence relation for the binomial coefficients to find all values $C(9, k)$, where $k = 0, 1, \ldots, 9$.

9. Find a recurrence relation and initial conditions that describe the following problem. In how many ways can a person climb a flight of stairs with n steps if the person can skip at most one step at a time?

10. Let b_n be the number of bit strings of length n with no consecutive 0's. Find a recurrence relation and initial conditions for b_n.

11. Let t_n be the number of ternary (base 3) numbers of length n with no consecutive 0's. Find a recurrence relation and initial conditions for t_n.

12. The Tower of Hanoi is a puzzle consisting of three pegs mounted on a base and n circular disks of increasing diameters with holes in their centers on one peg, with the largest disk on the bottom. The object of this game is to transfer all n disks one at a time to another peg. If a disk is on a peg, only a disk of smaller diameter can be placed on top of the other disk. Let t_n be the number of moves in which the n-disk game can be completed. The initial condition is $t_1 = 1$, since only one move is required to move one disk from peg 1 to peg 2.
    a. Find a recurrence relation for the sequence $t_1, t_2, t_3, \ldots$.
    b. Find an explicit formula for t_n.

13. Let t_n be the number of valid arithmetic expressions of length n, without parentheses, using the ten digits and the binary operation symbols $+, -, *, /$. Each valid arithmetic expression must begin and end with a digit; the binary operation symbols can never be adjacent; division by zero can never occur. For example, $79+3/9*5$ is a valid expression, but neither $79-3*$ nor $93/0+53$ is valid. Find a recurrence relation for t_n.

For Exercises 14–17 refer to the Fibonacci sequence in Example 9.1.8.

14. Prove each of the following.
  a. $F_1 + F_3 + \cdots + F_{2n-1} = F_{2n}$ **b.** $F_0 + F_2 + \cdots + F_{2n} = F_{2n+1} - 1$

15. Calculate $F_{n+2}F_n - (F_{n+1})^2$ for several values of n. Make a conjecture based on your calculations and prove that your conjecture is true.

16. Calculate $F_0 - F_1 + F_3 - \cdots + (-1)^n F_n$ for several values of n. Make a conjecture based on your calculations and prove that your conjecture is true. [*Hint:* Consider Exercise 14 above.]

17. Calculate $F_0^2 + F_1^2 + F_2^2 + \cdots + F_n^2$ for several values of n. Make a conjecture based on your calculations and prove that your conjecture is true. [*Hint:* $F_0^2 + F_1^2 + F_2^2 + \cdots + F_n^2$ is the product of two Fibonacci numbers.]

18. Here is an alternate form of the Fibonacci problem. Suppose each pair of rabbits born in a given month produces a pair of offspring in each of the next two months, then dies at the end of the third month. Let G_n be the number of pairs of rabbits alive at the end of the nth month. Find a recurrence relation for G_n.

19. Here is a sequence very similar to the Fibonacci sequence. The **Lucas sequence** is defined by: $L_2 = 1$, $L_3 = 3$, and $L_n = L_{n-1} + L_{n-2}$ for $n \geq 4$.
  a. Find $L_4, L_5, L_6,$ and L_7. **b.** Prove $L_n = F_n + F_{n-2}$ for $n = 2, 3, \ldots$.

20. Prove that the initial conditions $t_0 = a$ and $t_1 = b$ and the recurrence relation $t_n = t_{n-1} + t_{n-2}$ for $n = 2, 3, \ldots$ for the sequence t_n uniquely determine the sequence.

21. Here is an outline of a proof that the sum of the binomial coefficients along the diagonals of Pascal's triangle, running up from the left, are equal to the Fibonacci numbers. That is, we will prove:

$$C(n, 0) + C(n-1, 1) + C(n-2, 2) + \cdots + C(n-k, k) = F_{n+1}, \text{ where } k = \lfloor n/2 \rfloor$$

Verify steps 2–4. You may assume $C(i, j) = 0$ for $i < j$.
Proof
1. Let $t_{n+1} = C(n, 0) + C(n-1, 1) + C(n-2, 2) + \cdots + C(n-k, k)$, where $k = \lfloor n/2 \rfloor$.
2. $F_1 = t_1$ and $F_2 = t_2$.
3. $t_n = t_{n-1} + t_{n-2}$ for $n = 3, 4, \ldots$
4. Therefore, $t_{n+1} = F_{n+1}$ for $n = 0, 1, 2, \ldots$ by the result of Exercise 20.

Refer to Ackermann's function in Example 9.1.11 for Exercises 22–26.

22. Calculate $A(3, 3)$. **23.** Calculate $A(2, 5)$.

24. Prove $A(1, m) = m + 2 = 2 + (m + 3) - 3$ for $m = 0, 1, \ldots$.

25. Prove $A(2, m) = 2m + 3 = 2(m + 3) - 3$ for $m = 0, 1, \ldots$.

26. Prove $A(3, m) = 2^{m+3} - 3$ for $m = 0, 1, \ldots$.

27. Refer to Example 9.1.7. Let n be any positive integer. Then there is a positive integer r such that $2^{r-1} \leq n < 2^r$. Show that $T(n) = r$ and, therefore, $T(n) = 1 + \lfloor \log_2 n \rfloor$.

| PROGRAMMING
EXERCISES

28. Refer to the function Annuity in the Programming Notes. Write Annuity without using recursion. For example, you may want to use a while loop or a for loop.

29. Write functions in Pascal to calculate the Fibonacci numbers using:
   a. A loop structure and no recursion.   **b.** Recursion.

9.2
RECURRENCE RELATIONS AND THE CHARACTERISTIC EQUATION METHOD

There is no known general method for solving all recurrence relations. We will, however, consider two methods of solving certain recurrence relations. One is the characteristic equation method discussed in this section. It is easy to use but limited in scope. The other method makes use of generating functions and is more broadly applicable. We will investigate generating functions in Section 9.3. First, let us define some terminology.

DEFINITION

An equation of the form

$$c_0 a_n + c_1 a_{n-1} + c_2 a_{n-2} + \cdots + c_k a_{n-k} = f(n)$$

where $c_0, c_1, \ldots, c_k$ are constants, is called a **linear recurrence relation with constant coefficients**. The recurrence relation is said to have **order k** when $c_0 \neq 0$ and $c_k \neq 0$. When $f(n) = 0$, the equation is said to be **homogeneous**.

The first two examples we consider are from Section 9.1. They are both homogeneous linear recurrence relations with constant coefficients. Each of these examples will help to illustrate the terminology introduced above.

EXAMPLE 9.2.1 The recurrence relation $A_n = (1.07)A_{n-1}$ in Example 9.1.1 can be rewritten as $1A_n + (-1.07)A_{n-1} = 0$. It is a linear recurrence relation with constant coefficients $c_0 = 1$ and $c_1 = -1.07$. It is of order 1 since $c_0 \neq 0$ and $c_1 \neq 0$. This recurrence relation is homogeneous since the right side of the equation is 0.   ∎

EXAMPLE 9.2.2 In the Fibonacci sequence of Example 9.1.8, the two initial conditions are $F_0 = 0$, and $F_1 = 1$; and the recurrence relation is $F_n = F_{n-1} + F_{n-2}$ for $n \geq 2$. The recurrence relation $F_n - F_{n-1} - F_{n-2} = 0$ is a homogeneous linear recurrence relation of order 2 with coefficients $c_0 = 1$, $c_1 = -1$ and $c_2 = -1$.   ∎

What does a nonlinear and nonhomogeneous recurrence relation look like? The next recurrence relation is of order 2. However, it is neither linear nor homogeneous.

9.2 RECURRENCE RELATIONS AND THE CHARACTERISTIC EQUATION METHOD

EXAMPLE 9.2.3 The recurrence relation $a_n + 4(a_{n-1})^3 + 2a_{n-2} = 6n$ is not linear because of the cubed term, $(a_{n-1})^3$, and it is not homogeneous because $f(n) = 6n \neq 0$.

By Theorem 9.1.4 we know that the homogeneous linear recurrence relation $a_n - ra_{n-1} = 0$ with initial condition $a_0 = b$ has the solution $a_n = br^n$, where b and r are fixed positive numbers. In order to gain some insight into the solutions for linear recurrence relations in general, let us assume we do not know that $a_n = br^n$ is a solution for $a_n - ra_{n-1} = 0$. Rather, suppose we had guessed (by trial and error, and intuition) that the solution of this recurrence relation had to be some constant times a power of some number; in other words, we assume $a_n = kx^n$ for $n = 0, 1, 2, \ldots$.

Then we have $0 = a_n - ra_{n-1} = kx^n - rkx^{n-1} = kx^{n-1}(x - r)$. Hence, either $kx^{n-1} = 0$ for $n = 1, 2, 3, \ldots$ or $x - r = 0$. If $kx^{n-1} = 0$, then $a_{n-1} = 0$ for $n = 1, 2, 3, \ldots$ and, thus, $b = a_0 = 0$. This is contrary to our assumption that b is a positive number. Therefore, $x = r$. Now, $a_n = kr^n$. But $b = a_0 = kr^0 = k$. So, $k = b$. Therefore, $a_n = br^n$, as before.

This method is an example of the **characteristic equation method** of solving a homogeneous linear recurrence relation. The characteristic equation is $x - r = 0$.

We will use this method to solve homogeneous linear recurrence relations of order 2. The quadratic equation $c_0x^2 + c_1x + c_2 = 0$ is the **characteristic equation** for the homogeneous linear recurrence relation $c_0a_n + c_1a_{n-1} + c_2a_{n-2} = 0$.

Note that, if we assume that a solution for $c_0a_n + c_1a_{n-1} + c_2a_{n-2} = 0$ has the form $a_n = kx^n$ with $a_0 \neq 0$, then $c_0kx^n + c_1kx^{n-1} + c_2kx^{n-2} = 0$. So, $kx^{n-2}(c_0x^2 + c_1x + c_2) = 0$. Hence, $c_0x^2 + c_1x + c_2 = 0$. So, solving the characteristic equation will yield a solution for the homogeneous linear recurrence relation.

In order to find roots for some of the characteristic equations that follow, we need to recall the quadratic formula. The real roots for $ax^2 + bx + c = 0$, where $a \neq 0$ and $b^2 - 4ac \geq 0$, are

$$r = \frac{-b + \sqrt{b^2 - 4ac}}{2a} \quad \text{and} \quad s = \frac{-b - \sqrt{b^2 - 4ac}}{2a}$$

Let us use the characteristic equation method to solve the following recurrence relation.

EXAMPLE 9.2.4 Solve the recurrence relation $a_n - 3a_{n-1} + 2a_{n-2} = 0$ with initial conditions $a_0 = 1$ and $a_1 = 4$.

Solution The characteristic equation is $x^2 - 3x + 2 = 0$. Use either the quadratic formula or a simple factorization to get the roots $r = 2$ and $s = 1$. By using Theorem 9.2.5 or substitution, we see that $a_n = a \cdot 2^n + b \cdot 1^n$ is a solution to $a_n - 3a_{n-1} + 2a_{n-2} = 0$. To find a and b, we first use the initial conditions $a_0 = 1$ and $a_1 = 4$ to obtain two simultaneous equations. Then we solve the resulting system of equations for a and b. The solution is illustrated schematically:

$$\left.\begin{array}{l} a_0 = 1 \\ a_1 = 4 \end{array}\right\} \rightarrow \left.\begin{array}{l} a \cdot 2^0 + b \cdot 1^0 = 1 \\ a \cdot 2^1 + b \cdot 1^1 = 4 \end{array}\right\} \rightarrow \left.\begin{array}{l} a + b = 1 \\ 2a + b = 4 \end{array}\right\} \rightarrow \begin{array}{l} a = 3 \\ b = -2 \end{array}$$

Hence, $a = 3$ and $b = -2$ yields $a_n = 3 \cdot 2^n - 2$ as a solution. ∎

Here is a theorem we can use to solve recurrence relations like that of Example 9.2.4.

THEOREM 9.2.5 If the characteristic equation for $c_0 a_n + c_1 a_{n-1} + c_2 a_{n-2} = 0$ has distinct real roots r and s, then a solution for $c_0 a_n + c_1 a_{n-1} + c_2 a_{n-2} = 0$ is $a_n = ar^n + bs^n$. The constants a and b depend on the initial conditions.

Proof
$$\begin{aligned} c_0 a_n &+ c_1 a_{n-1} + c_2 a_{n-2} \\ &= c_0(ar^n + bs^n) + c_1(ar^{n-1} + bs^{n-1}) + c_2(ar^{n-2} + bs^{n-2}) \\ &= (c_0 ar^n + c_1 ar^{n-1} + c_2 ar^{n-2}) + (c_0 bs^n + c_1 bs^{n-1} + c_2 bs^{n-2}) \\ &= ar^{n-2}(c_0 r^2 + c_1 r + c_2) + bs^{n-2}(c_0 s^2 + c_1 s + c_2) \\ &= ar^{n-2} \cdot 0 + bs^{n-2} \cdot 0 = 0 \end{aligned}$$
∎

PRACTICE PROBLEM 1 Solve the recurrence relation $a_n - 4a_{n-1} - 21a_{n-2} = 0$.

EXAMPLE 9.2.6 **THE FIBONACCI SEQUENCE REVISITED** Recall from Example 9.1.8 (page 358) that $F_n - F_{n-1} - F_{n-2} = 0$, $F_0 = 0$, and $F_1 = 1$. The characteristic equation is $x^2 - x - 1 = 0$. By the quadratic formula, we find the distinct roots

$$r = \frac{1 + \sqrt{5}}{2} \qquad s = \frac{1 - \sqrt{5}}{2}$$

So, $F_n = ar^n + bs^n$. Using the initial conditions $a_0 = 0$ and $a_1 = 1$, we find $a = \sqrt{5}/5$ and $b = -\sqrt{5}/5$. Hence,

$$F_n = \frac{\sqrt{5}}{5} \left(\frac{1 + \sqrt{5}}{2} \right)^n - \frac{\sqrt{5}}{5} \left(\frac{1 - \sqrt{5}}{2} \right)^n$$
∎

PRACTICE PROBLEM 2 Do the algebra necessary to find the numbers r, s, a, and b in Example 9.2.6.

The ancient Greeks considered the **golden section** to be the most perfectly proportioned rectangle. The facade of the ancient Acropolis in Greece is in the shape of a golden section. A standard 35 mm slide is also approximately a golden section. The golden section is constructed as follows.

Take a 2×2 square and bisect the bottom side at point B. Place a compass so that one leg is at the point B and the other leg is on the upper right corner of the square. Rotate the compass, with the leg on the point B, to mark

9.2 RECURRENCE RELATIONS AND THE CHARACTERISTIC EQUATION METHOD

a point C on the bottom side extended. Construct a rectangle with line segment AC as the bottom side and line segment AD as the vertical side. See Figure 9.2.1.

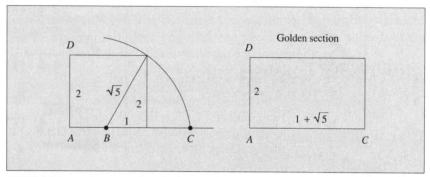

FIGURE 9.2.1

The segment AD has a height of 2 units. By the Pythagorean theorem, we see the line segment AC must have a length of $1 + \sqrt{5}$. Hence, the ratio of the sides of the golden section is $(1 + \sqrt{5})/2 \approx 1.6180$. This number, called the **golden ratio**, is one of the roots found above when we solved the recurrence relation for the Fibonacci sequence F_n.

Moreover, for large values of n, the sequence of fractions F_{n+1}/F_n is closely approximated by the golden ratio. Using terminology from calculus, we say the limit, as n approaches infinity, of F_{n+1}/F_n is equal to the golden ratio. In other words,

$$\lim_{n \to \infty} \frac{F_{n+1}}{F_n} = \frac{1 + \sqrt{5}}{2}$$

You are asked to prove that the limit of the ratio of successive Fibonacci numbers is equal to the golden ratio in Exercise Set 9.2 (see Exercise 21).

Before solving more recurrence relations, we prove an important fact about homogeneous recurrence relations.

THEOREM 9.2.7 If s_n and t_n are solutions of a linear homogeneous recurrence relation with constant coefficients, then $as_n + bt_n$, where a and b are arbitrary constants, is also a solution.

Proof We will give the proof for a linear homogeneous recurrence relation of degree 2. You will see that the proof applies to any linear homogeneous recurrence relation with constant coefficients. Assume s_n and t_n are solutions of $c_0 a_n + c_1 a_{n-1} + c_2 a_{n-2} = 0$. To prove $as_n + bt_n$ is also a solution, we substitute $a_n = as_n + bt_n$ in the left side of the recurrence relation to obtain

$c_0(as_n + bt_n) + c_1(as_{n-1} + bt_{n-1}) + c_2(as_{n-2} + bt_{n-2})$. Because the recurrence relation is linear,

$$c_0(as_n + bt_n) + c_1(as_{n-1} + bt_{n-1}) + c_2(as_{n-2} + bt_{n-2})$$
$$= a(c_0 s_n + c_1 s_{n-1} + c_2 s_{n-2}) + b(c_0 t_n + c_1 t_{n-1} + c_2 t_{n-2})$$

Then,

$$a(c_0 s_n + c_1 s_{n-1} + c_2 s_{n-2}) + b(c_0 t_n + c_1 t_{n-1} + c_2 t_{n-2}) = a \cdot 0 + b \cdot 0 = 0$$

since the recurrence relation is homogeneous. Hence, $as_n + bt_n$ is also a solution. ∎

So far we have shown how to solve linear homogeneous recurrence relations of degree 2 when their characteristic equations have distinct real roots. What if the characteristic equation has identical roots? The following example illustrates how the characteristic equation method is used in this case.

EXAMPLE 9.2.8 Solve the recurrence relation $a_n - 6a_{n-1} + 9a_{n-2} = 0$.

Solution The characteristic equation is $x^2 - 6x + 9 = 0$. Since $x^2 - 6x + 9 = (x-3)^2$, we have identical roots $r = 3$. We know 3^n is a solution. What about any other solutions? Of course $a \cdot 3^n$, where a is a constant, is also a solution. We verify that $n \cdot 3^n$ is also a solution as follows:

$$n \cdot 3^n - 6(n-1)3^{n-1} + 9(n-2)3^{n-2}$$
$$= (n-2)3^n + 2 \cdot 3^n - 6(n-2)3^{n-1} - 6 \cdot 3^{n-1} + 9(n-2)3^{n-2}$$
$$= (n-2)(3^n - 6 \cdot 3^{n-1} + 9 \cdot 3^{n-2}) + 2 \cdot 3^n - 6 \cdot 3^{n-1}$$
$$= (n-2)0 + 2 \cdot 3^n - 2 \cdot 3 \cdot 3^{n-1} = 0$$

Hence, by Theorem 9.2.7, $a \cdot 3^n + bn \cdot 3^n$ is a solution. ∎

The following theorem includes Example 9.2.8 as a special case.

THEOREM 9.2.9 The second-degree linear homogeneous recurrence relation $a_n - 2ra_{n-1} + r^2 a_{n-2} = 0$ has as a solution $ar^n + bnr^n$, where a and b are arbitrary constants depending on the initial conditions.

Proof By Theorem 9.2.7 it is sufficient to prove that nr^n is a solution. The computation in Example 9.2.8 is a special case of the proof needed here. The proof is left as an exercise. (See Exercise 19, Exercise Set 9.2.) ∎

Here is one more example to illustrate the case of a second-degree linear homogeneous recurrence relation that has a characteristic equation with identical real roots.

EXAMPLE 9.2.10 Solve $a_n + 4a_{n-1} + 4a_{n-2} = 0$ for $n = 2, 3, \ldots$ with initial conditions $a_0 = 5$ and $a_1 = 4$.

Solution The characteristic equation is $x^2 + 4x + 4 = 0$ with identical roots $r = -2$. Hence, the solution is $a(-2)^n + bn(-2)^n$. To find the constants a and b, we use the initial conditions $5 = a_0 = a$ and $4 = a_1 = -2a - 2b$. We obtain $a = 5$ and $b = -7$. Hence, the solution is $5(-2)^n - 7n(-2)^n$. ∎

In Example 9.1.9 we let t_n be the number of n-digit octal numbers with no consecutive 0's. We found the recurrence relation $t_n = 7t_{n-1} + 7t_{n-2}$. The initial conditions are $t_0 = 1$ and $t_1 = 8$. We will use the characteristic equation method to solve this recurrence relation.

EXAMPLE 9.2.11 Solve the recurrence relation $t_n - 7t_{n-1} - 7t_{n-2} = 0$ for $n = 2, 3, \ldots$ with initial conditions $t_0 = 1$ and $t_1 = 8$.

Solution The characteristic equation is $x^2 - 7x - 7 = 0$. By the quadratic formula we find the distinct roots

$$r = \frac{7 + \sqrt{77}}{2} \qquad s = \frac{7 - \sqrt{77}}{2}$$

So, $t_n = ar^n + bs^n$. Using the initial conditions $t_0 = 1$ and $t_1 = 8$, we have $a + b = 1$ and $ra + sb = 8$. Solving for a and b:

$$b = \frac{8 - r}{s - r} \qquad a = \frac{8 - s}{r - s}$$

Hence,

$$b = \frac{77 - 9\sqrt{77}}{154} \qquad a = \frac{77 + 9\sqrt{77}}{154}$$

Therefore,

$$t_n = \frac{77 + 9\sqrt{77}}{154} \cdot \left(\frac{7 + \sqrt{77}}{2}\right)^n + \frac{77 - 9\sqrt{77}}{154} \cdot \left(\frac{7 - \sqrt{77}}{2}\right)^n$$

∎

In this section we have used the characteristic equation method to solve second-degree linear homogeneous recurrence relations. Because we wish to restrict ourselves to the real numbers, we have investigated only those recurrence relations with real roots for the characteristic equation. The method we have outlined above can also be used to solve recurrence relations having characteristic equations with nonreal roots.

We can also use the characteristic equation method when the degree of the recurrence relation is greater than 2. However, it is usually much more difficult to solve a polynomial equation of degree 3 or greater. The next theorem regarding polynomial equations is very helpful for solving some equations of degree larger than 2.

THEOREM 9.2.12 Let c_i be an integer for all $i = 0, 1, 2, \ldots, n$ and assume $c_n \neq 0$. If r is a rational root of $x^n + c_1 x^{n-1} + c_2 x^{n-2} + \cdots + c_n = 0$, then r is an integer and r divides c_n.

The proof of Theorem 9.2.12 is left as an exercise. (See Exercise 20, Exercise Set 9.2.)

We have already seen that, if r is a root of multiplicity 2 for a quadratic characteristic equation, both r^n and nr^n are solutions of the recurrence relation. Also, it can be shown that, if r is a root of multiplicity m for the characteristic equation, then $r^n, nr^n, \ldots, n^{m-1}r^n$ are all solutions of the recurrence relation. Here is an example to illustrate the characteristic equation method for a recurrence relation of degree 5.

EXAMPLE 9.2.13 Solve $a_n - 10a_{n-1} + 39a_{n-2} - 74a_{n-3} + 68a_{n-4} - 24a_{n-5} = 0$.

Solution The characteristic equation is $x^5 - 10x^4 + 39x^3 - 74x^2 + 68x - 24 = 0$. By Theorem 9.2.12, the only possible rational roots for this equation must divide 24. The divisors of 24 are: $\pm 1, \pm 2, \pm 3, \pm 4, \pm 6, \pm 8, \pm 12$, and ± 24. It may be that none of these numbers are roots. Let $p(x) = x^5 - 10x^4 + 39x^3 - 74x^2 + 68x - 24$. If r is a root, then $p(r) = 0$. So let us substitute, as a possible root, one of the divisors listed. We try $r = 1$ first. It is easy to see that $p(1) = 0$. Hence, $x - 1$ is a factor of $p(x)$. If we divide $p(x)$ by $x - 1$, we obtain $p(x) = (x - 1)(x^4 - 9x^3 + 30x^2 - 44x + 24)$.

Any remaining roots of $p(x)$ must be roots of $q(x) = x^4 - 9x^3 + 30x^2 - 44x + 24$. Since $q(1) \neq 0$, 1 is not a multiple root. However, both 2 and 3 are roots since $q(2) = q(3) = 0$. If we divide $q(x)$ by $x - 3$, we obtain $q(x) = (x - 3)(x^3 - 6x^2 + 12x - 8)$. It is easy to check that 2 is a root of $x^3 - 6x^2 + 12x - 8$. Again dividing, we obtain $x^3 - 6x^2 + 12x - 8 = (x - 2)(x^2 - 4x + 4)$. But, $x^2 - 4x + 4 = (x - 2)^2$. Hence, the characteristic equation is $(x - 1)(x - 2)^3(x - 3) = 0$. The roots are 1, 2, and 3, and 2 is a root of multiplicity 3.

Therefore, the solution to the recurrence relation is

$$a_n = a \cdot 1^n + b \cdot 3^n + c \cdot 2^n + dn \cdot 2^n + gn^2 \cdot 2^n$$

where a, b, c, d, and g are constants that depend on the initial conditions. ∎

Next, we introduce a very efficient method for evaluating polynomials. This method, given by the mathematician William G. Horner (1786–1837), is called Horner's method.

Horner's Method for Evaluating a Polynomial

Consider the polynomial $p(x) = 2x^3 + 5x^2 - 3x + 7$. Suppose we want to evaluate $p(x)$ at $x = \pi$; in other words, calculate $p(\pi)$. The most familiar method is to "plug in" π for x and do the resulting arithmetic to obtain $p(\pi) = 2 \cdot \pi \cdot \pi \cdot \pi + 5 \cdot \pi \cdot \pi - 3 \cdot \pi + 7$. This evaluation technique requires six multiplications and three additions or subtractions.

Following is a recursive definition of a polynomial that can be used to give us a more efficient method for evaluating a polynomial.

9.2 RECURRENCE RELATIONS AND THE CHARACTERISTIC EQUATION METHOD

DEFINITION

A recursive definition of polynomials in x over a set of coefficients: Let C be a set of coefficients.

Basis Step ($n = 0$) Every element of C is a zero-degree polynomial in x.
Induction Step ($n \geq 1$) For any $(n-1)$th-degree polynomial $q(x)$ and any element $k \in C$, the expression $p(x) = xq(x) + k$ is an nth-degree polynomial.

EXAMPLE 9.2.14 Verify that the expression $p(x) = 2x^3 + 5x^2 - 3x + 7$ is a polynomial over the set of integers.

Solution Repeated factoring yields

$$p(x) = x(2x^2 + 5x - 3) + 7$$
$$= x[x(2x + 5) - 3] + 7$$
$$= x[x(2 \cdot x + 5) - 3] + 7$$

By the basis step, 2 is a zero-degree polynomial. So, $2x + 5$ is a first-degree polynomial by the induction step. Again by the induction step, $x(2x + 5) - 3 = 2x^2 + 5x - 3$ is a second-degree polynomial. Similarly, $p(x) = x(2x^2 + 5x - 3) + 7$ is a third-degree polynomial over the set of integers. ∎

If we use this recursive definition to evaluate $p(\pi)$, we obtain

$$p(\pi) = \pi[\pi(2 \cdot \pi + 5) - 3] + 7$$

Using the recursive definition of a polynomial to evaluate the polynomial at a specific value is **Horner's method**. Note that it requires just three multiplications and three additions or subtractions. Horner's method is a most efficient method for evaluating polynomials. Newton invented Horner's method before Horner but never published it.

PRACTICE PROBLEM 3 Use Horner's method to evaluate $q(\pi)$, where $q(x) = x^4 - 5x^2 + 7x + 9$.

EXERCISE SET 9.2

Solve each of the recurrence relations in Exercises 1–14 by the characteristic equation method.

1. $a_n - a_{n-1} - 2a_{n-2} = 0$ for $n = 2, 3, 4, \ldots$; $a_0 = 4$ and $a_1 = 5$
2. $a_n - 5a_{n-1} - 14a_{n-2} = 0$ for $n = 2, 3, 4, \ldots$; $a_0 = 1$ and $a_1 = 7$
3. $a_n - 4a_{n-1} + a_{n-2} = 0$ for $n = 2, 3, 4, \ldots$; $a_0 = 0$ and $a_1 = 2\sqrt{3}$
4. $a_n - 4a_{n-1} + 4a_{n-2} = 0$ for $n = 2, 3, 4, \ldots$; $a_0 = 3$ and $a_1 = 7$
5. $a_n + 10a_{n-1} + 25a_{n-2} = 0$ for $n = 2, 3, 4, \ldots$; $a_0 = 1$ and $a_1 = 3$
6. $a_n + a_{n-1} - 6a_{n-2} = 0$ for $n = 2, 3, 4, \ldots$; $a_0 = 5$ and $a_1 = 0$
7. $a_n - a_{n-1} - 6a_{n-2} = 0$ for $n = 2, 3, 4, \ldots$; $a_0 = 4$ and $a_1 = -3$

8. $a_n - 14a_{n-1} + 49a_{n-2} = 0$ for $n = 2, 3, 4, \ldots$; $a_0 = 2$ and $a_1 = 21$
9. $a_n - a_{n-1} - 5a_{n-2} - 3a_{n-3} = 0$ for $n = 3, 4, \ldots$; $a_0 = 2$, $a_1 = 0$, and $a_2 = 14$
10. $a_n - 3a_{n-2} + 2a_{n-3} = 0$ for $n = 3, 4, \ldots$; $a_0 = 6$, $a_1 = -4$, and $a_2 = 13$
11. $a_n - 5a_{n-1} + 3a_{n-2} + 9a_{n-3} = 0$ for $n = 3, 4, \ldots$; $a_0 = 2$, $a_1 = -1$, and $a_2 = 10$
12. $a_n + 4a_{n-1} - 3a_{n-2} - 18a_{n-3} = 0$ for $n = 3, 4, \ldots$; $a_0 = 3$, $a_1 = 3$, and $a_2 = 30$

The characteristic equations for Exercises 13 and 14 have nonreal (complex) roots. Recall that $i = \sqrt{-1}$.

13. $a_n + 2a_{n-1} + 2a_{n-2} = 0$ for $n = 2, 3, 4, \ldots$; $a_0 = 2$ and $a_1 = -2$
14. $a_n - 2a_{n-1} + 10a_{n-2} = 0$ for $n = 2, 3, 4, \ldots$; $a_0 = 2$ and $a_1 = 2$
15. Let t_n be the number of valid arithmetic expressions of length n, without parentheses, using the ten digits and the symbols $+, -, *, /$. Solve for t_n. First convince yourself that $t_n = 10t_{n-1} + 39t_{n-2}$ for $n = 2, 3, \ldots$.
16. Let t_n be the number of ternary (base 3) numbers of length n with no consecutive 0's. Solve for t_n.
17. Refer to the alternate form of the Fibonacci problem given in Exercise 18 of Exercise Set 9.1. Let $G_0 = a$ and $G_1 = b$.
   a. Solve for G_n. [First convince yourself that $G_n = 2G_{n-1} - G_{n-2}$ for $n = 2, 3, \ldots$.]
   b. Describe what happens when $a = b$.
18. Here is another alternate form of the Fibonacci problem. Suppose we have one pair of rabbits at the beginning of the first month. After one month of life, a pair of rabbits reproduces. For the second month and each month thereafter of a rabbit's life, every pair of rabbits produces *two* pairs of offspring. Assume no rabbits die. Let G_n be the number of pairs of rabbits alive at the end of n months.
   a. Find a recurrence relation for G_n.   b. Solve for G_n.
19. Finish the proof of Theorem 9.2.9.
20. Prove Theorem 9.2.12. That is, let c_i be an integer for all $i = 0, 1, 2, \ldots, n$ and assume $c_n \neq 0$. Prove: If r is a rational root of $x^n + c_1 x^{n-1} + c_2 x^{n-2} + \cdots + c_n = 0$, then r is an integer and r divides c_n. [*Hint*: Substitute $x = r$ in $x^n + c_1 x^{n-1} + c_2 x^{n-2} + \cdots + c_{n-1} x + c_n = 0$ and then subtract c_n from both sides of the equation.]
21. This exercise is for those who have had a semester of calculus. Refer to the Fibonacci sequence in Example 9.2.6. Prove
$$\lim_{n \to \infty} \frac{F_{n+1}}{F_n} = \frac{1 + \sqrt{5}}{2}$$
22. Use the recursive definition of a polynomial to verify that the expression $p(x) = x^3 + 3x^2 - 7x - 5$ is a polynomial over the integers.
23. Use the recursive definition of a polynomial to verify that the expression $p(x) = 3x^4 + x^2 - 2x + 4$ is a polynomial over the integers.
24. Use Horner's method to evaluate the polynomial $p(x) = x^3 + 3x^2 - 7x - 5$ at $x = 3$.
25. Use Horner's method to evaluate the polynomial $p(x) = 3x^4 + x^2 - 2x + 4$ at $x = \pi$.

■ PROGRAMMING EXERCISES

26. Write a computer program to implement Horner's method.
27. Write a computer program that uses Horner's method to convert a positive binary integer to a decimal integer.

9.3
RECURRENCE RELATIONS AND GENERATING FUNCTIONS

In Section 9.2 the solutions were not derived. That is, we took the "Let's try this... Gee, it works" approach. In what follows, we will use generating functions to show how the solutions we used earlier are derived.

Generating functions were invented by the mathematician Pierre-Simon Laplace (1749–1827). The idea of a generating function is simple but brilliant. Given a sequence of real numbers $(a_0, a_1, a_2, \ldots, a_n, \ldots)$, the associated infinite series $A(x) = a_0 + a_1 x + a_2 x^2 + \cdots + a_n x^n + \cdots$ is called the **generating function** of the given sequence (a_n).

The reason for replacing a sequence by a generating function is simplification. Operations performed on sequences correspond to simpler operations on generating functions. Generating functions are a useful tool for manipulating and solving sequences.

For the examples following, we will need the **geometric series**

$$\frac{1}{1-ax} = 1 + ax + a^2 x^2 + a^3 x^3 + \cdots = \sum_{j=0}^{\infty} a^j x^j$$

Notice this means that $1/(1-ax)$ is the generating function for the sequence $(1, a, a^2, a^3, \ldots)$.

Partial Fraction Decomposition

To apply the method of generating functions, we need a technique called partial fraction decomposition. In the next two examples, we will illustrate how **partial fraction decomposition** is used to write a quotient of polynomials as the sum of fractions.

EXAMPLE 9.3.1 Find constants a and b such that

$$\frac{1-x}{(1+x)(1-2x)} = \frac{a}{1-2x} + \frac{b}{1+x}$$

Solution First, write the sum on the right as one fraction with a common denominator:

$$\frac{1-x}{(1+x)(1-2x)} = \frac{a}{1-2x} + \frac{b}{1+x} = \frac{a(1+x) + b(1-2x)}{(1+x)(1-2x)}$$

When two quotients with the same denominator are equal, the numerators must also be equal. So, we equate the numerators to obtain

$$1 - x = a(1+x) + b(1-2x) \tag{1}$$

When two polynomials in x are equal, they must have the same value no matter what value is substituted for x. Therefore, the polynomials $1 - x$ and $a(1+x) + b(1-2x)$ in Equation (1) must be equal for *any* value of x that is substituted.

To find a and b we are wise to substitute numbers that will give us the easiest possible equations to solve. In this case, the substitutions $x = -1$ and $x = \frac{1}{2}$ are best because an unknown (either a or b) disappears after each substitution. Substituting $x = -1$ yields $2 = b(1 + 2)$. Solving for b gives $b = \frac{2}{3}$. Similarly, $x = \frac{1}{2}$ yields $\frac{1}{2} = a(\frac{3}{2})$. Solving, we get $a = \frac{1}{3}$. Therefore,

$$\frac{1-x}{(1+x)(1-2x)} = \frac{\frac{1}{3}}{1-2x} + \frac{\frac{2}{3}}{1+x}$$

The substitution method used above is sometimes called the **method of judicious substitution**.

When two polynomials in x are equal, the coefficients of corresponding powers of x are the same and the constant terms are the same. So, an alternate method of finding the constants a and b in Equation (1) is to equate coefficients of like powers of x on either side of

$$1 - x = a(1 + x) + b(1 - 2x) = (a + b) + (a - 2b)x$$

Hence, $a + b = 1$ and $a - 2b = -1$. Solving these two equations in two unknowns yields $a = \frac{1}{3}$ and $b = \frac{2}{3}$, as before. ∎

EXAMPLE 9.3.2 Find constants a, b, and c such that

$$\frac{1-11x}{(1+x)^2(1-2x)} = \frac{a}{1-2x} + \frac{b}{1+x} + \frac{c}{(1+x)^2}$$

As in Example 9.3.1, write the sum on the right as one fraction with a common denominator:

$$\frac{a}{1-2x} + \frac{b}{1+x} + \frac{c}{(1+x)^2} = \frac{a(1+x)^2 + b(1-2x)(1+x) + c(1-2x)}{(1+x)^2(1-2x)}$$

Equating the original numerator with the numerator we just found yields

$$1 - 11x = a(1+x)^2 + b(1-2x)(1+x) + c(1-2x)$$

Now we make three useful substitutions in this equation. Substituting $x = -1$ yields $12 = c(1 + 2)$. Hence, $c = 4$. Substituting $x = \frac{1}{2}$ yields $-\frac{9}{2} = a(\frac{3}{2})^2$. Hence, $a = -2$. Substituting $x = 0$ yields $1 = a + b + c = -2 + b + 4$. Hence, $b = -1$. Therefore,

$$\frac{1-11x}{(1+x)^2(1-2x)} = \frac{-2}{1-2x} + \frac{-1}{1+x} + \frac{4}{(1+x)^2}$$

Using the alternate method of finding the constants a, b, and c by equating coefficients of like powers of x, we obtain:

$$1 - 11x = a(1+x)^2 + b(1-2x)(1+x) + c(1-2x)$$
$$= (a + b + c) + (2a - b - 2c)x + (a - 2b)x^2$$

Hence,

$$a + b + c = 1 \quad \text{and} \quad 2a - b - 2c = -11 \quad \text{and} \quad a - 2b = 0$$

Solving these three equations in three unknowns yields $a = -2$, $b = -1$, and $c = 4$, as before. ∎

We now demonstrate the **generating function method** for solving a homogeneous linear recurrence relation of degree 2 with constant coefficients.

EXAMPLE 9.3.3 Let $a_0 = 1$, $a_1 = 4$, and $a_n - 5a_{n-1} + 6a_{n-2} = 0$. Suppose $A(x) = a_0 + a_1 x + a_2 x^2 + \cdots$ is the generating function of the sequence for which we are trying to find an explicit formula. In order to find an explicit formula for a_n, we will find an infinite series that equals $A(x)$ and then equate coefficients to solve for a_n. By the given recurrence relation, $a_2 - 5a_1 + 6a_0 = 0$, $a_3 - 5a_2 + 6a_1 = 0, \ldots$. In step 1 below we use these facts to get a finite sum, due to cancellation, after adding $A(x)$ to $-5x \cdot A(x)$ and to $6x^2 \cdot A(x)$.

Step 1. Use the recurrence relation to find $A(x)$ as a quotient of polynomials.

$$
\begin{aligned}
A(x) &= a_0 + a_1 x + a_2 x^2 + \cdots \\
-5x \cdot A(x) &= \quad\quad - 5a_0 x - 5a_1 x^2 - \cdots \\
6x^2 \cdot A(x) &= \quad\quad\quad\quad\quad + 6a_0 x^2 + \cdots \\
\hline
(1 - 5x + 6x^2)A(x) &= a_0 + (a_1 - 5a_0)x + (a_2 - 5a_1 + 6a_0)x^2 + \cdots \\
&= a_0 + (a_1 - 5a_0)x + 0 + \cdots \\
&= a_0 + (a_1 - 5a_0)x
\end{aligned}
$$

So,

$$A(x) = \frac{a_0 + (a_1 - 5a_0)x}{1 - 5x + 6x^2}$$

Step 2. Use the initial conditions to simplify the quotient:

$$A(x) = \frac{1 + (4 - 5 \cdot 1)x}{1 - 5x + 6x^2} = \frac{1 - x}{1 - 5x + 6x^2}$$

Step 3. Use the method of partial fractions to write $A(x)$ as a sum of linear quotients:

$$A(x) = \frac{1 - x}{(1 - 3x)(1 - 2x)} = \frac{a}{1 - 2x} + \frac{b}{1 - 3x}$$

Putting the right-hand side over a common denominator and equating numerators yields $1 - x = a(1 - 3x) + b(1 - 2x)$. Substituting $x = \frac{1}{2}$ and $x = \frac{1}{3}$, we get $a = -1$ and $b = 2$. So,

$$A(x) = \frac{2}{1 - 3x} - \frac{1}{1 - 2x}$$

Step 4. Write each linear quotient as a geometric series and combine terms:

$$
\begin{aligned}
A(x) &= 2(1 + 3x + 3^2 x^2 + \cdots) - (1 + 2x + 2^2 x^2 + \cdots) \\
&= (2 \cdot 1 - 1) + (2 \cdot 3 - 2)x + (2 \cdot 3^2 - 2^2)x^2 + \cdots \\
&\quad\quad\quad\quad\quad\quad\quad\quad\quad\quad\quad\quad\quad + (2 \cdot 3^n - 2^n)x^n + \cdots
\end{aligned}
$$

Step 5. Equate coefficients to find $a_n = 2 \cdot 3^n - 2^n$. ∎

PRACTICE PROBLEM 1 Do the algebra necessary to find a and b in step 3 of Example 9.3.3.

The method of generating functions illustrated in Example 9.3.3 can be used to solve many homogeneous linear recurrence relations with constant coefficients. Before going further with recurrence relations, we will investigate generating functions.

EXAMPLE 9.3.4 Find the generating function for the binomial coefficients.

Solution From the binomial theorem, it follows that

$$(1 + x)^n = C(n, 0) + C(n, 1)x + C(n, 2)x^2 + \cdots + C(n, n)x^n$$

Hence, $(1 + x)^n$ is the generating function for the sequence

$$(C(n, 0), C(n, 1), \ldots, C(n, n), 0, 0, \ldots)$$ ■

Here is a special case of the geometric series:

$$\frac{1}{1-x} = 1 + x + x^2 + x^3 + \cdots = \sum_{j=0}^{\infty} x^j$$

Hence, $1/(1 - x)$ is the generating function for the sequence $(1, 1, 1, 1, \ldots)$.

It will be useful to know something about the generating function $1/(1 - x)^2$. By differentiating both sides of the equation

$$\frac{1}{1-x} = 1 + x + x^2 + x^3 + \cdots$$

we obtain

$$\frac{1}{(1-x)^2} = 1 + 2x + 3x^2 + 4x^3 + \cdots$$

Hence, $1/(1 - x)^2$ is the generating function for the sequence $(1, 2, 3, 4, \ldots)$. Note that the infinite series $1 + x + x^2 + x^3 + \cdots$ was differentiated term by term.

Following is a list of the generating functions discussed so far. We write each generating function in its most useful form.

GENERATING FUNCTION	SEQUENCE
$\dfrac{1}{1 - ax}$	$(1, a, a^2, a^3, \ldots, a^n, \ldots)$
$\dfrac{1}{(1 - ax)^2}$	$(1, 2a, 3a^2, \ldots, (n + 1)a^n, \ldots)$
$(1 + x)^n$	$(C(n, 0), C(n, 1), \ldots, C(n, n), 0, 0, \ldots)$

THEOREM 9.3.5 Suppose $S(x)$ is the generating function for the sequence (s_n) and $T(x)$ is the generating function for the sequence (t_n). Then, $a \cdot S(x) + b \cdot T(x)$ is the generating function for the sequence $(as_n + bt_n)$.

Proof We have

$$S(x) = s_0 + s_1 x + s_2 x^2 + \cdots + s_n x^n + \cdots \quad \text{and}$$
$$T(x) = t_0 + t_1 x + t_2 x^2 + \cdots + t_n x^n + \cdots$$

So,

$$a \cdot S(x) + b \cdot T(x) = (as_0 + bt_0) + (as_1 + bt_1)x + \cdots + (as_n + bt_n)x^n + \cdots \quad \blacksquare$$

EXAMPLE 9.3.6 Use the method of generating functions to solve:

$$a_n - 4a_{n-1} + 4a_{n-2} = 0; \quad a_0 = 1, \quad a_1 = 1$$

Solution Let $A(x) = a_0 + a_1 x + a_2 x^2 + \cdots$ be the generating function of the sequence for which we are trying to find an explicit formula. We will proceed as in Example 9.3.3.

Step 1. Use the recurrence relation to find $A(x)$ as a quotient of polynomials:

$$\begin{aligned}
A(x) &= a_0 + a_1 x + a_2 x^2 + \cdots \\
-4x \cdot A(x) &= -4a_0 x - 4a_1 x^2 - \cdots \\
4x^2 \cdot A(x) &= 4a_0 x^2 + \cdots \\
\hline
(1 - 4x + 4x^2) A(x) &= a_0 + (a_1 - 4a_0)x
\end{aligned}$$

So,

$$A(x) = \frac{a_0 + (a_1 - 4a_0)x}{1 - 4x + 4x^2}$$

Step 2. Use the initial conditions to simplify the quotient:

$$A(x) = \frac{1 + (1 - 4 \cdot 1)x}{1 - 4x + 4x^2} = \frac{1 - 3x}{1 - 4x + 4x^2}$$

Step 3. Use the method of partial fractions to write $A(x)$ as a sum of linear quotients:

$$A(x) = \frac{1 - 3x}{(1 - 2x)^2} = \frac{a}{1 - 2x} + \frac{b}{(1 - 2x)^2}$$

Putting the right-hand side over a common denominator and equating numerators yields $1 - 3x = a(1 - 2x) + b$. Substituting $x = \frac{1}{2}$ and $x = 0$, we get $b = -\frac{1}{2}$ and $a = \frac{3}{2}$. So,

$$A(x) = \frac{\frac{3}{2}}{1 - 2x} - \frac{\frac{1}{2}}{(1 - 2x)^2}$$

Step 4. We conclude from Theorem 9.3.5 that $A(x)$ is the generating function for the sequence $((\frac{3}{2})2^n - (\frac{1}{2})(n + 1)2^n)$.

Step 5. Hence, $a_n = (\frac{3}{2})2^n - (\frac{1}{2})(n + 1)2^n$. An easy simplification yields

$$a_n = 3 \cdot 2^{n-1} - (n + 1)2^{n-1} = 2^n - n \cdot 2^{n-1} \quad \blacksquare$$

Note that we made no assumptions about the solution we were seeking in Example 9.3.6. In other words, we *derived* the solution. Similarly, we derived the solution in Example 9.3.3.

Let us use the generating function method to derive the solution for an arbitrary second-degree recurrence relation with a characteristic equation that has distinct real roots.

THEOREM 9.3.7 The solution for $a_n - (r+s)a_{n-1} + rsa_{n-2} = 0$ is $a_n = ar^n + bs^n$, where a and b are arbitrary constants.

Proof Step 1. Use the recurrence relation to find $A(x)$ as a quotient of polynomials:

$$A(x) = a_0 + a_1 x + a_2 x^2 + \cdots$$
$$-(r+s)x \cdot A(x) = \quad - (r+s)a_0 x - (r+s)a_1 x^2 - \cdots$$
$$rsx^2 \cdot A(x) = \quad\quad\quad\quad\quad\quad\quad\quad\quad rsa_0 x^2 + \cdots$$
$$\overline{[1 - (r+s)x + rsx^2]A(x) = a_0 + [a_1 - (r+s)a_0]x}$$

So,

$$A(x) = \frac{a_0 + [a_1 - (r+s)a_0]x}{1 - (r+s)x + rsx^2}$$

Step 2. Let $k = a_1 - (r+s)a_0$, and factor the denominator to obtain

$$A(x) = \frac{a_0 + kx}{(1 - rx)(1 - sx)}$$

Step 3. Use the method of partial fractions to write $A(x)$ as a sum of linear quotients:

$$A(x) = \frac{a}{1 - rx} + \frac{b}{1 - sx}$$

Step 4. By Theorem 9.3.5, $A(x)$ is the generating function for the sequence $(ar^n + bs^n)$. ∎

Similarly, the generating function method can be used to derive the solution for an arbitrary second-degree recurrence relation with a characteristic equation that has identical real roots.

THEOREM 9.3.8 The solution for $a_n - 2ra_{n-1} + r^2 a_{n-2} = 0$ is $a_n = ar^n + bnr^n$, where a and b are arbitrary constants.

The proof of Theorem 9.3.8 is left as an exercise. (See Exercise 15.)

Although some of the problems you will be solving can be done more easily by the characteristic equation method, it is important that you gain familiarity with both methods.

9.3 RECURRENCE RELATIONS AND GENERATING FUNCTIONS

More on Generating Functions

There are many properties and applications of generating functions; however, it is beyond the scope of this book to discuss them all. Please see the References for other examples of the use of generating functions. We will show a couple of examples illustrating the power of generating functions. Generating functions are particularly useful for deriving identities.

Let us reconsider the binomial coefficients. Recall that

$$(1 + x)^n = C(n, 0) + C(n, 1)x + C(n, 2)x^2 + \cdots + C(n, n)x^n$$

Substituting $x = 1$ yields the identity

$$2^n = C(n, 0) + C(n, 1) + C(n, 2) + \cdots + C(n, n)$$

Differentiating both sides of the equation $(1 + x)^n = C(n, 0) + C(n, 1)x + C(n, 2)x^2 + \cdots + C(n, n)x^n$, we obtain

$$n(1 + x)^{n-1} = C(n, 1) + 2 \cdot C(n, 2)x + \cdots + n \cdot C(n, n)x^{n-1}$$

Substituting $x = 1$ yields the identity

$$n \cdot 2^{n-1} = C(n, 1) + 2 \cdot C(n, 2) + \cdots + n \cdot C(n, n)$$

Consider what happens when we multiply two generating functions. For example, $(1 + x)^2(1 + x)^3 = (1 + 2x + x^2)(1 + 3x + 3x^2 + x^3)$. If we just place the coefficients of matching powers in the same column, we can write the multiplication process in much the same way as we multiply decimal numbers. The powers of x act as placeholders just as we use the powers of 10 in decimal multiplication. There is no need to carry numbers to the next column when multiplying generating functions.

```
      1  3  3  1                 (1 + 2x + x²)(1 + 3x + 3x² + x³) =
         1  2  1                      1 + 3x +  3x² +   x³
      1  3  3  1              +           2x +  6x² +  6x³ + 2x⁴
      2  6  6  2              +                  x² +  3x³ + 3x⁴ + x⁵
         1  3  3  1              = 1 + 5x + 10x² + 10x³ + 5x⁴ + x⁵
      1  5  10 10  5  1
```

In the multiplication, the fourth column from the left $(1 + 6 + 3 = 10)$ is formally written

$$C(2, 0) \cdot C(3, 3) + C(2, 1) \cdot C(3, 2) + C(2, 2) \cdot C(3, 1) = C(5, 3)$$

This is a special case of the next theorem concerning generating functions.

THEOREM 9.3.9 Suppose $S(x)$ is the generating function for the sequence (s_n) and $T(x)$ is the generating function for the sequence (t_n). Then $S(x) \cdot T(x)$ is the generating function for the sequence (u_n), where $u_n = s_0 t_n + s_1 t_{n-1} + \cdots + s_n t_0$.

Proof This theorem is proved by multiplying the series $S(x)$ by the series $T(x)$ and noting that the coefficient of the x^n term in the product is

$$s_0 t_n + s_1 t_{n-1} + \cdots + s_n t_0 \qquad \blacksquare$$

EXAMPLE 9.3.10 Derive the identity

$$C(n, 0)^2 + C(n, 1)^2 + \cdots + C(n, n)^2 = C(2n, n)$$

Solution We know $(1 + x)^n (1 + x)^n = (1 + x)^{2n}$. Therefore,

$$C(n, 0) \cdot C(n, n) + C(n, 1) \cdot C(n, n-1) + \cdots + C(n, n) \cdot C(n, 0) = C(2n, n)$$

by Theorem 9.3.9. Since $C(n, k) = C(n, n - k)$, we have

$$\sum_{j=0}^{n} C(n, j)^2 = C(2n, n) \qquad \blacksquare$$

EXAMPLE 9.3.11 Let a_n be the number of octal strings of length n with the property that the digits 5 and 7 always precede the other six digits. For example, 5753046 is legal but 7345310 is not. Find a generating function for a_n.

Solution Let us look at strings of length 4 to see the pattern. Denote the number of characters in the string that are either 5 or 7 by j:

$j = 0$: 6^4 strings of length 4
$j = 1$: $2 \cdot 6^3$ strings of length 4
$j = 2$: $2^2 \cdot 6^2$ strings of length 4
$j = 3$: $2^3 \cdot 6$ strings of length 4
$j = 4$: 2^4 strings of length 4

Hence, there are a total of $2^0 \cdot 6^4 + 2 \cdot 6^3 + 2^2 \cdot 6^2 + 2^3 \cdot 6 + 2^4 \cdot 6^0$ strings of length 4. Since this pattern holds for arbitrary n, the generating function is

$$\frac{1}{1 - 2x} \cdot \frac{1}{1 - 6x}$$

by Theorem 9.3.9. $\qquad \blacksquare$

Remarks on Generating Functions

If you have studied infinite series, you may be wondering why we are not concerned about the convergence of each generating function. It is true that all the generating functions we have used are convergent for real numbers x such that $|x| < 1$. But we need not consider the variable x in a generating function as a number. The variable x in each generating function has acted only as a placeholder.

We are interested in generating functions as formal sums to be manipulated. So far we have multiplied generating functions by constants, added them together, and differentiated them. We have also multiplied a generating function by the variable x and multiplied two generating functions together. Since a generating function is also an infinite series, these manipulations of generating functions are also manipulations of infinite series. Each manipulation of generating functions we use in this section can be justified as valid in the theory of infinite series.

Our interest is in the application of generating functions to the solution of recurrence relations and the derivation of identities. Here is another example in that spirit.

Multiplying both sides of $1/(1-x)^2 = 1 + 2x + 3x^2 + 4x^3 + \cdots$ by x yields a new identity $x/(1-x)^2 = x + 2x^2 + 3x^3 + 4x^4 + \cdots$. In other words, $x/(1-x)^2$ is the generating function for the sequence $(0, 1, 2, 3, \ldots)$.

PRACTICE PROBLEM 2   Differentiate both sides of $x/(1-x)^2 = x + 2x^2 + 3x^3 + 4x^4 + \cdots$ and thereby derive a generating function for the sequence $(1^2, 2^2, 3^2, \ldots)$.

EXERCISE SET 9.3

In Exercises 1–4 find the partial fraction decomposition of the given quotient of polynomials.

1. $\dfrac{2 - 3x}{(1 + 5x)(1 - x)}$

2. $\dfrac{1 - x}{(1 + x)(1 - 2x)^2}$

3. $\dfrac{4x + 1}{x^2 - x - 2}$

4. $\dfrac{1 - 7x}{6x^2 - 5x + 1}$

Solve each of the recurrence relations in Exercises 5–14 by the generating function method.

5. $a_n - a_{n-1} - 2a_{n-2} = 0$ for $n = 2, 3, 4, \ldots$; $a_0 = 4$ and $a_1 = 5$
6. $a_n - 5a_{n-1} - 14a_{n-2} = 0$ for $n = 2, 3, 4, \ldots$; $a_0 = 1$ and $a_1 = 7$
7. $a_n - 4a_{n-1} + a_{n-2} = 0$ for $n = 2, 3, 4, \ldots$; $a_0 = 0$ and $a_1 = 2\sqrt{3}$
8. $a_n - 4a_{n-1} + 4a_{n-2} = 0$ for $n = 2, 3, 4, \ldots$; $a_0 = 3$ and $a_1 = 7$
9. $a_n + 10a_{n-1} + 25a_{n-2} = 0$ for $n = 2, 3, 4, \ldots$; $a_0 = 1$ and $a_1 = 3$
10. $a_n - 6a_{n-1} + 9a_{n-2} = 0$ for $n = 2, 3, 4, \ldots$; $a_0 = 2$ and $a_1 = 21$
11. $a_n - 8a_{n-1} + 16a_{n-2} = 0$ for $n = 2, 3, 4, \ldots$; $a_0 = 3$ and $a_1 = 32$
12. $a_n + a_{n-1} - 6a_{n-2} = 0$ for $n = 2, 3, 4, \ldots$; $a_0 = 5$ and $a_1 = 0$
13. $a_n - a_{n-1} - 6a_{n-2} = 0$ for $n = 2, 3, 4, \ldots$; $a_0 = 4$ and $a_1 = -3$
14. $a_n - 14a_{n-1} + 49a_{n-2} = 0$ for $n = 2, 3, 4, \ldots$; $a_0 = 2$ and $a_1 = 21$
15. Prove Theorem 9.3.8.
16. Find a closed-form expression for a_n of Example 9.3.11.
17. Let a_n be the number of octal strings of length n with the property that the digits 0, 1, and 2 always precede the other five digits. For example, 1103746 is legal but 2145310 is not.
    a. Find a generating function for a_n.   b. Find a closed-form expression for a_n.

18. Let a_n be the number of positive decimal integers of length n with the property that the digits 0, 1, and 2 always precede the other seven digits. For example, 1103946 is legal but 2145310 is not.
    a. Find a generating function for a_n. b. Find a closed-form expression for a_n.

In Exercises 19–21 find the sequences for which each of the following functions is the generating function.

19. $\dfrac{1}{1+x}$ 20. $\dfrac{1}{(1-x)^3}$ 21. $\dfrac{x}{(1-x)^3}$

22. Determine the generating function for the sequence $(1, -1, 1, -1, \ldots)$.
23. Determine the generating function for the sequence $(1, -2, 3, -4, \ldots)$.
24. Determine the generating function for the sequence

    $$(C(n, 0), -C(n, 1), C(n, 2), -C(n, 3), \ldots, (-1)^n \cdot C(n, n), 0, 0, \ldots)$$

25. Prove that the generating function for the sequence $(C(2, 2), C(3, 2), C(4, 2), C(5, 2), \ldots)$ is $1/(1-x)^3$.

Find a closed-form expression for each of the sums in Exercises 26–28.

26. $\displaystyle\sum_{j=0}^{n} 2^j \cdot C(n, j), \quad n > 1$ 27. $\displaystyle\sum_{j=0}^{n} j \cdot 2^j \cdot C(n, j), \quad n > 1$ 28. $\displaystyle\sum_{j=0}^{n} j^2 \cdot C(n, j)$

9.4
APPLICATION: DIVIDE AND CONQUER ALGORITHMS

Many sorting and searching algorithms are classified as divide and conquer algorithms because they accomplish the task by dividing a given problem into smaller but similar problems. For example, to sort a list of names alphabetically, we might first divide the list into sublists corresponding to the first letter of each name. Some of the most efficient algorithms known, including mergesort and quicksort, are divide and conquer algorithms. Mergesort is described in Exercise 8. Both mergesort and quicksort are discussed in detail in *Data Structures and Program Design* by Robert Kruse.

Special recurrence relations, called **divide and conquer relations**, are used to analyze the time complexity of divide and conquer algorithms. First we will analyze a specific divide and conquer algorithm using the techniques of this chapter to solve the corresponding recurrence relation. In Exercise Set 9.4 we give several divide and conquer relations and their solutions.

ALGORITHM 9.4.1   **ALGORITHM HIGHLOW**   This algorithm provides a method for searching a list of numbers and finding the largest number and the smallest number. Suppose we are given a list of n numbers. Algorithm HighLow is a recursive algorithm that divides each list of numbers successively in half and calls itself to act on the smaller lists.

Let A be a list $A(1), A(2), \ldots, A(n)$ of n numbers, with $n \geq 1$. This recursive algorithm finds the largest number H and the smallest number L in A. Initially $H = 0 = L$.

HighLow(A, H, L)
  If n = 1, then H := the number in A and L := H.   (*Basis*)
  Let A_1 be the array $A(1), \ldots, A(\lceil n/2 \rceil)$ and   (*Divide array in two halves*)
    A_2 be the array $A(\lceil n/2 \rceil + 1), \ldots, A(n)$.
  Call HighLow(A_1, H_1, L_1) and   (*Solve smaller problems*)
    call HighLow(A_2, H_2, L_2).
  H := max(H_1, H_2) and L := min(L_1, L_2).   (*Find largest and smallest element*)

  Return H and return L.   (*Return largest and smallest element*)

Let a_n be the number of comparisons done to find the largest and smallest element. We will use a_n as a measure of time complexity. How can we find a recurrence relation for a_n?

For simplicity, we assume n is a power of 2 and $n \geq 2$; hence, $n = 2^r$ for some $r = 1, 2, \ldots$. Note that, if A has $n = 2^r$ numbers, then both A_1 and A_2 have $n/2 = 2^{r-1}$ numbers. Since one pair of numbers requires exactly one comparison, $a_2 = 1$ is the initial condition.

If we have already determined the largest and smallest numbers in each half of the sequence A of n numbers, then it will require exactly two more comparisons to determine the largest and smallest numbers in A. Hence, the recurrence relation is

$$a_n = 2a_{n/2} + 2 \quad \text{for } n = 4, 8, \ldots$$

In order to solve this recurrence relation using the techniques of this chapter, we make a change of variable. Since $n = 2^r$, we let $b_r = a_{2^r} = a_n$. So, the recurrence relation we will solve is

$$b_r = 2b_{r-1} + 2 \quad \text{for } r = 2, 3, \ldots$$

with initial condition $b_1 = 1$.

Let $g(x) = b_0 + b_1 x + b_2 x^2 + \cdots$ be the generating function for (b_r). The recurrence relation yields $b_r - 2b_{r-1} = 2$ for $r = 2, 3, \ldots$. The algebra used to manipulate the generating function will be easier if we set $b_1 - 2b_0 = 2$. Since $b_1 = 1$, we set $b_0 = -\frac{1}{2}$. The recurrence relation becomes $b_r - 2b_{r-1} = 2$ for $r = 1, 2, 3, \ldots$, with initial condition $b_0 = -\frac{1}{2}$.

This is not a homogeneous recurrence relation because the right side is not equal to zero. However, we will proceed as we did when solving homogeneous recurrence relations. First we calculate $g(x) - 2x \cdot g(x)$:

$$
\begin{array}{rl}
g(x) = & b_0 + b_1 x + b_2 x^2 + \cdots + b_n x^n + \cdots \\
-2x \cdot g(x) = & \quad - 2b_0 x - 2b_1 x^2 - \cdots - 2b_{n-1} x^n - \cdots \\
\hline
(1-2x)g(x) = & b_0 + (b_1 - 2b_0)x + (b_2 - 2b_1)x^2 + \cdots + (b_n - 2b_{n-1})x^n + \cdots \\
= & -\tfrac{1}{2} + \quad 2x \quad + \quad 2x^2 \quad + \cdots + \quad 2x^n \quad + \cdots
\end{array}
$$

Hence,

$$(1 - 2x)g(x) = -\frac{1}{2} + \frac{2x}{1 - x}$$

Combining terms and dividing both sides by $1 - 2x$ we obtain

$$g(x) = \frac{5x - 1}{2(1 - x)(1 - 2x)} = \frac{a}{1 - x} + \frac{b}{1 - 2x}$$

where a and b are constants.

By the method of partial fractions, we find $a = -2$ and $b = \frac{3}{2}$. Therefore,

$$g(x) = \frac{-2}{1 - x} + \frac{\frac{3}{2}}{1 - 2x}$$

Hence, $b_r = (\frac{3}{2})2^r - 2$. Since $n = 2^r$ and $b_r = a_n$, we have

$$a_n = \tfrac{3}{2}n - 2 \qquad \blacksquare$$

The recurrence relation for the algorithm HighLow is an example of a class of recurrence relations called divide and conquer relations. They frequently arise in the study of recursive algorithms and they are usually of the form

$$a_n = ka_{n/d} + f(n)$$

where k and d are constants and f is some function of n.

For example, the maximum number of comparisons needed to sort a list of n elements using the recursive algorithm mergesort is given by the recurrence relation: $a_n = 2a_{n/2} + (n - 1)$ for $n = 2, 3, \ldots$, with $a_1 = 0$. (See Exercises 7 and 8, Exercise Set 9.4.)

In Exercise Set 9.4, several forms of divide and conquer relations are given along with their solutions.

EXERCISE SET 9.4

For all of the following exercises, d is assumed to be a positive integer larger than 1 and $n = d^r$ for $r = 0, 1, 2, \ldots$. The letters b, c, and k designate arbitrary constants.

In Exercises 1 and 2 use the method of generating functions to find the given solution. [Hint: Make a change of variable and let $b_r = a_{d^r} = a_n$. You may use the following fact about logarithms: $n^{\log_d k} = k^{\log_d n}$.]

RECURRENCE RELATION	SOLUTION
1. $a_n = a_{n/d} + b;\quad a_1 = b$	$b(\log_d n + 1)$
2. $a_n = ka_{n/d} + b;\quad k \neq 1, a_1 = b$	$\dfrac{b(kn^{\log_d k} - 1)}{k - 1}$

In Exercises 3–6 use mathematical induction to verify the given solution. It is helpful, though not necessary, to make a change of variable and work with $b_r = a_{d^r} = a_n$.

RECURRENCE RELATION	SOLUTION
3. $a_n = da_{n/d} + b$	$cn + \dfrac{b}{1 - d}$
4. $a_n = ka_{n/d} + bn;\quad k < d$	$\dfrac{bdn}{d - k}$
5. $a_n = da_{n/d} + bn$	$bn(\log_d n)$

	RECURRENCE RELATION	SOLUTION
6.	$a_n = ka_{n/d} + bn; \quad k > d$	$c(n^{\log_d k}) + \left(\dfrac{bd}{d-k}\right)d^r$

7. Assume n is a power of 2. Solve the recurrence relation $a_n = 2a_{n/2} + (n-1)$ with $a_1 = 0$, for the mergesort by filling in the details for the following steps.
   a. Write $n = 2^r$ and change variables to obtain $b_r = a_n$ with $b_0 = 0$.
   b. Note that $b_r - 2b_{r-1} = 2^r - 1$.
   c. Verify that $2^j b_{r-j} - 2^{j+1} b_{r-j-1} = 2^r - 2^j$ for $j = 0, 1, \ldots, r-1$.
   d. Write the sum
   $$(b_r - 2b_{r-1}) + (2b_{r-1} - 2^2 b_{r-2}) + (2^2 b_{r-2} - 2^3 b_{r-3}) + \cdots + (2^{r-1}b_1 - 2^r b_0)$$
   in two ways to obtain $b_r = r \cdot 2^r - (1 + 2 + 2^2 + \cdots + 2^{r-1})$.
   e. Therefore, $b_r = r \cdot 2^r - (2^r - 1)$ and $a_n = n \log_2 n - (n-1)$.

8. Mergesort is a recursive sorting algorithm that divides an unsorted list into two halves, sorts each half separately, and then merges the sorted lists together to obtain a single sorted list. For simplicity, we assume that the number of items in the original list is some power of 2. Let a_n be the maximum number of comparisons needed to sort a list with n elements. Prove $a_n = 2a_{n/2} + (n-1)$ for $n = 2, 3, \ldots$, with $a_1 = 0$. [*Hint:* The key step in this proof is to show that the maximum number of comparisons needed to merge two sorted lists each with $n/2$ elements into a sorted list with n elements is $n-1$.]

KEY TERMS

Recurrence relation
Initial conditions
Closed-form expression
Solve a recurrence relation
Solution by iteration
Geometric progression
Arithmetic progression
Fibonacci sequence
Double recurrence
Ackermann's function
Linear recurrence relation with constant coefficients
Recurrence relation of order k

Homogeneous recurrence relation
Characteristic equation method
Characteristic equation
Golden section
Golden ratio
Recursive definition of polynomials
Horner's method
Generating function
Geometric series
Partial fraction decomposition
Method of judicious substitution
Generating function method
Divide and conquer relation

REVIEW EXERCISES

In Exercises 1 and 2 find the partial fraction decomposition.

1. $\dfrac{7 - 3x}{1 - x^2}$

2. $\dfrac{-9}{x^2 + x - 2}$

Solve each of the recurrence relations in Exercises 3–11 by both the characteristic equation method and the generating function method.

3. $a_n = 7a_{n-1} - 10a_{n-2}$ for $n \geq 2$; $a_0 = 3$ and $a_1 = 15$
4. $a_n = 4a_{n-1} + 21a_{n-2}$ for $n \geq 2$; $a_0 = 0$ and $a_1 = 1$

5. $a_n = -6a_{n-1} - 9a_{n-2}$ for $n \geq 2$; $a_0 = 1$ and $a_1 = -2$

6. $a_n = 10a_{n-1} + 29a_{n-2}$ for $n \geq 2$; $a_0 = 0$ and $a_1 = 1$

7. $a_n = 10a_{n-1} + 29a_{n-2}$ for $n \geq 3$; $a_1 = 10$ and $a_2 = 100$

8. $a_n = a_{n-1} + 4a_{n-2} - 4a_{n-2}$ for $n \geq 3$; $a_0 = 7$, $a_1 = 5$, and $a_2 = 13$

9. $a_n = 3a_{n-1} + 9a_{n-2} + 5a_{n-3}$ for $n \geq 3$; $a_0 = 4$, $a_1 = 0$, and $a_2 = 32$

10. $a_n = 2a_{n-1} + 12a_{n-2} + 14a_{n-3} + 5a_{n-4}$ for $n \geq 4$;
    $a_0 = 4$, $a_1 = 0$, $a_2 = 34$, and $a_3 = 110$

11. $a_n = 6a_{n-1} - 13a_{n-2} + 12a_{n-3} - 4a_{n-4}$ for $n \geq 4$;
    $a_0 = 6$, $a_1 = 12$, $a_2 = 31$, and $a_3 = 82$

12. A decaying radioactive substance loses 12% of its mass every hour. Let r_n be the amount present at the end of n hours. Assume $r_0 = b$.
    a. Find r_1 and r_2.   **b.** Find a recurrence relation for r_n.
    c. Solve the recurrence relation for r_n using iteration.

13. Suppose there are n lines in the Euclidean plane with no two lines parallel and no three lines intersecting in a point. Let t_n, for $n = 1, 2, 3, \ldots$, be the number of regions into which the plane is divided by the n lines.
    a. Find t_1, t_2, t_3, t_4, and t_5.   **b.** Find a recurrence relation for t_n.
    c. Solve the recurrence relation for t_n using iteration.

14. Consider the set $A_n = \{1, 2, 3, \ldots, n\}$. Let p_n, for $n = 1, 2, \ldots$, be the number of ways in which A_n can be partitioned into two nonempty disjoint subsets.
    a. Find p_1, p_2, and p_3.   **b.** Find a recurrence relation for p_n.
    c. Solve the recurrence relation for p_n.

15. Suppose there are n hundreds of thousands of dollars worth of houses to be sold. There is one type of house selling for $100,000 and six types of houses selling for $200,000. Let h_n be the number of permutations of the houses sold. We assume that types of houses are distinguishable, but houses within each type are not. For example, $h_3 = 13$ since either three $100,000 houses, or one of the six $200,000 houses and a $100,000 house, or a $100,000 house and one of the six $200,000 houses could be sold.
    a. Find h_1 and h_2.   **b.** Find a recurrence relation for h_n.
    c. Solve the recurrence relation for h_n.

16. This exercise is similar to Exercise 15. Suppose there are seven types of $200,000 houses, six types of $300,000 houses, and no $100,000 houses.
    a. Find h_1, h_2, and h_3.   **b.** Find a recurrence relation for h_n.
    c. Solve the recurrence relation for h_n.

17. Let F_n be the nth Fibonacci number. Prove $F_0^2 + F_1^2 + F_2^2 + \cdots + F_n^2 = F_n F_{n+1}$ for $n = 0, 1, 2, \ldots$.

REFERENCES

Beckenbach, E. F. (Ed.). *Applied Combinatorial Mathematics*. New York: Wiley, 1964.

Bogart, K. P. *Introductory Combinatorics*. Marshfield, MA: Pitman, 1983.

Hamming, R. W. *Methods of Mathematics Applied to Calculus, Probability, and Statistics*. Englewood Cliffs, NJ: Prentice-Hall, 1985.

Kruse, R. *Data Structures and Program Design*, 2nd ed. Englewood Cliffs, NJ: Prentice-Hall, 1987.

Tucker, A. *Applied Combinatorics*, 2nd ed. New York: Wiley, 1984.

CHAPTER 10

AUTOMATA THEORY AND FORMAL LANGUAGES

Automata theory and formal languages are particularly important in the definition of computer programming languages and the study of compilers. Automata and formal languages are closely related. In this chapter we discuss a special kind of automaton called a finite state recognizer. The relationship between finite state recognizers and a type of formal language called a regular language will be studied.

In Section 10.4 we discuss another type of formal language, called a context-free language. Context-free languages are very useful for defining and analyzing programming languages. A programming language such as Pascal is also an example of a formal language.

Formal languages are most commonly used in computer programming and mathematical logic. Programming languages are similar to ordinary languages, such as English, but are specially constructed with precise rules that can be automated in a machine. This process requires automatic translation from the formal programming language to machine language.

Automatic translation of text from one language to another has been a major goal of computer scientists and linguists. This goal has motivated the study of formal languages.

APPLICATION

Compilers are computer programs written to perform two tasks. The first task, syntax checking or parsing, is to check that strings of symbols are legal strings in the language. The second task is to translate statements written in a formal language into another formal language such as assembly language or machine language.

Lexical analysis is an important part of parsing. We will see how finite state machines are used in the design of a compiler's lexical analyzer. Finite state machines are one of several kinds of automata that are useful to computer scientists. The application in Section 10.5 describes how another, more powerful type of automaton, called a push-down automaton, can be applied to the task of syntax checking.

10.1
FINITE STATE RECOGNIZERS AND REGULAR EXPRESSIONS

The most interesting formal languages admit infinitely many strings of symbols or characters. So providing a finite specification for an infinite set of objects is a problem of language theory. We will discuss the two most common methods of finitely specifying a given language. The **generative method** makes use of a grammar to specify a given language. Grammars are introduced in Section 10.2. The second method, called the **automata method**, uses finite state machines (or finite automata) to decide whether a given string belongs to a language. We will discuss this method first.

As we will see in Section 10.2, a grammar generates a language. Writing a grammatically correct sentence in English means generating an acceptable string in English. When we use our knowledge of English to determine whether a sentence is correctly written, we are acting as language recognizers. We either accept or reject the sentence as a valid English sentence. This method for specifying languages is the automata method. The recognizers for formal languages are called automata.

The Automata Method

Before giving formal definitions let us see how a simple recognizer might work. In Example 10.1.1 we will design an automaton to check the parity of bit strings.

It is important to detect errors that can occur when data are stored or transmitted. Data are usually stored or transmitted in groups of bits, called **words**. To facilitate error detection, an extra bit, called a **parity bit**, is frequently included in each word. For example, if each word consists of eight bits (seven data bits plus one parity bit), the parity bit might be chosen so that all words have an odd number of 1's. Each word is said to have **odd parity** in this case. When this is done, a single bit error that occurs later in a word is easily detected since the parity of that word will no longer be odd. Of course, we could just as well choose to give each word an even number of 1's (**even parity**). Here are several examples of words having odd parity:

DATA BITS (seven)	PARITY BIT	WORD
0010100	1	00101001
0100101	0	01001010
1111100	0	11111000

Two bit errors occurring in the same word will, of course, defeat a single bit parity scheme. If errors are expected frequently, then more sophisticated checking techniques must be devised. One such technique is called a **parity checker**.

EXAMPLE 10.1.1   Design a parity checker.

10.1 FINITE STATE RECOGNIZERS AND REGULAR EXPRESSIONS

Solution The parity checker will read a string of 0's and 1's and determine whether the string contains an odd number of 1's. Let k be the number of 1's in the given string.

The parity checker we will design is an example of a finite state recognizer. As the definition following this example makes clear, finite state recognizers have only a fixed finite number of states. In the finite state recognizer of this example, states will be used to keep track of the number of 1's in the string as it is being read. The initial state is designated s_0. In the initial state s_0, the number of 1's in the string being processed is zero. Hence, we want s_0 to symbolize the total number of 1's as even. We need only two states: s_0 for even k, and s_1 for odd k.

A finite state recognizer can be easily represented by a digraph called a **state graph**. The state graph for the parity checker is pictured in Figure 10.1.1. The vertices of this state graph (that is, the labeled circles) are the states of the finite state recognizer; the edges (arcs) specify the next-state transition for a given input.

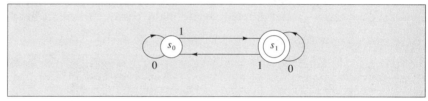

FIGURE 10.1.1

The double boundary around the vertex labeled s_1 designates s_1 as a recognizing state. As we will see, that means this finite state recognizer accepts strings with an odd number of 1's. ∎

DEFINITION

A **finite state recognizer** (FSR) is denoted by $M = (S, V, g, F, s_0)$, where

S is a finite set of **states**,
V is a finite set of **input symbols**,
$g: S \times V \to S$ is the **next-state function**,
$F \subseteq S$ is the set of **recognizing states**, and
$s_0 \in S$ is a designated state called the **initial state**.

A finite state recognizer is a special kind of finite state machine. Finite state machines are the most fundamental and simple of all automata models. They are frequently used in the design of compilers for computer languages such as Pascal. Finite state machines model computers that have a fixed and finite amount of memory, accept strings of symbols from a set of symbols called an **alphabet**, and produce output signals.

The finite state machines we will consider all have a very small number of states. Since states act as memory, the fewer machine states, the smaller the machine memory. Modern computers can have millions of bytes of main memory and billions of bytes of secondary storage. Even though a finite state machine must have a finite number of states, that number can be huge.

Let V be a set of characters. A **string** over V is a finite sequence of characters from V. Strings will be written with characters adjacent to one another. Finite state recognizers, which we study in this section, indicate whether a particular string is acceptable or not. Any string acceptable to a given machine may be considered syntactically correct.

Suppose we decide that strings of a's and b's with one a followed by any number (possibly none) of a's, followed by two b's, in turn followed by any number (possibly none) of b's are syntactically correct. We will write the set of such strings as $aa*bbb*$, where $a*$ means any number of a's and $b*$ means any numbers of b's. The notation $aa*bbb*$ is an example of *a regular expression*. Regular expressions are discussed at the end of this section.

The finite state recognizer pictured in Figure 10.1.2 is designed to check strings of a's and b's and accept those, and only those, strings that are in the set $aa*bbb*$. We formally describe this FSR in Example 10.1.2.

FIGURE 10.1.2

EXAMPLE 10.1.2 Design an FSR to accept those, and only those, strings in the set $aa*bbb*$.

Solution Each state is the label for a vertex. Recognizing states are customarily pictured as vertices with double boundaries. To make directed graphs easier to read, duplicate arcs with different labels will be drawn only once with all labels entered as a set of labels. So,

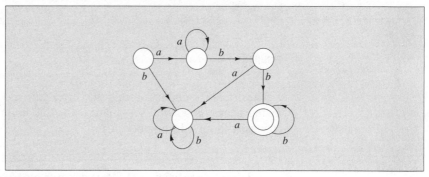

Figure 10.1.3 shows a state graph with vertices labeled properly.

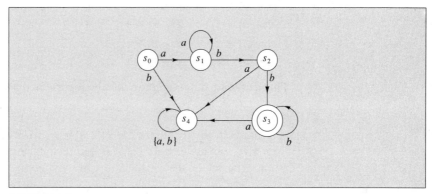

FIGURE 10.1.3

How does a finite state recognizer respond to a string of inputs? Assume the FSR always starts in state s_0. Strings will be processed or read left to right. For example, if this FSR gets the input string *abaa*, it will start in state s_0 and move consecutively to s_1, s_2, s_4, and s_4 as it reads *a, b, a, a* in order from left to right. Pictorially, we have

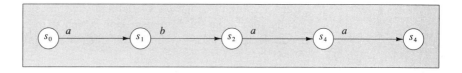

The next-state function is shown by placing an arc labeled x from state s to state $g(s, x)$:

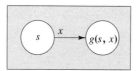

For example, we can see from the state graph in Figure 10.1.3 that $g(s_0, b) = s_4$, $g(s_0, a) = s_1$, and $g(s_2, b) = s_3$.

Finally, here is the FSR formally described. Refer to Figure 10.1.3. The FSR is $M = (S, V, g, F, s_0)$, with $S = \{s_0, s_1, s_2, s_3, s_4\}$, $V = \{a, b\}$, $F = \{s_3\}$,

TABLE 10.1.1

g	a	b
s_0	s_1	s_4
s_1	s_1	s_2
s_2	s_4	s_3
s_3	s_4	s_3
s_4	s_4	s_4

and $g: S \times V \to S$ defined by

$g(s_0, a) = s_1 \qquad g(s_0, b) = s_4$

$g(s_1, a) = s_1 \qquad g(s_1, b) = s_2$

$g(s_2, a) = s_4 \qquad g(s_2, b) = s_3$

$g(s_3, a) = s_4 \qquad g(s_3, b) = s_3$

$g(s_4, a) = s_4 \qquad g(s_4, b) = s_4$

The next-state function g can also be represented as a table. See Table 10.1.1.

A finite state recognizer $M = (S, V, g, F, s_0)$ **accepts** a string of symbols from V if, starting in state s_0, M is in a recognizing state after processing the string. The set of strings that an FSR accepts is the set **recognized** by the FSR. The FSR of Example 10.1.2 recognizes the set $aa*bbb*$.

Designing an FSR to Check Syntax

Checking for syntax errors is one task a compiler performs. One reason finite state recognizers are important to computer science is their utility in designing the compiler's syntax checker. When the syntax checker part is designed for a language, syntax rules and, by implication, acceptable strings are given. The designer will create a finite state recognizer, which is simulated in a program, that recognizes syntactically correct strings. The choice of recognizing states (the set F) depends on which strings are to be accepted. The examples in this section are in the spirit of design for syntax checking.

EXAMPLE 10.1.3   **PASCAL IDENTIFIERS**   Design an FSR to recognize the set of Pascal identifiers.

Solution   Recall that a Pascal identifier is a string of letters and digits that must begin with a letter. Let V be the alphabet (set of symbols) available on a given computer, $L = \{a, b, c, \ldots, z\}$, and $D = \{0, 1, 2, \ldots, 9\}$. The state graph shown in Figure 10.1.4 is an FSR to recognize the set of Pascal identifiers.

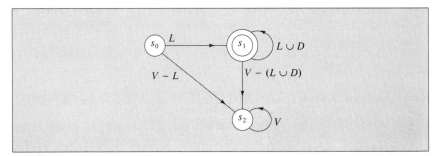

FIGURE 10.1.4

10.1 FINITE STATE RECOGNIZERS AND REGULAR EXPRESSIONS

An automated teller machine used by bank customers for deposits and withdrawals is linked to the bank's central computing system, which has stored each customer's personal identification number. When a card is inserted, the customer is asked to enter an identification number. The machine communicates this entry to the bank's central computer, which either accepts or rejects the number based on stored data.

EXAMPLE 10.1.4 **AUTOMATED TELLER MACHINE** Design an automated teller machine.

Solution   For simplicity, we will not distinguish between the automated teller machine and the bank's central computer. We will design a finite state recognizer that has alphabet $D = \{0, 1, 2, \ldots, 9\}$ and accepts only the string 73 (your identification number). The states will be s_0 (waiting for first input), s_1 (waiting for second input after correct first input), s_2 (correct first two inputs), and s_3 (incorrect string). Clearly, $F = \{s_2\}$. Figure 10.1.5 shows the state graph.

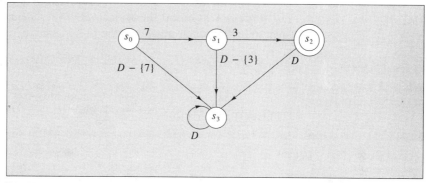

FIGURE 10.1.5

A state like s_3 in Example 10.1.4 is frequently called a **dump state** or **error state**.

Regular Expressions and Regular Sets

The set $aa*bbb*$ of strings recognized by the FSR of Example 10.1.2 contains strings of a's and b's beginning with one a, followed by any number of a's, followed by two b's, then followed by any number of b's. The string $aa*bbb*$ is an example of a regular expression. The set designated by $aa*bbb*$ is an example of a regular set. We will define *regular expression* and *regular set* shortly.

The following theorem, proved by the logician Stephen Kleene in 1956, closely links regular sets to finite state recognizers. We state Kleene's Theorem without proof.

THEOREM 10.1.5 **KLEENE'S THEOREM** Every set recognized by an FSR is a regular set. Conversely, for every regular set, there is an FSR that recognizes it.

We assume V is a nonempty finite set of characters. As usual, the empty set is written $\emptyset$. The **empty string** over V is the string with no characters, denoted by λ.

DEFINITION

> **Regular expressions** over V must satisfy:
>
> 1. $\emptyset$ and λ are both regular expressions.
> 2. Each element of V is a regular expression.
> 3. If A and B are regular expressions, then so are AB, $A \vee B$, and A^*.

When symbols are placed adjacent to one another, as in AB, the symbols are said to be **concatenated**.

As usual, parentheses may be used for clarity but can be omitted in many cases by observing the following hierarchical rules. Parenthesized expressions are of highest precedence; they are applied first. In the absence of parentheses, the hierarchical order is: * is applied first, concatenation second, and $\vee$ third. Operations at the same level of precedence are applied left to right. The following diagram shows the precedence hierarchy.

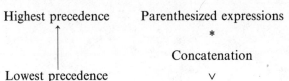

So,

$ab \vee a$ means $(ab) \vee a$, not $a(b \vee a)$
ab^* means $a(b^*)$, not $(ab)^*$
$a \vee b^*$ means $a \vee (b^*)$, not $(a \vee b)^*$
$a \vee b \vee c$ means $(a \vee b) \vee c$

When A and B are sets, the set AB is the set of all concatenations ab, where a is any element of A and b is any element of B. In other words, $AB = \{ab : a \in A \text{ and } b \in B\}$. For example, if $A = \{1, 2\}$ and $B = \{x, 1, \lambda\}$, then $AB = \{1x, 11, 1, 2x, 21, 2\}$.

DEFINITION

> Regular expressions represent sets, called **regular sets**, by the following rules:
>
> 1. $\emptyset$ represents the empty set $\emptyset$, and λ represents $\{\lambda\}$. *Note:* $\emptyset \neq \lambda$.
> 2. For each element $x \in V$, the variable x represents the singleton set $\{x\}$.
> 3. AB represents the set $AB = \{ab : a \in A \text{ and } b \in B\}$; $A \vee B$ represents the union $(A \cup B)$ of the sets A and B; A^* represents the set of all strings of symbols from A.

10.1 FINITE STATE RECOGNIZERS AND REGULAR EXPRESSIONS

A Pascal identifier begins with a letter followed by any number of letters or digits. So, if L represents the set of alphabetic letters and D represents the set of digits, then $L(L \vee D)^*$ is the regular expression representing the set of Pascal identifiers.

By definition we consider the regular expression representing a set to be identified with that set. So, for $d \in V$, we write d as a regular expression for the regular set $\{d\}$. This convention is helpful but may seem confusing at first. It is worth the effort to understand its ramifications. The following example will help make the notation clear.

EXAMPLE 10.1.6

REGULAR EXPRESSION	REGULAR SET
$\emptyset$	$\emptyset$
a	$\{a\}$
$\emptyset^*$	$\{\lambda\}$
λ	$\{\lambda\}$
λ^*	$\{\lambda\}$
b^*	$\{\lambda, b, bb, bbb, \ldots\}$
ab^*	$\{a, ab, abb, \ldots\}$
ab	$\{ab\}$
$a \vee b$	$\{a, b\}$
$(a \vee b)^*$	$\{\lambda, a, b, ab, ba, \ldots\} = \{a, b\}^*$
$a \vee b^*$	$\{a, \lambda, b, bb, \ldots\}$

∎

An integer in Pascal is a string of digits with at least one digit that may be preceded by a sign $(+, -)$. How can we use a regular expression to specify the set of all integers in Pascal? If D represents the set of digits, the regular expression for integers is $(+ \vee - \vee \lambda)DD^*$.

PRACTICE PROBLEM 1 Decimal numbers in Pascal may be preceded by a sign $(+, -)$, and the decimal point must be preceded by at least one digit and followed by at least one digit. Write a regular expression for the set of all decimal numbers in Pascal.

By the result of Practice Problem 1 and Kleene's Theorem (Theorem 10.1.5), there exists a finite state recognizer that recognizes the set of all decimal numbers in Pascal.

EXAMPLE 10.1.7 Design an FSR that recognizes the set of all decimal numbers in Pascal.

Solution Let $V = D \cup \{+, -, .\}$, where $D = \{0, 1, \ldots, 9\}$. Figure 10.1.6 shows the state graph. Note that s_5 is a dump state.

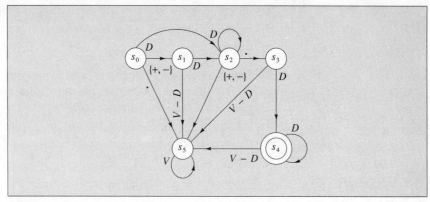

FIGURE 10.1.6

Since every regular expression represents a set, each of the following identities can be proved using the pick-a-point method. Several of the proofs are left as exercises.

REGULAR EXPRESSION IDENTITIES

a. $A\emptyset = \emptyset A = \emptyset$
b. $A\lambda = \lambda A = A$
c. $A \vee \emptyset = \emptyset \vee A = A$
d. $A \vee \lambda = \lambda \vee A$, but $A \vee \lambda \neq A$ in general
e. $A \vee B = B \vee A$, but $AB \neq BA$ in general
f. $(A \vee B) \vee C = A \vee (B \vee C)$ and $(AB)C = A(BC)$ Associative laws
g. $A \vee A = A$, but $AA \neq A$ in general
h. $A(B \vee C) = AB \vee AC$ and $(B \vee C)A = BA \vee CA$ Distributive laws

---

Identities b and c show that a given regular expression can be written in more than one way.

PRACTICE PROBLEM 2   Convince yourself that $(a \vee b)^* = (a^*b^*)^*$.

The associative laws for regular expressions justify writing $A \vee B \vee C$ and ABC without confusion.

■ PROGRAMMING NOTES

We have already mentioned that FSR's are useful in designing the syntax checking part (**lexical analyzer**) of a compiler. So, in order to use an FSR as part of a compiler, it is necessary to translate it into some language.

The following is a Pascal computer program fragment that implements the FSR of Example 10.1.2. In other words, it recognizes *aa*bbb**.

10.1 FINITE STATE RECOGNIZERS AND REGULAR EXPRESSIONS

For the program fragment, we will assume that V is the set $\{a, b\}$ and the variable ch is a defined variable of type ['a', 'b']. The variable State takes on the values S0, S1, S2, S3, and S4.

```
State := S0;
while not eoln do
    begin
        read(ch);
        case State of
            S0: if ch = 'a' then State := S1
                else State := S4;
            S1: if ch = 'a' then State := S1
                else State := S2;
            S2: if ch = 'b' then State := S3
                else State := S4;
            S3: if ch = 'b' then State := S3
                else State := S4;
            S4:                                    (*Stay There*)
        end                                        (*Of case*)
    end;                                           (*Of while*)
    if State = S3 then
        writeln ('String accepted.')
    else
        writeln ('String not accepted.')
end;
```

As this example indicates, it is relatively easy to translate an FSR into a high-level language like Pascal.

EXERCISE SET 10.1

For Exercises 1–4 write, as concisely as possible, a regular expression for the set recognized by the FSR represented by the given state graph. Let $V = \{0, 1\}$.

1.
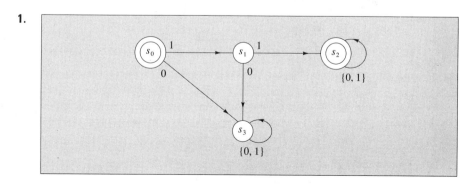

2.

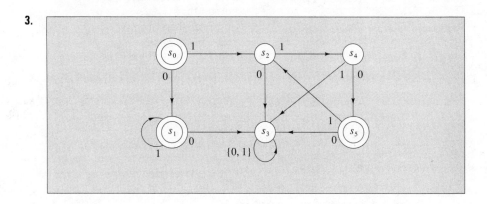

3.

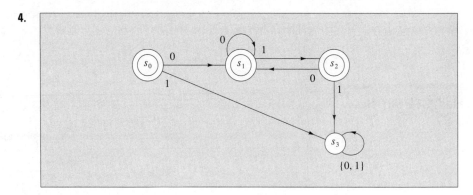

4.

5. Write a regular expression for the set recognized by the FSR of the given example.
   a. Example 10.1.3   **b.** Example 10.1.4
6. Do the given strings belong to the given sets?
   a. 0011, 01001;   0*10*01(01 ∨ λ)   **b.** 00011, 01001;   (01 ∨ 00)*01*
7. Do the given strings belong to the given sets?
   a. *abc*, *abacb*;   (*ac*b*)*   **b.** *ab*, *abc*, *acb*, *acb*^{2};   (*ab** ∨ *c*b*)*

8. Write a regular expression for each of the following sets.
   a. $\{101, 10001, 1000001, \ldots\}$ b. $\{11, 111, 1111, \ldots\} \cup \{\lambda, 0, 00, 000, \ldots\}$
9. Write a regular expression for each of the following sets.
   a. $\{11, 10001, 10000001, \ldots\}$ b. $\{\lambda, 111, 1111, \ldots\}\{0, 00, 000, \ldots\}$
10. Write a regular expression for the set of strings of 0's and 1's starting with a 1 and ending with 00.
11. Write a regular expression for the set of strings of 0's and 1's with an odd number of 1's.
12. Write a regular expression for the set of strings of 0's and 1's with an even number of 0's.
13. Design an FSR to recognize each of the following.
    a. $\emptyset$ b. λ
14. Design an FSR to recognize each of the following.
    a. $a \vee b$ b. ab c. $a*$

In Exercises 15–20 represent as a state graph an FSR that accepts strings of 0's and 1's with k 0's satisfying the given condition.

15. $k \equiv 0 \pmod 2$ 16. $k \equiv 1 \pmod 3$
17. k is not divisible by 3 18. $k \equiv 3 \pmod 4$
19. k is not divisible by 4 20. $k \equiv 2 \pmod 5$ or $k \equiv 4 \pmod 5$

Use pick-a-point proofs to prove the identities in Exercises 21–25.

21. $A(B \vee C) = AB \vee AC$
22. a. $A\emptyset = \emptyset A = \emptyset$ b. $A\lambda = \lambda A = A$
23. a. $A \vee \emptyset = \emptyset \vee A = A$ b. $A \vee \lambda = \lambda \vee A$
    c. Give an example to show that $A \vee \lambda \neq A$ in general.
24. a. $A \vee B = B \vee A$ b. Give an example to show that $AB \neq BA$ in general.
25. a. $A \vee A = A$ b. Give an example to show that $AA \neq A$ in general.

■ PROGRAMMING EXERCISES

26. Write a program, in some programming language, that implements the FSR of Example 10.1.3.
27. Write a program, in some programming language, that implements the FSR of Example 10.1.4.
28. Design an FSR to recognize the set of valid decimal numbers in exponential form for Pascal. Write a program fragment that implements the FSR you designed.

10.2
PHRASE STRUCTURE GRAMMARS

Data are represented in computers by strings of symbols from a given alphabet. For a particular computer the alphabet is the character set available. In general an **alphabet** is a finite nonempty set of characters.

EXAMPLE 10.2.1 **a.** $\{a, b, c, \ldots, z\}$ is the alphabet of lowercase English letters.
b. $\{0, 1\}$ is the alphabet of binary digits.
c. $\{a, x, 3, \#, \$\}$ is also an alphabet. ∎

Strings over an Alphabet

Let V be an alphabet. Recall that a string (or word) over V is a finite sequence of characters from V; strings are written with their characters adjacent to one another; the empty string over any alphabet is the string with no characters, denoted by λ.

EXAMPLE 10.2.2 **a.** $abcb$ is a string over $\{a, b, c, \ldots, z\}$.
b. 01001 is a string over $\{0, 1\}$.
c. $01a1$ is not a string over $\{0, 1\}$.
d. λ is a string over any alphabet. ∎

Strings over $\{0, 1, 2, \ldots, 9\}$ are usually interpreted as nonnegative decimal integers. The strings over $\{0, 1\}$ are also frequently interpreted as nonnegative integers called binary numbers. If we consider 010 and 10 as binary numbers, then $010 = 10$. However, $010 \neq 10$ when 010 and 10 are considered as strings.

For any alphabet V, the symbol V^* denotes the set of all strings over V. Note that V^* is an infinite set since $\lambda \in V^*, a \in V^*, aa \in V^*, \ldots$ if $a \in V$. Two strings w and v in V^* may be concatenated to form a new string $wv \in V^*$.

The **length** of a string w, written length(w), is the number of adjacent characters in w. The length(w) is 0 if and only if $w = \lambda$. Also, length(wv) = length(w) + length(v). If strings w and v have equal length and the characters of w are exactly the same as those of v when read left to right, they are said to be **equal strings**. An important fact about the empty string is that $\lambda w = w\lambda = w$, for all strings w.

EXAMPLE 10.2.3 The set $\{0, 1\}^*$ of all strings over $\{0, 1\}$ is the set of all binary numbers. Different strings can denote the same binary numbers. The different strings 010, 10, and 00010 all denote the same binary number. ∎

Formal Languages

How do formal languages differ from other languages? One dictionary defines a language as "a body of words and systems for their use common to a people of the same community or nation, the same geographical region, or the same cultural tradition." An alternate meaning is "any set or system of symbols as used in a more or less uniform fashion by a number of people, who are thus enabled to communicate intelligibly with one another." A language that fits either of the above descriptions is known as a natural language.

A natural language for a given community is a complex, evolving system. The rules by which sentences in the language are formed are often ambiguous and imprecise. Natural languages have a simple objective. They support the exchange of ideas and information among community members. A natural language has rules that have evolved naturally over time.

On the other hand, formal languages are highly artificial; they are created and precisely formulated for a specific purpose. Any changes in a formal language are deliberate and consistent. The rules are created as the language is designed. These rules are precise and simpler than those of most natural languages.

To understand the basic concepts underlying all formal languages, we will study some simple formal languages. Let V be a given alphabet. Any subset L of V^* is called a **language** over V.

EXAMPLE 10.2.4   **a.** $L = \{0, 00, 01, \lambda\}$ is a language over $\{0, 1\}$.
b. $L = \{1, 11, 111, \ldots\}$ is also a language over $\{0, 1\}$. ∎

Most languages over an alphabet V are infinite subsets of V^*. As we have noted, in order to work with languages, finite representations must be found for them.

Here is an example of a finite representation for a simple language.

EXAMPLE 10.2.5 $L = \{1, 11, 111, \ldots\} = \{w \in \{0, 1\}^* : w \neq \lambda \text{ and } w \text{ contains no } 0\}$ ∎

PRACTICE PROBLEM 1   Consider $L = \{a, ab, abb, abbb, \ldots\}$.

a. Assume V is a two-element set and L is a language over V. Specify the set V.
b. Find a string over V that is not in L.
c. Find a finite representation for L.

For most languages of interest, the technique used in Example 10.2.5 is not very useful for finitely representing the language. In Section 10.1 we developed a very concise representation (a regular expression) of the language in Example 10.2.5 and many others similar to it. Recall that $\{1, 11, 111, \ldots\} = 11^*$.

Another useful and complementary approach to constructing a finite representation for an infinite language is to find a finite set of generating rules that determine a given language. Let us look at how this might work for the language of Example 10.2.5.

Generating Languages

To generate $\{1, 11, 111, \ldots\}$ we need a starting point, called the start symbol, and a finite set of rules that produce the strings, and only the strings, in $\{1, 11, 111, \ldots\}$. Such a system is called a grammar.

DEFINITION

> A **grammar** or **phrase structure grammar** is denoted by $G = (V, T, @, P)$, where
>
> V is the alphabet,
> T is the set of **terminals**, $T \subset V$,
> $V - T$ is the set of **nonterminals**,
> $@$ is the **start symbol**, and $@ \in V - T$, and
> P is the finite set of **productions**.

A production is of the form $w \to v$, where $w, v \in V^*$ and w contains at least one nonterminal. Productions can be considered replacement rules or rewriting rules. So, $w \to v$ means "w can be replaced by v." Note that using a production with λ on the right, such as $w \to \lambda$, has the effect of simply deleting w.

A grammar makes use of a set of rewriting rules, called productions, to specify acceptable strings. A string will be considered acceptable whenever it can be generated by applying a sequence of productions in which the first production is applied to the start symbol. Productions are a formal set of rules used to generate all, and only, the acceptable strings in the language.

The syntax of a language is determined by its grammar. **Syntax** is the structure of a language, whereas **semantics** concerns meaning. For a programming language, syntax determines whether or not a program compiles. The semantics of a programming language defines a program's output for specified input.

Here is a simple grammar to generate the language L of Example 10.2.5. We will use it to illustrate how productions work as rules to generate a given language.

EXAMPLE 10.2.6 The grammar is $G = (V, T, @, P)$; $V = \{@, 0, 1\}$ is the alphabet, $T = \{0, 1\}$ is the set of terminals, $V - T = \{@\}$ is the set of nonterminals. There are two productions: $@ \to 1$, $@ \to 1@$.

Suppose we want to produce the string 111 in L. We will write the start symbol $@$ first:

	STRING
	$@$
1. Then use $@ \to 1@$ to produce	$1@$
2. Then use $@ \to 1@$ to produce	$11@$
3. Then use $@ \to 1$ to produce	111

At each step a production is used to rewrite a portion of the string. For example, in step 2 we rewrote the $@$ in $1\underline{@}$ as $1@$ to obtain $11\underline{@}$.

Notice that any sequence of productions, starting with $@$, will terminate if the string consists of only 1's. If we had included productions involving

10.2 PHRASE STRUCTURE GRAMMARS

0's, then any sequence of productions would terminate when the string consisted only of 0's and 1's. That is why the set $T = \{0, 1\}$ is called the set of terminals.

Strings in the language to be generated are strings of terminals. This is the key point about terminals. When designing a grammar to generate a given language, the set of terminals must include all characters that are in strings of the language.

To make the idea of a grammar more clear, consider the following simplified set of English grammar rules.

1. A sentence @ is made up of a noun phrase Q followed by a verb R.
2. A noun phrase is formed by an article A followed by a noun N.
3. "The" is an article.
4. "Deer" is a noun.
5. "Graze" is a verb.

The preceding rules can be written as a grammar.

EXAMPLE 10.2.7 The grammar is $G = (V, T, @, P)$; with $V = \{@, Q, R, A, N, \text{The}, \text{Deer}, \text{Graze}\}$, $T = \{\text{The}, \text{Deer}, \text{Graze}\}$, $P = \{@ \rightarrow QR, Q \rightarrow AN, A \rightarrow \text{The}, N \rightarrow \text{Deer}, R \rightarrow \text{Graze}\}$.

The words "The," "Deer," and "Graze" are treated as single characters in this example. They are chosen as terminals because we want strings in the language generated to consist of only those characters. The set of nonterminals contains the start symbol and the symbols Q, R, A, and N, which stand for the syntactic categories "noun phrase," "verb," "article," and "noun."

An alternate and useful method of listing productions is to list them without set braces, numbered for ease of reference. So, in Example 10.2.7 we write:

P: 1. $@ \rightarrow QR$ 2. $Q \rightarrow AN$ 3. $A \rightarrow \text{The}$ 4. $N \rightarrow \text{Deer}$ 5. $R \rightarrow \text{Graze}$

We will refer to Example 10.2.7 later in this section. Meanwhile, we need some terminology to examine how languages are generated by their respective grammars.

Assume $q, v, w, r \in V^*$. Whenever $x = qwr$ then w is a **substring** of x. It is also true that r and q are substrings of x when $x = qwr$. Suppose $w \rightarrow v$ is a production from some grammar. Then $w \rightarrow v$ can be used to replace w, as a substring of $x = qwr$, by v. The newly produced string is $qvr = y$. We say y is **directly derived** from x, written $x \Rightarrow y$. We call $x \Rightarrow y$ a **direct derivation**. A direct derivation uses *one* production to rewrite a substring of a given string to obtain another string. In Example 10.2.6, we can use the production $@ \rightarrow 1@$ to replace $@$ in $00@0$ by $1@$ to obtain $001@0$. This replacement is written $00@0 \Rightarrow 001@0$.

For $x, y \in V^*$, y can be **derived** from x (written $x \stackrel{*}{\Rightarrow} y$, with an asterisk above the arrow) when $x = y$ or when there is a finite sequence of one or more direct derivations:

$$x \Rightarrow z_1, \quad z_1 \Rightarrow z_2, \quad \ldots, \quad z_n \Rightarrow y$$

The sequence of direct derivations that show $x \stackrel{*}{\Rightarrow} y$ is called a **derivation**. As we will see, more than one sequence of derivations could yield $x \stackrel{*}{\Rightarrow} y$. Usually a derivation ends when no more productions can be applied; this occurs when the derived string consists of terminals only.

In Example 10.2.6, we have @ $\stackrel{*}{\Rightarrow}$ 111 since @ $\Rightarrow$ 1@, 1@ $\Rightarrow$ 11@, and 11@ $\Rightarrow$ 111. We will write such a sequence of direct derivations more efficiently as

$$@ \Rightarrow 1@ \Rightarrow 11@ \Rightarrow 111$$

DEFINITION

The **language $L(G)$ generated by a grammar** G is the set

$$L(G) = \{w \in T^* : @ \stackrel{*}{\Rightarrow} w\}$$

So, $L(G)$ is the set of strings of *terminals* that can be derived from the start symbol @. Because of this definition we are most interested in strings derived from the start symbol. Hence, unless otherwise noted, all derivations will begin with the start symbol.

To show that 1111 is in the language generated by the grammar G of Example 10.2.6, we need to show that @ $\stackrel{*}{\Rightarrow}$ 1111 using the productions of G. Here is that derivation:

$$@ \Rightarrow 1@ \Rightarrow 11@ \Rightarrow 111@ \Rightarrow 1111$$

For the grammar G of Example 10.2.6, the language $L(G) = \{1, 11, 111, \ldots\}$. A useful notation for this example and others is to let 1^k mean $11\ldots 1$ (k times), where k is a positive integer. We also define $1^0 = \lambda$. So, $L(G) = \{1^k : k \text{ is a positive integer}\}$ in Example 10.2.6.

PRACTICE PROBLEM 2 Write a recursive definition of the string a^k, where a is an arbitrary character and $k = 0, 1, 2, \ldots$.

The restricted English grammar of Example 10.2.7 yields $L(G) = \{\text{TheDeerGraze}\}$. So there is only one string, consisting of three characters, in this language. Note that there is more than one way to show @ $\stackrel{*}{\Rightarrow}$ TheDeerGraze.

PRACTICE PROBLEM 3 Refer to Example 10.2.7. Find at least two different derivations (in other words, sequences of direct derivations) that show @ $\stackrel{*}{\Rightarrow}$ TheDeerGraze.

10.2 PHRASE STRUCTURE GRAMMARS

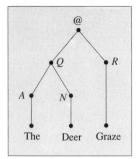

FIGURE 10.2.1

For the grammar of Example 10.2.7, a **parse tree** can be used to illustrate $@ \stackrel{*}{\Rightarrow} \text{TheDeerGraze}$. Each node represents a single element of V. The start symbol is the label for the root of the parse tree. The characters on the right of each direct derivation are the labels on the direct descendants of the node, which is labeled by each character on the left of that direct derivation. For example, the direct derivation $Q \Rightarrow AN$ is illustrated by the parse tree in Figure 10.2.1. Parse trees will also be discussed in Section 10.4.

The following grammar G' generates the same language as the one generated by the grammar G of Example 10.2.6. However, parse trees cannot be used to illustrate every derivation in G'.

EXAMPLE 10.2.8 Consider $G' = (V, T, @, P)$; with $V = \{@, 1\}$, $T = \{1\}$, and P: $@ \rightarrow 1$, $@ \rightarrow @11$, $@1 \rightarrow 1$.

To see why a parse tree cannot be used to illustrate certain derivations, consider $@ \Rightarrow @11 \Rightarrow 11$. Because the production $@1 \rightarrow 1$ is used in this derivation, 1 would be the direct descendant of both $@$ and 1 in any "parse tree." But this contradicts the definition of a tree. ∎

PRACTICE PROBLEM 4 Refer to Example 10.2.8.

a. Show that $@ \stackrel{*}{\Rightarrow} 1$ and $@ \stackrel{*}{\Rightarrow} 11$ and $@ \stackrel{*}{\Rightarrow} 111$.
b. Show that $@ \stackrel{*}{\Rightarrow} 1111$ and $@ \stackrel{*}{\Rightarrow} 11111$.
c. How would you show that $@ \stackrel{*}{\Rightarrow} 1^{2n}$ and $@ \stackrel{*}{\Rightarrow} 1^{2n+1}$, where $n \geq 1$?
d. Convince yourself that a parse tree cannot be used to illustrate $@ \stackrel{*}{\Rightarrow} 11$.

Here is another grammar, G'', for practice in deriving strings. The grammar G'' of Example 10.2.9 is different from the grammars G and G' of Example 10.2.6 and Example 10.2.8, respectively. However, G'' shares a most important property with G and G'. We will see what that property is in Practice Problem 5.

EXAMPLE 10.2.9 Consider $G'' = (V, T, @, P)$; with $V = \{@, A, 1\}$, $T = \{1\}$, and P: $@ \rightarrow A1$, $A \rightarrow A1$, $A \rightarrow 1$, $A \rightarrow \lambda$.

The derivation $@ \stackrel{*}{\Rightarrow} 111$ is shown by

$@ \Rightarrow A1 \Rightarrow A11 \Rightarrow A111 \Rightarrow \lambda 111 = 111$ or $@ \Rightarrow A1 \Rightarrow A11 \Rightarrow 111$ ∎

PRACTICE PROBLEM 5 Refer to Example 10.2.9.

a. Show that $@ \stackrel{*}{\Rightarrow} 1$ and $@ \stackrel{*}{\Rightarrow} 1111$.
b. Convince yourself that G'' generates the same language as the grammars G of Example 10.2.6 and G' of Example 10.2.8.

Designing a Grammar to Generate a Given Language

To design a grammar that generates a given language, we need to develop intuition for what is needed. First, the start symbol is required as one of the nonterminals. Second, terminals for the strings in the language are needed. Here is a simple example to illustrate the process.

EXAMPLE 10.2.10 Find a grammar G such that $L(G) = \{(01)^n : n \in \mathbb{N}\}$.

Solution We require @ $\in V$ and $0, 1 \in T$. Since $\lambda \in L(G)$, we use @ $\to \lambda$ as one of the productions. After trying reasonable possibilities, we should see that @ $\to$ 01@ will be useful. Here is one solution: Let $V = \{@, 0, 1\}$, $T = \{0, 1\}$, and $P:$ @ $\to \lambda$, @ $\to$ 01@. ∎

PRACTICE PROBLEM 6 Refer to Example 10.2.10. Find a solution with a different set of productions.

EXERCISE SET 10.2

1. Consider $G = (V, T, @, P)$; with $V = \{@, Q, R, A, N, \text{The}, \text{Student}, \text{Child}, \text{Plays}, \text{Studies}\}$, $T = \{\text{The}, \text{Student}, \text{Child}, \text{Plays}, \text{Studies}\}$, and

   $P:$ 1. @ $\to QR$ 2. $Q \to AN$ 3. $A \to$ The   4. $N \to$ Student   5. $N \to$ Child
   6. $R \to$ Plays   7. $R \to$ Studies

   a. Show @ $\stackrel{*}{\Rightarrow}$ TheChildPlays.
   b. Draw a parse tree for the derivation in part a.
   c. Convince yourself that TheChildStudiesStudent is not an element of $L(G)$.

2. Consider $G = (V, T, @, P)$; with $V = \{@, Q, R, A, N, S, M, \text{The}, \text{Woman}, \text{Man}, \text{Works}, \text{Sings}, \text{Sweetly}, \text{Diligently}\}$, $T = \{\text{The}, \text{Woman}, \text{Man}, \text{Works}, \text{Sings}, \text{Sweetly}, \text{Diligently}\}$, and

   $P:$ 1. @ $\to QR$ 2. $Q \to AN$ 3. $R \to SM$ 4. $A \to$ The   5. $N \to$ Woman
   6. $N \to$ Man   7. $S \to$ Works   8. $S \to$ Sings   9. $M \to$ Sweetly
   10. $M \to$ Diligently

   a. Show @ $\stackrel{*}{\Rightarrow}$ TheWomanWorksDiligently.
   b. Show @ $\stackrel{*}{\Rightarrow}$ TheManSingsSweetly.
   c. Draw the parse trees for the derivations in parts a and b.
   d. Convince yourself that TheManWomanSings is not a member of $L(G)$.

3. Here is a grammar to generate arithmetic expressions with addition and multiplication. Let $G = (V, T, @, P)$; with $V = \{@, I, x, +, *,), (\}$, $T = \{x, +, *,), (\}$, and

   $P:$ 1. @ $\to$ @ + @   2. @ $\to$ @ * @   3. @ $\to$ (@)   4. @ $\to I$ 5. $I \to x$

   a. Show @ $\stackrel{*}{\Rightarrow} (x + x) * x$.
   b. Show $x * x + x \in L(G)$.
   c. Draw parse trees for the derivations in parts a and b.

4. Here is a grammar to generate arithmetic expressions with subtraction and division. Let $G = (V, T, @, P)$; with $V = \{@, I, y, -, /,), (\}$, $T = \{y, -, /,), (\}$, and

   $P:$ 1. @ $\to$ @ - @   2. @ $\to$ @/@   3. @ $\to$ (@)   4. @ $\to I$ 5. $I \to y$

   a. Show @ $\stackrel{*}{\Rightarrow} (y - y)/y$.
   b. Show $(y - y)/(y/y) \in L(G)$.

c. Convince yourself that $(y + y)/y$ is not a member of $L(G)$.
d. Draw parse trees for the derivations in parts a and b.

5. Consider $G = (V, T, @, P)$; with $V = \{@, A, a, b, c\}$, $T = \{a, b, c\}$, and

   P: 1. $@ \to a@$ 2. $@ \to bA$ 3. $A \to bA$ 4. $A \to c$

   a. Show $@ \stackrel{*}{\Rightarrow} a^2b^3c$.
   b. Draw a parse tree for the derivation in part a.
   c. Show $b^5c \in L(G)$.
   d. If $a^n b^m c^j \in L(G)$, what conditions must n, m, and j satisfy?

6. Consider $G = (V, T, @, P)$; with $V = \{@, A, 0, 1\}$, $T = \{0, 1\}$, and

   P: 1. $@ \to 0A$ 2. $A \to \lambda$ 3. $A \to 0A1$

   a. Show $@ \stackrel{*}{\Rightarrow} 00011$.
   b. Show $0 \in L(G)$.
   c. If $0^n 1^m \in L(G)$, what conditions must n and m satisfy?

In Exercises 7–15 find a grammar G that generates the given language L(G).

7. $L(G) = \{0^n 1^n : n \geq 1\}$
8. $L(G) = \{0^n 1^n : n \in \mathbb{N}\}$
9. $L(G) = \{0^m 1^n : m, n \in \mathbb{N}\}$
10. $L(G) = \{0^m 1^n : m \geq 1 \text{ and } n \geq 0\}$
11. $L(G) = \{a^{n-1} c b^n : n \geq 1\}$
12. $L(G) = \{a^n b^{n+2} : n \in \mathbb{N}\}$
13. $L(G) = \{a^n b^m : m \in \mathbb{N} \text{ and } (n = 1 \text{ or } n = 2)\}$
14. $L(G) = \{a^n b^m : n, m \in \mathbb{N} \text{ and } m \equiv 1 \pmod 3\}$
15. $L(G) = \{x^{n+3} a^n : n \in \mathbb{N}\} \cup \{z^{n+2} a^{n+1} : n \in \mathbb{N}\}$

10.3
REGULAR GRAMMARS AND REGULAR LANGUAGES

Regular grammars are grammars with restrictions on the type of productions allowed.

DEFINITION

A grammar $G = (V, T, @, P)$ is a **regular grammar** if every production is one of the following three types:

$@ \to \lambda$
$N \to t$ with $N \in V - T$ and $t \in T$
$N \to tM$ with $N, M \in V - T$ and $t \in T$

Suppose G is a regular grammar with $V = \{@, A, B, a, b\}$ and $T = \{a, b\}$. Then, $@ \to \lambda$, $@ \to a$, $A \to b$, $B \to bA$ could be productions, but $A \to \lambda$, $B \to A$, $A \to aBb$ would not be legal productions.

DEFINITION

A language generated by a regular grammar is called a **regular language**.

Generating Regular Languages

Suppose we want to generate the set $L = \{a^n b^m : n \geq 1 \text{ and } m \geq 2\}$. To generate a string in L, we can follow the rule: First write one letter a followed by any number of a's; then write two b's followed by any number of b's.

Example 10.3.1 gives a regular grammar that generates L.

EXAMPLE 10.3.1 Consider $G = (V, T, @, P)$; with $V = \{@, A, B, a, b\}$; $T = \{a, b\}$, and

$P:\quad @ \to aA,\quad A \to aA,\quad A \to bB,\quad B \to bB,\quad B \to b$ ∎

PRACTICE PROBLEM 1 Convince yourself that $L(G) = \{a^n b^m : n \geq 1 \text{ and } m \geq 2\}$ in Example 10.3.1.

Regular languages are useful for analyzing certain parts of computer programming languages. Here is a regular grammar that generates the set of identifiers in Pascal.

EXAMPLE 10.3.2 Show that the set of identifiers in Pascal is a regular language.

Solution Consider $G = (V, T, @, P)$; with $V = \{@, I, a, b, \ldots, z, 0, 1, \ldots, 9\}$, $T = \{a, b, \ldots, z\} \cup \{0, 1, \ldots, 9\}$, and

$P:\quad @ \to a,\quad @ \to b,\quad \ldots,\quad @ \to z$
$@ \to aI,\quad @ \to bI,\quad \ldots,\quad @ \to zI$
$I \to aI,\quad I \to bI,\quad \ldots,\quad I \to zI$
$I \to 0I,\quad I \to 1I,\quad \ldots,\quad I \to 9I$
$I \to a,\quad I \to b,\quad \ldots,\quad I \to z$
$I \to 0,\quad I \to 1,\quad \ldots,\quad I \to 9$ ∎

PRACTICE PROBLEM 2 Refer to Example 10.3.2.

a. Does every string in $L(G)$ start with a letter?
b. Show $ad\,3 \in L(G)$.

We discussed in Section 10.2 the need for concise representations of languages. Regular languages have particularly concise finite representations because they are regular sets.

The language generated in Example 10.3.1 consists of strings $w \in \{a, b\}^*$ beginning with an a, followed by any number of a's (possibly zero), then two b's, followed by any number of b's (possibly zero). Hence, we can write $L(G) = aa^*bbb^*$ to represent this language.

We know that every regular grammar generates a regular expression. Moreover, given a regular expression, there is a regular grammar to generate it.

EXAMPLE 10.3.3 Find a regular grammar to generate $a^3 a^* b^2$.

Solution Let $G = (V, T, @, P)$; with $V = \{@, A, B, C, E, a, b\}$, $T = \{a, b\}$, and

$P: \quad @ \to aC, \quad C \to aA, \quad A \to aA, \quad A \to aE, \quad E \to bB, \quad B \to b$ ∎

PRACTICE PROBLEM 3 Convince yourself that G generates $a^3a^*b^2$ in Example 10.3.3.

Constructing a Regular Grammar from a Given FSR

ALGORITHM 10.3.4 This algorithm constructs a regular grammar G that generates the language recognized by a given finite state recognizer M.

Let M be given and let S be the set of states for M.

Define G by $G = (V, T, @, P)$, where the set of terminals T is the set of input symbols, $@ = s_0$, and $V = S \cup T$. The nonterminals in G are the states in M.

The productions are given by the following rules:

1. If M has a directed arc labeled x from state s to state s', then $s \to xs'$ is a production.
2. If M has a directed arc labeled x from any state s to any recognizing state, then $s \to x$ is a production.
3. If s_0 is a recognizing state, then $s_0 \to \lambda$ is a production.

EXAMPLE 10.3.5 Use Algorithm 10.3.4 to construct a grammar G that generates the language aa^*bbb^*.

Solution First recall the FSR of Example 10.1.2 shown in Figure 10.1.3. Let $G = (V, T, @, P)$; with $T = \{a, b\}$, $@ = s_0$, $V = \{s_0, s_1, s_2, s_3, s_4, a, b\}$, and

$P: \quad s_0 \to as_1, \quad s_0 \to bs_4$
$ s_1 \to as_1, \quad s_1 \to bs_2$
$ s_2 \to as_4, \quad s_2 \to bs_3, \quad s_3 \to as_4, \quad s_3 \to bs_3$
$ s_4 \to as_4, \quad s_4 \to bs_4, \quad s_2 \to b, \quad s_3 \to b$ ∎

PRACTICE PROBLEM 4 Convince yourself that $L(G) = aa^*bbb^*$ in Example 10.3.5. For example, do the derivation $s_0 \stackrel{*}{\Rightarrow} aabbb$ in G, and then verify that $aabbb$ is accepted by M.

Recall that the regular grammar G of Example 10.3.1 also generated aa^*bbb^*. Comparing the grammar of Example 10.3.5 with that of Example 10.3.1 shows that Algorithm 10.3.4 may yield more productions than necessary to generate a given language. This inefficiency is a small price to pay for an algorithm that guarantees a regular grammar that generates the language recognized by *any* given FSR.

PRACTICE PROBLEM 5 Which productions listed in Example 10.3.5 can be deleted from G, with the grammar still yielding $L(G) = aa^*bbb^*$?

Proving that a Language Is Not Regular

It can be proved that every set of strings recognized by some FSR is a regular expression. Conversely, for every regular expression, there is an FSR that recognizes it. Recall that every regular expression represents a regular language, and conversely. Therefore, an effective method to prove that a language is not regular is to prove that certain strings in the language are not acceptable by any FSR.

Let us prove that the set of all well-balanced parentheses is not a regular language. Recall the definition of a string of well-balanced parentheses. As the string is read left to right, the number of left parentheses must be at least as great as the number of right parentheses. When we are finished reading, the number of left parentheses must equal the number of right parentheses.

THEOREM 10.3.6 The set of all well-balanced parentheses is not a regular language.

Proof We give a proof by contradiction. Suppose there is an FSR that recognizes the set of all well-balanced parentheses. This FSR must have a finite number of states; say it has m states. The string $(^{m+1})^{m+1}$ is well-balanced, so the given FSR must accept it. While processing the left parentheses of $(^{m+1})^{m+1}$, the FSR has to be in some state s more than once. This follows from the pigeonhole principle and the fact that the FSR has m states to process $m + 1$ left parentheses. So, it takes some finite number of left parentheses to go from s back to s. Let us say $(^k$ will do it. See Figure 10.3.1. Then it is not difficult to see that the FSR must also accept $(^{k+m+1})^{m+1}$. But this is a contradiction since $(^{k+m+1})^{m+1}$ is not well-balanced.

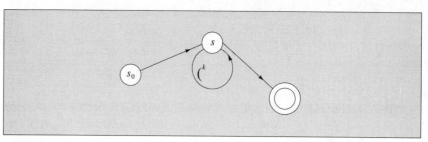

FIGURE 10.3.1

The proof of Theorem 10.3.6 can be modified to show that $\{a^n b^n : n \in \mathbb{N}\}$ is not regular. This problem is included in Exercise Set 10.3 (Exercise 17).

Regular Expression Identities

Recall the list of identities given in Section 10.1. We repeat it here for easy reference.

> **REGULAR EXPRESSION IDENTITIES**
>
> a. $A\varnothing = \varnothing A = \varnothing$
> b. $A\lambda = \lambda A = A$
> c. $A \vee \varnothing = \varnothing \vee A = A$
> d. $A \vee \lambda = \lambda \vee A$, but $A \vee \lambda \neq A$ in general
> e. $A \vee B = B \vee A$, but $AB \neq BA$ in general
> f. $(A \vee B) \vee C = A \vee (B \vee C)$ and
> $(AB)C = A(BC)$ Associative laws
> g. $A \vee A = A$, but $AA \neq A$ in general
> h. $A(B \vee C) = AB \vee AC$ and
> $(B \vee C)A = BA \vee CA$ Distributive laws
>
> **MORE REGULAR EXPRESSION IDENTITIES**
>
> i. $(A^*)^* = A^*$
> j. $AA^* = A^*A$
> k. $AA^* \vee \lambda = A^*$
> l. $A(BA)^* = (AB)^*A$
> m. $(A \vee B)^* = (A^* \vee B^*)^*$
> n. $(A \vee B)^* = (A^*B^*)^*$
> o. $(A \vee B)^* = A^*(BA^*)^*$

It is possible to prove Identities a–o by considering the expressions as sets and using pick-a-point proofs. Here is a proof of Identity n.

Proof
1. First we prove $(A \vee B)^* \subseteq (A^*B^*)^*$: Let $w \in (A \vee B)^*$. Then w is a string of symbols from A or B. So, $w = a_1b_1 \ldots a_nb_n$, where a_i is either a symbol from A or $a_i = \lambda$ and b_i is either a symbol from B or $b_i = \lambda$. Hence, $a_i \in A^*$ and $b_i \in B^*$. Therefore, $w \in (A^*B^*)^*$.
2. Prove $(A^*B^*)^* \subseteq (A \vee B)^*$: Let $w \in (A^*B^*)^*$. Then w is a string of strings from A or B. So, $w = a_1b_1 \ldots a_nb_n$, where a_i is an element of A^* and b_i is an element of B^*. Hence, a_i is a string of symbols from A and b_i is a string of symbols from B. Therefore, $w \in (A \vee B)^*$. ■

PROGRAMMING NOTES

The UNIX system is a family of computer operating systems developed by AT&T's Bell Laboratories. UNIX has been implemented on many different computers including micros, minis, and mainframes.

To manage files effectively, an operating system needs an efficient search capability. **Grep** is a text file utility in UNIX that searches for text patterns in files. Grep is an acronym standing for "global regular expression print." Both grep and the more powerful extended grep (**egrep**) use regular expressions to specify string patterns.

Expressions available in grep include:

- a where a is any single character that does not otherwise have special meaning in UNIX
- $E*$ specifying zero or more occurrences of the regular expression E
- AB specifying two regular expressions A and B concatenated

Egrep includes all expressions available in grep as well as:

- $A|B$ specifying either A or B

In addition, egrep allows specified regular expressions to be enclosed in parentheses.

The use of regular expressions gives UNIX a very powerful pattern matching capability.

EXERCISE SET 10.3

1. Define $G = (V, T, @, P)$ by $V = \{@, A, B, a, b\}$, $T = \{a, b\}$, and

    P: $@ \to \lambda$, $@ \to aA$, $A \to aB$, $B \to bB$, $B \to b$

    a. Which of the following strings belong to $L(G)$?

    $\lambda, \quad a, \quad ab, \quad a^3b, \quad a^2b^2a, \quad a^2b^2$

    b. If $a^n b^m \in L(G)$, what conditions must the integers m and n satisfy?

    c. Write $L(G)$ as a regular expression.

2. Define $G = (V, T, @, P)$ by $V = \{@, B, a, b, c\}$, $T = \{a, b, c\}$, and

    P: $@ \to a@$, $@ \to bB$, $B \to bB$, $B \to c$

    a. Which of the following strings belong to $L(G)$?

    $\lambda, \quad a, \quad abc, \quad a^2bc, \quad ab^2c, \quad abc^2$

    b. If $a^n b^m c^j \in L(G)$, what conditions must the integers j, m, and n satisfy?

    c. Write $L(G)$ as a regular expression.

    In Exercises 3–6 find a regular grammar G that generates the given language $L(G)$.

3. $L(G) = aa*b*$
4. $L(G) = ab*$
5. $L(G) = aa*bbc*$
6. $L(G) = a* \vee b*$

    Simplify each of the regular expressions in Exercises 7–10.

7. $(\lambda \vee 0* \vee 1*)*$
8. $(0*1)* \vee (1*0)*$
9. $((0*1)*(1*0*)*)*$
10. $(0 \vee 1)*0(0 \vee 1)*$

    For Exercises 11–16 use Algorithm 10.3.4 to define a regular grammar G such that $L(G)$ is the set recognized by the given FSR.

11. The FSR of Exercise 1, Exercise Set 10.1.

12. The FSR of Exercise 2, Exercise Set 10.1.
13. The FSR of Exercise 3, Exercise Set 10.1.
14. The FSR of Exercise 4, Exercise Set 10.1.
15. The FSR of Example 10.1.1.
16. The FSR of Example 10.1.3.
17. Prove that the set $\{a^n b^n : n \in \mathbb{N}\}$ is not regular. [*Hint:* Use an argument similar to that used in Theorem 10.3.6.]

Use pick-a-point proofs to prove the identities in Exercises 18–23.

18. $(A^*)^* = A^*$
19. $AA^* = A^*A$
20. $AA^* \vee \lambda = A^*$
21. $A(BA)^* = (AB)^*A$
22. $(A \vee B)^* = (A^* \vee B^*)^*$
23. $(A \vee B)^* = A^*(BA^*)^*$

10.4

CONTEXT-FREE GRAMMARS AND PARSE TREES

Context-free grammars are important for specifying and analyzing programming languages. A context-free grammar has the property that the left side of every production consists of a single nonterminal.

DEFINITION

A grammar $G = (V, T, @, P)$ is a **context-free grammar** if every production is of the form $N \to w$, with $N \in V - T$ and $w \in V^*$.

So every production of a context-free grammar has a single nonterminal on the left side.

A language is **context-free** if it is generated by a context-free grammar. Note that a regular grammar is also a context-free grammar. Hence, every regular language is context-free. However, as we will see, not every context-free language is regular.

A string in the language $L(G)$ for a context-free grammar G is generated by the following rules:

1. Start with @.
2. Find a nonterminal in the current string that appears on the left side of a production.
3. Replace an occurrence of this nonterminal by the string that appears on the right side of the production.
4. Repeat steps 2 and 3 until no nonterminal can be found that appears on the left side of a production.

Recall that the grammar of Example 10.2.8 contains the production $@1 \to 1$. Because the left side of this production is not a single nonterminal, the grammar is not context-free. As we saw in Example 10.2.8, any derivation using the production $@1 \to 1$ cannot be illustrated by a parse tree.

The fact that the left side of every production in a context-free grammar consists of a single nonterminal means that every derivation in a context-free grammar can be illustrated by a parse tree. Parse trees simplify the analysis of derivations and decisions concerning which strings are in the language. This is one reason context-free grammars are so important.

Noam Chomsky invented phrase structure grammars in the mid-1950s. Phrase structure grammars include context-free grammars and context-sensitive grammars. The terms *context-free* and *context-sensitive* limit the types of productions allowed in each grammar. For example, $A \to aBbb$ is a context-free production, and it means A can be replaced by $aBbb$ no matter where A occurs in a string—in other words, regardless of context. An example of a context-sensitive production is $dAD \to daBbD$. It means A can be replaced by aBb whenever A has d on its left and D on its right. In other words, A can be replaced by aBb in the context of d on the left and D on the right.

Chomsky developed phrase structure grammars to study structural aspects of natural languages. Although context-free grammars are the most important grammars for the study of formal languages, they were found to be inadequate for understanding the syntax of natural languages.

A Context-Free Language that Is Not Regular

By Exercise 17 in Exercise Set 10.3, $\{a^n b^n : n \in \mathbb{N}\}$ is not regular.

EXAMPLE 10.4.1 Show that $\{a^n b^n : n \in \mathbb{N}\}$ is a context-free language that is not regular.

Solution Consider the context-free grammar $G = (V, T, @, P)$; with $V = \{@, a, b\}$, $T = \{a, b\}$, and

P: $@ \to \lambda$, $@ \to a@b$

We show $@ \stackrel{*}{\Rightarrow} a^2 b^2$ by the derivation $@ \Rightarrow a@b \Rightarrow aa@bb \Rightarrow aabb$. ∎

PRACTICE PROBLEM 1 Verify that $L(G) = \{a^n b^n : n \in \mathbb{N}\}$ in Example 10.4.1.

We have already seen in Section 10.2 that the same language can be generated by different grammars.

EXAMPLE 10.4.2 Consider $G' = (V, T, @, P)$; with $V = \{@, A, a, b\}$, $T = \{a, b\}$, and

P: $@ \to \lambda$, $@ \to ab$, $@ \to aAb$, $A \to aAb$, $A \to ab$ ∎

PRACTICE PROBLEM 2 Verify that $L(G') = L(G)$, where G is the grammar of Example 10.4.1 and G' is the grammar of Example 10.4.2.

The Membership Problem

So far, we have discussed the question of how to find a grammar to generate a given language. Another important question is: For a given grammar or

10.4 CONTEXT-FREE GRAMMARS AND PARSE TREES

language, is there an effective procedure to decide whether a given string is in the language?

Specifically, for a given context-free grammar G and a string $w \in V^*$, is there an algorithm to decide whether $w \in L(G)$? This is the **membership problem** for context-free languages, and the answer is yes. We now show why. We assume all grammars are context-free, unless explicitly stated otherwise.

The empty string λ plays a crucial role in this discussion. A **λ-production** is any production with λ on the right side. For example, $@ \to \lambda$ is a λ-production. Hence, any grammar that generates a language containing λ must have a λ-production.

Let us examine how λ affects derivations.

EXAMPLE 10.4.3 Find a grammar that generates $L = \{1^m d 0^n 1^n : m, n \in \mathbb{N}\}$, where d is a given terminal.

Solution **a.** The productions that generate L are:

P: $@ \to AdB$, $A \to \lambda$, $A \to 1A$, $B \to \lambda$, $B \to 0B1$

This set of productions includes two λ-productions.

b. If we try to find a set of productions without λ-productions to generate L, we see that more productions are needed. Such a set is given in the following, alternate solution to Example 10.4.3.

P: $@ \to AdB$, $@ \to d$, $@ \to 1d$, $@ \to d01$, $A \to 1$, $A \to 1A$,
$B \to 01$, $B \to 0B1$

Figure 10.4.1 shows a parse tree for the derivation $@ \overset{*}{\Rightarrow} 1d$ in each case.

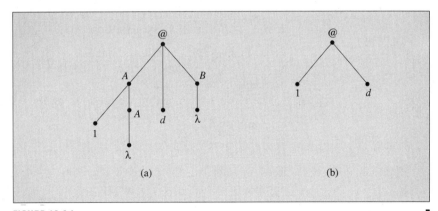

FIGURE 10.4.1

PRACTICE PROBLEM 3 Find a grammar G such that $L(G) = \{1^m 0^n 1^n : m, n \in \mathbb{N}\}$. [*Hint:* Note that $\lambda \in L(G)$ and d is not needed as a terminal. You can make very simple changes to the productions in Example 10.4.3 to get two different solutions.]

From Example 10.4.3, it is clear that λ-productions can be useful for reducing the number of productions needed. On the other hand, their inclusion in the list of productions requires that we allow λ as a label for nodes in parse trees. More crucial is the unfortunate possibility of the strings contracting during a derivation. This is illustrated by $AdB \Rightarrow dB \Rightarrow d$ using productions from Example 10.4.3. This kind of derivation can occur whenever there are productions, such as $A \to \lambda$ and $B \to \lambda$, with right sides shorter than left sides.

A grammar in which all productions have a right side at least as long as the left side is called a **noncontracting grammar**. In particular, any grammar without λ-productions is noncontracting. The following theorem is the reason why a noncontracting grammar is desirable. In Theorem 10.4.4, the grammar G need *not* be context-free.

THEOREM 10.4.4 If G is a noncontracting grammar, then there is an algorithm for deciding whether a given nonempty string w is a member of $L(G)$.

Proof Let $G = (V, T, @, P)$ be noncontracting, and let $w \in V^*$ with $w \neq \lambda$. Then, length$(w) = n > 0$. Define a sequence of sets S_i by: $S_0 = \{@\}$ and $S_k = S_{k-1} \cup \{v \in V^* : \text{length}(v) \leq n \text{ and } u \Rightarrow v \text{ for some } u \in S_{k-1}\}$. Hence, S_k includes all strings in V^* of length no more than n that are derivable from $@$ in no more than k steps.

Since G is noncontracting, it is clear that $w \in L(G)$ if and only if $w \in S_i$ for some i. We will show that one of the sets S_i includes all the strings derivable from $@$ and of length less than or equal to n. Two easily verified facts will help.

1. V^* has a finite number of strings of length less than or equal to n since V is finite.
2. $S_{k-1} \subseteq S_k$ for all k, by definition of S_k.

From Facts 1 and 2 it follows that $S_{j-1} = S_j$ for some j. But this also means $S_j = S_{j+1} = \cdots$. So, all strings of length less than or equal to n that are derivable from $@$ are in S_j.

Therefore, the algorithm we are looking for consists of two steps.

1. Compute S_j.
2. If $w \in S_j$, then $w \in L(G)$ else $w \notin L(G)$. ∎

PRACTICE PROBLEM 4 Refer to Fact 1 in the proof of Theorem 10.4.4. Let Card$(V) = m$. Find the number of strings in V^* of length less than or equal to n.

Theorem 10.4.4 is important and does not depend on the grammar being context-free. However, it does not seem to guarantee a similar algorithm for context-free grammars because, as we have seen, context-free grammars may not be noncontracting. Fortunately, there is a bridge for this conceptual gap.

THE DISPENSABILITY OF λ-PRODUCTIONS

If G is any context-free grammar, then there is a context-free grammar G' that has no λ-productions and for which

$$L(G') = L(G) - \{\lambda\}$$

For a proof of this fact, see *Elementary Computability, Formal Languages, and Automata* by Robert McNaughton.

One other gap remains: how to decide whether or not $\lambda \in L(G)$.

ALGORITHM 10.4.5 This algorithm determines whether or not $\lambda \in L(G)$ for any context-free grammar G.

Underline any nonterminal that is on the left side of a λ-production.

Repeat until no production has a nonunderlined nonterminal on the left and all underlined nonterminals are on the right.

Underline every nonterminal not already underlined that appears on the left side of a production with a right side consisting entirely of underlined nonterminals.

If @ is underlined, then $\lambda \in L(G)$ else $\lambda \notin L(G)$.

To see how Algorithm 10.4.5 works, refer to Example 10.4.3. The productions are $P: @ \to AdB, A \to \lambda, A \to 1A, B \to \lambda, B \to 0B1$. We first underline A and B since A and B appear on the left side of a λ-production. This yields

$$@ \to A d\underline{B}, \underline{A} \to \lambda, \underline{A} \to 1\underline{A}, \underline{B} \to \lambda, \underline{B} \to 0\underline{B}1$$

The only production to which the repeat loop can possibly be applied is $@ \to A d\underline{B}$, but the right side does not consist of only underlined nonterminals. So, @ is not underlined and, hence, λ is not in $L(G)$.

If we change the production set slightly to get $P: @ \to AB, A \to \lambda, A \to 1A, B \to \lambda, B \to 0B1$, it is easy to verify that Algorithm 10.4.5 will lead to the conclusion $\lambda \in L(G)$.

Here then is the main theorem of this section.

THEOREM 10.4.6 If G is a context-free grammar, then there is an algorithm to decide whether a given string w is a member of $L(G)$.

Proof If $w = \lambda$, then use Algorithm 10.4.5.

Assume $w \neq \lambda$. If G is noncontracting, use the algorithm of Theorem 10.4.4. Otherwise, find a context-free grammar G' that has no λ-productions and such that $L(G') = L(G) - \{\lambda\}$. Now apply the algorithm of Theorem 10.4.4 to G'. ∎

Parsing and Ambiguity

We have used the word *parse* several times. When applied to natural languages, **parse** means "describe grammatically by giving the parts of speech and the syntactic relations." Parsing is best illustrated by diagrams. We will use parse trees for languages generated by context-free grammars. For natural languages another kind of diagram is frequently used. We now consider a simple example illustrating the parsing of a sentence in English.

The sentence "He is using a fishing pole" can be diagrammed as follows:

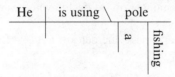

Both "a" and "fishing" modify "pole" in this sentence.

To see why it is impossible to parse sentences in a natural language like English without taking meaning and context into consideration, we will consider three more sentences. First we diagram the following two sentences, with particular attention to the modifiers.

1. She walks through the park with a fishpond.
2. She walks through the park with a fishing pole.

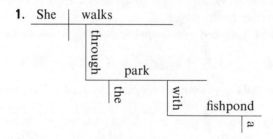

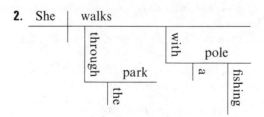

Now consider the sentence "She walks through the park with flowers." How do we diagram it? It is reasonable to conclude that "with flowers" might modify "park" or it might modify "walks." So it could be diagrammed like Sentence 1 or like Sentence 2. Here are the diagrams for each case:

3.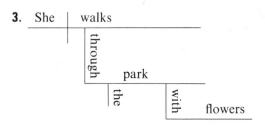

4.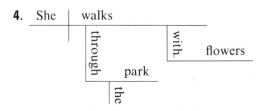

This sentence is structurally ambiguous; it has two distinct meanings. The meaning depends on knowledge not contained in the sentence itself. That is, the meaning of the sentence depends on context. We may even have to ask the author of the sentence what it means.

It appears inevitable that ambiguities will occur in natural languages because they are so complex. It is desirable that a formal language be as free of ambiguity as possible. To understand ambiguity in formal languages and how to avoid it, if possible, we will look at several examples. In what follows we will consider formal languages only.

First, we need to describe what is meant by the term *parse* as applied to formal languages generated by a grammar. To parse a string in a given language, we will examine the derivation of the string. If a given string has more than one distinct derivation, we will consider the grammar to be ambiguous. But how can we tell if two derivations are distinct? As before, we use diagrams. The next example illustrates these ideas.

EXAMPLE 10.4.7 Construct an ambiguous grammar.

Solution Let $G = (V, T, @, P)$; where $V = \{@, b, x\}$, $T = \{b, x\}$, and $P: @ \to @b@$, $@ \to x$.

It seems we can derive xbx from $@$ in two distinct ways:

1. $@ \Rightarrow @b@ \Rightarrow xb@ \Rightarrow xbx$
2. $@ \Rightarrow @b@ \Rightarrow @bx \Rightarrow xbx$

Both of these derivations yield $@ \overset{*}{\Rightarrow} xbx$. Should they be considered distinct?

We can use diagrams to decide whether the two derivations are distinct. Because we are considering a formal language, the diagrams will be parse

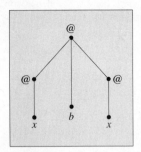

FIGURE 10.4.2

trees. Both derivations 1 and 2 can be pictured as the parse tree in Figure 10.4.2.

The bottom vertices of a tree are called **leaves**. Each vertex must be labeled by λ or by a single symbol from a given alphabet. The **yield** of a parse tree is the string obtained by concatenating the symbols that label the leaves from left to right. The yield of this parse tree is xbx.

Since the parse trees for the two derivations yielding $@ \stackrel{*}{\Rightarrow} xbx$ are the same, the derivations are not considered distinct. Another way to see this is to note that, if we require each direct derivation to be applied to the leftmost nonterminal first, it would make derivation 1 the only valid derivation. Such a derivation is called a **leftmost derivation**.

Now consider the following two derivations of $xbxbx$ from $@$:

3. $@ \Rightarrow @b@ \Rightarrow xb@ \Rightarrow xb@b@ \Rightarrow xbxb@ \Rightarrow xbxbx$
4. $@ \Rightarrow @b@ \Rightarrow @b@b@ \Rightarrow xb@b@ \Rightarrow xbxb@ \Rightarrow xbxbx$

Are these two derivations really the same? They are both valid leftmost derivations. The parse trees are shown in Figure 10.4.3.

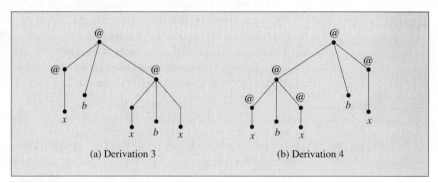

(a) Derivation 3   (b) Derivation 4

FIGURE 10.4.3

The parse trees are different. So, derivations 3 and 4 are considered distinct. Hence, the grammar G is said to be ambiguous. ∎

A context-free grammar G is an **ambiguous grammar** if there is a string with more than one leftmost derivation—in other words, more than one parse tree.

It is often possible to find another grammar that is not ambiguous and that generates the same language as generated by an ambiguous grammar.

EXAMPLE 10.4.8 Find a context-free grammar that is not ambiguous and that generates the language generated by the grammar of Example 10.4.7.

Solution Let $G' = (V, T, @, P)$; with $V = \{@, A, b, x\}$, $T = \{b, x\}$, and $P: @ \to @bA$, $@ \to A$, $A \to x$. ∎

10.4 CONTEXT-FREE GRAMMARS AND PARSE TREES

PRACTICE PROBLEM 5 Convince yourself that $L(G) = L(G')$, where G is the grammar of Example 10.4.7 and G' is the grammar of Example 10.4.8. Is G' ambiguous? Try to find more than one leftmost derivation $@ \stackrel{*}{\Rightarrow} xbxbx$ in G'. Draw the parse tree for any derivation you find. Convince yourself that G' is not ambiguous.

Here is an ambiguous context-free grammar that generates arithmetic expressions with minus signs and the numbers 2 and 5.

EXAMPLE 10.4.9 Let $G = (V, T, @, P)$; with $V = \{@, B, Y, -, 2, 5\}$, $T = \{-, 2, 5\}$, and

P: $@ \to @B@$, $@ \to Y$, $B \to -$, $Y \to 2$, $Y \to 5$

There are two distinct leftmost derivations for $@ \stackrel{*}{\Rightarrow} 5 - 2 - 2$:

1. $@ \Rightarrow @B@ \Rightarrow YB@ \Rightarrow 5B@ \Rightarrow 5 - @ \Rightarrow 5 - @B@ \Rightarrow 5 - YB@ \Rightarrow 5 - 2B@ \Rightarrow 5 - 2 - @ \Rightarrow 5 - 2 - Y \Rightarrow 5 - 2 - 2$

2. $@ \Rightarrow @B@ \Rightarrow @B@B@ \Rightarrow YB@B@ \Rightarrow 5B@B@ \Rightarrow 5 - @B@ \Rightarrow 5 - YB@ \Rightarrow 5 - 2B@ \Rightarrow 5 - 2 - @ \Rightarrow 5 - 2 - Y \Rightarrow 5 - 2 - 2$

Because context-free grammars are used to analyze syntax, it is important to avoid ambiguity. How does one interpret a string like $5 - 2 - 2$? Does it mean $5 - (2 - 2) = 5$ or $(5 - 2) - 2 = 1$? Choosing derivation 1 is the same as interpreting $5 - 2 - 2$ as $5 - (2 - 2)$, and choosing derivation 2 is picking the interpretation $(5 - 2) - 2$. To see this clearly, draw the two parse trees. ∎

The grammar G' defined in Example 10.4.10 is unambiguous, and it generates the same language as G defined in Example 10.4.9.

EXAMPLE 10.4.10 Let $G' = (V, T, @, P)$; with $V = \{@, A, B, Y, -, 2, 5\}$, $T = \{-, 2, 5\}$, and

P: $@ \to @BA$, $@ \to A$, $A \to Y$, $B \to -$, $Y \to 2$, $Y \to 5$

Here is the only leftmost derivation for $@ \stackrel{*}{\Rightarrow} 5 - 2 - 2$ in G':

$@ \Rightarrow @BA \Rightarrow @BABA \Rightarrow ABABA \Rightarrow YBABA \Rightarrow 5BABA \Rightarrow 5 - ABA \Rightarrow 5 - YBA \Rightarrow 5 - 2BA \Rightarrow 5 - 2 - A \Rightarrow 5 - 2 - Y \Rightarrow 5 - 2 - 2$ ∎

Syntax Rules and Language Specification

When checking for syntax errors, a compiler must decide which strings in a language are syntactically correct. Syntax rules are given by the language specifications, which are known before the compiler is designed. One reason why context-free grammars are important in computer science is their usefulness in generating only the syntactically correct strings of a given language. The syntax rules tell us what strings the language must contain. The grammar is then designed to generate exactly those strings in the language.

Here is an example of an unambiguous grammar that generates a part of many programming languages. The language generated by this grammar consists of all syntactically correct strings over a given alphabet.

EXAMPLE 10.4.11 Let $G = (V, T, @, P)$; with $V = \{@, E, F,), (, +, \cdot, x, 1, 2\}$, $T = \{), (, +, \cdot, x, 1, 2\}$, and

P: $@ \to @ + E$, $@ \to E$, $E \to E \cdot F$, $E \to F$, $F \to (@)$, $F \to x1$, $F \to x2$

Because $x1$ and $x1 \cdot (x1 + x2)$ are in $L(G)$, they are, by definition, syntactically correct arithmetic expressions. To show that an expression such as $2 \cdot x1$ is not syntactically correct with respect to this grammar requires that we show that there is no derivation $@ \stackrel{*}{\Rightarrow} 2 \cdot x1$. Parse trees are useful in this type of verification, and most texts in compiler theory discuss parsing and parse trees. ∎

It would be nice if all languages could be generated by context-free grammars. Unfortunately, this is not the case. It can be proved, using techniques beyond the scope of this text, that there is no context-free grammar G such that $L(G) = \{a^n b^n c^n : n > 0\}$. Hence, $\{a^n b^n c^n : n > 0\}$ is not a context-free language.

Fortunately, almost all the syntax of any given programming language can be specified by a context-free grammar. However, some programming languages, when generally defined and not limited by a given implementation, are not context-free. One example is Pascal. Pascal allows identifiers of arbitrary length. It can be shown that the problem of checking that identifiers have been declared in Pascal is equivalent to the problem of specifying $\{wdw : w \in \{0, 1\}^*\}$, where d is a given terminal. It can also be shown that $\{wdw : w \in \{0, 1\}^*\}$ is not context-free. See Exercise 19, Exercise Set 10.4, for the grammar that generates $\{wdw : w \in \{0, 1\}^*\}$.

■ **PROGRAMMING NOTES**

The If . . . then . . . else Ambiguity The most well-known ambiguity for a programming language is the "dangling else" ambiguity that can occur in a programming language having an if . . . then . . . else construction.

Suppose the grammar defining the language has the productions:

P: $A \to$ if B then A else A

$A \to$ if B then A

$A \to a$

$B \to b$

This grammar is ambiguous because

if b then if b then a else a

can be interpreted as either

1. if *b* then (if *b* then *a* else *a*)
2. if *b* then (if *b* then *a*) else *a*

Figure 10.4.4 shows the parse trees for the derivations yielding the preceding interpretations.

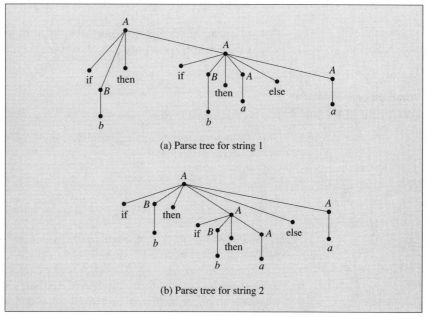

FIGURE 10.4.4

Eliminating this ambiguity is usually accomplished by choosing the interpretation illustrated by the parse tree in Figure 10.4.4(a). That is, each "else" is associated with the nearest "then" to the left of it.

PRACTICE PROBLEM 6 Find the derivations illustrated by the parse trees in Figures 10.4.4(a) and 10.4.4(b).

PRACTICE PROBLEM 7 Write a computer program in Pascal by which you can determine whether interpretation 1 or interpretation 2 is used for Pascal. [*Hint:* You might use a statement like: If (x > 2) then if (x = 3) then write('2') else write('3'). Assuming interpretation 1 is the one chosen for Pascal, what would you expect to see printed for each of the three assignments: x := 1, x := 2, x := 3?]

Pascal has chosen interpretation 1 as the syntactically correct one. In order for you to implement interpretation 2, you must write:

```
if (x > 2) then
    begin
        if (x = 3) then
            write('2')
    end
else
    write('3')
```

For this program fragment, the three different assignments $x := 1$, $x := 2$, $x := 3$ will yield the outputs 3, 3, 2, respectively.

EXERCISE SET 10.4

For the following exercises $j, m, n \in \mathbb{N}$.

1. This grammar generates well-balanced parentheses. Let $G = (V, T, @, P)$; with $V = \{@, A,), (\}$, $T = \{), (\}$, and

   $P: @ \to \lambda, \quad @ \to A, \quad A \to (A), \quad A \to AA, \quad A \to ()$

   a. Show $@ \stackrel{*}{\Rightarrow} (())()$ with two different derivations and draw the parse trees.
   b. Which strings in $L(G)$ can be derived using derivations of three or fewer direct derivations?
   c. Find all strings in $L(G)$ of length less than or equal to 3.
   d. Find all strings in $\{w \in V^* : @ \stackrel{*}{\Rightarrow} w\}$ of length less than or equal to 3.

2. Do each of the following for the grammar of Exercise 1.
   a. Show $@ \stackrel{*}{\Rightarrow} ()()(())$ with two different derivations and draw the parse trees.
   b. Which strings in $L(G)$ can be derived using derivations of five or fewer direct derivations?
   c. Find all strings in $L(G)$ of length less than or equal to 5.
   d. Find all strings in $\{w \in V^* : @ \stackrel{*}{\Rightarrow} w\}$ of length less than or equal to 5.

3. Let $G = (V, T, @, P)$; with $V = \{@, A, 0, 1\}$, $T = \{0, 1\}$, and $P: @ \to A$, $A \to 1A1$, $A \to 0$.
   a. Show $@ \stackrel{*}{\Rightarrow} 1^3 0 1^3$ and draw the parse tree.
   b. If $1^m 0 1^n \in L(G)$, what conditions must m and n satisfy?
   c. Find all strings in $L(G)$ of length less than or equal to 9.

4. Let $G = (V, T, @, P)$; with $V = \{@, A, 0, 1\}$, $T = \{0, 1\}$, and $P: @ \to AA$, $A \to 1A1$, $A \to 0$.
   a. Show $@ \stackrel{*}{\Rightarrow} 1^3 0 1^5 0 1^2$.
   b. If $1^m 0 1^n 0 1^j \in L(G)$, what conditions must m, n, and j satisfy?
   c. Find all strings in $L(G)$ of length less than or equal to 9.

5. Let $G = (V, T, @, P)$; with $V = \{@, A, 0, 1\}$, $T = \{0, 1\}$, and $P: @ \to AA$, $A \to A1$, $A \to 1A$, $A \to 0$.
   a. Show $1^3 0 1^2 0 1^4 \in L(G)$.
   b. If $1^m 0 1^n 0 1^j \in L(G)$, what conditions must m, n, and j satisfy?
   c. Find all strings in $L(G)$ of length less than or equal to 5.

6. **a.** Find a context-free, but not regular, grammar G such that $L(G) = a(ab)^*$.
   b. Find a regular grammar G' such that $L(G') = a(ab)^*$.

7. **a.** Find a context-free, but not regular, grammar G such that $L(G) = a^*(bb)^*$.
   b. Find a regular grammar G' such that $L(G') = a^*(bb)^*$.

Exercises 8–11 refer to the proof of Theorem 10.4.4.

8. Let $n = 3$. For the grammar of Exercise 1, find the sets $S_0, S_1, S_2, \ldots$.
9. Let $n = 5$. For the grammar of Exercise 3, find the sets $S_0, S_1, S_2, \ldots$.
10. Let $n = 7$. For the grammar of Exercise 4, find the sets $S_0, S_1, S_2, \ldots$.
11. Let $n = 4$. For the grammar of Exercise 5, find the sets $S_0, S_1, S_2, \ldots$.

In Exercises 12–15 find a context-free grammar G that generates the given language $L(G)$.

12. $L(G) = \{0^n 1^n : n > 0\}$
13. $L(G) = \{01^n : n \in \mathbb{N}\}$
14. $L(G) = \{ba^n b^2 : n > 0\}$
15. $L(G) = \{a^2 b^n c^2 : n \in \mathbb{N}\}$

16. The following grammar generates arithmetic expressions with the symbols $+$ and $*$ and the letters a and b. Let $G = (V, T, @, P)$; with $V = \{@, Y, +, *, a, b\}$, $T = \{+, *, a, b\}$, and

    P: $@ \to @+@$, $@ \to @*@$, $@ \to Y$, $Y \to a$, $Y \to b$

    Show that G is ambiguous by finding two distinct leftmost derivations for $a*b+b$. Draw the parse trees for the two derivations.

17. The following unambiguous grammar generates arithmetic expressions with the symbols $+$ and $*$ and the letters a and b. Let $G = (V, T, @, P)$; with

    $V = \{@, E, F,), (, +, *, a, b\}$, $T = \{), (, +, *, a, b\}$

    and

    P: 1. $@ \to @+E$ 2. $@ \to E$ 3. $E \to E*F$ 4. $E \to F$ 5. $F \to (@)$
    6. $F \to a$ 7. $F \to b$

    a. Find the leftmost derivation for $a*b+b$.
    b. Find the leftmost derivation for $a*(b+a)$.

18. For $w \in V^*$, the symbol w^r denotes w in reverse order. So, if $w = 1011$, then $w^r = 1101$. Supply a recursive definition of w^r.

19. For $w \in V^*$, the symbol w^r denotes w in reverse order.
    a. Let $V = \{@, 0, 1, d\}$. Find a context-free grammar G such that $L(G) = \{wdw^r : w \in \{0, 1\}^*\}$, where d is a given terminal.
    b. It can be proved that there is no context-free grammar G such that $L(G) = \{wdw : w \in \{0, 1\}^*\}$. Convince yourself that the following grammar generates this language. Let $V = \{@, A, B, J, 0, 1, d\}$, $T = \{0, 1, d\}$, and

    P: $@ \to d$, $@ \to 0d0$, $@ \to 1d1$, $@ \to 0AJ$, $@ \to 1BJ$, $J \to 0AJ$,
    $J \to 1BJ$, $BJ \to J1$, $AJ \to J0$, $0J \to 0d$, $A0 \to 0A$, $A1 \to 1A$,
    $1J \to 1d$, $B0 \to 0B$, $B1 \to 1B$

    In particular, show $@ \overset{*}{\Rightarrow} 110d110$.

20. Consider $G = (V, T, @, P)$; with $V = \{@, a, b\}$, $T = \{a, b\}$, and

$P:\ @ \to \lambda,\ @ \to a@b$

Use mathematical induction to prove $\{a^k b^k : k \in \mathbb{N}\} \subseteq L(G)$.

21. Consider $G = (V, T, @, P)$; with $V = \{@, a, b\}$, $T = \{a, b\}$, and

$P:\ @ \to \lambda,\ @ \to ab@$

Use mathematical induction to prove $\{(ab)^k : k \in \mathbb{N}\} \subseteq L(G)$.

22. Consider $G' = (V, T, @, P)$; with $V = \{@, A, a, b\}$, $T = \{a, b\}$, and

$P:\ @ \to \lambda,\ @ \to ab,\ @ \to aAb,\ A \to aAb,\ A \to ab$

Use mathematical induction to prove $\{a^k b^k : k \in \mathbb{N}\} \subseteq L(G')$.

There is a useful shorthand, called **Backus-Naur Form (BNF)**, *for listing a large set of productions in a context-free grammar. For the productions of the regular grammar in Example 10.3.2, BNF is:*

$@ ::= a|b|\ldots|z|aI|bI|\ldots|zI$
$I ::= aI|bI|\ldots|zI|0I|1I|\ldots|9I$
$I ::= a|b|\ldots|z|0|1|\ldots|9$

The vertical lines in BNF are read "or." So, for example, $@ ::= a|aI|zI$ means a or aI or zI can be directly derived from @. BNF can be used for any context-free grammar.

23. Use BNF to list the productions in the grammar of Exercise 1.
24. Use BNF to list the productions in the grammar of Exercise 19b.

10.5

APPLICATION: PUSH-DOWN AUTOMATA AND STACKS

We have shown how finite state machines are used for syntax checking. But we also know that finite state machines are limited to recognizing regular languages. Some parts of Pascal and other programming languages—for example, identifiers and numerals—can be defined by a regular grammar. But most constructions, such as expressions and statements, can only be defined by a context-free grammar. So, we need a method for recognizing context-free languages. Push-down automata recognize context-free languages.

In Theorem 10.3.6 we proved that a finite state machine is not powerful enough to recognize the set of all well-balanced parentheses. Exercise 1 of Exercise Set 10.4 demonstrates that the set of well-balanced parentheses is a context-free language. In this application we will describe a push-down automaton that recognizes the set of all well-balanced parentheses.

The reason a finite state machine cannot recognize the set of all well-balanced parentheses is that it cannot "remember" the left or right parentheses it has processed. The design of finite state recognizers for regular languages depends on the fact that regular grammars have only productions of the type $N \to t$ or $N \to tM$. A finite state machine mimics such a production with the transition "When in state N, read t and move to state M."

10.5 APPLICATION: PUSH-DOWN AUTOMATA AND STACKS

Context-free grammars can have productions of the type $N \to tMw$. If we try to mimic this production in a finite state machine by the rule "When in state N, read t and move to state M," then there is no way to remember that w needs to be on the right to match the t on the left. A push-down automaton uses a stack or **push-down store** as auxiliary memory to keep track of such things. For example, we imitate the production $N \to tMw$ in a push-down automaton by the rule "When in state N, read t, put w on the stack, and move to state M."

A **stack** is a linearly ordered list for which all insertions and deletions are made from one end, called the **top**, of the list. A string such as *abb* on the stack is read left to right as top down. Stacks are often likened to a pile of dinner plates. When the plates are individually placed on and taken from the pile, the last one placed on the pile will be the first one taken from the pile. A stack, like the dinner-plate pile, is a **last-in-first-out (lifo) structure**.

DEFINITION

> A **push-down automaton (PDA)** consists of:
>
> a finite set of **states** S,
> a set F of **recognizing states**,
> an **initial state** s_0,
> an alphabet V of the **input symbols**,
> an alphabet W of the **stack symbols**,
> a special stack symbol β to designate the **bottom** of the stack, and
> a set of **transitions**.

The transitions will be written as 5-tuples from $S \times (V \cup \{\lambda\}) \times W \times S \times W^*$. Informally, a transition such as (s_1, a, b, s_2, w) means that, when the automaton is in state s_1 and it reads the input symbol a with b on top of the stack, then it moves to state s_2 and replaces the symbol b by the string w on top of the stack. A transition such as $(s_1, \lambda, b, s_2, w)$ is interpreted to mean the PDA can move from state s_1 to state s_2 and replace b by the string w without reading any symbols on the input string.

A PDA reads the input string left to right. Once a PDA has read and acted on a symbol of the input string, it continues reading to the right; in other words, it cannot read any part of the string a second time. The actions of a PDA depend on the state of the PDA, on the leftmost symbol of the unread input string, and on the symbol at the top of the stack.

To **pop** a symbol from the top of the stack is to remove a symbol from the top of the stack; to add a symbol to the top of the stack is to **push** it. For example, the transition (s_1, a, b, s_2, c) pops b and pushes c, the transition (s_1, a, c, s_2, bc) pushes b, and the transition $(s_1, a, b, s_2, \lambda)$ pops b.

Let us consider a PDA with an input alphabet $\{0, 1\}$, a stack alphabet $\{a, b\}$, and a set of states $\{s_0, s_1, s_2\}$. Suppose this PDA has been processing an input string of 0's and 1's. Further, suppose the PDA has reached the stage in which it is in state s_1, the unread input string is 1011, and the stack contains *abbaa*. If the transition $(s_1, 1, a, s_2, baa)$ is in the PDA, it will result in the PDA moving to state s_2 with an unread input string of 011 and a stack containing *baabbaa*. See Figure 10.5.1.

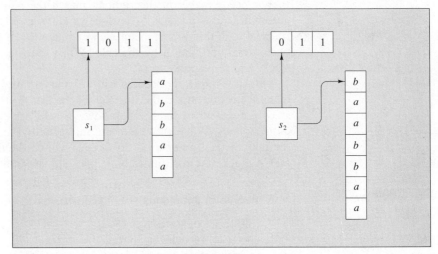

FIGURE 10.5.1   A result of transition $(s_1, 1, a, s_2, baa)$

Before processing a string, an automaton is in the initial state s_0, and the stack contains only β. A push-down automaton **accepts** a string if, after processing the string, the automaton is in a recognizing state *and* the stack is empty. The set **recognized** by the automaton is the set of all strings accepted by the automaton.

As an application of push-down automata, in Example 10.5.1 we design a PDA to recognize the set of well-balanced parentheses. The design of this PDA is based on the fact that a string of parentheses is well-balanced if and only if, while reading the string left to right, the number of right parentheses never exceeds the number of left parentheses and, when we are finished reading the string, the number of right parentheses equals the number of left parentheses.

EXAMPLE 10.5.1   **WELL-BALANCED PARENTHESES CHECKER**   Design a well-balanced parentheses checker.

Solution   To keep count of left parentheses versus right parentheses, the PDA pushes a 1 on the stack whenever a left parenthesis is read and pops a 1 (if possible) whenever a right parenthesis is read. If a right parenthesis is read with only the bottom symbol on the stack, then the number of right parentheses exceeds the number of left parentheses and the PDA will not finish processing the string, indicating that the parentheses are not well-balanced.

We let $S = \{s_0, s_1\}$, $F = \{s_1\}$, $V = \{(,)\}$, and $W = \{1, \beta\}$, where β is the symbol for the bottom of the stack. The set of transitions is:

10.5 APPLICATION: PUSH-DOWN AUTOMATA AND STACKS

1. $(s_0, (, \beta, s_1, 1\beta)$
2. $(s_1,), 1, s_1, \lambda)$
3. $(s_1, (, 1, s_1, 11)$
4. $(s_1, (, \beta, s_1, 1\beta)$
5. $(s_1, \lambda, \beta, s_1, \lambda)$

Table 10.5.1 shows how the string (())() is processed and accepted. According to the last line in Table 10.5.1, the unread input string is empty, the stack is empty, and the automaton is in a recognizing state. Therefore, (())() is accepted. It is instructive to do a similar table for the string ()). If you do so, you will note that, when the unread input string is), the automaton is in state s_1 and the stack contains β. No further moves are possible and the string is therefore not accepted.

TABLE 10.5.1

TRANSITION	STATE	UNREAD INPUT	STACK
—	s_0	(())()	β
1	s_1	())()	1β
3	s_1	))()	11β
2	s_1	)()	1β
2	s_1	()	β
4	s_1	)	1β
2	s_1	λ	β
5	s_1	λ	λ

To implement a push-down automaton as part of a compiler written in Pascal, procedures that manipulate stacks must be defined. A stack is an example of a user-defined data structure in Pascal. Any book on data structures, such as *Data Structures and Program Design* by Robert Kruse, discusses stacks. Stacks are commonly implemented in Pascal with linked lists. In order to manipulate stacks, procedures to implement pop and push are needed.

Stacks are very useful for manipulating and evaluating algebraic expressions. For this reason they are often used to write computer programs that act as calculators. Stacks are also used in some hand-held calculators. If you have ever used a hand-held calculator that works with reverse Polish notation (RPN), then you have worked with a stack. When the keystrokes [5] [enter] are executed on an RPN calculator, the number 5 is pushed on a stack. If 5 is at the top of the stack and the keystrokes [3] [+] are executed, the number 5 is popped, the operation + is carried out to obtain 3 + 5, and the result 8 is pushed on the stack.

Several popular minicomputers, including the HP 3000, use stacks as part of their basic architecture. So, stacks are important in the study of computer architecture as well as data structures.

Perspective: Automata and Formal Languages

Finite state machines are the least powerful automata. Three other kinds of automata are commonly discussed. They are, in increasing order of their power as recognizers, push-down automata, linear bounded automata, and Turing machines. The Turing machine is our best model of a computer, though the model assumes infinite memory. Each type of automaton corresponds to a specific type of set (the language) recognized.

The four types of languages are named Type 3 (regular), Type 2 (context-free), Type 1 (noncontracting or context-sensitive), and Type 0 (phrase structure). As we have seen, Type 3 languages correspond to finite state recognizers, and Type 2 languages correspond to push-down automata. Type 1 languages correspond to linear bounded automata, and Type 0 to Turing machines.

With appropriate care in defining Type 2 languages so that they are generated by noncontracting context-free grammars, we have

Type 3 $\subset$ Type 2 $\subset$ Type 1 $\subset$ Type 0

These languages form the **Chomsky hierarchy**. See Figure 10.5.2.

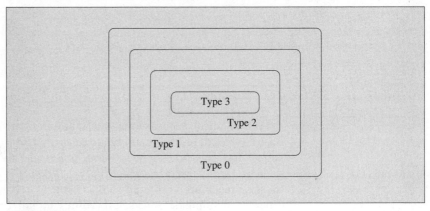

FIGURE 10.5.2

As we have seen, there are context-free languages that are not regular and context-sensitive languages that are not context-free. There are also phrase-structure languages that are not context-sensitive. So, all the set inclusions illustrated in Figure 10.5.2 are proper set inclusions.

EXERCISE SET 10.5

1. Refer to Exercise 19 of Exercise Set 10.4. Design a push-down automaton to recognize the language $L = \{wdw^r : w \in \{0, 1\}^*\}$, where d is a given terminal. Recall that w^r is w in reverse order.

2. It can be proved that there is no push-down automaton that recognizes the language $L = \{wdw : w \in \{0, 1\}^*\}$, where d is a given terminal. Contrast this result with that of Exercise 1, and informally explain why there is no way to design a push-down automaton to recognize L.
3. Design a push-down automaton to recognize the language $L = \{a^n b^n : n \geq 1\}$.
4. Design a push-down automaton to recognize the language

$$L = \{w : w \in \{a, b\}^* \text{ and } w \text{ has the same number of } a\text{'s and } b\text{'s}\}$$

5. Since every regular language is a context-free language, every regular language has a PDA that recognizes it. In other words, for every finite state recognizer, there is a push-down automaton that recognizes the same language. Explain how to design a PDA to emulate a given FSR.

PROGRAMMING EXERCISE

6. Write a computer program to implement the well-balanced parentheses checker.

KEY TERMS

Generative method
Automata method
Words
Parity bit
Odd parity
Even parity
Parity checker
State graph
Finite state recognizer
States
Input symbols
Next-state function
Recognizing states
Initial state
Alphabet
String
Acceptance of a
  string
Recognition of a
  set of strings
Dump state
Error state
Empty string
Regular expressions
Concatenation
  (of strings)
Regular sets
Lexical analyzer

Length (of a string)
Equal strings
Language
Grammar
Phrase structure
  grammar
Terminals
Nonterminals
Start symbol
Productions
Syntax
Semantics
Substring
Directly derived
Direct derivation
Derived
Derivation
Language generated
  by a grammar
Parse tree
Regular grammar
Regular language
Grep
Egrep
Context-free grammar
Context-free language
Membership problem
λ-Production

Noncontracting
  grammar
Parse
Leaves
Yield
Leftmost derivation
Ambiguous grammar
Stack
Push-down store
Top (of stack)
Last-in-first-out
  (lifo) structure
Push-down
  automaton
States (for PDA)
Initial state (for PDA)
Input symbols
  (for PDA)
Stack symbols
Bottom (of stack)
Transitions
Pop
Push
Accepted string
  (for PDA)
Recognized set of
  strings (for PDA)
Chomsky hierarchy

REVIEW EXERCISES

1. **a.** Design an FSR that recognizes $a^* \vee a^3b^*$.
   b. Find a regular grammar G such that $L(G) = a^* \vee a^3b^*$.
   c. Find a context-free grammar G' that is not regular such that $L(G') = a^* \vee a^3b^*$.

2. **a.** Design an FSR that recognizes $a^*b \vee b^2b^*$.
   b. Find a regular grammar G such that $L(G) = a^*b \vee b^2b^*$.
   c. Find a context-free grammar G' that is not regular such that $L(G') = a^*b \vee b^2b^*$.

3. Write a regular expression for the set recognized by the FSR represented by the state graph in the figure.

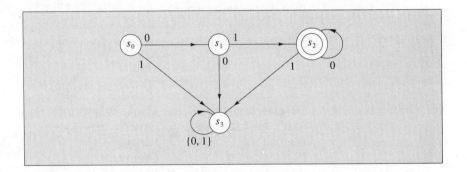

4. Let $L = \{a^n b^m : n, m \in \mathbb{N} \text{ and } m \equiv 1 \pmod{3}\}$. There are two methods for proving L is a regular language. What are they? Use both methods to prove L is a regular language.

5. Let $L = \{a^n b^n : n \in \mathbb{N} \text{ and } n \equiv 1 \pmod{3}\}$. There is one method discussed in this chapter for proving L is a context-free language. What is it? Use the method to prove L is a context-free language.

6. Let $L = \{a^n b^n : n \in \mathbb{N} \text{ and } n \equiv 1 \pmod{3}\}$. There is one method discussed in this chapter for proving L is not a regular language. What is it? Use the method to prove L is not a regular language.

7. Consider $G = (V, T, @, P)$; with $V = \{@, A, 0, 1\}$, $T = \{0, 1\}$, and $P: @ \to A$, $@ \to 1@1$, $A \to 0A$, $A \to 0$.
   a. Show $@ \stackrel{*}{\Rightarrow} 1^3 0^2 1^3$ and draw the parse tree.
   b. If $1^k 0^m 1^n \in L(G)$, what condition must k, m, and n satisfy?
   c. Find all strings in $L(G)$ of length less than or equal to 5.

8. Consider $G = (V, T, @, P)$; with $V = \{@, a, b\}$, $T = \{a, b\}$, and $P: @ \to \lambda$, $@ \to a@b^3$. Use mathematical induction to prove $\{a^k b^{3k} : k \in \mathbb{N}\} \subseteq L(G)$.

9. Let F be the set of all arithmetic expressions of the form: addition of two nonnegative integers or subtraction of two nonnegative integers.
   a. Design an FSR that recognizes F.
   b. Write a regular expression for the set F.

10. Prove $A(BA)^* = (AB)^*A$.

11. Let A be a regular set, and let $A^r = \{w^r : w \in A\}$, where w^r is w in reverse order. Prove A^r is a regular set.

12. In standard Pascal, a **comment** is any string of characters beginning with the character {, ending with the character }, and having no } between { and }. Let L be the set of strings that are valid comments.
   a. Write L as a regular expression.
   b. Design an FSR that recognizes L.
   c. Find a regular grammar that generates L.

13. Many Pascal compilers allow the characters (* as an alternate to { and *) as an alternate to } for comments. (See Exercise 12 above.) Let L be the set of all strings that are valid comments when the compiler allows comments to be enclosed by the pair { and }, or the pair (* and *), or the pair { and *), or the pair (* and }.
   a. Design an FSR that recognizes L.
   b. Find a regular grammar that generates L.

14. The cruise control on many new cars is designed to maintain a steady speed without the driver's foot on the accelerator. It has an off/on switch to activate it. Once the cruise control is activated, a "set speed" switch is turned on when the desired speed is attained. If the driver touches the accelerator, brake, or clutch while the cruise control is maintaining the preset speed, the cruise control loses control of the speed. It remembers the preset speed and, when the driver is ready, a "resume" switch can be used. There are also controls for speeding up and slowing down while the cruise control is controlling the car's speed. Design an FSR that implements the cruise control described above. You may assume the starting state is "cruise control off" and the only recognizing state is "cruise control maintaining preset speed."

REFERENCES

Aho, A. V., and J. D. Ullman. *Principles of Compiler Design.* Reading, MA: Addison-Wesley, 1977.

Backhouse, R. C. *Syntax of Programming Languages: Theory and Practice.* London: Prentice-Hall International, 1979.

Barnier, W. *Finite State Machines as Recognizers.* Arlington, MA: UMAP Module 671, Consortium for Mathematics and Its Applications, 1986.

Barrett, W. A., and J. D. Crouch. *Compiler Construction: Theory and Practice.* Chicago: Science Research Associates, 1979.

Beckman, F. S. *Mathematical Foundations of Programming.* Reading, MA: Addison-Wesley, 1981.

Denning, P. J., J. B. Dennis, and J. E. Qualitz. *Machines, Languages, and Computation.* Englewood Cliffs, NJ: Prentice-Hall, 1978.

Hayes, B. "Computer Recreations." *Scientific American*, 249(6) (December 1983): 19.

Hopcroft, J. E., and J. D. Ullman. *Introduction to Automata Theory, Languages, and Computation.* Reading, MA: Addison-Wesley, 1979.

Kruse, R. *Data Structures and Program Design,* 2nd ed. Englewood Cliffs, NJ: Prentice-Hall, 1987.

Lewis, H. R., and C. H. Papadimitriou. *Elements of the Theory of Computation.* Englewood Cliffs, NJ: Prentice-Hall, 1981.

Lewis, P. M., II, D. J. Rosenkrantz, and R. E. Stearns. *Compiler Design Theory.* Reading, MA: Addison-Wesley, 1976.

Manna, Z. *Mathematical Theory of Computation.* New York: McGraw-Hill, 1974.

McCulloch, W. S., and W. Pitts. "A Logical Calculus of the Ideas Immanent in Nervous Activity." *Bulletin of Mathematical Biophysics,* 5 (1943): 115–133.

McNaughton, R. *Elementary Computability, Formal Languages, and Automata.* Englewood Cliffs, NJ: Prentice-Hall, 1982.

Minsky, M. *Computation: Finite and Infinite Machines.* Englewood Cliffs, NJ: Prentice-Hall, 1967.

Pollack, S. V. (Ed.). *Studies in Computer Science.* Washington, DC: The Mathematical Association of America, 1982.

SOLUTIONS TO SELECTED PRACTICE PROBLEMS

CHAPTER 1

Section 1.1

1. a. $F = \{0, 1, 2, 3, \ldots, 18, 19\}$ and $G = \{-2, -1, 0, 1, 2, 3, 4, 5\}$ **b.** Card$(F) = 20$ and Card$(G) = 8$
2. a. Since $x^2 - 5x + 6 = 0$ can be rewritten as $(x - 2)(x - 3) = 0$, $S = \{2, 3\}$. Hence, $S = T$.
   b. The statement $\emptyset \subseteq A$ is true for any set A; $B \subseteq A$ since every element in B is in A.
3. $A \cap \emptyset = \emptyset$ for any set A; $B \cup \emptyset = B$ for any set B; $A \cap B = \{d, f\}$; $A \cup B = \{a, b, c, d, e, f, 1, 2\}$
4. a. $A \cup B \cup C = \{a, b, c, 3, 4, 5, 6\}$ **b.** $A \cap B \cap C = \{a, 3\}$
5. a. $A' = \{-5, -4, -3, -2, 1, 3, 5\}$; $B' = \{-5, -4, -3, -2, -1, 0, 5\}$; $A \cup B = \{-1, 0, 1, 2, 3, 4\}$; $A \cap B = \{2, 4\}$
   b. $A - B = \{-1, 0\}$; $B - A = \{1, 3\}$; $A - B \neq B - A$ **c.** $\{-1, 0\} \subseteq \{-1, 0, 1, 2, 3, 4\}$
6. a. Card(COLORS) $= 5$
   b. $\{1, 3\}, \{1, 5\}, \{1, 7\}, \{1, 9\}, \{1, 11\}, \{3, 5\}, \{3, 7\}, \{3, 9\}, \{3, 11\}, \{5, 7\}, \{5, 9\}, \{5, 11\}, \{7, 9\}, \{7, 11\}, \{9, 11\}$
7. In Pascal notation, $B * C = [2, 4]$, $B - A = [2, 4]$, $B - C = [1, 3, 5]$, $B + C = [1, 2, 3, 4, 5, 6, 8]$.

Section 1.2

1. The four disjoint sets in terms of A, B, $'$, and $\cap$ are $A \cap B$, $A \cap B'$, $A' \cap B$, and $A' \cap B'$, as shown below.

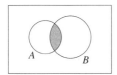
Shaded region is $A \cap B$

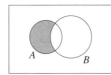
Shaded region is $A \cap B'$

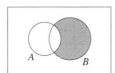
Shaded region is $A' \cap B$

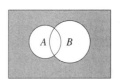
Shaded region is $A' \cap B'$

2.

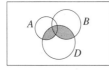
$D \cup (A \cap B)$

$D \cap (A \cup B)$

a. We conjecture that $D \cap (A \cup B) \subseteq D \cup (A \cap B)$.

A-1

2. b. $D \cup (A \cap B) = (A \cap B \cap D) \cup (A \cap B \cap D') \cup (A \cap B' \cap D) \cup (A' \cap B \cap D) \cup (A' \cap B' \cap D)$
   $D \cap (A \cup B) = (A \cap B \cap D) \cup (A \cap B' \cap D) \cup (A' \cap B \cap D)$

3. a. $A \cup B = \{a, b, c, d, e, f\}$ and $B \cup C = \{a, d, e, f, g, k\}$
   b. $(A \cup B) \cup C = \{a, b, c, d, e, f, g, k\}$ and $A \cup (B \cup C) = \{a, b, c, d, e, f, g, k\}$. Therefore, $(A \cup B) \cup C = A \cup (B \cup C)$.
   c. $A \cap B = \{a, d\}$ and $B \cap C = \{d, f\}$
   d. $(A \cap B) \cap C = \{d\}$ and $A \cap (B \cap C) = \{d\}$. Hence, $(A \cap B) \cap C = A \cap (B \cap C)$.

4. a. $A \times B = \{(a, 3), (a, 4), (b, 3), (b, 4), (c, 3), (c, 4), (d, 3), (d, 4), (e, 3), (e, 4)\}$
   b. $B \times A = \{(3, a), (3, b), (3, c), (3, d), (3, e), (4, a), (4, b), (4, c), (4, d), (4, e)\}$
   c. $A \times B \neq B \times A$ since $(a, 3) \in A \times B$ and $(a, 3) \notin B \times A$ **d.** $\text{Card}(A \times B) = 10 = \text{Card}(B \times A)$

5. a. $P(\emptyset) = \{\emptyset\}$ and $\text{Card}(P(\emptyset)) = 1$
   b. $P(A) = \{\emptyset, \{1\}, \{2\}, \{3\}, \{4\}, \{1, 2\}, \{1, 3\}, \{1, 4\}, \{2, 3\}, \{2, 4\}, \{3, 4\}, \{1, 2, 3\}, \{1, 2, 4\}, \{1, 3, 4\}, \{2, 3, 4\}, A\}$ and $\text{Card}(P(A)) = 16$
   c. Since $A \times \emptyset = \emptyset$, $\text{Card}(A \times \emptyset) = 0$.   **d.** $\text{Card}(A \times A) = 16$

6.

S	Card(S)	Card($P(S)$)	Card($S \times S$)
$\emptyset$	0	1	0
$\{a\}$	1	2	1
$\{a, b\}$	2	4	4
$\{a, b, c\}$	3	8	9

Section 1.5

1. The statement $\emptyset \subseteq A$ is vacuously true because a pick-a-point proof of this statement begins with the hypothesis $x \in \emptyset$. But $x \in \emptyset$ is false. Hence, $x \in \emptyset \to x \in A$ is true.

2. a. False   **b.** True   **c.** False

3. Proof: Assume $a < b$ and $u < 0$. Then $a + u < b + u$ by Rule A3. Similarly, $u + b < 0 + b$ by Rule A3. But $u + b = b + u$ and $0 + b = b$. Hence, $b + u < b$. Since $a + u < b + u$ and $b + u < b$, it follows that $a + u < b$ by Rule A1.

CHAPTER 2

Section 2.1

1. a. $((v_3, v_5), (v_5, v_1), (v_1, v_2), (v_2, v_3))$ is a cycle of length 4   **b.** $((v_4, v_3) (v_3, v_5), (v_5, v_5), (v_5, v_1), (v_1, v_4))$ is a circuit of length 5
2. $(v_1, v_2, v_3, v_4, v_2, v_5, v_1)$ is a circuit. This circuit is not a cycle, and it contains no self-loop.

3. $((v_2, v_3), (v_3, v_1), (v_1, v_4))$, $((v_2, v_3), (v_3, v_5), (v_5, v_1), (v_1, v_4))$

Section 2.2

1. $(1, 1, 2)$ does not belong to R because $1^2 + 1^2 \neq 2^2$; $(3, 4, 5) \in R$; $(4, 5, 5) \notin R$; $(0, 5, 5) \in R$; $(8, 6, 10) \in R$
2. P_{12}(Schedule)

COURSE	CREDIT
C.S. 101	4
C.S. 102	3
C.S. 408	3
C.S. 514	3

3. a. $R = \{(1, 1), (1, 3), (2, 1), (2, 2), (2, 3), (3, 2), (3, 3), (4, 4)\}$ **b.** $S = \{(1, 1), (2, 1), (3, 1), (3, 2), (3, 3)\}$ **c.** $T = \{(1, 1), (2, 2)\}$
4. a. Not reflexive since there is no edge from vertex 4 to vertex 4.
  b. Not symmetric since there is an edge from 4 to 2 but not from 2 to 4.
  c. Not antisymmetric since there are edges from 2 to 3 and from 3 to 2.
  d. Not transitive since there is an edge from 4 to 2, an edge from 2 to 3, but no edge from 4 to 3.
5. (c, c)   **6. c.** (a, b) $\in R$ and (b, d) $\in R$ but (a, d) $\notin R$
7. a. For a relation R, W is the reflexive closure of R whenever:
  1. $R \subseteq W$;
  2. W is reflexive; and
  3. If V is any reflexive relation and $R \subseteq V$, then $W \subseteq V$.
b. $R \cup \{(a_1, a_1), (a_2, a_2), \ldots, (a_n, a_n)\}$

Section 2.3

2. First, R is a subset of R^+. Since there is a path from 1 to 3 and a path from 1 to 4, (1, 3) and (1, 4) are elements of R^+. There are no other pairs in R^+.
3. $R^2 = \{(a, a), (a, b), (b, a), (b, b), (c, c)\}$, $R^2 \circ R = \{(a, a), (a, b), (b, a), (b, b), (c, c)\}$, $R \circ R^2 = \{(a, a), (a, b), (b, a), (b, b), (c, c)\}$. Therefore, $R \circ R^2 = R^2 \circ R$.
4. $R \circ R = \{(1, 4), (2, 3), (3, 3), (4, 4)\}$, $R \circ S = \{(1, 3), (2, 2), (4, 3)\}$, and $S \circ R = \{(1, 1), (3, 4)\}$

Section 2.4

1. $M = \begin{array}{c} \\ v_1 \\ v_2 \\ v_3 \\ v_4 \\ v_5 \end{array} \begin{array}{c} \begin{array}{ccccc} v_1 & v_2 & v_3 & v_4 & v_5 \end{array} \\ \begin{bmatrix} 0 & 1 & 0 & 1 & 0 \\ 0 & 1 & 1 & 0 & 0 \\ 1 & 1 & 0 & 0 & 1 \\ 0 & 0 & 1 & 0 & 1 \\ 1 & 0 & 0 & 0 & 1 \end{bmatrix} \end{array}$ **2.** $M_R = \begin{array}{c} \\ a \\ b \\ c \end{array} \begin{array}{c} \begin{array}{ccc} a & b & c \end{array} \\ \begin{bmatrix} 1 & 1 & 1 \\ 0 & 1 & 0 \\ 0 & 1 & 1 \end{bmatrix} \end{array}$

The relation R is reflexive since every entry on the main diagonal of M_R is 1. Fold the matrix M_R over along the main diagonal. We see that R is not symmetric since not all of the entries off the main diagonal match each other. R is antisymmetric since every "1" off the main diagonal is matched by a "0."

3. $P \oplus Q = \begin{bmatrix} 1+1 & 0+1 \\ 0+1 & 0+0 \end{bmatrix} = \begin{bmatrix} 1 & 1 \\ 1 & 0 \end{bmatrix}$ **4.** $P \otimes Q = \begin{bmatrix} 1 & 0 \\ 1 & 1 \\ 1 & 1 \end{bmatrix}$

Here is the computation for the entry in row 3, column 2 of $P \otimes Q$:

$$\begin{bmatrix} 1 & 0 & 1 \\ 1 & 1 & 0 \\ 0 & 1 & 1 \end{bmatrix} \otimes \begin{bmatrix} 1 & 0 \\ 1 & 1 \\ 1 & 0 \end{bmatrix} = \begin{bmatrix} 1 & 0 \\ 1 & 1 \\ 1 & 0 \cdot 0 + 1 \cdot 1 + 1 \cdot 0 \end{bmatrix}$$

Section 2.5

2. $M_{R^+} = \begin{bmatrix} 1 & 1 & 0 & 0 \\ 1 & 1 & 0 & 0 \\ 1 & 1 & 0 & 0 \\ 1 & 1 & 1 & 1 \end{bmatrix}$ **3.** $P_3[1, 3] = 1$ and $P_3[4, 5] = 1$

4. Algorithm Reach
  REACH := MATRIX ∨ IDENTITY
  for k = 1 to n do
    for i = 1 to n do
      for j = 1 to n do
        REACH [i, j] := REACH [i, j] + (REACH [i, k] · REACH [k, j])   (*Where + and · are Boolean addition and
  Return REACH.   Boolean multiplication respectively.*)

CHAPTER 3

Section 3.1

1. R is reflexive since each person has the same last name as himself or herself; R is symmetric since, if person x has the same last name as person y, then person y has the same last name as person x; R is transitive since, if person x has the same last name as person y and person y has the same last name as person z, then person x has the same last name as person z. Thus, by definition, R is an equivalence relation.
2. Let $A = \{1, 2, 3, 4, 5, 6, 7, 8, 9, 10\}$. The relation $\equiv$ is reflexive since $m - m = 0 = 4 \cdot 0$ and $0 \in \mathbb{Z}$. The relation $\equiv$ is symmetric for, if $n - m = 4k$, then $m - n = 4(-k)$. The relation $\equiv$ is transitive since, if $m \equiv n$ and $n \equiv p$, then $n - m = 4k$ and $p - n = 4j$ for some $k, j \in \mathbb{Z}$. But $p - m = (p - n) + (n - m) = 4j + 4k = 4(j + k)$. Hence, $p - m$ is divisible by 4 and, therefore, $m \equiv p$. The equivalence classes are: $\{1, 5, 9\}, \{2, 6, 10\}, \{3, 7\}, \{4, 8\}$.
3. The reflexive property is used in the proof of parts 1 and 3. The symmetric and transitive properties are used in the proof of part 2.
4. **a.** For example, $\{1, 2, 3\}$ and $\{4\}$ are not comparable.
   b. Both $\varnothing$ and $\mathbb{Z}$ are comparable to every other member in $P(\mathbb{Z})$, and these are the only two such members.

Section 3.2

1.   2. The poset $\mathbb{Z}$ with the standard order has neither a maximal nor a minimal element.

3. L_1: $e \prec a \prec b \prec c \prec d \prec f \prec g$; L_2: $a \prec b \prec c \prec d \prec e \prec f \prec g$
4. **a.** The lower bounds of $\{d, f\}$ are a, c, and d. The upper bound of $\{d, f\}$ is f.
   b. The lower bound of $\{c, d, e\}$ is c. The upper bound of $\{c, d, e\}$ does not exist.
   c. There are no lower bounds or upper bounds of $\{a, b\}$.
5. **a.** The greatest lower bound of $\{d, f\}$ is d. The least upper bound of $\{d, f\}$ is f.
   b. The greatest lower bound of $\{c, d, e\}$ is c. The least upper bound of $\{c, d, e\}$ does not exist.
   c. There is no greatest lower bound or least upper bound of $\{a, b\}$.

Section 3.3

1. The digraphs are shown below. The relation f is not a function since the points c and d in D have no images under f. The relation g is a function where $g(a) = 1, g(b) = 1, g(c) = 2$, and $g(d) = 3$.

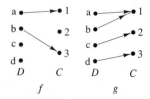

2. a. The table for $\chi_{A \cup B}$ is:

x	u_1	u_2	u_3	u_4	u_5
$\chi_{A \cup B}(x)$	1	1	1	0	1

The bit string for $A \cup B$ is 11101. We see that $\chi_{A \cup B}(x) = 1$ when $\chi_A(x) = 1$ or $\chi_B(x) = 1$.

c. The table for $\chi_{A'}$ is:

x	u_1	u_2	u_3	u_4	u_5
$\chi_{A'}(x)$	1	0	0	1	0

The bit string for A' is 10010. We see that $\chi_{A'}(x) = 1$ when $\chi_A(x) = 0$.

b. The table for $\chi_{A \cap B}$ is:

x	u_1	u_2	u_3	u_4	u_5
$\chi_{A \cap B}(x)$	0	0	1	0	1

The bit string for $A \cap B$ is 00101. We observe that $\chi_{A \cap B}(x) = 1$ when $\chi_A(x) = 1$ and $\chi_B(x) = 1$.

The table for $\chi_{B'}$ is:

x	u_1	u_2	u_3	u_4	u_5
$\chi_{B'}(x)$	0	1	0	1	0

The bit string for B' is 01010. We note that $\chi_{B'}(x) = 1$ when $\chi_B(x) = 0$.

Section 3.4

1. Assume $g(x) = \log_2 x = \log_2 w = g(w)$. Then, $x = 2^{\log_2 x} = 2^{\log_2 w} = w$. Therefore, g is 1–1.

2. a. Let $w \in \mathbb{R}^+$. We want $2^x = w$ for some x. Since we know $2^{\log_2 w} = w$, we choose $x = \log_2 w$. Then, $f(x) = w$. Therefore, f is onto.

b. Because $5 \in \mathbb{N}$ has no preimage, h is not onto.

3. The four additional permutations are: $\begin{pmatrix} 1 & 2 & 3 \\ 1 & 3 & 2 \end{pmatrix}$, $\begin{pmatrix} 1 & 2 & 3 \\ 2 & 1 & 3 \end{pmatrix}$, $\begin{pmatrix} 1 & 2 & 3 \\ 3 & 1 & 2 \end{pmatrix}$, $\begin{pmatrix} 1 & 2 & 3 \\ 3 & 2 & 1 \end{pmatrix}$.

4. We are given

$$f = \begin{pmatrix} 1 & 2 & 3 \\ 2 & 3 & 1 \end{pmatrix} \text{ and } g = \begin{pmatrix} 1 & 2 & 3 \\ 2 & 1 & 3 \end{pmatrix}$$

Since $(f \circ g)(1) = f(g(1)) = f(2) = 3$, $(f \circ g)(2) = f(g(2)) = f(1) = 2$, and $(f \circ g)(3) = f(g(3)) = f(3) = 1$, we have

$$g \circ f = \begin{pmatrix} 1 & 2 & 3 \\ 3 & 2 & 1 \end{pmatrix}$$

5. We will show that $g \circ f = 1_\mathbb{R}$ and $f \circ g = 1_\mathbb{R}$. For each $x \in \mathbb{R}$, we have

$$(g \circ f)(x) = g(f(x)) = g(8x + 1) = \frac{(8x + 1) - 1}{8} = x = 1_\mathbb{R}(x)$$

and

$$(f \circ g)(x) = f(g(x)) = f\left(\frac{x-1}{8}\right) = 8\left(\frac{x-1}{8}\right) + 1 = x - 1 + 1 = x = 1_\mathbb{R}(x)$$

Hence, g is the inverse of f, and f is the inverse of g.

6. We have

$$(f \circ f^{-1})(x) = f(f^{-1}(x)) = f\left(\frac{x-5}{3}\right) = 3\left(\frac{x-5}{3}\right) + 5 = x - 5 + 5 = x$$

and

$$(f^{-1} \circ f)(x) = f^{-1}(f(x)) = f^{-1}(3x + 5) = \frac{(3x + 5) - 5}{3} = \frac{3x}{3} = x$$

Therefore, $(f \circ f^{-1})(x) = x = (f^{-1} \circ f)(x)$.

7. We have

$(f \circ f^{-1})(1) = f(f^{-1}(1)) = f(3) = 1 \quad (f \circ f^{-1})(2) = f(f^{-1}(2)) = f(1) = 2 \quad (f \circ f^{-1})(3) = f(f^{-1}(3)) = f(2) = 3$

Hence, $\qquad\qquad\qquad\qquad$ Similarly,

$$f \circ f^{-1} = \begin{pmatrix} 1 & 2 & 3 \\ 1 & 2 & 3 \end{pmatrix} \qquad g^{-1} \circ g = \begin{pmatrix} 1 & 2 & 3 \\ 1 & 2 & 3 \end{pmatrix}$$

CHAPTER 4
Section 4.1

1. Since $A = \{1, 2, 3, 4, 5, 6, 7\}$, $B = \{4, 7, 10, 12, 13\}$, and $C = \{2, 4, 6\}$, we have
   $A \cup B = \{1, 2, 3, 4, 5, 6, 7, 10, 12, 13\}$ $A \cap B = \{4, 7\}$ $A \cup C = \{1, 2, 3, 4, 5, 6, 7\}$ $A \cap C = \{2, 4, 6\}$
   $B \cup C = \{2, 4, 6, 7, 10, 12, 13\}$ $B \cap C = \{4\}$
   a. Card$(A \cup B) = 10$ Card$(A \cup C) = 7$ Card$(B \cup C) = 7$
   b. Card$(A \cap B) = 2$ Card$(A \cap C) = 3$ Card$(B \cap C) = 1$
   c. Card$(A \cup B) = 10 = 7 + 5 - 2 = $ Card$(A) + $ Card$(B) - $ Card$(A \cap B)$
   d. Card$(A \cup C) = 7 = 7 + 3 - 3 = $ Card$(A) + $ Card$(C) - $ Card$(A \cap C)$
   e. Card$(B \cup C) = 7 = 5 + 3 - 1 = $ Card$(B) + $ Card$(C) - $ Card$(B \cap C)$

2. Let P be the set of all juniors in our sample who can write computer programs in Pascal. Let F be the set of all juniors in our sample who can write computer programs in FORTRAN. Thus,
   $$\text{Card}(P \cup F) = \text{Card}(P) + \text{Card}(F) - \text{Card}(P \cap F) = 800 + 500 - 325 = 975$$
   Hence, there are $1000 - 975 = 25$ juniors who can write computer programs in neither Pascal nor FORTRAN.

3. $8 \cdot 10^6$

4. a. 7, 8, 6, 9, 3, 3, 3, 2, 2, 3, 2, 1, 2, 1, 1
   b. Card$(A \cup B \cup C \cup D) = 7 + 8 + 6 + 9 - 3 - 3 - 3 - 2 - 2 - 3 + 2 + 1 + 2 + 1 - 1 = 19$

5.
S	Card(S)	Card$(P(S))$	Card$(S \times S)$
$\emptyset$	0	$2^0 = 1$	$0 = 0 \cdot 0$
$\{a\}$	1	$2^1 = 2$	$1 = 1 \cdot 1$
$\{a, b\}$	2	$2^2 = 4$	$4 = 2 \cdot 2$
$\{a, b, c\}$	3	$2^3 = 8$	$9 = 3 \cdot 3$

Section 4.2

1. The arrangement of A corresponding to the bijective function f from A to A is $f(a)f(b)f(c)f(d)f(e)$, which is bcdea.
2. The set of letters in NUMBER is $\{N, U, M, B, E, R\}$, and Card$\{N, U, M, B, E, R\} = 6$. By Theorem 4.2.3, the number of permutations is $6! = 720$.
3. Since there are 4 choices for the first element, 3 choices for the second element, and 2 choices for the third element, there are $4 \cdot 3 \cdot 2 = 24$ 3-permutations.
4. There are 26 letters. By Theorem 4.2.6, the number of 3-letter initials with distinct letters is $P(26, 3) = 26 \cdot 25 \cdot 24$. There are also $P(26, 3)$ 1–1 functions from $\{0, 1, 2\}$ to the set of 26 letters of the alphabet.
5. There are 9 choices for the first digit since 0 cannot be used; there are 8 choices for the second digit, since 0 can be used but the second digit cannot be equal to the first digit or the fourth digit. There is 1 choice for the third digit, and there are 8 choices for the fourth digit. Hence, the number of passwords is $9 \cdot 8 \cdot 1 \cdot 8 = 576$.
6. There are five odd integers in $D = \{0, 1, 2, 3, 4, 5, 6, 7, 8, 9\}$. Hence, the required number is 5^7.   7. 6!3!

Section 4.3

1. $\{1, 2, 3\}, \{1, 2, 4\}, \{1, 3, 4\}, \{2, 3, 4\}$ 2. $C(25, 3) \cdot C(20, 2) = 437{,}000$
3. $(a + b)^6 = C(6, 0)a^{6-0}b^0 + C(6, 1)a^{6-1}b^1 + C(6, 2)a^{6-2}b^2 + C(6, 3)a^{6-3}b^3 + C(6, 4)a^{6-4}b^4 + C(6, 5)a^{6-5}b^5 + C(6, 6)a^{6-6}b^6$
   $= a^6 + 6a^5b + 15a^4b^2 + 20a^3b^3 + 15a^2b^4 + 6ab^5 + b^6$

4. $(3+4)^5 = 7^5$ **5.**
$$
\begin{array}{ll}
n=0 & 1=2^0 \\
n=1 & 2=2^1 \\
n=2 & 4=2^2 \\
n=3 & 8=2^3 \\
n=4 & 16=2^4 \\
n=5 & 32=2^5 \\
\vdots & \vdots
\end{array}
$$
The sum of the coefficients in row n of Pascal's triangle is 2^n.

6. $(x+y)^4 = x^4 + 4x^3y + 6x^2y^2 + 4xy^3 + y^4$ **7.** $n=4, r=8$, and $C(n-1+r, r) = C(11, 8) = \dfrac{11!}{8!3!} = 165$

Section 4.4

1. a. $\Pr(\{a\}) = \frac{1}{5}$ since we have an equiprobable space with 5 simple events. **b.** $\Pr(\{a, b\}) = \frac{2}{5}$ **c.** $\Pr(\{c, d, e\}) = \frac{3}{5}$
2. $E_3 = \{(1,2), (2,1)\}$, $E_7 = \{(1,6), (2,5), (3,4), (4,3), (5,2), (6,1)\}$, $E_9 = \{(3,6), (4,5), (5,4), (6,3)\}$; $\Pr(E_3) = \frac{2}{36} = \frac{1}{18}$, $\Pr(E_7) = \frac{6}{36} = \frac{1}{6}$, $\Pr(E_9) = \frac{4}{36} = \frac{1}{9}$
3. $\Pr(\emptyset) = 0$ since $\Pr(\emptyset) = \Pr(S') = 1 - P(S) = 1 - 1 = 0$
4. Let A be the event of obtaining at least one tail in the last two tosses. Then, A' is the event of obtaining no tail in the last two tosses, and $A' = \{HHH, THH\}$. Hence, $\Pr(A) = 1 - \Pr(A') = 1 - \frac{2}{8} = \frac{3}{4}$.

CHAPTER 5

Section 5.1

1. a. False **b.** True **2.**

Steps: $(p \wedge \neg q) \to (p \vee q)$, 3 2 4 2

p	q	$(p \wedge \neg q)$	$\to$	$(p \vee q)$
T	T	F F T	T	
T	F	T T T	T	
F	T	F F T	T	
F	F	F T T	F	

Steps: 1 1 3 2 4 2

Section 5.2

1.

p	q	$\neg(p \vee q)$	$\neg p \wedge \neg q$
T	T	F	F
T	F	F	F
F	T	F	F
F	F	T	T

2.

p	q	$[p \wedge (p \to q)]$	$\to$	q
T	T	T	T	T
T	F	F	T	F
F	T	F	T	T
F	F	F	T	F

3.

p	q	$(p \to q) \wedge (q \to p)$
T	T	T T T
T	F	F F T
F	T	T F F
F	F	T T T

4. We need only show that OR can be written in terms of NOT and AND. By De Morgan's law and the rule of double negation, $p \vee q \equiv \neg(\neg p \wedge \neg q)$.

Section 5.3

1. a. Assume $mn < 100$. Suppose it is not the case that $m < 10$ or $n < 10$. Then, $m \geq 10$ and $n \geq 10$. Hence, $mn \geq 100$. This contradicts our assumption.
b. We prove: If it is not the case that $m < 10$ or $n < 10$, then it is not the case that $mn < 100$. Assume it is not the case that $m < 10$ or $n < 10$. Then, $m \geq 10$ and $n \geq 10$. Hence, $mn \geq 100$. Therefore, it is not the case that $mn < 100$.

2. Let x and y be two integers. We prove: xy is even if and only if x is even or y is even. Proof: ($\rightarrow$) Assume xy is even. Suppose it is not the case that x is even or y is even. Then, x is odd and y is odd. Hence, $x = 2j + 1$ and $y = 2k + 1$ for some integers j and k. But this yields $xy = 2m + 1$, where $m = 2(2jk + j + k)$, and hence xy is odd. Contradiction. ($\leftarrow$) Assume x is even or y is even. If x is even, then $x = 2k$ for some integer k. Hence, $xy = 2m$, and xy is even. The proof is similar when y is even.

3. Reasons in order: definition of difference set, commutative law, distributive law, commutative law, definition of difference set

Section 5.4

1. Basis step ($n = 1$): The sum of the first (one) summand on the left is 1, and $1 = 2^1 - 1$. Induction hypothesis: Assume $1 + 2 + 2^2 + \cdots + 2^{k-1} = 2^k - 1$. Hence, $1 + 2 + 2^2 + \cdots + 2^{k-1} + 2^k = 2^k - 1 + 2^k = 2 \cdot 2^k - 1 = 2^{k+1} - 1$.
2. Basis step ($n = 3$): $2 \cdot 3 + 1 = 7 \leq 8 = 2^3$. Induction hypothesis: Assume $2k + 1 \leq 2^k$. Then, $2(k + 1) + 1 = 2k + 1 + 2$. Hence, $2(k + 1) + 1 \leq 2^k + 2$ by the induction hypothesis. But, $2^k + 2 \leq 2^k + 2^k$ for $k \geq 1$. Using the fact that $2^k + 2^k = 2 \cdot 2^k = 2^{k+1}$, we conclude that $2(k + 1) + 1 \leq 2^{k+1}$.

Section 5.5

1. The quantified statement is "u is a greatest element for a set A of real numbers if and only if $u \in A \wedge \forall x(x \in A$ implies $x \leq u)$."
2. The statement $\forall x\, \exists y\, (x + y = 0)$ is true because, for $U = \mathbb{Z}$, for all x there is an additive inverse $-x$ such that $x + (-x) = 0$. The statement $\exists y\, \forall x\, (x + y = 0)$ is false because, when y is chosen first, there exists an element x such that $x + y \neq 0$. For example, $x = 1 - y$ will make $x + y \neq 0$.
3. The statement $\neg\, \exists x\, (p(x))$ is read "There is no element x such that $p(x)$ is true." And $\forall x\, (\neg p(x))$ is read "For every x, $p(x)$ is false."
4. Let $U = \{u, v\}$. Now we write Rule Q4: $\neg(p(u) \vee p(v))$ means $\neg\, \exists x\, (p(x))$ means $\forall x(\neg p(x))$ means $\neg p(u) \wedge \neg p(v)$.
5. $\neg(A \subseteq B) \leftrightarrow \neg\, \forall u\, (u \in A \rightarrow u \in B) \leftrightarrow \exists u\, (u \in A \wedge u \notin B)$
6. **a.** $\neg\, \exists x\, \exists p\, (x + p = 0) \leftrightarrow \forall x\, [\neg\, \exists p\, (x + p = 0)] \leftrightarrow \forall x\, \forall p\, (x + p \neq 0)$
   c. $\neg\, \exists y\, \forall x\, (x + y = 0) \leftrightarrow \forall y\, [\neg\, \forall x\, (x + y = 0)] \leftrightarrow \forall y\, \exists x\, (x + y \neq 0)$
   Statements a and b are true. The negations of c and d are also true.
7. $\neg\, \exists x\, (p(x) \wedge o(x)) \leftrightarrow \forall x\, \neg(p(x) \wedge o(x)) \leftrightarrow \forall x\, (\neg p(x) \vee \neg o(x)) \leftrightarrow \forall x\, (p(x) \rightarrow \neg o(x))$
8. Let R be a relation on a set A. The universe $U = A$.
   a. R is reflexive. $\qquad \forall x\, ((x, x) \in R)$
   R is not reflexive. $\qquad \exists x\, ((x, x) \notin R)$
   b. R is symmetric. $\qquad \forall x\, \forall y\, ((x, y) \in R$ implies $(y, x) \in R)$
   R is not symmetric. $\qquad \exists x\, \exists y\, ((x, y) \in R \wedge (y, x) \notin R)$
   c. R is antisymmetric. $\qquad \forall x\, \forall y\, ((x, y) \in R \wedge (y, x) \in R$ implies $x = y)$
   R is not antisymmetric. $\qquad \exists x\, \exists y\, ((x, y) \in R \wedge (y, x) \in R \wedge x \neq y)$
   d. R is transitive. $\qquad \forall x\, \forall y\, \forall z\, ((x, y) \in R \wedge (y, z) \in R$ implies $(x, z) \in R)$
   R is not transitive. $\qquad \exists x\, \exists y\, \exists z\, ((x, y) \in R \wedge (y, z) \in R \wedge (x, z) \notin R)$
9. **a.** $\bigcap\{A, B\} = \{x : (\forall Y \in \{A, B\})(x \in Y)\} = \{x : x \in A \wedge x \in B\} = A \cap B$
   b. $\bigcup\{A, B\} = \{x : (\exists X \in \{A, B\})(x \in X)\} = \{x : x \in A \vee x \in B\} = A \cup B$
10. Prove: $D \cup \bigcap\Omega = \bigcap\{D \cup A : A \in \Omega\}$. Proof: Let $x \in D \cup \bigcap\Omega$. Then $x \in D$ or $x \in \bigcap\Omega$. Hence, $x \in D$ or $x \in A$ for all $A \in \Omega$. So $x \in D \cup A$ for all $A \in \Omega$. Therefore, $x \in \bigcap\{D \cup A : A \in \Omega\}$. Hence, $D \cup \bigcap\Omega \subseteq \bigcap\{D \cup A : A \in \Omega\}$. Reverse each of the steps above to prove the other set inclusion.

CHAPTER 6

Section 6.1

1. The primes between 20 and 50 are 23, 29, 31, 37, 41, 43, and 47.
2. Euclid's proof for Theorem 6.1.5: Assume there are a finite number of primes: $p_0, p_1, \ldots, p_k$. We need only find one more prime to obtain a contradiction. Let
$$m = 1 + p_0 \cdot p_1 \cdot \cdots \cdot p_k$$
Now prove m has a prime factor different from any of $p_0, p_1, \ldots, p_k$.

3. a. $q = 16$ and $r = 3$ **b.** $q = 0$ and $r = 17$ **c.** $q = -5$ and $r = 2$ **d.** $q = -7$ and $r = 2$
4. Let $m = 2$ and $n = -3$. We have $1 = 20m + 13n = 20 \cdot 2 + 13(-3)$.
5. a. 29 div 3 = 9   **b.** 29 mod 3 = 2   **c.** -21 div 4 = -5 **d.** -21 mod 4 = 3

Section 6.2

1. gcd(48, 18) = 6 because:
$$48 = 2 \cdot 18 + 12 \quad 18 = 1 \cdot 12 + 6 \quad 12 = 2 \cdot 6 + 0$$
2. $155 = 7 \cdot 20 + 15 \quad 20 = 1 \cdot 15 + 5 \quad 15 = 3 \cdot 5 + 0$
Hence, gcd(155, 20) = 5.
$$5 = 20 - 1 \cdot 15 = 20 - (155 - 7 \cdot 20) = (-1)155 + 8 \cdot 20$$
Hence, $m = -1$ and $n = 8$.
3. Factoring yields $155 = 5 \cdot 31$ and $20 = 2^2 \cdot 5$. Hence, lcm(155, 20) = $2^2 \cdot 5 \cdot 31$ and gcd(155, 20) = 5. So, $155 \cdot 20 = 5(2^2 \cdot 5 \cdot 31) = $ gcd(155, 20) $\cdot$ lcm(155, 20).

Section 6.3

1. a. $\frac{2043}{3} = 681 + 0$; hence, $2043 \equiv 0 \pmod{3}$ **b.** $\frac{759}{8} = 94 + \frac{7}{8}$; hence, $759 \equiv 7 \pmod{8}$
2. a. $[8] +_{12} [7] = [3]$ **b.** $[1] +_5 [3] = [4]$ **c.** $[3] +_5 [4] = [2]$
3. a. $(10011010)_2 = 2 + 8 + 16 + 128 = 154$
   b. $169 = 128 + 41 = 128 + 32 + 9 = 128 + 32 + 8 + 1$; hence, $169 = (10101001)_2$
5. $784 = 2 \cdot 392 + 0 \quad 392 = 2 \cdot 196 + 0 \quad 196 = 2 \cdot 98 + 0 \quad 98 = 2 \cdot 49 + 0 \quad 49 = 2 \cdot 24 + 1 \quad 24 = 2 \cdot 12 + 0$
$12 = 2 \cdot 6 + 0 \quad 6 = 2 \cdot 3 + 0 \quad 3 = 2 \cdot 1 + 1 \quad$ Hence, $784 = (1100010000)_2$.

CHAPTER 7

Section 7.1

1. Axiom 4a, $x + 0 = x$: Since $x = 1$, we have $x + 0 = 1 + 0 = 1 = x$ by the operation $+$ defined on $B = \{0, 1\}$.
Axiom 5b, $x \cdot x' = 0$: Since $x = 1$, we have $x \cdot x' = 1 \cdot 1' = 1 \cdot 0 = 0$ by the operations $\cdot$ and $'$ defined on $B = \{0, 1\}$ in Example 7.1.2.

2.

S_1	S_2	$L = S_1 \cdot S_2$
0	0	0
0	1	0
1	0	0
1	1	1

We have $L = S_1 \cdot S_2$ as shown in the table. Note that $L = 1$ only when both $S_1 = 1$ and $S_2 = 1$; $L = 0$ otherwise.

3. $P(U) = \{\emptyset, \{a\}, \{b\}, U\}$. We verify Boolean laws 5a and 5b for set theory for $X = \emptyset, \{a\}, \{b\}, U$ in the table. Hence, Boolean laws 5a and 5b for set theory hold.

X	X'	$X \cup X' = U$	$X \cap X' = \emptyset$
$\emptyset$	U	$\emptyset \cup U = U$	$\emptyset \cap U = \emptyset$
$\{a\}$	$\{b\}$	$\{a\} \cup \{b\} = U$	$\{a\} \cap \{b\} = \emptyset$
$\{b\}$	$\{a\}$	$\{b\} \cup \{a\} = U$	$\{b\} \cap \{a\} = \emptyset$
U	$\emptyset$	$U \cup \emptyset = U$	$U \cap \emptyset = \emptyset$

4. Reasons, in order: identity law, complement law, distributive law, commutative law, assumption, complement law, commutative law, distributive law, assumption, identity law

5. 1a. $\text{lcm}(x, y) = \text{lcm}(y, x)$ **1b.** $\gcd(x, y) = \gcd(y, x)$
2a. $\text{lcm}(\text{lcm}(x, y), z) = \text{lcm}(x, \text{lcm}(y, z))$ **2b.** $\gcd(\gcd(x, y), z) = \gcd(x, \gcd(y, z))$
3a. $\text{lcm}(x, \gcd(y, z)) = \gcd(\text{lcm}(x, y), \text{lcm}(x, z))$ **3b.** $\gcd(x, \text{lcm}(y, z)) = \text{lcm}(\gcd(x, y), \gcd(x, z))$
4a. $\text{lcm}(x, 1) = x$ **4b.** $\gcd(x, 14) = x$
5a. $\text{lcm}(x, 14/x) = 14$ **5b.** $\gcd(x, 14/x) = 1$

6. In $(D_{12}, +, \cdot, ', 1, 12)$, Axiom 5a $(x + x' = 1)$ is $x + 12/x = 12$. But, for $x = 2$, $x + x' = 2 + 12/2 = 2 + 6 = \text{lcm}(2, 6) = 6 \neq 12$.

Section 7.2

1.

x	f(x)	x	g(x)	x	h(x)	x	k(x)
0	1	0	0	0	0	0	1
1	0	1	1	1	0	1	1

2. a. $\text{Card}(B) = 2$, $\text{Card}(B^3) = 2^3 = 8$ **b.** $2^{2^3} = 2^8 = 256$

3.

x	y	z	g(x, y, z) = yx + xz' + yy'
0	0	0	0
0	0	1	0
0	1	0	0
0	1	1	0
1	0	0	1
1	0	1	0
1	1	0	1
1	1	1	1

4.

x	y	f(x) = xx'
0	0	0
0	1	0
1	0	0
1	1	0

5.

x	y	z	x(y + z)	xy + xz	xy + xz + xx'
0	0	0	0	0	0
0	0	1	0	0	0
0	1	0	0	0	0
0	1	1	0	0	0
1	0	0	0	0	0
1	0	1	1	1	1
1	1	0	1	1	1
1	1	1	1	1	1

Since the last three columns have identical values, the three Boolean expressions define the same Boolean function from $\{0, 1\}^3$ to $\{0, 1\}$.

6. $(x + y)'(y' + x) + x' = x'y'(y' + x) + x' = x'y'y' + x'y'x + x' = x'y' + x'xy' + x' = x'y' + 0 \cdot y' + x' = x'y' + x' = x'(y' + 1) = x' \cdot 1 = x'$

7. There are 2^n minterms.

8. First we define the Boolean function by $(x + y)(x' + y')$ in a truth table.

x	y	f(x, y) = (x + y)(x' + y')
0	0	0
0	1	1
1	0	1
1	1	0

From the truth table, we obtain the minterms $x'y$ and xy'. Hence, $(x + y)(x' + y')$ is equivalent to the standard sum of products $x'y + xy'$.

9. The standard sum of products that defines g is $x'y'z + xy'z + xyz'$.

10. We need only show that $+$ can be written in terms of $'$ and $\cdot$. By De Morgan's law and the involution law, $x + y = (x' \cdot y')'$.

Section 7.3

1.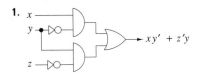

2. $(x + (y + z)')'$ **3.** $(x + y')(x'y)'$

Section 7.4

1. $0 \leq x$ follows from $x \cdot 0 = 0$ **2.** The atoms are 3, 7, and 11.
3. Proof by contradiction: Suppose $a_1 \neq a_2$. Then, since $a_1 \leq a_2$ and a_2 is an atom, $a_1 = 0$. Contradiction.
4. Atom$[\{a, c\}] = \{\{a\}, \{c\}\}$

CHAPTER 8
Section 8.1

1. One possible answer is Monday: Grading Policy, Parking, Public Relations; Wednesday: Course Review, Space, Scholarship; Friday: Travel.
2. a and b, a and e, b and d, b and f, c and d, c and e, c and f, c and g, d and e, d and f, e and g, f and e, f and g
3. $C(n, 2) = n(n - 1)/2$
4. We use $\approx$ to denote "connected to." Let $G = (V, E)$ be a graph, and let v, w, and u be arbitrary vertices of G. By definition, $v \approx w$ whenever there is a path from v to w or when $v = w$. Now we have $v \approx v$ since $v = v$ for every vertex v.

Assume v and w are distinct and $v \approx w$. Then, $w \approx v$ since a path from w to v can be obtained from a path from v to w by reversing the sequence of edges of the path from v to w. The case when $v = w$ is easy.

Assume v, w, and u are distinct. If $v \approx w$ and $w \approx u$, then $v \approx u$ since a path from v to w and a path from w to u can be concatenated to form a path from v to u. If $v = w$, $v = u$, or $w = u$, the proof is immediate.

Hence, $\approx$ is reflexive, symmetric, and transitive.
5. One possible answer is shown in the following figure.

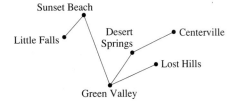

Section 8.2

1. Necessary and sufficient conditions for a connected multigraph to contain an Eulerian path from vertex v to a different vertex w: v and w are the only two vertices of odd degree.

2.

3. One solution: (Boston, Atlanta, Dallas, Chicago, Cleveland, Boston) for a total of 3858 miles

A-12 SOLUTIONS TO SELECTED PRACTICE PROBLEMS

4. Assume G is a complete graph with $m+1$ vertices. There are $m+1$ choices for the first vertex of a Hamiltonian cycle. After choosing this first vertex, we have m choices for the next vertex. Then, we have $m-1$ choices, and so on. Hence, there are $(m+1)m(m-1)\cdots 1 = (m+1)!$ Hamiltonian cycles.
5. The closest-neighbor cycle starting at Boston is (Boston, Cleveland, Chicago, Atlanta, Dallas, Boston) for a total of 4326 miles. No, this cycle is not the cycle of minimum weight.

Section 8.3

1. The level of e is 4; the level of k is 4. There is one vertex at level 5.
2. a. The leaves are b, f, g, d, and h.   b. The internal vertices are a, c, and e.   c. The siblings of c are b, d, and e.
3. a. By Theorem 8.3.2, we have $n = mt + 1$. Hence, $t = (n-1)/m$.
   b. By Theorem 8.3.3, $k = (m-1)t + 1 = (m-1)[(n-1)/m] + 1 = [(m-1)n - (m-1)]/m + 1 = [(m-1)n - m + 1 + m]/m = [(m-1)n + 1]/m$.
4. a. By Practice Problem 3b, the 5-ary tree has $[(m-1)n + 1]/m = [(5-1)61 + 1]/5 = 245/5 = 49$ leaves.
   b. By Theorem 8.3.2, the number n of vertices in a full 7-ary tree must satisfy the equation $n = 7t + 1$ for some integer t. Hence, $n - 1$ must be a multiple of 7. Since $61 - 1$ is not a multiple of 7, we cannot form a full 7-ary tree with 61 vertices. To solve this problem, we form a full 7-ary tree T with q vertices, where q is the smallest integer larger than 61 such that $q - 1$ is a multiple of 7. For this problem, $q = 64$ since 63 is a multiple of 7. Since $[(m-1)n + 1]/m = [(7-1)64 + 1]/7 = 55$, then T has 55 leaves. But T has $3 = 64 - 61$ more leaves than needed for the actual telephone tree of 61 vertices. Hence, 52 members will not have to make a phone call. The actual telephone tree has 52 leaves, and it is a 7-ary tree but not a full 7-ary tree.
5. 0011101; 110111
6. Since each edge corresponds to a telephone call, we will find the number of edges. By Theorem 8.3.14, the number of edges is $n - 1 = 61 - 1 = 60$. Hence, the number of telephone calls is 60. This problem can also be solved by noting that every group member except the president must receive a telephone call.

Section 8.4

1.
2.

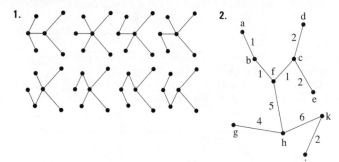

3. Same as the minimal spanning tree shown in solution 2 above.

CHAPTER 9

Section 9.1

1. We are given $a_n = a_{n-1} + 5$. So, $a_0 = 9, a_1 = 9 + 5, a_2 = 9 + 2 \cdot 5, \ldots, a_n = 9 + n \cdot 5, \ldots$, by iteration. Prove $a_n = 9 + n \cdot 5$ for $n = 0, 1, 2, 3, \ldots$. Proof: The basis step is true since $a_0 = 9 = 9 + 0 \cdot 5$. For the induction hypothesis, we assume $a_k = 9 + k \cdot 5$. Then $a_{k+1} = a_k + 5 = (9 + k \cdot 5) + 5 = 9 + (k+1)5$.
2. We are given that $b_1 = 0$ and $b_n = b_{n-1} + (n-1)$ for $n = 2, 3, \ldots$. Prove $b_n = 1 + 2 + \cdots + (n-1)$ for $n = 1, 2, 3, \ldots$. Proof: The basis step is true since $b_1 = 0$ and $0 = 1 + 2 + \cdots + (n-1)$ for $n = 1$. For the induction hypothesis, we assume $b_k = 1 + 2 + \cdots + (k-1)$. By the given recurrence relation, $b_{k+1} = b_k + [(k+1) - 1]$. Therefore, $b_k = 1 + 2 + \cdots + (k-1) + [(k+1) - 1]$.

3. $F_{13} = 233$ **4.** $A(2, 2) = 7$
5. Proof: We are given NFactorial(0) = 1 and we know $1 = 0!$. Hence, the basis step follows. For the induction hypothesis, we assume NFactorial(k) = k!. The function NFactorial yields NFactorial(n) = $n \cdot$ NFactorial(n − 1). Therefore, NFactorial(k + 1) = (k + 1) $\cdot$ NFactorial(k) = (k + 1)k! = (k + 1)!.
6. Recall: Annuity(20) = $102,320

Section 9.2

1. The characteristic equation is $x^2 - 4x - 21 = 0$. Since $x^2 - 4x - 21 = (x - 7)(x + 3)$, the roots of the characteristic equation are $r = 7$ and $s = -3$. Therefore, the solution to the recurrence relation is $a_n = a \cdot 7^n + b(-3)^n$.
2. To find r and s, use the quadratic formula with $a = 1$, $b = -1$, and $c = -1$. To find the constants a and b, use the initial conditions to obtain

$$0 = a[(1 + \sqrt{5})/2]^0 + b[(1 - \sqrt{5})/2]^0 = a + b$$
$$1 = a[(1 + \sqrt{5})/2]^1 + b[(1 - \sqrt{5})/2]^1 = a[(1 + \sqrt{5})/2] + b[(1 - \sqrt{5})/2]$$

Hence, $b = -a$. Substituting in the second equation gives

$$1 = a[(1 + \sqrt{5})/2] - a[(1 - \sqrt{5})/2] = \sqrt{5}a$$

Therefore, $a = 1/\sqrt{5} = \sqrt{5}/5$ and $b = -1/\sqrt{5} = -\sqrt{5}/5$.
3. $q(\pi) = \pi^4 - 5\pi^2 + 7\pi + 9 = \pi[\pi^3 - 5\pi + 7] + 9 = \pi[\pi(\pi^2 - 5) + 7] + 9 = \pi[\pi(\pi \cdot \pi - 5) + 7] + 9$

Section 9.3

1. We have $1 - x = a(1 - 3x) + b(1 - 2x)$ for all x. Substituting $x = 1/2$ yields $1 - 1/2 = a(1 - 3/2) + b(1 - 2/2) = -a/2$. So, $1/2 = -a/2$. Hence, we get $a = -1$. Substituting $x = 1/3$ yields $1 - 1/3 = a(1 - 3/3) + b(1 - 2/3)$. So, $2/3 = b/3$. Hence, $b = 2$.
2. Differentiating both sides of $x/(1 - x)^2 = x + 2x^2 + 3x^3 + 4x^4 + \cdots$ yields $(1 + x)/(1 - x)^3 = 1 + 2^2x + 3^2x^2 + 4^2x^2 + \cdots$. Hence, the generating function for the sequence $(1^2, 2^2, 3^2, \ldots)$ is $(1 + x)/(1 - x)^3$.

CHAPTER 10

Section 10.1

1. $(+ \vee - \vee \lambda)DD^*.DD^*$
2. Both sets $(a \vee b)^*$ and $(a^*b^*)^*$ contain all possible strings of a's and b's.

Section 10.2

1. a. $V = \{a, b\}$ **b.** bab is one string not in L
   c. $L = \{w \in \{a, b\}^* : w$ consists of one a followed by any number (possibly zero) of b's$\}$
2. $a^k = \lambda$ for $k = 0$ and $a^k = aa^{k-1}$ for $k \geq 1$
3. @ $\Rightarrow$ QR $\Rightarrow$ QGraze $\Rightarrow$ ANGraze $\Rightarrow$ TheNGraze $\Rightarrow$ TheDeerGraze;
    @ $\Rightarrow$ QR $\Rightarrow$ ANR $\Rightarrow$ ANGraze $\Rightarrow$ ADeerGraze $\Rightarrow$ TheDeerGraze
4. a. @ $\Rightarrow$ 1, @ $\Rightarrow$ @11 $\Rightarrow$ 11, @ $\Rightarrow$ @11 $\Rightarrow$ 111
   b. @ $\Rightarrow$ @11 $\Rightarrow$ @1111 $\Rightarrow$ 1111, @ $\Rightarrow$ @11 $\Rightarrow$ @1111 $\Rightarrow$ 11111
   c. Show @ $\stackrel{*}{\Rightarrow} 1^{2n}$ by using the production @ $\to$ @11 n times, then use @1 $\to$ 1. Show @ $\stackrel{*}{\Rightarrow} 1^{2n+1}$ by using the production @ $\to$ @11 n times, then use @ $\to$ 1.
5. a. @ $\Rightarrow$ A1 $\Rightarrow$ λ1 = 1, @ $\Rightarrow$ A1 $\Rightarrow$ A11 $\Rightarrow$ A111 $\Rightarrow$ 1111
6. Let $V = \{@, A, 0, 1\}$, $T = \{0, 1\}$, and $P: @ \to \lambda, @ \to 0A, A \to 1@$.

Section 10.3

2. a. Yes   **b.** @ $\Rightarrow$ aI $\Rightarrow$ adI $\Rightarrow$ ad3   **4.** $s_0 \Rightarrow as_1 \Rightarrow aas_1 \Rightarrow aabs_2 \Rightarrow aabbs_3 \Rightarrow aabbb$
5. Delete all productions with s_4.

Section 10.4

1. To derive $a^n b^n$ for $n = 1, 2, \ldots$, use $@ \stackrel{*}{\Rightarrow} a^n @ b^n \Rightarrow a^n b^n$.
2. First show how to derive $a^n b^n$ for $n = 0, 1, 2$. To derive $a^n b^n$ for $n > 2$, use the derivation $@ \Rightarrow aAb \stackrel{*}{\Rightarrow} a^{n-1} A b^{n-1} \Rightarrow a^n b^n$.
3. One set of productions is $P: @ \to AB, A \to \lambda, A \to 1A, B \to \lambda, B \to 0B1$.
4. $1 + m + m^2 + \cdots + m^n$ **5.** G' is not ambiguous.
6. Figure 10.4.4(a): $A \Rightarrow \text{if}B\text{then}A \Rightarrow \text{if}b\text{then}A \Rightarrow \text{if}b\text{then if}B\text{then}A\text{else}A \stackrel{*}{\Rightarrow} \text{if}b\text{then if}b\text{then}a\text{else}a$
   Figure 10.4.4(b): $A \Rightarrow \text{if}B\text{then}A\text{else}A \Rightarrow \text{if}b\text{then}A\text{else}A \Rightarrow \text{if}b\text{then if}B\text{then}A\text{else}A \stackrel{*}{\Rightarrow} \text{if}b\text{then if}b\text{then}a\text{else}a$
7. Nothing is printed; nothing is printed; 2 is printed.

SOLUTIONS FOR THE EXERCISES

CHAPTER 1

Section 1.1

1. a. True   **b.** False   **c.** True   **d.** False   **e.** False   **f.** True   **3.** $A = \{-5, 5\}$
5. a. $\{N, O\}$ **b.** $\{A, D, M\}$ **c.** $\{A, C, E, H, I, M, S, T\}$ **d.** $\{A, C, E, R\}$ **7. a.** 0   **b.** 2   **c.** 0   **d.** 11
9. a. True   **b.** False   **c.** True   **d.** False   **e.** True   **11. a.** True   **b.** False   **c.** False   **d.** True
13. $\emptyset, \{a\}$ **15. a.** True   **b.** False   **c.** True   **d.** False   **e.** False   **f.** True
17. a. $A \cup B = \{2, 6, 9, 10\}$ **b.** $A \cap B = \{6, 10\}$ **c.** $A \cup B' = \{0, 1, 2, 3, 4, 5, 6, 7, 8, 10\}$ **d.** $B \cap A' = \{9\}$ **e.** No
19. a. $A \cap B = \{1\}$ **b.** $A \cup B = \{1, 2\}$ **c.** $A - B = \{2\}$ **d.** $B - A = \emptyset$
21. a. {Peter, Martha, David}; intersection
  b. {Peter, Joan, Eva, Jean, Mark, Steve, Martha, Bruce, David, Anastasia, Anna, Daisy, Patty}; union
23. a. $B \subseteq A$ **b.** $A \subseteq B$ **c.** A can be any subset of $\mathbb{R}$.   **d.** $A \subseteq B$
25. a. $K' \cap J$ **b.** J' **c.** $I \cap J$ **d.** $I \cup K$

Section 1.2

1. a. $B' = \{x : x > 4\}$ **b.** $A \cup B = \{x : x \leq 4 \text{ or } x = 7\}$ **c.** $A \cup D = \{-1, 1, 2, 4, 7, 9\}$ **d.** $D - B = \{7, 9\}$
  e. $A \cap D = \{7\}$ **f.** $A \cap B = \{-1\}$
3. a. $D' = \{x : x \leq 5\}$ **b.** $D - A = D$ **c.** $P(A) = \{\emptyset, \{1\}, \{3\}, \{5\}, \{1, 3\}, \{1, 5\}, \{3, 5\}, A\}$ **d.** $A \cup B = \{1, 3, 5\}$
  e. A **f.** $A \times A = \{(1, 1), (1, 3), (1, 5), (3, 1), (3, 3), (3, 5), (5, 1), (5, 3), (5, 5)\}$
5. a. $1 \in A$; $\{1\} \subseteq A$; $\{b\} \in A$ **b.** $P(A) = \{\emptyset, \{1\}, \{\{b\}\}, \{\{1, b\}\}, \{1, \{b\}\}, \{1, \{1, b\}\}, \{\{b\}, \{1, b\}\}, A\}$
7. a. $\{\emptyset\}$ **b.** $\{\emptyset\}$ **c.** $\{\{\emptyset\}\}$ **9. a.** $P(A) = \{\emptyset, A\}$ **b.** $P(\{A\}) = \{\emptyset, \{A\}\}$
11. a. True
  b. False. For example, $A = \{1, 2, 3\}$, $B = \{4\}$, and $\text{Card}(A - B) = 3$, which is not equal to $\text{Card}(A) - \text{Card}(B) = 2$.
13. a.   **b.**   **15. a.**   **b.**

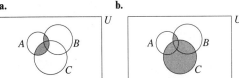
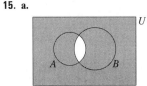
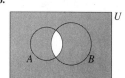

17. a. False   **b.** False
19. False. For example, $A = \{a, b, c, d, e\}$, $B = \{f, g\}$, $C = \{a\}$, $D = \{b\}$; $(A \cup B) \cap (C \cup D) = \{a, b\} \neq \{a\} = (A \cap C) \cup (B \cap D)$.

21. a. $C \times D = \{(1, 8), (2, 8), (3, 8), (4, 8)\}$ **b.** $D \times C = \{(8, 1), (8, 2), (8, 3), (8, 4)\}$ **c.** $\text{Card}(C \times D) = 4 = \text{Card}(D \times C)$
  d. $C - D = \{1, 2, 3, 4\}$, $D - C = \{8\}$; therefore, $C - D \neq D - C$.
23. a. From $u - 7 = 10$, $v + 3 = -4$, we obtain $u = 17$ and $v = -7$.
  b. From $3 = 2u - v$, $u + 8v = 6$, we obtain $u = \frac{30}{17}$ and $v = \frac{9}{17}$. **c.** From $u + v = 9$, $v = 3$, we obtain $v = 3$ and $u = 6$.
  d. From $u + v = 8$, $3v = 9$, we obtain $v = 3$ and $u = 5$.
25. a. $A \Delta B = (-1, 0] \cup [1, 2)$ **b.** $(-\infty, -\sqrt{2}) \cup (-\sqrt{2}, -1) \cup (1, \sqrt{2}) \cup (\sqrt{2}, \infty)$
27. a. [i, j, k, l, n, m, n, p] **b.** [x, w] **c.** [6, 9, 21, 30]

Section 1.3

1. Begin
   Taxableincome := Income $-$ 10000
   Tax := (0.03)(Taxableincome)
   Return Tax
  End

3. Begin
   Sum := A(1)
   For j = 2 to n do
   Sum := Sum + A(j)
   Return Sum
  End

5. Begin
   Sum := A(1)
   For j = 2 to n do
   Sum := Sum + A(j)
   Average := Sum/n
   Return Average
  End

7. Begin
   Sum := 0
   For k = 0 to n $-$ 1 do
   Sum := 2 $\cdot$ k + 1 + Sum
   Return Sum^2
  End

9. Begin
   Sum := 0
   For k = 1 to n do
   Sum := Sum + k^2
   Return Sum
  End

11. Begin
   Number := a
   While Number > 0 do
   Number := Number $-$ 2
   If Number = 0 then return "true" else return "false"
  End

13. This algorithm finds the smaller of two numbers. **15.** This algorithm computes the product of $A(1), A(2), \ldots, A(n)$.
17. The algorithm computes $S = 2^n - 1$.
19. $J = 7, A = 3$ **21.** $J = 7, A = 3$ **23.** $J = 7, A = 10$ **25.** $J = 3, k = 5$

Section 1.4

1. a. To show, if $x \in A$, then $x \in A \cup B$ **b.** A and B are arbitrary sets. **c.** Definition of union of two sets
  d. If $x \in A$, then $x \in A$ or $x \in B$.
3. a. To show, if $\sqrt{xy} = (x + y)/2$, then $x = y$ **b.** x and y are nonnegative numbers, and $\sqrt{xy} = (x + y)/2$. **c.** None
  d. Facts of algebra and equivalent equations.
5. a. If the right triangle is isosceles, then the area of the triangle is equal to $c^2/4$.
  b. a, b, and c are the sides and the hypotenuse, respectively, of a right triangle. Also, the triangle is isoceles.
  c. Definition of a right triangle **d.** The Pythagorean Theorem, facts of algebra, and formula of area of a triangle
7. a. To show, if x and y are odd integers, then $x + y$ is even **b.** x and y are odd integers.
  c. Definitions of odd and even integers **d.** Facts of algebra
9. a. To show $x + 1/x \geq 2$ for any positive real number x **b.** x is a positive real number.
  c. Definition of a positive real number **d.** Facts of algebra and inequalities
11. a. To show $\sin x \leq x$ for all nonnegative real numbers **b.** x is a nonnegative real number.
  c. Definition of nondecreasing function
  d. Theorem stating that, if derivative of a function is nonnegative, then the function is nondecreasing; $x' = 1$,
    $(\sin x)' = \cos x$, and $\sin 0 = 0$
13. a. To show, if a^2 is an even integer, then a is an even integer. **b.** a^2 is an even integer.
  c. Definitions of even and odd integers **d.** The fact that the product of two odd integers is odd

Section 1.5

1. a. True **b.** True **3. a.** True **b.** False
5. If we assert "This sentence is false" to be true, then this sentence is false. If we assert "This sentence is false" to be false,
  then this sentence is true.

7. F, T, T, T, F, F, T, T, F   **9.** T, yes   **11.** T, no

13. Prove: $-1 < 0$. Proof: By A2, exactly one of the following is true: $1 < 0$, $0 < 1$, or $1 = 0$. By A5, we have $1 > 0$. Hence, $1 < 0$ and $1 = 0$ are not true. By A2, exactly one of the following is true: $-1 < 0$, $0 < -1$, or $-1 = 0$. If $0 < -1$, then $0 + 1 < -1 + 1$, or $1 < 0$, by A3. This contradicts the fact that $1 < 0$ is not true. If $-1 = 0$, then $-1 + 1 = 0 + 1$, or $0 = 1$, by A3. But $0 = 1$ is not true since $1 = 0$ is not true from above. Hence $-1 < 0$.

15. Prove: $v < 0$ if and only if $0 < -v$. Proof: ($\rightarrow$) Assume $v < 0$. Adding $-v$ to both sides of $v < 0$ yields $0 < -v$ by A3. ($\leftarrow$) Assume $0 < -v$. Adding v to both sides of $0 < -v$ yields $v < 0$ by A3.

17. Prove: If $x > 1$, then $x^2 > 1$. Proof: Assume $x > 1$. By A5, we have $1 > 0$. Since $x > 1$ and $1 > 0$, we have $x > 0$ by A1. Multiplying both sides of $x > 1$ by x yields $x^2 > x$ by A4. Since $x^2 > x$ and $x > 1$, it follows that $x^2 > 1$ by A1.

19. Prove: If $0 < x$ and $x < y$, then $x^2 < y^2$. Proof: Since $0 < x$ and $x < y$, it follows that $0 < y$ by A1. Multiplying by x on both sides of $x < y$ yields $x^2 < xy$ by A4. Multiplying by y on both sides of $x < y$ yields $xy < y^2$. Since $x^2 < xy$ and $xy < y^2$, it follows that $x^2 < y^2$ by A1.

Chapter 1 Review

1. a, c, d, and f are true.   **3. a.** $A' = \{1, 2, 3\}$ **b.** $A - U = \emptyset$ **c.** $U - A' = A$ **d.** $A \cap U = A$ **e.** $\emptyset$

5. a. $P(\emptyset) = \{\emptyset\}$ $P(A) = \{\emptyset, A\}$ $P(B) = \{\emptyset, \{b\}, \{c\}, B\}$
  b. $P(C) = \{\emptyset, \{d\}, \{e\}, \{f\}, \{d, e\}, \{d, f\}, \{e, f\}, C\}$; $\text{Card}(P(C)) = 8$
  c. $P(D) = \{\emptyset, \{g\}, \{h\}, \{i\}, \{j\}, \{g, h\}, \{g, i\}, \{g, j\}, \{h, i\}, \{h, j\}, \{i, j\}, \{g, h, i\}, \{g, h, j\}, \{g, i, j\}, \{h, i, j\}, D\}$; $\text{Card}(P(D)) = 16$
  d. $B \times C = \{(b, d), (b, e), (b, f), (c, d), (c, e), (c, f)\}$; $\text{Card}(B \times C) = 6$
  e. $A \times P(B) = \{(a, \emptyset), (a, \{b\}), (a, \{c\}), (a, B)\}$; $\text{Card}(A \times P(B)) = 4$

7. a and c

9. a. True for all sets A, B, and C.
  b. False. For example, let $U = \{A, 1, 2, 3, 4\}$, $A = \{a\}$, $B = \{A, 1, 2\}$, and $C = \{A, 1, 2, 3\}$. Then $A \in B$ and $B \subseteq C$, but it is not true that $A \subseteq C$.

11. Suppose A is a sequence $A(1), A(2), \ldots, A(n)$ of real numbers.
  Begin
    Product := $A(1)$
    For j = 2 to n do
      Product := (Product)($A(j)$)
    If Product ≥ 0 then Squareroot := SQRT(Product) else return
      "Square root not real."
  End

13. Proof: From Exercise 24 in Section 1.1, $A \triangle B = \{x : x \in A \text{ or } x \in B \text{ but not both}\}$. First we show $A \triangle B \subseteq (A \cup B) - (A \cap B)$. Pick $x \in A \triangle B$. Then $x \in A$ or $x \in B$ but not both. This means that $x \in A \cup B$ and $x \notin (A \cap B)$. Hence, $A \triangle B \subseteq (A \cup B) - (A \cap B)$. Now we show $(A \cup B) - (A \cap B) \subseteq A \triangle B$. Pick $x \in (A \cup B) - (A \cap B)$. Therefore, $x \in A \cup B$ and $x \notin A \cap B$. In other words, $x \in A$ or $x \in B$ but not both.

15. Prove: If $x > 1$, then $x < x^2$. Proof: Since $x > 1$ and $1 > 0$, it follows that $x > 0$ by A1 of Section 1.5. Multiplying both sides of $x > 1$ by x yields $x^2 > x$ by A4 of Section 1.5.

17. Prove: $(x - y)^2 = 0$ if and only if $x = y$. Proof: ($\rightarrow$) If $a^2 = 0$ then $a = 0$. From $(x - y)^2 = 0$, we have $x - y = 0$. Hence, adding y to both sides of $x - y = 0$ yields $x = y$. ($\leftarrow$) If $x = y$, then $x - y = 0$. Since $x - y = 0$, we can multiply both sides of $x - y = 0$ by $x - y$ to obtain $(x - y)^2 = 0$.

CHAPTER 2

Section 2.1

1. $V = \{\text{Village, River, Field, Pond, Well}\}$
  $E = \{(\text{Village, River}), (\text{River, Village}), (\text{River, Field}), (\text{Field, Village}), (\text{Pond, Field}), (\text{Village, Well}), (\text{Well, Village}), (\text{Well, Pond})\}$

3. Let v = village, r = river, f = field, p = pond, and w = well.   **a.** $\{w, p, f, v, r\}$ **b.** (w, p, f), (w, v, r, f), (w, v, w, p, f)

5. a. [graph with vertices a, b, c, d; loops at a and d] **b.** ((a, b), (b, d), (d, c)) **c.** {a, b, c, d}

7. a. [graph with vertices a, b, c, d] **b.** {a, b, c} **c.** No

9. a. $(v_2, v_3, v_4, v_6, v_2)$ **b.** $(v_1, v_2, v_3, v_4, v_6, v_1)$ **c.** $(v_4, v_6, v_1, v_6, v_4)$
11. a. $(v_1, v_2, v_3, v_4, v_1)$ **b.** $\{v_6\}$ **c.** $(v_2, v_3, v_4, v_1, v_2)$
13. True. Suppose there is a path P from v to w that is not simple. Eliminating any self-loops, unnecessary edges, and unnecessary vertices from closed paths in P, we obtain a simple path from v to w.

Section 2.2

1. $R = \{(1, 4), (1, 9), (1, 21), (1, 25), (2, 4), (3, 9), (3, 21)\}$
3. a. $R = \{(1, 1), (1, 2), (1, 3), (1, 4), (2, 2), (2, 3), (2, 4), (3, 3), (3, 4), (4, 4)\}$
  b. R is reflexive since $(1, 1), (2, 2), (3, 3)$, and $(4, 4)$ are in R. R is not symmetric since $(1, 3) \in R$ and $(3, 1) \notin R$. R is antisymmetric since R does not contain (x, y) and (y, x) for distinct x and y in A. Finally, we note that, if $(x, y) \in R$ and $(y, z) \in R$, then $(x, z) \in R$ for all x, y, z in A. Hence, R is transitive.
5. a. $R = \{(2, 6), (3, 9)\}$ **b.** $S = \{(2, 9), (2, 10), (2, 13), (3, 9), (3, 10), (3, 13)\}$
7. a. $T = \{(2, 6), (2, 10), (2, 12), (3, 9)\}$ **b.** $S = \{(2, 6), (2, 10), (2, 12), (3, 6), (3, 9), (3, 12)\}$
9. a. $R = \{(1, 1), (1, 2), (1, 5), (1, 10), (2, 2), (2, 10), (5, 5), (5, 10), (10, 10)\}$ **b.**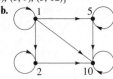

  c. R is reflexive since $(a, a) \in R$ for each $a \in A$. R is not symmetric since $(1, 2)$ is in R but $(2, 1)$ is not in R. R is antisymmetric since there are no distinct $x, y \in A$ such that both (x, y) and (y, x) are in R. Finally, R is transitive since, for all x, y, z in A, if (x, y) and (y, z) are both in R, then (x, z) is in R.
11. a. The transitive closure of R is the set $\{(a, b), (b, c), (c, c), (a, a), (a, c)\}$.
  b. The reflexive closure of R is the set $\{(a, b), (b, c), (c, c), (a, a), (b, b), (d, d)\}$.
13. a. {(Cathy, 110, Tango, Brown), (Connie, 100, Swing, Red), (Sally, 115, Madison, Blonde), (Barbara, 98, Cha-cha, Black), (John, 135, Fox trot, Black)}

  b. P_{134}(Personal)   **c.** P_{13}(Personal) contains the first two columns of the table in part b.

NAME	FAVORITE DANCE	HAIR COLOR
Cathy	Tango	Brown
Connie	Swing	Red
Sally	Madison	Blonde
Barbara	Cha-cha	Black
John	Fox trot	Black

15. The transitive closure of R is {(c, b), (b, c), (j, l), (l, j), (b, j), (j, b), (c, c), (b, b), (j, j), (l, l), (c, j), (c, l), (b, l), (j, c), (l, c), (l, b)}.
17. $R = \{(1, 2), (2, 1), (2, 3), (3, 2)\}$ **19.** $T = \{(c, c), (d, d)\}$ **21. a.** $(1, 1) \notin R$ **b.** $(1, 2) \in R$ and $(2, 4) \in R$ but $(1, 4) \notin R$

Section 2.3

1. To find an ordered pair in $R \circ S$, we use the following graphical representation of R and S.

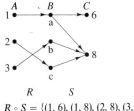

$R \circ S = \{(1, 6), (1, 8), (2, 8), (3, 8)\}$

3. $R \circ S = \{(1, 1), (1, 5), (a, a), (a, b), (b, b), (5, 5)\}$

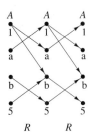

5. a. Since $(1, 2)$ and $(2, 3)$ are in R but $(1, 3)$ is not in R, R is not transitive.
  b. Let $S = \{(1, 2), (1, 3), (1, 4), (2, 3), (2, 4)\}$, and let R^+ be the transitive closure of R. We will show that $S = R^+$. Since $R \subseteq S$ and S is transitive, we have $R^+ \subseteq S$ by definition of the transitive closure of R. To show that $S \subseteq R^+$: By Theorem 2.3.3a, $R \subseteq R^+$. Hence, $(1, 2)$, $(2, 3)$, and $(2, 4)$ are in R^+. Since $(1, 2)$ and $(2, 3)$ are in R^+ and R^+ is transitive, then $(1, 3) \in R^+$. Similarly, $(1, 4) \in R^+$. Hence, $S \subseteq R^+$. Therefore, $R^+ = S$, and S is the transitive closure of R.
7. $R^+ = R^1 \cup R^2 \cup R^3$

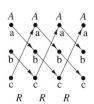

$R^2 = \{(a, c), (b, a), (c, b)\}$, $R^3 = \{(a, a), (b, b), (c, c)\}$. Hence,
$R^+ = \{(a, b), (b, c), (c, a), (a, c), (b, a), (c, b), (a, a), (b, b), (c, c)\} = A \times A$.

9. a. Yes   **b.** Yes   **c.** Yes   **d.** 2, 4, 5
11. $R = \{(1, 2), (2, 3)\}$, $R^2 = \{(1, 3)\}$, $R^3 = \emptyset$. Thus, R^2 is not a subset of R.
13. a. R is a binary relation on a set A.   **b.** To show R^+ is transitive
  c. Definitions of R^+, binary relation, subset, and transitive

15. a. 367 for leap years, otherwise 366   **b.** 11
17. Let $A = \{1, 2\}$ and let $R = \{(2, 1)\}$. Note that $R \circ R = \emptyset$ and R is not a subset of $R \circ R$.
19. Suppose $(j, x) \in R$ for each j. Since R is symmetric, we have $(x, j) \in R$ for each j. By transitivity, we have $(j, j) \in R$ for each j. Hence, R is reflexive.
21. *Hint:* Remove cycles and repeated vertices.

Section 2.4

1. a. $M_R = \begin{array}{c} \\ a \\ b \\ c \\ d \end{array} \begin{array}{c} a\ b\ c\ d \\ \begin{bmatrix} 1 & 1 & 0 & 0 \\ 0 & 0 & 0 & 1 \\ 1 & 0 & 1 & 0 \\ 0 & 0 & 1 & 1 \end{bmatrix} \end{array}$ **b.** $M_R \oplus M_R = \begin{bmatrix} 1 & 1 & 0 & 0 \\ 0 & 0 & 0 & 1 \\ 1 & 0 & 1 & 0 \\ 0 & 0 & 1 & 1 \end{bmatrix}$

3. a. $P \oplus Q = \begin{bmatrix} 1 & 1 & 1 \\ 1 & 1 & 1 \\ 1 & 1 & 0 \end{bmatrix}$ $P \otimes Q = \begin{bmatrix} 0 & 1 & 1 \\ 0 & 1 & 1 \\ 1 & 1 & 0 \end{bmatrix} \otimes \begin{bmatrix} 1 & 1 & 0 \\ 1 & 0 & 1 \\ 0 & 1 & 0 \end{bmatrix} = \begin{bmatrix} 1 & 1 & 1 \\ 1 & 1 & 1 \\ 1 & 1 & 1 \end{bmatrix}$ $Q \otimes P = \begin{bmatrix} 0 & 1 & 1 \\ 1 & 1 & 1 \\ 0 & 1 & 1 \end{bmatrix}$

$P^2 = \begin{bmatrix} 0 & 1 & 1 \\ 0 & 1 & 1 \\ 1 & 1 & 0 \end{bmatrix} \otimes \begin{bmatrix} 0 & 1 & 1 \\ 0 & 1 & 1 \\ 1 & 1 & 0 \end{bmatrix} = \begin{bmatrix} 1 & 1 & 1 \\ 1 & 1 & 1 \\ 0 & 1 & 1 \end{bmatrix}$ $Q^2 = \begin{bmatrix} 1 & 1 & 0 \\ 1 & 0 & 1 \\ 0 & 1 & 0 \end{bmatrix} \otimes \begin{bmatrix} 1 & 1 & 0 \\ 1 & 0 & 1 \\ 0 & 1 & 0 \end{bmatrix} = \begin{bmatrix} 1 & 1 & 1 \\ 1 & 1 & 0 \\ 1 & 0 & 1 \end{bmatrix}$

b. $R = \{(1, 2), (1, 3), (2, 2), (2, 3), (3, 1), (3, 2)\};\ S = \{(1, 1), (1, 2), (2, 1), (2, 3), (3, 2)\}$
c. Since $M_{R \circ S} = M_R \otimes M_S = P \otimes Q$ and $M_{S \circ R} = Q \otimes P$, we obtain $R \circ S = A \times A$, and $S \circ R = \{(1, 2), (1, 3), (2, 1), (2, 2), (2, 3), (3, 2), (3, 3)\}$.
d. $R^2 = \{(1, 1), (1, 2), (1, 3), (2, 1), (2, 2), (2, 3), (3, 2), (3, 3)\}$

5. a. $M_R = \begin{array}{c} \\ a \\ b \\ c \\ d \end{array} \begin{array}{c} a\ b\ c\ d \\ \begin{bmatrix} 0 & 1 & 0 & 0 \\ 1 & 0 & 1 & 0 \\ 0 & 0 & 0 & 1 \\ 0 & 0 & 0 & 0 \end{bmatrix} \end{array}$; $(M_R)^2 = \begin{bmatrix} 1 & 0 & 1 & 0 \\ 0 & 1 & 0 & 1 \\ 0 & 0 & 0 & 0 \\ 0 & 0 & 0 & 0 \end{bmatrix}$; $(M_R)^3 = \begin{bmatrix} 0 & 1 & 0 & 1 \\ 1 & 0 & 1 & 0 \\ 0 & 0 & 0 & 0 \\ 0 & 0 & 0 & 0 \end{bmatrix}$; $(M_R)^4 = \begin{bmatrix} 1 & 0 & 1 & 0 \\ 0 & 1 & 0 & 1 \\ 0 & 0 & 0 & 0 \\ 0 & 0 & 0 & 0 \end{bmatrix}$;

b. $M_{R^+} = \begin{bmatrix} 1 & 1 & 1 & 1 \\ 1 & 1 & 1 & 1 \\ 0 & 0 & 0 & 1 \\ 0 & 0 & 0 & 0 \end{bmatrix}$ **c.** From M_{R^+}, we obtain $R^+ = \{(a, a), (a, b), (a, c), (a, d), (b, a), (b, b), (b, c), (b, d), (c, d)\}$.

7. a. $M_R = \begin{bmatrix} 1 & 1 & 1 & 1 & 1 \\ 0 & 1 & 1 & 1 & 1 \\ 0 & 0 & 1 & 1 & 1 \\ 0 & 0 & 0 & 1 & 1 \\ 0 & 0 & 0 & 0 & 1 \end{bmatrix}$ **b.** No   **c.** Yes   **d.** $M_{R^+} = M_R$ **e.** $R^+ = R$

9. a. $M_R = \begin{bmatrix} 0 & 1 & 1 & 0 \\ 0 & 0 & 0 & 0 \\ 0 & 0 & 0 & 1 \\ 0 & 0 & 0 & 0 \end{bmatrix}$; $M_S = \begin{bmatrix} 0 & 0 & 0 & 0 \\ 1 & 1 & 0 & 1 \\ 0 & 0 & 0 & 1 \\ 0 & 0 & 0 & 0 \end{bmatrix}$

b. $R \cup S = \{(1, 2), (1, 3), (2, 1), (2, 2), (2, 4), (3, 4)\};\ M_{R \cup S} = \begin{bmatrix} 0 & 1 & 1 & 0 \\ 1 & 1 & 0 & 1 \\ 0 & 0 & 0 & 1 \\ 0 & 0 & 0 & 0 \end{bmatrix}$

c. $\begin{bmatrix} 0 & 1 & 1 & 0 \\ 0 & 0 & 0 & 0 \\ 0 & 0 & 0 & 1 \\ 0 & 0 & 0 & 0 \end{bmatrix} \oplus \begin{bmatrix} 0 & 0 & 0 & 0 \\ 1 & 1 & 0 & 1 \\ 0 & 0 & 0 & 1 \\ 0 & 0 & 0 & 0 \end{bmatrix} = \begin{bmatrix} 0 & 1 & 1 & 0 \\ 1 & 1 & 0 & 1 \\ 0 & 0 & 0 & 1 \\ 0 & 0 & 0 & 0 \end{bmatrix}$

11. b. $R^2 = \{(a, a), (a, b), (a, c), (a, d), (b, e), (d, c)\}$; $R^3 = \{(a, a), (a, b), (a, c), (a, d), (a, e), (d, e)\}$
13. a. $R = \{(a, b), (c, d)\}$; find M_R from R **b.** $S = \{(a, a), (b, b), (c, c), (d, d), (a, b), (b, a), (b, c), (c, b)\}$; find M_S from S
15. The (2, 5) entry in $(M_R)^3$ is 1 since there is a path of length 3 from 2 to 5.   **17.** Routine verifications
19. For example, $M = \begin{bmatrix} 1 & 1 \\ 0 & 0 \end{bmatrix}$ and $T = \begin{bmatrix} 1 & 0 \\ 1 & 0 \end{bmatrix}$.

Section 2.5

1. a. $M_R = \begin{array}{c} \\ v_1 \\ v_2 \end{array} \begin{array}{c} v_1 \quad v_2 \\ \begin{bmatrix} 0 & 1 \\ 1 & 0 \end{bmatrix} \end{array}$

b. $P_0 = M_R = \begin{bmatrix} 0 & 1 \\ 1 & 0 \end{bmatrix}$

To find P_1, we use $k = 1$ in (1) on page 80 and follow the method in Example 2.5.3:

$P_1[1, 1] = P_0[1, 1] + (P_0[1, 1] \cdot P_0[1, 1]) = 0 + (0 \cdot 0) = 0$
$P_1[1, 2] = P_0[1, 2] + (P_0[1, 1] \cdot P_0[1, 2]) = 1 + (0 \cdot 1) = 1$
$P_1[2, 1] = P_0[2, 1] + (P_0[2, 1] \cdot P_0[1, 1]) = 1 + (1 \cdot 0) = 1$
$P_1[2, 2] = P_0[2, 2] + (P_0[2, 1] \cdot P_0[1, 2]) = 0 + (1 \cdot 1) = 1$

Hence, $P_1 = \begin{bmatrix} 0 & 1 \\ 1 & 1 \end{bmatrix}$.

To find P_2, we use $k = 2$ in (1) on page 80.

$P_2[1, 1] = P_1[1, 1] + (P_1[1, 2] \cdot P_1[2, 1]) = 0 + (1 \cdot 1) = 1$
$P_2[1, 2] = P_1[1, 2] + (P_1[1, 2] \cdot P_1[2, 2]) = 1 + (1 \cdot 1) = 1$
$P_2[2, 1] = P_1[2, 1] + (P_1[2, 2] \cdot P_1[2, 1]) = 1 + (1 \cdot 1) = 1$
$P_2[2, 2] = P_1[2, 2] + (P_1[2, 2] \cdot P_1[2, 2]) = 1 + (1 \cdot 1) = 1$

Hence, $P_2 = \begin{bmatrix} 1 & 1 \\ 1 & 1 \end{bmatrix}$.

c. $(M_R)^2 = \begin{bmatrix} 0 & 1 \\ 1 & 0 \end{bmatrix} \otimes \begin{bmatrix} 0 & 1 \\ 1 & 0 \end{bmatrix} = \begin{bmatrix} 1 & 0 \\ 0 & 1 \end{bmatrix}$

d. We note that P_1 and P_2 are not powers of M_R. In particular, $P_1 \neq (M_R)^2$, and $P_2 \neq (M_R)^2$.

e. $M_{R^+} = P_2 = \begin{bmatrix} 1 & 1 \\ 1 & 1 \end{bmatrix}$, and $R^+ = \{(v_1, v_1), (v_1, v_2), (v_2, v_1), (v_2, v_2)\}$

f. $M_{R^+} = M_R \oplus (M_R)^2 = \begin{bmatrix} 0 & 1 \\ 1 & 0 \end{bmatrix} \oplus \begin{bmatrix} 1 & 0 \\ 0 & 1 \end{bmatrix} = \begin{bmatrix} 1 & 1 \\ 1 & 1 \end{bmatrix}$ $R^+ = \{(v_1, v_1), (v_1, v_2), (v_2, v_1), (v_2, v_2)\} = V \times V$.

Same result from part e.

3. a. $M_{R^*} = \begin{bmatrix} 1 & 1 \\ 1 & 1 \end{bmatrix}$ **b.** $R^* = V \times V$

5. a.

5. b. We find paths of length 2 and obtain $R \circ R = \{(v_2, v_2), (v_2, v_4), (v_3, v_1), (v_3, v_3), (v_3, v_4), (v_4, v_1), (v_4, v_2), (v_4, v_3), (v_4, v_4)\}$.

$$(M_R)^2 = \begin{bmatrix} 0 & 0 & 0 & 0 \\ 1 & 0 & 1 & 0 \\ 0 & 1 & 0 & 1 \\ 1 & 0 & 1 & 1 \end{bmatrix} \otimes \begin{bmatrix} 0 & 0 & 0 & 0 \\ 1 & 0 & 1 & 0 \\ 0 & 1 & 0 & 1 \\ 1 & 0 & 1 & 1 \end{bmatrix} = \begin{bmatrix} 0 & 0 & 0 & 0 \\ 0 & 1 & 0 & 1 \\ 1 & 0 & 1 & 1 \\ 1 & 1 & 1 & 1 \end{bmatrix}$$

c. $M_{R^*} = \begin{bmatrix} 1 & 0 & 0 & 0 \\ 1 & 1 & 1 & 1 \\ 1 & 1 & 1 & 1 \\ 1 & 1 & 1 & 1 \end{bmatrix}$ **d.** $M_{R^*} = I \vee P_4 = \begin{bmatrix} 1 & 0 & 0 & 0 \\ 1 & 1 & 1 & 1 \\ 1 & 1 & 1 & 1 \\ 1 & 1 & 1 & 1 \end{bmatrix}$ after finding P_4.

7. a. 1   **b.** 1   **c.** 0   **d.** 1   **e.** 0   **f.** 0
9. Follow Example 2.5.9.

a. $M_R = \begin{bmatrix} 1 & 1 & 1 \\ 0 & 1 & 1 \\ 1 & 0 & 0 \end{bmatrix}$ **b.** $S_0 = \begin{bmatrix} 0 & 1 & 1 \\ 300 & 0 & 1 \\ 1 & 300 & 0 \end{bmatrix}$; $S_1 = \begin{bmatrix} 0 & 1 & 1 \\ 300 & 0 & 1 \\ 1 & 2 & 0 \end{bmatrix}$; $S_2 = \begin{bmatrix} 0 & 1 & 1 \\ 300 & 0 & 1 \\ 1 & 2 & 0 \end{bmatrix}$; $S_3 = \begin{bmatrix} 0 & 1 & 1 \\ 2 & 0 & 1 \\ 1 & 2 & 0 \end{bmatrix}$

11. Follow Algorithm Newpath.   **13.** Since $M_R = M_{R^+} = P_n$, we have $M_R = P_0 = P_i$, for $i = 1, \ldots, n$.
15. a. $R = \{(2, 1), (3, 2), (4, 3)\}$ $R^+ = \{(2, 1), (3, 1), (3, 2), (4, 1), (4, 2), (4, 3)\}$
b. $R^* = \{(1, 1), (2, 1), (2, 2), (3, 1), (3, 2), (3, 3), (4, 1), (4, 2), (4, 3), (4, 4)\}$. Note that $x R^* y$ if and only if $y \leq x$.

Section 2.6

1. $\begin{bmatrix} 0 & 1 & 0 & 0 & 0 & 0 \\ 0 & 0 & 0 & 1 & 0 & 0 \\ 0 & 1 & 0 & 0 & 1 & 1 \\ 0 & 0 & 1 & 0 & 0 & 0 \\ 0 & 1 & 0 & 1 & 0 & 1 \\ 1 & 1 & 0 & 0 & 0 & 0 \end{bmatrix}$ **3.** There is a cycle at each vertex.

Chapter 2 Review

1. Let $A = \{1, 2, 3, 4\}$.
a. $R = \{(1, 1), (2, 2), (3, 3), (4, 4), (1, 2), (2, 3)\}$ **b.** $R = \{(1, 2)\}$ **c.** $R = \{(1, 1), (2, 2), (3, 3), (4, 4), (1, 2), (2, 3), (1, 3)\}$
3. R is reflexive. R is not symmetric since (a, d) is in R but (d, a) is not in R. R is not transitive. R is antisymmetric.

5. a. $M_R = \begin{bmatrix} 0 & 1 & 0 & 0 & 0 \\ 0 & 0 & 1 & 0 & 0 \\ 0 & 0 & 0 & 1 & 0 \\ 0 & 0 & 0 & 0 & 1 \\ 0 & 0 & 0 & 0 & 0 \end{bmatrix}$ **b.** $(M_R)^2 = \begin{bmatrix} 0 & 0 & 1 & 0 & 0 \\ 0 & 0 & 0 & 1 & 0 \\ 0 & 0 & 0 & 0 & 1 \\ 0 & 0 & 0 & 0 & 0 \\ 0 & 0 & 0 & 0 & 0 \end{bmatrix}$ **c.** $P_3 = \begin{bmatrix} 0 & 1 & 1 & 1 & 0 \\ 0 & 0 & 1 & 1 & 0 \\ 0 & 0 & 0 & 1 & 0 \\ 0 & 0 & 0 & 0 & 1 \\ 0 & 0 & 0 & 0 & 0 \end{bmatrix}$

d. $x R^+ y$ if and only if $x < y$
7. a. Yes.
b. No. For example, $A = \{1, 2, 3\}$, $R = \{(1, 2), (2, 1)\}$, and $S = \{(1, 3), (3, 1)\}$. The relation $R \circ S = \{(2, 3)\}$ is not symmetric.
c. No. For example, $A = \{1, 2, 3\}$; the relations $R = \{(1, 1), (2, 3)\}$ and $S = \{(1, 2), (3, 3)\}$ are transitive. The relation $R \circ S = \{(1, 2), (2, 3)\}$ is not transitive.
9. R is transitive if and only if $R^2 \subseteq R$. Also, $R^2 \subseteq R$ if and only if $R^2 \cup R = R$. The result follows.
11. a. $T^+ = \{(1, 1), (1, 3), (3, 1), (3, 3), (1, 5), (5, 1), (5, 5), (2, 4), (4, 2), (2, 2), (4, 4), (3, 5), (5, 3)\}$
b. $T^* = T^+$. We see $x T^* y$ if and only if $x - y$ is divisible by 2.

CHAPTER 3

Section 3.1

1.

3. If $x\ R\ y$ and $y\ R\ z$, then $x\ R\ z$ since, if x and y produce the same output from the same input and y and z produce the same output from the same input, then x and z produce the same output from the same input. Hence, R is transitive. Similarly, we can show that R is reflexive and symmetric.

5.

7. a. $P(A) = \{\emptyset, \{a\}, \{b\}, \{c\}, \{d\}, \{a, b\}, \{a, c\}, \{a, d\}, \{b, c\}, \{b, d\}, \{c, d\}, \{a, b, c\}, \{a, c, d\}, \{a, b, d\}, \{b, c, d\}, A\}$
b. R is reflexive since $x \subseteq x$ for any x in $P(A)$. R is antisymmetric since $x \subseteq y$ and $y \subseteq x$ imply $x = y$ for all x and y in $P(A)$. R is transitive since $x \subseteq y$ and $y \subseteq z$ imply $x \subseteq z$ for all x, y, and z in $P(A)$.

9. No. For example, $l \perp m$ and $m \perp n$ but l is parallel to n or l and n are the same line.

11. a. The relation $\equiv$ is reflexive since $m - m$ is divisible by 3 for any m in $\mathbb{Z}$. The relation $\equiv$ is symmetric for, if $m - n = 3k$ for some $k \in \mathbb{Z}$, then $n - m = 3(-k)$ for all m and n in $\mathbb{Z}$. The relation $\equiv$ is transitive since $m - n = 3k$ and $n - p = 3j$ for some k and j in $\mathbb{Z}$ imply $m - p = m - n + n - p = 3k + 3j = 3(k + j)$ for all m, n, and p in $\mathbb{Z}$. Hence, $\equiv$ is an equivalence relation on $\mathbb{Z}$.
b. There are three equivalence classes: $[0] = \{\ldots, -3, 0, 3, 6, \ldots\}$, $[1] = \{\ldots, -5, -2, 1, 4, \ldots\}$, and $[2] = \{\ldots, -1, 2, 5, \ldots\}$.

13. a. R is clearly reflexive and symmetric. Also, R is transitive, since, if x and y are in the same row and y and z are in the same row, then x and z are in the same row, for all x, y, and z in A.
b. There are four equivalence classes since there are four rows in M.

15. a. R is clearly reflexive. Also, R is symmetric for, if x and y have at least one character in common, then y and x have at least one character in common.
b. R is not an equivalence relation because R is not transitive. For example, consider $x = $ number, $y = $ big, and $z = $ go. Then x and y have a character in common, and y and z have a character in common, but x and z have no character in common.

17. a. R is reflexive since $\text{Card}(A) = \text{Card}(A)$ for any A in $P(D)$. R is symmetric since $\text{Card}(A) = \text{Card}(B)$ implies $\text{Card}(B) = \text{Card}(A)$ for all A and B in $P(D)$. R is transitive since $\text{Card}(A) = \text{Card}(B)$ and $\text{Card}(B) = \text{Card}(C)$ imply $\text{Card}(A) = \text{Card}(C)$ for all A, B, and C in $P(D)$.
b. $[\emptyset] = \{\emptyset\}$, $[D] = \{D\}$ **c.** $[\{4\}] = [\{0\}, \{1\}, \{2\}, \ldots, \{9\}]$
d. $[\{4, 7\}]$ contains all members in $P(D)$ containing two elements in D.

19. a. R is reflexive and symmetric since multiplication in $\mathbb{Z}$ is commutative. To show R is transitive, suppose $(m, n)\ R\ (p, q)$ and $(p, q)\ R\ (r, s)$. Then $mq = np$ and $ps = qr$. It follows that $n = mq/p$ and $s = qr/p$. Hence, $ms = nr$ and, therefore, $(m, n)\ R\ (r, s)$.

21. a. $T^+ = \{(n, m) : m = n + 2k \text{ for } k \text{ in } \mathbb{Z}\}$
b. $T^* = \{(n, m) : m \equiv n \pmod{2} \text{ for } n \text{ and } m \text{ in } \mathbb{Z}\}$; T^* is the well-known equivalence relation $\equiv$ modulo 2 on $\mathbb{Z}$.

Section 3.2

1. a. Maximal element: e; Minimal elements: a and c   **b.** Greatest element: e; Least element: none
3. a. Maximal element: 12; Minimal elements: 2, 3   **b.** Greatest element: 12; Least element: none

5. Refer to the Hasse diagram in Figure 3.1.4.
   a. Maximal element: {a, b, c}; Minimal element: ∅   b. Greatest element: {a, b, c}; Least element: ∅
7. The set B is $[\sqrt{5}, 3)$.   a. It is routine to show that $(B, \leq)$ is a poset.   b. $\text{glb}(B) = \sqrt{5}$ c. $\text{lub}(B) = 3$
9. Proof: Suppose r and s are both least elements for $(A, \preceq)$. Since $r \in A$ and s is a least element of A, we have $s \preceq r$. Similarly, $r \preceq s$. Since $s \preceq r$ and $r \preceq s$ and $\preceq$ is antisymmetric, it follows that $r = s$.
11. $a \prec b \prec c \prec d \prec f \prec e \prec g$ is one answer.
13. $1 \prec 2 \prec 3 \prec 5 \prec 6 \prec 10 \prec 15 \prec 30$ is one solution.
15. The $(1, 3), (2, 3), (4, 3), (5, 3), \ldots, (n, 3)$ entries of column 3 of M_R must be zero.
17. The $(2, 1)$ entry of M_R is 1, and the $(3, 1)$ entry of M_R is 1. In addition, if the $(2, i)$ entry is 1 and the $(3, i)$ entry is 1, then the $(1, i)$ entry must be 1.

Section 3.3

1. Yes, f is a function from A to B.
3. a. Domain of f is $\{x : x \in \mathbb{R} \text{ and } x \neq -3\}$ b. Domain of g is $\{x : x \in \mathbb{R} \text{ and } x \geq -8\}$

5. a. The table for χ_A is

x	u_1	u_2	u_3	u_4
$\chi_A(x)$	1	1	0	0

The bit string for A is 1100.

b. The table for χ_A is

x	u_1	u_2	u_3	u_4
χ_A	0	0	1	0

The bit string for A is 0010.

c. The table for χ_A is

x	u_1	u_2	u_3	u_4
χ_A	1	1	1	1

The bit string for A is 1111.

d. The table for χ_A is

x	u_1	u_2	u_3	u_4
χ_A	0	0	0	0

The bit string for A is 0000.

7. $s_n = \begin{cases} n & \text{if } n \text{ is even and } n \geq 0 \\ 1/n & \text{if } n \text{ is odd and } n \geq 1 \end{cases}$

9. a. $g(1) = [\dot{1}], g(-1) = [2], g(18) = [0]$ b. $\{x : g(x) = [0]\} = \{\ldots, -6, -3, 0, 3, 6, \ldots\}$
11. a. 2   b. $3 + 8 = 11$ c. -2 d. $-2 - (-3) = 1$
13. a. $\preceq$ is a partial ordering since $\subseteq$ is a partial ordering on $P(S)$.

b. 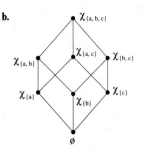   c. Identical diagrams except for labels

15. a. $(F, \preceq)$ is a poset since $\preceq$ is reflexive, antisymmetric, and transitive.
  b. No, since $\preceq$ is not symmetric. Let $f:\mathbb{R} \to \mathbb{R}$ be defined by $f(x) = 0$ for every $x \in \mathbb{R}$. Let $g:\mathbb{R} \to \mathbb{R}$ be defined by $g(x) = x^2$ for every $x \in \mathbb{R}$. Clearly, $f \preceq g$ since $f(x) \leq g(x)$ for every $x \in \mathbb{R}$. But it is not true that $g \preceq f$.
17. R is reflexive on D since $f(x) = f(x)$. R is symmetric on D since $f(x) = f(y)$ implies $f(y) = f(x)$. R is transitive since $f(x) = f(y)$ and $f(y) = f(z)$ imply $f(x) = f(z)$. Hence, R is an equivalence relation on D.
19. x is a minimal element of A.

Section 3.4

1. f is a function; it is not injective.

3. $f = \begin{pmatrix} 1 & 2 & 3 & 4 \\ 2 & 1 & 4 & 3 \end{pmatrix}$, $g = \begin{pmatrix} 1 & 2 & 3 & 4 \\ 1 & 3 & 4 & 2 \end{pmatrix}$ **a.** $f \circ g = \begin{pmatrix} 1 & 2 & 3 & 4 \\ 2 & 4 & 3 & 1 \end{pmatrix}$, $g \circ f = \begin{pmatrix} 1 & 2 & 3 & 4 \\ 3 & 1 & 2 & 4 \end{pmatrix}$

  b. $f^{-1} = \begin{pmatrix} 1 & 2 & 3 & 4 \\ 2 & 1 & 4 & 3 \end{pmatrix}$. Since $(f \circ f^{-1})(1) = 1$, $(f \circ f^{-1})(2) = 2$, $(f \circ f^{-1})(3) = 3$, and $(f \circ f^{-1})(4) = 4$, we have $f \circ f^{-1} = \begin{pmatrix} 1 & 2 & 3 & 4 \\ 1 & 2 & 3 & 4 \end{pmatrix}$. Similarly, we find $f^{-1} \circ f = \begin{pmatrix} 1 & 2 & 3 & 4 \\ 1 & 2 & 3 & 4 \end{pmatrix}$.

  c. $g^{-1} = \begin{pmatrix} 1 & 2 & 3 & 4 \\ 1 & 4 & 2 & 3 \end{pmatrix}$. Since $(g \circ g^{-1})(1) = 1$, $(g \circ g^{-1})(2) = 2$, $(g \circ g^{-1})(3) = 3$, and $(g \circ g^{-1})(4) = 4$, we have $g \circ g^{-1} = \begin{pmatrix} 1 & 2 & 3 & 4 \\ 1 & 2 & 3 & 4 \end{pmatrix}$. Similarly, we find $g^{-1} \circ g = \begin{pmatrix} 1 & 2 & 3 & 4 \\ 1 & 2 & 3 & 4 \end{pmatrix}$.

5. $f:\mathbb{R} \to \mathbb{R}$, $f(x) = 2 - 5x$. Assume f is bijective.
  $v = f(u) = 2 - 5u$ $v = 2 - 5u$
  Solving for u in terms of v: $u = (2 - v)/5$. Hence, $f^{-1}(v) = (2 - v)/5$, and $f^{-1}(x) = (2 - x)/5$. Therefore
  $(f \circ f^{-1})(x) = f\left(\dfrac{2-x}{5}\right) = 2 - 5\left(\dfrac{2-x}{5}\right) = x = f^{-1}(2 - 5x) = (f^{-1} \circ f)(x)$

7. Since the codomain of g is the domain of f, then $f \circ g$ is well-defined. Explicitly, $(f \circ g)(n) = f(g(n)) = f(2 - n) = 3(2 - n) - 2 = 4 - 3n$, or $(f \circ g)(n) = 4 - 3n$. But $g \circ f$ is not well-defined.

9. Since the codomain of g is the domain of f, $f \circ g$ is well-defined. Explicitly, $(f \circ g)(n) = f(g(n)) = f(2^n - 1) = \log_2[(2^n - 1) + 1] = \log_2 2^n = n$. But $g \circ f$ is not well-defined.

11. Since the codomain of g is not a subset of the domain of f, $f \circ g$ is not well-defined. But $g \circ f$ is well-defined since the codomain of f is a subset of the domain of g. Explicitly, $(g \circ f)(n) = g(f(n)) = g(2 - n^2) = 2 - n^2 + 7 = 9 - n^2$.

13. $(p \circ s)(x) = p(s(x)) = p(x + 1) = (x + 1) - 1 = x$; $(s \circ p)(x) = s(p(x)) = s(x - 1) = (x - 1) + 1 = x$. Therefore, $p \circ s = s \circ p = 1_{\mathbb{Z}}$. Hence, p and s are inverses of each other.

15. a. $h(x) = 2^x$ is a unary operation on $\mathbb{N}$.
  b. $h(x) = \log_2(x + 1)$ is not a unary operation on $\mathbb{N}$. For example, $h(2) = \log_2(2 + 1) = \log_2 3$, which is not in $\mathbb{N}$.

17. a. No. For example, $k(1, 1) = \frac{1}{2}$; but $\frac{1}{2}$ is not an integer.   **b.** Yes
  c. Yes, $x^2 + y^2 - xy \geq x^2 + y^2 - 2xy = (x - y)^2 \geq 0$.

19. a. Define $f:D \to C$ by $f(1) = 2$, $f(2) = 3$, $f(3) = 1$.   **b.** Such a function does not exist.
  c. Such a function does not exist.
  d. Define $f:D \to C$ by $f(1) = 2$, $f(2) = 2$, $f(3) = 2$. The function f is neither 1–1 nor onto.

21. a. Define $f:\mathbb{N} \to \mathbb{Z}$ by

$$f(n) = \begin{cases} n/2 & \text{if } n \text{ is even} \\ -(n + 1)/2 & \text{if } n \text{ is odd} \end{cases}$$

  b. Define $f:\mathbb{N} \to \mathbb{Z}$ by $f(n) = n$ for each $n \in \mathbb{N}$. The function f is 1–1 but not onto.
  c. Define $f:\mathbb{N} \to \mathbb{Z}$ by

$$f(n) = \begin{cases} 0 & \text{if } n = 0 \\ (n - 2)/2 & \text{if } n \text{ is even and } n \geq 2 \\ -(n + 1)/2 & \text{if } n \text{ is odd} \end{cases}$$

  The function f is onto but not 1–1.
  d. Define $f:\mathbb{N} \to \mathbb{Z}$ by $f(n) = 0$. The function f is neither 1–1 nor onto.

23. There are (7)(1024) bits in the computer memory. By Exercise 22, at least one location has more than $[((7)(1024) - 1)/7] = 1023$ bits. Hence, there is a location with at least 1024 bits.

Section 3.5

1. a. $n = 10$, $n \log_2 n = 33.2$; $33.2/10^9 = 3.32 \cdot 10^{-8}$ second $= 3.32 \cdot 10^{-8}/10^{-6}$ $\mu s = 3.32 \cdot 10^{-2}$ μs
   b. $n = 32$, $n \log_2 n = 160$; $160/10^9 = 1.6 \cdot 10^{-7}$ second $= 1.6 \cdot 10^{-7}/10^{-6}$ $\mu s = 1.6 \cdot 10^{-1}$ $\mu s = 0.16$ μs
   c. $n = 100$, $n \log_2 n = 664$; $664/10^9 = 6.64 \cdot 10^{-7}$ second $= 6.64 \cdot 10^{-7}/10^{-6}$ $\mu s = 6.64 \cdot 10^{-1}$ $\mu s = 0.66$ μs
3. a. $n = 10$, $n^2 = 100$; $100/10^9 = 10^{-7}$ second $= 10^{-7}/10^{-6}$ $\mu s = 10^{-1}$ μs
   b. $n = 32$, $n^2 = 1024$; $1024/10^9 = 1.024 \cdot 10^{-6}$ second $= 1.024$ μs
   c. $n = 100$, $n^2 = 10^4$; $10^4/10^9 = 10^{-5}$ second $= 10$ μs
5. $n + (n-1) + (n-2) + \cdots + 2 + 1 = n^2/2 + n/2$ **7.** $(n-1) + (n-2) + \cdots + 1 = (n^2/2 + n/2) - n = n^2/2 - n/2$

Chapter 3 Review

1. a. $B' = \{x : x > 8\}$ **b.** $A \cup D = \{-3, 1, 3, 5, 7, 9\}$ **c.** $A \cap D = \{7, 9\}$
   d. $A \cup B = \{\ldots, -3, -2, -1, 0, 1, 2, 3, 4, 5, 6, 7, 8, 9\}$ **e.** $D - B = \{9\}$ **f.** $A \cap B = \{-3, 7\}$
   g. $A \cap (B \cup D) = \{-3, 7, 9\}$ **h.** $(A \cap B) \cup D = \{-3, 1, 3, 5, 7, 9\}$
   i. $P(A) = \{\emptyset, \{-3\}, \{7\}, \{9\}, \{-3, 7\}, \{-3, 9\}, \{7, 9\}, A\}$ **j.** $A \times \emptyset = \emptyset$
   k. $A \times D = \{(-3, 1), (-3, 3), (-3, 5), (-3, 7), (-3, 9), (7, 1), (7, 3), (7, 5), (7, 7), (7, 9), (9, 1), (9, 3), (9, 5), (9, 7), (9, 9)\}$
3. a. False. For example, let $A = \{3, 4\}$, $B = \{3, 4\}$, and $D = \{4\}$. Then $D \cap (A \cup B) = \{4\} \neq \{3, 4\} = D \cup (A \cap B)$.
   b. True   **c.** False. For example, let $A = \{1, 3\}$, $B = \{3\}$, and $C = \{3\}$. Then $(A - B) - C = \{1\} \neq \{1, 3\} = A - (B - C)$.
5. a. f is bijective.   **b.** g is neither 1–1 nor onto.   **c.** j is neither 1–1 nor onto.   **d.** k is onto but not 1–1.
7. a. 42   **b.** Maximal element: 42   **c.** Greatest element: 42
                 Minimal element: 1   Least element: 1

9. f is not a function. It is a partial function with a domain of definition $\{a, b, c, e\}$.
11. $f(x) = 3 - x^5$, $v = f(u) = 3 - u^5$. Solving for u in terms of v: $u = \sqrt[5]{3 - v}$. Hence, $f^{-1}(v) = \sqrt[5]{3 - v}$ and $f^{-1}(x) = \sqrt[5]{3 - x}$.
Now, $(f \circ f^{-1})(x) = f(f^{-1}(x)) = f(\sqrt[5]{3 - x}) = 3 - (\sqrt[5]{3 - x})^5 = 3 - (3 - x) = x$, and $(f^{-1} \circ f)(x) = f^{-1}(f(x)) = f^{-1}((3 - x)^5) = \sqrt[5]{3 - (3 - x^5)} = \sqrt[5]{x^5} = x$. Therefore, $f \circ f^{-1} = 1_\mathbb{R} = f^{-1} \circ f$.
13. a. No, $2^{-1} = \frac{1}{2} \notin \mathbb{N}$ **b.** Yes
15. a. Assume $R = S$. Suppose $(x, y) \in R$. By definition of S, we have $(y, x) \in S$. Since $S = R$, it follows that $(y, x) \in R$. Since $(x, y) \in R$ implies $(y, x) \in R$, then R is symmetric.
   b. Assume R is reflexive, antisymmetric, and transitive. Since $(x, x) \in R$, we have $(x, x) \in S$ for each $x \in A$. Hence, S is reflexive. Suppose $(x, y) \in S$ and $(y, x) \in S$. By definition of S, we have $(y, x) \in R$ and $(x, y) \in R$. Since R is antisymmetric, we have $x = y$. Hence, S is antisymmetric. Suppose $(x, y) \in S$ and $(y, z) \in S$. By definition, we have $(y, x) \in R$ and $(z, y) \in R$. Hence, $(z, x) \in R$ and $(x, z) \in S$. Therefore, S is transitive.
17. Proof: There are four cases.
   Case 1: Assume x and y are both even. Then $f(x + y) = 1$ since $x + y$ is even. Also, $f(x) = 1$ and $f(y) = 1$.
   Case 2: Assume x is even and y is odd. Then $f(x + y) = -1$ since $x + y$ is odd. Also, $f(x) = 1$ and $f(y) = -1$.
   Case 3: Assume x and y are both odd. Then $f(x + y) = 1$ since $x + y$ is even. Also, $f(x) = -1$ and $f(y) = -1$.
   Case 4: Assume x is odd and y is even. Then $f(x + y) = -1$ since $x + y$ is odd. Also, $f(x) = -1$ and $f(y) = 1$. Hence, $f(x + y) = f(x) \cdot f(y)$ is all cases.

CHAPTER 4

Section 4.1

1. $45 \cdot 62 = 2790$ **3.** 2^{16} **5.** $30 \cdot 29 = 870$ **7.** $7 \cdot 11 \cdot 3 + 4 \cdot 11 \cdot 3 + 4 \cdot 7 \cdot 3 + 4 \cdot 7 \cdot 11 + 4 \cdot 7 \cdot 11 \cdot 3 = 1679$
9. **a.** $\text{Card}(A \times B) = 6 \cdot 3 = 18$ **b.** $\text{Card}(A \times A) = 36$ **c.** $\text{Card}(A^B) = 6^3$ **d.** $\text{Card}(P(A)) = 2^6$
   e. $\text{Card}(A \cup B) = 9$
11. **a.** 2^9 **b.** 5^9 **13.** $(2 \cdot 4)^{11}$ **15.** 2^8
17. Let A_1, A_2, and A_3 denote the sets of car owners, home owners, and computer owners, respectively.
   a. $\text{Card}(A_1 \cup A_2 \cup A_3) = 90 + 40 + 35 - 32 - 21 - 26 + 17 = 103$ **b.** 17 **c.** $35 - 21 - 26 + 17 = 5$
19. Let m be the desired number of integers between 1 and 500, inclusive.
   a. $m = 250$
   b. Since there are $\lfloor 500/2 \rfloor = 250$ multiples of 2, $\lfloor 500/3 \rfloor = 166$ multiples of 3, and $\lfloor 500/6 \rfloor = 83$ multiples of 6, we have $m = 250 + 166 - 83 = 333$.
   c. $m = \lfloor 500/2 \rfloor + \lfloor 500/3 \rfloor + \lfloor 500/7 \rfloor - \lfloor 500/6 \rfloor - \lfloor 500/14 \rfloor - \lfloor 500/21 \rfloor + \lfloor 500/42 \rfloor = 357$
   d. Using the formula preceding Example 4.1.7, we find $m = 385$.
21. Let B_k = The set of all integers with kth digit equal to k for $k = 1, 2, 3$.
   $\text{Card}(B_1 \cup B_2 \cup B_3) = 10 \cdot 10 + 9 \cdot 10 + 9 \cdot 10 - 10 - 10 - 9 + 1 = 252$.
23. First show $A = (A - B) \cup (A \cap B)$. Let $x \in A$. Then, $(x \in A$ and $x \notin B)$ or $(x \in A$ and $x \in B)$. Hence, $A \subseteq (A - B) \cup (A \cap B)$. Since $A - B \subseteq A$ and $A \cap B \subseteq A$, we have $(A - B) \cup (A \cap B) \subseteq A$. Since $A - B$ and $A \cap B$ are disjoint sets, we have $\text{Card}(A) = \text{Card}(A - B) + \text{Card}(A \cap B)$.
25. $\text{Card}(S) = \text{Card}(B^A) = \text{Card}(B)^{\text{Card}(A)} = 2^{2^n}$ **27. a.** m^n **b.** 2^{nm} **c.** n^m
29. We must include the set $S = \{(m, k): m = 1, \ldots, n$ for some $k \in A\}$ in the binary relation. There are 2^{n^2-n} ways to choose subsets from the $n^2 - n$ remaining ordered pairs in $A \times A$ to adjoin to S to form a desired binary relation. Hence, there are 2^{n^2-n} binary relations on A with the required property.

Section 4.2

1. 123, 132, 213, 231, 312, 321 **3.** $5! = 120$ **5.** $5 \cdot 4 \cdot 3 = 60$ **7. a.** $5 \cdot 4 \cdot 3 = 60$ **b.** $5! = 120$
9. $26 + 26 \cdot 36 + 26 \cdot 36^2 + 26 \cdot 36^3 + 26 \cdot 36^4 + 26 \cdot 36^5$ **11.** 8^6 **13.** 2^7 **15.** 2^8 **17. a.** 7 **b.** $128 - 96 = 32$
19. **a.** $8! = 40{,}320$ **b.** $3!3!2!3! = 432$ **21. a.** $7! = 5040$ **b.** $3!3!3!1 = 216$ **c.** $2 \cdot 7 \cdot 3! \cdot 3! = 504$ **23.** $5!3! = 720$
25. **a.** $5! = 120$ **b.** 0 **c.** $5 \cdot 4 \cdot 3 = 60$ **27.** $8^5 - 8 \cdot 7 \cdot 6 \cdot 5 \cdot 4 = 26{,}048$

Section 4.3

1. $C(4, 3) = \dfrac{4!}{3!1!} = 4$ **3. a.** $C(8, 5) = 56$ **b.** $C(n - 1 + r, r) = C(12, 5) = 792$
5. **a.** $P(5, 3) = \dfrac{5!}{2!} = 60$ **b.** $C(5, 3) = 10$ **7.** $C(7, 2) = 21$
9. **a.** $23 \cdot 22 \cdot 21 = P(23, 3) = 10{,}626$ **b.** $C(23, 3) = 1771$
11. **a.** $C(28, 5) = 98{,}280$ **b.** $C(26, 5) = 65{,}780$ **c.** $C(26, 3) + C(27, 4) + C(27, 4) = 37{,}700$ **13.** $C(11, 4) = 330$
15. $n = 3, r = 7, C(n - 1 + r, r) = C(9, 7) = 36$ **17.** $n = 4, r = 7, C(n - 1 + r, r) = C(10, 7) = 120$
19. **a.** $11 \cdot 10 \cdot 9 \cdot 8 \cdot 7 \cdot 6 = P(11, 6) = 332{,}640$ **b.** 11^6
21. **a.** 1, 7, 21, 35, 35, 21, 7, 1 **b.** $(a + b)^7 = a^7 + 7a^6b + 21a^5b^2 + 35a^4b^3 + 35a^3b^4 + 21a^2b^5 + 7ab^6 + b^7$
   c. $(1 + 1)^7 = 2^7$
23. **a.** 7^{23} **b.** $(7 + 1)^{23} = 8^{23}$
25. $C(5, 5) + C(6, 5) + C(7, 5) = C(6, 6) + C(6, 5) + C(7, 5)$ Since $C(5, 5) = 1 = C(6, 6)$
   $ = C(7, 6) + C(7, 5)$ Since $C(6, 6) + C(6, 5) = C(7, 6)$
   $ = C(8, 6)$ Since $C(7, 6) + C(7, 5) = C(8, 6)$
27. **a.** $C(n, r) = \dfrac{n!}{r!(n - r)!} = \dfrac{n!}{(n - r)![n - (n - r)]!} = C(n, n - r)$
   b. $\dfrac{n}{r} \cdot C(n - 1, r - 1) = \dfrac{n}{r} \cdot \dfrac{(n-1)!}{(r-1)![(n-1)-(r-1)]!} = \dfrac{n}{r} \cdot \dfrac{(n-1)!}{(r-1)!(n-r)!} = \dfrac{n!}{r!(n-r)!} = C(n, r)$

29. From Exercise 27b, we have $C(n, 1) = n \cdot C(n - 1, 0)$, $2 \cdot C(n, 2) = n \cdot C(n - 1, 1)$, $3 \cdot C(n, 3) = n \cdot C(n - 1, 2), \ldots$, $n \cdot C(n, n) = n \cdot C(n - 1, n - 1)$. Hence, $C(n, 1) + 2 \cdot C(n, 2) + \cdots + n \cdot C(n, n) = n \cdot C(n - 1, 0) + n \cdot C(n - 1, 1) + \cdots + n \cdot C(n - 1, n - 1) = n[C(n - 1, 0) + C(n - 1, 1) + \cdots + C(n - 1, n - 1)] = n \cdot 2^{n-1}$ by Exercise 28.

Section 4.4

1. **a.** $\{x_0\}, \{x_3\}$ **b.** $\{x_1, x_4\}$ **c.** $\Pr(S_i) = \frac{1}{5}$; $\Pr(\{x_0, x_2, x_4\}) = \frac{3}{5}$
3. **a.** $\Pr(E_0) = \frac{1}{7}$, $\Pr(E_1) = \frac{2}{7}$, $\Pr(E_2) = \frac{1}{7}$, $\Pr(E_3) = \frac{2}{7}$, $\Pr(E_4) = \frac{1}{7}$ **b.** $\frac{2}{7}$ **c.** $\frac{5}{7}$
5. **a.** {HHHH, HHHT, HHTH, HHTT, HTHH, HTHT, HTTH, HTTT, THHH, THHT, THTH, THTT, TTHH, TTHT, TTTH, TTTT}
   b. $\frac{1}{16}$ **c.** $\frac{5}{16}$
7. **a.** abc, cde, bcd **b.** $\frac{1}{60}$ **c.** $\frac{12}{60} = \frac{1}{5}$ **d.** $\frac{3}{5}$
9. **a.** $\Pr(A \cup B) = .5 + .3 - .2 = .6$ **b.** $\Pr(B') = .7$ **c.** $\Pr(A - B) = \Pr(A) - \Pr(A \cap B) = .5 - .2 = .3$
   d. $\Pr(B - A) = \Pr(B) - \Pr(A \cap B) = .3 - .2 = .1$
11. No, since $\Pr(B - A) \neq \Pr(B) - \Pr(A \cap B)$.
13. **a.** $\dfrac{C(6, 2)}{C(26, 2)} = .0462$ **b.** $\dfrac{C(18, 2)}{C(26, 2)} = .4708$ **c.** $\dfrac{C(2, 2)}{C(26, 2)} = .0031$
15. **a.** $\{a, b, c\}, \{c, d\}, \{a, e\}$ **b.** $\dfrac{1}{2^5}$ **c.** $\dfrac{C(5, 3)}{2^5} = \dfrac{10}{32} = \dfrac{5}{16}$ **d.** $\dfrac{C(5, 3) + C(5, 4) + C(5, 5)}{2^5} = \dfrac{16}{32} = \dfrac{1}{2}$
17. **a.** $\dfrac{C(4, 1) \cdot C(4, 1)}{C(52, 2)} = \dfrac{16}{26 \cdot 51} = \dfrac{8}{663}$ **b.** $\dfrac{C(13, 2)}{C(52, 2)} = \dfrac{78}{26 \cdot 51} = \dfrac{1}{17}$ **c.** $\dfrac{C(4, 1) \cdot C(4, 1)}{C(52, 2)} + \dfrac{C(13, 2)}{C(52, 2)} - \dfrac{1}{C(52, 2)}$
19. **a.** $\dfrac{C(4, 2) \cdot C(4, 2) \cdot C(44, 1)}{C(52, 5)}$ **b.** $\dfrac{C(13, 2) \cdot C(4, 2) \cdot C(4, 2) \cdot C(44, 1)}{C(52, 2)}$
21. $\Pr(E_n) = 1 - \Pr(E_n') = 1 - \dfrac{12 \cdot 11 \cdots [12 - (n - 1)]}{12^n}$ 23. $\dfrac{10 \cdot 9 \cdot 8 \cdot 7 \cdot 6}{10^5}$
25. Use Algorithm 4.4.2.
    a. Since $S = S_1 \cup \cdots \cup S_n$, we have $\Pr(S) = \Pr(S_1) + \cdots + \Pr(S_n) = 1$.
    b. For any event E in S, $E = S_{i_1} \cup \cdots \cup S_{i_k}$, where $S_{i_1}, \ldots, S_{i_k}$ are among $S_1, \ldots, S_n$. Hence, $\Pr(E) = \Pr(S_{i_1}) + \cdots + \Pr(S_{i_k}) \leq 1$. Since each $\Pr(S_{i_j}) \geq 0$, we have $\Pr(E) \geq 0$. Hence, $0 \leq \Pr(E) \leq 1$.
    c. If A and B are mutually exclusive events, the simple events in A are distinct from the simple events in B. The simple events in A and those in B are the simple events in $A \cup B$. Hence, $\Pr(A \cup B) = \Pr(A) + \Pr(B)$.

Section 4.5

1. Cat(0) = 1, Cat(1) = 1, Cat(2) = 2, Cat(4) = 14
3. **a.** Yes; push 4, push 3, pop 3, pop 4, push 2, push 1, pop 1, pop 2
   b. Yes; push 4, pop 4, push 3, push 2, push 1, pop 1, pop 2, pop 3   **c.** No   **5. a.** $14 = \text{Cat}(4)$
7. $\text{Cat}(n - 1) = C(2(n - 1), n - 1) - C(2(n - 1), n - 2) = C(2n - 2, n - 1) - \dfrac{n - 1}{n} \cdot C(2n - 2, n - 1) = \dfrac{1}{n} \cdot C(2n - 2, n - 1)$

Chapter 4 Review

1. **a.** $10 + 90 + 900 + 9000 + 90000 = 100000$ **b.** $\sum_{i=0}^{5} C(10, i)$ or $\dfrac{2^{10} + C(10, 5)}{2}$
3. **a.** $n = 5, r = 20, C(n - 1 + r, r) = C(24, 20) = 10{,}626$ **b.** $C(20, 10) \cdot C(5 - 1 + 10, 10) = C(20, 10) \cdot C(14, 10)$
5. $n!$ **7.** $2^n - 2$ (two functions are not onto)   **9. a.** $\dfrac{C(13, 5)}{C(52, 5)} = .000495$ **b.** $\dfrac{C(4, 1) \cdot C(13, 5)}{C(52, 5)} = .00198$
11. Let $n = C(50, 2)$. **a.** $\dfrac{C(3, 2)}{n} = .00245$ **b.** $\dfrac{C(45, 2)}{n} = .808$ **c.** $\dfrac{C(2, 2)}{n} = .000816$ **d.** $\dfrac{C(45, 1) \cdot C(5, 1)}{n} = .184$
13. 5   **15. a.** $P(15, 8)$ **b.** 5^8
17. **a.** 11^7 since each distinct object can be placed in 11 different boxes.   **b.** $C(n - 1 + r, r) = C(17, 7) = 19{,}448$

CHAPTER 5
Section 5.1

19. $C(m + n - 1, n)$ because this is the same as the number of bit strings with n 1's and $m - 1$ 0's.

21. We seek the number of bit strings with $m - 1$ 0's (as markers), n 1's, and no 0's on either end and no adjacent 0's. There are $n - 1$ spaces between n 1's in which to place $m - 1$ 0's, and no space contains more than one 0. This number is $C(n - 1, m - 1)$.

1. If the irises are not blooming, then it is winter.
The irises are not blooming if and only if it is winter.
p true, q true

3. If the irises are blooming, then the irises are blooming or it is winter.
If the irises are blooming, then the irises are blooming and it is winter.
p true, q false

5. a. $r \wedge (q \vee p)$ **b.** $q \to (\neg p \vee \neg r)$ **7.** $\underset{3}{(p \to \underset{2}{\neg q}) \wedge r \to q}$ **9.** $\underset{2}{(p \to q)} \to [\underset{3}{(p \wedge r)} \to \underset{4}{(q \wedge r)}]$

11. $[\underset{2}{(p \to q)} \wedge \underset{3}{(r \to s)}] \to [\underset{5}{(p \vee r)} \to \underset{7}{(q \vee s)}]$

13. a.

p	q	$p \wedge q \to p$
T	T	T
T	F	T
F	T	T
F	F	T

b.

p	q	$p \to p \vee q$
T	T	T
T	F	T
F	T	T
F	F	T

15. a.

p	q	$p \wedge (\neg p \vee q) \to q$
T	T	T
T	F	T
F	T	T
F	F	T

b.

p	q	$\neg p \wedge (p \to q) \to q$
T	T	T
T	F	T
F	T	T
F	F	F

17.

p	q	r	$(p \to q) \to [(p \vee r) \to (q \vee r)]$
T	T	T	T
T	T	F	T
T	F	T	T
T	F	F	T
F	T	T	T
F	T	F	T
F	F	T	T
F	F	F	T

19.

p	q	r	$(p \vee q \to r) \to (p \to r) \wedge (q \to r)$
T	T	T	T
T	T	F	T
T	F	T	T
T	F	F	T
F	T	T	T
F	T	F	T
F	F	T	T
F	F	F	T

21.

p	q	r	$(p \wedge q \to r) \to (p \to r) \vee (q \to r)$
T	T	T	T
T	T	F	T
T	F	T	T
T	F	F	T
F	T	T	T
F	T	F	T
F	F	T	T
F	F	F	T

23.

p	q	r	$\neg(p \wedge \neg q) \vee (r \to q)$
T	T	T	T
T	T	F	T
T	F	T	F
T	F	F	T
F	T	T	T
F	T	F	T
F	F	T	T
F	F	F	T

25.

a.
p	$p \veebar p$
T	F
F	F

b.
p	$p \veebar \neg p$
T	T
F	T

27.

a.
p	q	r	$p \veebar (q \wedge r)$
T	T	T	F
T	T	F	T
T	F	T	T
T	F	F	T
F	T	T	T
F	T	F	F
F	F	T	F
F	F	F	F

b.
p	q	r	$(p \veebar q) \wedge (p \veebar r)$
T	T	T	F
T	T	F	F
T	F	T	F
T	F	F	T
F	T	T	T
F	T	F	F
F	F	T	F
F	F	F	F

29. $(p \wedge \neg q) \vee (\neg p \wedge q)$ **31.** Both James and Amy are truth tellers.

Section 5.2

1.

a.
p	q	$q \to p$
T	T	T
T	F	T
F	T	F
F	F	T

b.
p	q	$\neg q \to \neg p$
T	T	T
T	F	F
F	T	T
F	F	T

3.
p	q	$\neg(p \wedge q)$	$\neg p \vee \neg q$
T	T	F	F
T	F	T	T
F	T	T	T
F	F	T	T

5.
p	$p \vee \neg p$
T	T
F	T

7.
p	q	r	$p \wedge r \to q$	$p \to (r \to q)$
T	T	T	T	T
T	T	F	T	T
T	F	T	F	F
T	F	F	T	T
F	T	T	T	T
F	T	F	T	T
F	F	T	T	T
F	F	F	T	T

9.
p	q	r	$p \vee q \to r$	$(p \to r) \wedge (q \to r)$
T	T	T	T	T
T	T	F	F	F
T	F	T	T	T
T	F	F	F	F
F	T	T	T	T
F	T	F	F	F
F	F	T	T	T
F	F	F	T	T

11, 13, 15, and **17.** The truth tables all have T's in the final column.
19. b. $[p \wedge (\neg p \vee q) \to q] \equiv [p \wedge (p \to q) \to q]$. Hence, $p \wedge (\neg p \vee q) \to q$ is a tautology because modus ponens is a tautology.
21. a. p false, q true, and r false is one counterexample.   **b.** p false, q false, and r true is one counterexample.
23. a. $p \uparrow p \equiv \neg(p \wedge p) \equiv \neg p$ **b.** $(p \uparrow p) \uparrow (q \uparrow q) \equiv (\neg p) \uparrow (\neg q) \equiv \neg(\neg p \wedge \neg q) \equiv p \vee q$
c. $\neg(\neg p \vee \neg q) \equiv \neg(\neg p) \wedge \neg(\neg q) \equiv p \wedge q$
25. *Hint:* Construct the appropriate truth tables.   **27.** *Hint:* Construct the appropriate truth tables.

Section 5.3

1. *Hint:* Do a direct proof of an implication. First, assume p. Deduce $\neg q$ and then conclude s.
3. *Hint:* Note that $r \to s$ is logically equivalent to the desired conclusion. Use a direct proof to prove $r \to s$.
5. *Hint:* Assume r. Use the premises to derive n, then $\neg p$, then $\neg s$.
7. a. Proof: If $x = 2k$ and $y = 2m$, where k and m are some integers, then $xy = 2k \cdot 2m = 2(2km)$. Hence, xy is even.
b. Proof: If $x = 2k + 1$ and $y = 2m + 1$, where k and m are some integers, then
$x + y = 2k + 1 + 2m + 1 = 2k + 2m + 2 = 2(k + m + 1)$. Hence, $x + y$ is even.
9. a. Proof: Pick any $b \in B$. Then, $b \in B$ or $b \in A$. Hence, $b \in A \cup B$.
b. Proof: $x \in (A')' \leftrightarrow \neg(x \in A') \leftrightarrow \neg[\neg(x \in A)] \leftrightarrow x \in A$
c. Proof: $x \in A \cap B \leftrightarrow x \in A$ and $x \in B \leftrightarrow x \in B$ and $x \in A \leftrightarrow x \in B \cap A$
11. a. Proof: We prove $(A \cup B) \times C \subseteq (A \times C) \cup (B \times C)$. Assume $(a, c) \in (A \cup B) \times C$. Then, $(a \in A$ or $a \in B)$ and $c \in C$. Hence, $(a \in A$ and $c \in C)$ or $(a \in B$ and $c \in C)$. Therefore, $(a, c) \in (A \times C) \cup (B \times C)$. Reverse the steps above to prove $(A \times C) \cup (B \times C) \subseteq (A \cup B) \times C$.
b. The proof is similar to that in part a.

13. a. Proof: Assume $(x, y) \in (A \times B) \cup (C \times D)$. Case 1: Assume $(x, y) \in (A \times B)$. Then, $x \in A$ and $y \in B$. Hence, $x \in A \cup C$ and $y \in B \cup D$. Therefore, $(x, y) \in (A \cup C) \times (B \cup D)$. Case 2 is similar.
  b. One counterexample is: $A = \{a\}$, $B = \{b\}$, $C = \{c\}$, and $D = \emptyset$.

15. Proof: Assume $f^{-1}(u) = f^{-1}(v)$. Then $(f \circ f^{-1})(u) = (f \circ f^{-1})(v)$. Therefore, $u = v$, since $f \circ f^{-1} = 1_C$. To show f^{-1} is onto, pick any $d \in D$. Then we have $f(d) = c \in C$ and $f^{-1}(c) = f^{-1}(f(d)) = d$. Hence, f^{-1} is bijective.

17. a. Proof: $x \in D - (A \cap B) \leftrightarrow x \in D$ and $\neg(x \in A \cap B) \leftrightarrow x \in D$ and $\neg(x \in A$ and $x \in B) \leftrightarrow x \in D$ and $(x \notin A$ or $x \notin B)$
$\leftrightarrow (x \in D$ and $x \notin A)$ or $(x \in D$ and $x \notin B) \leftrightarrow x \in (D - A) \cup (D - B)$
  b. Proof: $(D - A) \cap (D - B) = (D \cap A') \cap (D \cap B') = D \cap A' \cap B' = D \cap (A' \cap B')'' = D \cap (A'' \cup B'')' =$
$D \cap (A \cup B)' = D - (A \cup B)$

19. Proof: $(A \cap B) \cup (C - A) \cup (B \cap C) = (A \cap B) \cup (C \cap A') \cup (B \cap C) = (A \cap B) \cup (C \cap A') \cup [(B \cap C) \cap (A \cup A')] =$
$(A \cap B) \cup (C \cap A') \cup [(B \cap C \cap A) \cup (B \cap C \cap A')] = (A \cap B) \cup (A \cap B \cap C) \cup (C \cap A') \cup (B \cap C \cap A')$. Now use the fact that $X \cup (X \cap Y) = X$ to obtain the result.

21. Proof: Assume f is an equivalence relation on A. Then f is reflexive and so $(x, x) \in f$ for all $x \in A$. Hence, $f(x) = x$ for all $x \in A$. Therefore, $f = 1_A$.

23. Use De Morgan's law to prove $A' \cup B = U \leftrightarrow A \cap B' = \emptyset$. Prove: If $A \subseteq B$, then $A \cap B' = \emptyset$. Suppose $A \cap B' \neq \emptyset$. Then there is an element $x \in A \cap B'$. But then $x \in A$ and $x \notin B$, contradicting $A \subseteq B$. Prove: If $A' \cup B = U$, then $A \subseteq B$. Assume $A' \cup B = U$ and let $x \in A$. Since $x \in U$, it follows from $A' \cup B = U$ that $x \in A' \cup B$. Hence, $x \in B$, since $x \notin A'$. Therefore, $A \subseteq B$.

25. Let d_i be the digit assigned to the ith segment. Suppose $d_1 + d_2 < 8$, $d_2 + d_3 < 8, \ldots, d_9 + d_{10} < 8$, $d_{10} + d_1 < 8$. Adding, we obtain $2(d_1 + d_2 + \cdots + d_{10}) < 80$. Hence, $0 + 1 + 2 + \cdots + 9 < 40$. But, $0 + 1 + 2 + \cdots + 9 = 9 \cdot 10/2 = 45$. We have a contradiction.

Section 5.4

1. The basis step is: $A \cap (B_1) = A \cap B_1 = (A \cap B_1)$. The induction step is: $A \cap (B_1 \cup \cdots \cup B_k \cup B_{k+1}) =$
$A \cap [(B_1 \cup \cdots \cup B_k) \cup B_{k+1}] = [A \cap (B_1 \cup \cdots \cup B_k)] \cup (A \cap B_{k+1}) = (A \cap B_1) \cup \cdots \cup (A \cap B_k) \cup (A \cap B_{k+1})$.

3. The induction step is: $D - (A_1 \cup \cdots \cup A_k \cup A_{k+1}) = D - [(A_1 \cup \cdots \cup A_k) \cup A_{k+1}] =$
$[D - (A_1 \cup \cdots \cup A_k)] \cap (D - A_{k+1}) = (D - A_1) \cap \cdots \cap (D - A_k) \cap (D - A_{k+1})$.

5. Induction hypothesis: Assume $1 + 2 + \cdots + k = \dfrac{k(k+1)}{2}$. Induction step:

$$1 + 2 + \cdots + k + (k+1) = \frac{k(k+1)}{2} + (k+1) = \frac{k(k+1) + 2(k+1)}{2} = \frac{(k+1)(k+2)}{2}$$

7. Induction hypothesis: Assume $1 + \dfrac{1}{2} + \cdots + \dfrac{1}{2^k} = 2 - \dfrac{1}{2^k}$. Induction step:

$$1 + \frac{1}{2} + \cdots + \frac{1}{2^k} + \frac{1}{2^{k+1}} = 2 - \frac{1}{2^k} + \frac{1}{2^{k+1}} \text{ (By induction hypothesis)}$$
$$= 2 + \frac{-2 + 1}{2^{k+1}}$$
$$= 2 - \frac{1}{2^{k+1}}$$

9. Induction step: $(k+1)^2 = k^2 + (2k+1) < 2^k + 2^k = 2^{k+1}$

11. To finish the induction step, it is necessary to prove $\dfrac{2k(k+1) + 1}{(k+1)^2} \leq \dfrac{2(k+1)}{k+2}$. Use an indirect proof for this.

13. Induction step:
$1 \cdot 1! + 2 \cdot 2! + \cdots + k \cdot k! + (k+1) \cdot (k+1)! = (k+1)! - 1 + (k+1) \cdot (k+1)! = (k+2) \cdot (k+1)! - 1 = (k+2)! - 1$

15. Induction step: $M_{R^{k+1}} = M_{R \circ R^k} = M_R(M_{R^k}) = M_R(M_R)^k = (M_R)^{k+1}$

17. $\dfrac{1}{1 \cdot 2} + \dfrac{1}{2 \cdot 3} + \cdots + \dfrac{1}{n(n+1)} = \dfrac{n}{n+1}$, $n = 1, 2, 3, \ldots$

$\dfrac{1}{1 \cdot 2} + \dfrac{1}{2 \cdot 3} + \cdots + \dfrac{1}{k(k+1)} + \dfrac{1}{(k+1)(k+2)} = \dfrac{k}{k+1} + \dfrac{1}{(k+1)(k+2)} = \dfrac{k(k+2) + 1}{(k+1)(k+2)} = \dfrac{(k+1)^2}{(k+1)(k+2)} = \dfrac{k+1}{k+2}$

19. Induction step: $D^{k+1}(e^{bx}) = D[D^k(e^{bx})] = D[b^k e^{bx}] = b^k D[e^{bx}] = b^{k+1} e^{bx}$

21. Use mathematical induction to prove that $n \in A$ for $n = 0, 1, 2, \ldots$.   **23.** The induction step from $k = 1$ is not valid.

Section 5.5

1. a. $\emptyset, \{\frac{3}{5}\}$ **b.** $\{0, 1, 2\}, \{x : x < \frac{7}{3}\} = (-\infty, \frac{7}{3})$ **c.** $\mathbb{N}, \mathbb{R}$ **d.** $\emptyset, \{x : x = \pi/2 + 2m\pi, m \in \mathbb{Z}\}$
3. a. False **b.** True **c.** True **d.** True
5. a. $\exists x\, (x^2 + 2x - 3 \neq 0)$ **b.** $\forall x\, (x^2 + 2x - 3 \neq 0)$ **c.** $\exists x\, (\sin(x + \pi/2) \neq \cos x)$ **d.** $\forall x\, (\sin(x + \pi/2) \neq \cos x)$
7. a. $\neg \forall x\, \forall y\, [(x < y) \to (x^2 < y^2)] \leftrightarrow \exists x\, \neg \forall y\, [(x < y) \to (x^2 < y^2)] \leftrightarrow \exists x\, \exists y\, [(x < y) \land \neg(x^2 < y^2)] \leftrightarrow \exists x\, \exists y\, [(x < y) \land (y^2 \leq x^2)]$
b. It is not the case that some prime numbers are odd.
$\leftrightarrow \neg \exists x\, [p(x) \land o(x)] \leftrightarrow \forall x\, \neg [p(x) \land o(x)] \leftrightarrow \forall x\, [\neg p(x) \lor \neg o(x)] \leftrightarrow \forall x\, [p(x) \to \neg o(x)] \leftrightarrow$ No prime numbers are odd.
9. a. $\exists z\, (\forall w > 0)(0 \leq z < w \land z \neq 0)$, false **b.** $\forall m\, \exists x\, [x/(|x| + 1) \geq m]$, false
11. a. f is 1-1 if $(\forall u \in D)(\forall w \in D)(f(u) = f(w) \to u = w)$ **b.** f is onto if $(\forall z \in C)(\exists x \in D)(f(x) = z)$
13. a. $\forall x\, [g(x) \to c(x)],\ \exists x\, [g(x) \land \neg c(x)]$ **b.** $\exists x\, [g(x) \land c(x)],\ \forall x\, [g(x) \to \neg c(x)]$
15. a. $\exists M\, \exists Q\, (M \otimes Q = 0 \land M \neq 0 \land Q \neq 0)$ **b.** $\exists M\, \exists Q\, (M \otimes Q \neq Q \otimes M)$

$$M = \begin{bmatrix} 1 & 0 \\ 0 & 0 \end{bmatrix},\ Q = \begin{bmatrix} 0 & 0 \\ 0 & 1 \end{bmatrix} \qquad M = \begin{bmatrix} 1 & 1 \\ 0 & 0 \end{bmatrix},\ Q = \begin{bmatrix} 0 & 0 \\ 0 & 1 \end{bmatrix}$$

17. Proof:
$x \in D - \bigcap \Omega \leftrightarrow x \in D \land \neg(\forall A \in \Omega)(x \in A) \leftrightarrow x \in D \land (\exists A \in \Omega)(x \in A') \leftrightarrow (\exists A \in \Omega)(x \in D - A) \leftrightarrow x \in \bigcup \{D - A : A \in \Omega\}$
19. Proof: Pick $t = 1$ and $m = 0$. **21.** Proof: Pick $t = 1$ and $m = 3$.
23. Proof: Suppose $2^n < tn^2$ for some fixed t and m, and all $n > m$. We may assume $t < n$, since n can be arbitrarily large. Hence, $2^n < n^3$ for all $n > m$. We have a contradiction since it can be proved, using mathematical induction, that $2^n \geq n^3$ for $n > 9$.
25. Proof: Assume there exists m_1 and t_1 such that $f(n) \leq t_1 \cdot g(n)$ for all $n > m_1$ and there exists m_2 and t_2 such that $g(n) \leq t_2 \cdot h(n)$ for all $n > m_2$. Pick $t = t_1 t_2$ and $m = \max(m_1, m_2)$. Hence, $f(n) \leq t \cdot h(n)$ for all $n > m$.
27. Proof: Assume there exists m_1 and t_1 such that $f(n) \leq t_1 \cdot h(n)$ for all $n > m_1$ and there exists m_2 and t_2 such that $g(n) \leq t_2 \cdot h(n)$ for all $n > m_2$. Pick $t = at_1 + bt_2$ and $m = \max(m_1, m_2)$. Hence, $a \cdot f(n) + b \cdot g(n) \leq t \cdot h(n)$ for all $n > m$.

Section 5.6

1. Proof: We obtain clauses $\neg t$ and p by negating the conclusion. The first two premises are the clauses $\neg p \lor m$ and $t \lor \neg m$. The resolvent of these two clauses is $\neg p \lor t$. Using $\neg t$ we obtain $\neg p$. Hence, $p \land \neg p$ is a contradiction.
3. Proof: Negating the conclusion yields $\neg s$ and $\neg p \lor r$. The resolvent of the first two premises is $\neg r \lor s$. Resolving $\neg p \lor r$ with $s \lor \neg r$ yields $\neg p \lor s$. Using $\neg s$ we obtain $\neg p$. Hence, $\neg p \land p$ is a contradiction.
5. Proof: Negating the conclusion yields p and $\neg s$. The resolvent of p and the first premise is $\neg q$. The resolvent of $\neg q$ and the third premise is s. Hence, $\neg s \land s$ is a contradiction.

Chapter 5 Review

1. False (see page 190).
3. The induction step requires that the following be verified:
$$\frac{3}{4} - \frac{1}{2(k+1)} - \frac{1}{2(k+2)} + \frac{1}{(k+2)^2 - 1} = \frac{3}{4} - \frac{1}{2(k+2)} - \frac{1}{2(k+3)}$$
5. a. Assume R is transitive. Let $(x, y) \in R \circ R$. Then there is an element z such that $(x, z) \in R$ and $(z, y) \in R$. Hence, $(x, y) \in R$. Therefore, $R \circ R \subseteq R$.
b. Assume $R \circ R \subseteq R$. Let $(x, z) \in R$ and $(z, y) \in R$. Then $(x, y) \in R \circ R$. Hence, $(x, y) \in R$. Therefore, R is transitive.
7. Assume R is transitive. We need only prove that $R^+ \subseteq R$ since $R \subseteq R^+$. Let $(a, b) \in R^+$. Then there is a path in R from a to b. Use the result of Exercise 14 in Exercise Set 5.4 to conclude that $(a, b) \in R$.
9. Assume $R \subseteq S$. Let $(x, y) \in R \circ T$. Then, $(x, z) \in R$ and $(z, y) \in T$ for some $z \in A$. Hence, $(x, z) \in S$ and $(z, y) \in T$. Therefore, $(x, y) \in S \circ T$.
11. $(A - B) \cup (B - A) = (A \cap B') \cup (B \cap A') = (A \cup B) \cap (A \cup A') \cap (B' \cup B) \cap (A' \cup B') = (A \cup B) \cap (A' \cup B') = (A \cup B) \cap (A \cap B)' = (A \cup B) - (A \cap B) = A \Delta B$
13. Proof: Suppose $n(n - 1)/2 \leq tn$ for some fixed t and m, and all $n > m$. Hence, $(n - 1)/2 < t$. So, $n \leq 2t + 1$. But this contradicts the fact that the natural numbers are not bounded above.
15. Proof: Pick $t = 1$ and $m = 0$.

17. Assume $q(0)$ and, for all $k \in \mathbb{N}$, $q(j)$, for all $j = 0, 1, \ldots, k \to q(k + 1)$. Let $p(m)$ be the statement $(\forall j = 0, 1, \ldots, m)\, q(j)$, where $m \in \mathbb{N}$. Since $q(0)$, we have $p(0)$. The induction hypothesis is $p(k)$. Hence, $(\forall j = 0, 1, \ldots, k)\, q(j)$. By our initial assumption, $q(k + 1)$. Therefore, $p(k + 1)$, since $(\forall j = 0, 1, \ldots, k + 1)\, q(j)$. By the principle of mathematical induction, $p(n)$ for all $n \in \mathbb{N}$. Note that $p(m)$ implies $q(m)$. Therefore, $q(n)$ for all $n \in \mathbb{N}$.
19. Suppose there is a nonempty subset A of $\mathbb{N}$ with no least element. Then, $0 \in A'$ since $0 \in A$ implies A has a least element. Assume $i \in A'$ for $i = 0, 1, 2, \ldots, k$. Then, $k + 1 \in A'$ since $k + 1 \in A$ means $k + 1$ is the least element of A. By the principle of strong induction, $n \in A'$ for all $n \in \mathbb{N}$. Hence, $A = \emptyset$, contradicting the initial assumption. Therefore, every nonempty subset of $\mathbb{N}$ has a least element and the well-ordering principle holds.
21. Proof: The functions g and h have the same domain B. Let $x \in B$. Then, $f(g(x)) = (f \circ g)(x) = (f \circ h)(x) = f(h(x))$. Hence, $g(x) = h(x)$ since f is 1–1.

CHAPTER 6
Section 6.1

1. **a.** $210 = 2 \cdot 3 \cdot 5 \cdot 7$ **b.** $1024 = 2^{10}$ 3. Proof: $a = a \cdot 1$
5. Proof: Assume $d\,|\,a$ and $d\,|\,b$. Then, $a = dm$ and $b = dn$. Hence, $a + b = dm + dn = d(m + n)$. Therefore, $d\,|\,(a + b)$.
7. Proof: For $n = 1$, we have $2^{2n} - 1 = 2^2 - 1 = 3$. Induction hypothesis: Assume $3\,|\,(2^{2k} - 1)$. Then, $2^{2k} - 1 = 3m$. Hence, $2^{2(k+1)} - 1 = 4 \cdot 2^{2k} - 1 = 3 \cdot 2^{2k} + (2^{2k} - 1) = 3(2^{2k} + m)$. Therefore, $3\,|\,(2^{2(k+1)} - 1)$.
9. Proof: For $n = 0$, we have 3 divides 0. Induction hypothesis: Assume $(k^3 + 2k)$ is divisible by 3. Induction step: Prove that $(k + 1)^3 + 2(k + 1)$ is divisible by 3. We have
$$(k + 1)^3 + 2(k + 1) = k^3 + 3k^2 + 3k + 1 + 2k + 2 = k^3 + 2k + 3k^2 + 3k + 3$$
$$= 3m + 3(k^2 + k + 1) \quad \text{By induction hypothesis}$$
$$= 3(m + k^2 + k + 1)$$
11. By Theorem 6.1.4, $n = 1 + p^2$ has a prime factor. If this prime factor is p, then $p\,|\,n - p^2$. Hence, $p\,|\,1$, contradicting the fact that p is a prime.
13. By the division algorithm (Algorithm 6.1.6), $a = vq + r$ with $0 \leq r < v$. Then, $a = (am + pn)q + r$. So, $r = (1 - mq)a + (-nq)p$. Hence, $r = 0$ since $r < v$. Therefore, $v\,|\,a$.
15. Suppose $\sqrt{p} = m/n$, where m and n are positive integers with no common factors. Then $pn^2 = m^2$. Hence, $p\,|\,m^2$. By Theorem 6.1.8, $p\,|\,m$. So, $p^2\,|\,m^2$. Hence, $pn^2 = p^2k$ and $n^2 = pk$. Therefore, n^2, and hence n, has a prime factor p. This contradicts the fact that m and n have no common factors. Hence, $\sqrt{p}$ is an irrational number.
17. Suppose $\log_2 p = m/n$, where m and n are positive integers with no common factors. Then, $2^{m/n} = p$. Hence, $2^m = p^n$, contradicting the Fundamental Theorem of Arithmetic (Theorem 6.1.10).
19. Prove the contrapositive. Assume n is not prime and let $n = ab$. Then,
$2^n - 1 = 2^{ab} - 1 = (2^a)^b - 1 = (2^a - 1)[(2^a)^{b-1} + (2^a)^{b-2} + \cdots + 1]$. Hence, $2^n - 1$ is not prime.
21. Induction hypothesis: Assume $x^k - y^k = (x - y)m$. Then,
$x^{k+1} - y^{k+1} = xx^k - yy^k = (x - y + y)x^k - yy^k = (x - y)x^k + yx^k - yy^k = (x - y)x^k + y(x - y)m = (x - y)(x^k + ym)$.
23. Assume $a = dq + r_1$, with $0 \leq r_1 < d$, and $q = ks + r_2$, with $0 \leq r_2 < k$. Then, $a = d(ks + r_2) + r_1 = (kd)s + (dr_2 + r_1)$. Also, $0 \leq dr_2 + r_1$. We need to show $dr_2 + r_1 < kd$. We have $r_1 \leq d - 1$ and $r_2 \leq k - 1$. Hence, $dr_2 + r_1 \leq (k - 1)d + d - 1 = kd - 1$. Therefore, $dr_2 + r_1 < kd$.

Section 6.2

1. **a.** $\gcd(29, 11) = 1 = (-3)29 + 8 \cdot 11$ **b.** $\gcd(32, 76) = 4 = 3 \cdot 76 + (-7)32$
3. **a.** $\text{lcm}(40, 68) = 680$ **b.** $\text{lcm}(175, 450) = 3150$ 5. **a.** $\text{lcm}(21, 96) = 7 \cdot 96 = 672$ **b.** $\gcd(21, 96) = 3$
7. **a.** No, by Corollary 6.2.5   **b.** Yes, by Corollary 6.2.5
9. **a.** $\gcd(10, 33) = 1 = 10 \cdot 10 + (-3)33$
   b. Alternately fill the 10 cc beaker and pour it into the 33 cc beaker until full. Empty the 33 cc beaker. Pour what is left in the 10 cc beaker to the 33 cc beaker. Repeat until the 10 cc beaker has been filled and poured into the 33 cc beaker 10 times. There will be 1 cc of liquid in the 10 cc beaker.
11. Assume $g = \gcd(a, b)$ and $g\,|\,w$. By Theorem 6.2.4, $g = ax + by$ for some integers x and y. But $w = gm$. Hence, $w = a(xm) + b(ym) = (am)x + (bm)y$.
13. The only divisors of p are p and 1. Hence, $\gcd(a, p) = p$ or $\gcd(a, p) = 1$.

15. This follows from Corollary 6.2.5 since $g|1$ if and only if $g = 1$.
17. Assume $\gcd(a, b) = c$, $a = cd$, and $b = cf$. Then, $c = am + bn = cdm + cfn$. Hence, $1 = dm + fn$, and $\gcd(d, f) = 1$.
19. Assume $\gcd(a, b) = 1$, $ad|c$, $bd|c$. Then, $abd|bc$ since $ad|c$, and $abd|ac$ since $bd|c$. Now $c = acm + bcn$ since $1 = am + bn$. Therefore, $abd|c$.
21. Let $u = \gcd(a + b, a - b)$ and $v = \gcd(a, b)$. Then, u divides $2a = (a + b) + (a - b)$, and u divides $2b = (a + b) - (a - b)$. Since $a + b$ is odd, u is odd. By Exercise 18, u is a common divisor of a and b. Hence, $u|v$. But v divides both $a + b$ and $a - b$. Hence, $v|u$. Therefore, $u = v$.
23. Since a is a common divisor of a and ab, and a is the largest of all divisors of a, then $\gcd(a, ab) = a$. By Theorem 6.2.8, $a(ab) = \gcd(a, ab) \cdot \text{lcm}(a, ab)$. It follows that $\text{lcm}(a, ab) = ab$.
25. **a.** $e = 1$
    b. No such element exists. If $\gcd(a, f) = a$ for all a, then $a|f$ for all a. But this contradicts the fact that $\mathbb{Z}$ is not bounded above.

Section 6.3

1. **a.** 3 **b.** 2 **3. a.** 1 **b.** 1
5. Assume $x - u = md$ and $y - v = nd$. Then, $xy = (u + md)(v + nd) = uv + kd$. Hence, $xy - uv = kd$.
7. **a.** 11 **b.** 25 **9. a.** 78 **b.** 413 **11.** $(11111)_2 = 31$ **13.** $(1000000)_2 = 64$ **15.** $(101010)_2$
17. $(1101001)_2$ **19.** Conjecture is false. For example, $2 \cdot 3 \equiv 0 \pmod 6$, but neither 2 nor 3 is 0 mod 6.
21. Assume $a^2 \equiv b^2 \pmod p$. Then, $p|(a^2 - b^2)$ and $a^2 - b^2 = (a + b)(a - b)$. By Theorem 6.1.8, $p|(a + b)$ or $p|(a - b)$. Therefore, $a \equiv -b \pmod p$ or $a \equiv b \pmod p$.
23. Assume $m|n$ and $a - b = nk$. Then, $n = jm$ and $a - b = jkm$. Therefore, $a \equiv b \pmod m$.

Section 6.4

1. $s = 3$
3. Given $n = 5 \cdot 2 = 10$ and $r = 7$, the receiver calculates $s = 3$. The sender transmits $C = 8$, since $8 \equiv 2^7 \pmod{10}$. Since $8^3 \equiv 2 \pmod{10}$, the receiver recovers the message $M = 2$.
5. $9699690 = 2 \cdot 3 \cdot 5 \cdot 7 \cdot 11 \cdot 13 \cdot 17 \cdot 19$
7. Assume $a \equiv b \pmod p$ and $a \equiv b \pmod q$. Then, $p|(a - b)$ and $q|(a - b)$. Hence, $a - b = pk$ and $q|pk$. Since p and q are unequal primes, $q|k$ by Theorem 6.1.8. Therefore, $a - b = pqj$ and $a \equiv b \pmod{pq}$.

Chapter 6 Review

1. **a.** $\text{lcm}(2 \cdot 23, 2^3 \cdot 3 \cdot 5) = 2^3 \cdot 3 \cdot 5 \cdot 23$ **b.** $\gcd(46, 120) = 2$ **3.** $(1000010)_2$ **5.** $(111111)_2$
7. Induction step: $8^{k+1} - 3^{k+1} = 8 \cdot 8^k - 3 \cdot 3^k = (5 + 3)8^k - 3 \cdot 3^k = 5 \cdot 8^k + 3 \cdot 8^k - 3 \cdot 3^k = 5 \cdot 8^k + 3(8^k - 3^k)$
9. Assume $\gcd(a, b) = 1$, $a|c$, and $b|c$. Then, $1 = ma + nb$, $c = ka$, and $b|ka$. Hence, $b|k$ by Exercise 18 of Section 6.2. Therefore, $c = ka = (jb)a = j(ab)$ and $ab|c$.
11. $n^5 - n = (n^3 - n)(n^2 + 1)$. By Exercise 10, we have $6|(n^3 - n)$. By Exercise 9, we need only show $5|(n^5 - n)$. Finish the proof by considering the cases n is congruent to i mod 5 for $i = 0, 1, 2, 3, 4$.
13. Use Corollary 6.1.9 or a proof by mathematical induction. **15.** Use Exercise 14 and Theorem 6.2.8.

CHAPTER 7

Section 7.1

1. **a.** Yes **b.** No; let $X = \{1, 2\}$ **c.** No; let $x = 0$
3. **a.** No. One counterexample: $\text{lcm}(2, 3) = 6$ and $6 \notin S$. **b.** Yes
    c. No. One counterexample: $\{1, 3, 5, \ldots\} \cap \{2, 4, 6, \ldots\} = \emptyset$.
5. **a.** $(x + y')x' = x'y'$ **b.** $(x \cdot 1 + 0)x' = 0$ **c.** If $x + y = 0$, then $x = 0$ and $y = 0$. **7. a.** No **b.** No
9. **a.** Yes **b.** Yes **c.** No. One reason: $2 \cdot 2' = 2 \cdot 2 = \gcd(2, 2) = 2$
11. **a.** $x + xy = x \cdot 1 + xy = x(1 + y) = x \cdot 1 = x$ **b.** $(x + y)(x + y') = x + yy' = x + 0 = x$
13. **a.** $x + x = (x + x)1 = (x + x)(x + x') = x + xx' = x + 0 = x$ **b.** This proof is the dual of the proof in part a.

15. a. Suppose y is an element such that $x + y = x$ for all x. In particular, $0 + y = 0$. But $0 + y = y + 0$, and $y + 0 = y$. Therefore, $y = 0$.
   b. Suppose there is an element u such that $xu = x$ for all x. Then, in particular, $1 \cdot u = 1$. Hence, $u = u \cdot 1 = 1 \cdot u = 1$.
17. a. For any z, there exists a unique z' satisfying $z + z' = 1$ and $zz' = 0$. Since $x' + x = 1$ and $x'x = 0$, $x = (x')'$.
   b. Since $0 = 1'$, $0' = (1')' = 1$.
19. a. $(x + y)(x' + y) = (y + x)(y + x') = y + xx' = y + 0 = y$ **b.** $(xy)(x' + y) = xyx' + xyy = x'xy + xy = 0 \cdot y + xy = xy$
21. a. $(x + y + z)' = [(x + y) + z]' = (x + y)'z' = x'y'z'$ **b.** $(xyz)' = x' + y' + z'$ **23.** $xy + xy' + x'y + x'y' = x + x' = 1$
25. $xyz + xy'z + xy'z' + xyz' = xz + xz' = x$
27. $(x + xy' + z') + xz = x'(x' + y)z + xz = [x'(x' + y) + x]z = [x'x' + x'y + x]z = [x' + x + x'y]z = 1 \cdot z = z$
29. $(x'yw + x'z'w + x'y'w' + yz' + x'z)' = (x'yw + x'z'w + x'z + x'y'w' + yz' + x'z)' =$
$(x'yw + x'z'w + x'z(w + w') + x'y'w' + yz' + x'z)' = (x'yw + x'z'w + x'zw + x'y'w' + yz' + x'z + x'zw')' =$
$(x'yw + x'w + x'y'w' + yz' + x'z)' = (x'yw + x'w + x'w(y + y') + x'y'w' + yz' + x'z)' =$
$(x'w + x'wy + x'y'w + x'y'w' + yz' + x'z)' = [(x'w + x'wy) + x'y'(w + w') + yz' + x'z]' =$
$(x'w + x'y' + yz' + x'z)' = (x' + yz')' = x(y' + z)$

Section 7.2

1. 1   **3.** $xy + xy' + x'y$ **5.**

x	y	$f(x, y)$
0	0	0
0	1	1
1	0	0
1	1	1

   7. b. $(x + y)'z = x'y'z$ by De Morgan's law

9. a. $(x + y')x = xy + xy'$ **b.** $(x + y')(x' + y) = xy + x'y'$
11. a. $xyz + xyz' + x'yz + x'y'z$
   b. $(xy')'(x + z) = (x' + y)(x + z) = x'x + x'z + xy + yz = xy + x'z + yz = xy(z + z') + x'z(y + y') + yz(x + x') =$
   $xyz + xyz' + x'yz + x'y'z + xyz + x'yz = xyz + xyz' + x'yz + x'y'z$
13. a. $xyz + xyz' + xy'z + xy'z' + x'y'z$
   b. $x + z(x + y') = x(y + y')(z + z') + zx + zy' = x(y + y')(z + z') + xz(y + y') + (x + x')y'z =$
   $xyz + xyz' + xy'z + xy'z' + x'y'z$
15. $x'y$ **17.** $xy'z + x'yz + x'y'z + x'yz' + x'y'z'$ **19. a.** $E' = xyz + xyz'$ **b.** $E = (x' + y' + z')(x' + y' + z)$
21. a. $x' + y$ **b.** $(x' + y' + z)(x + y + z')$
23. Proof of transitivity: Assume $E \approx F$ and $F \approx G$. Then E and F define the same Boolean function, and F and G define the same Boolean function. Hence, E and G define the same Boolean function. Therefore, $E \approx G$.
25. a. $x \veebar x = xx' + x'x = 0 + 0 = 0$
   b. $x \veebar x' = x(x')' + x'x' = xx + x'x' = x + x' = 1$
   c. $(x \veebar y)' = (xy' + x'y)' = (x' + y)(x + y') = x'x + x'y' + xy + yy' = xy + x'y'$. It is easy to verify that $x' \veebar y = x \veebar y' = xy + x'y'$.

Section 7.3

1. a. y' **b.** $yz'u + yzu' + xyu'$ or $yz'u + yzu' + xyz'$ **c.** $xyz' + x'y'u' + yz'u'$ or $xyz' + x'y'u' + x'z'u'$ **d.** $xzu + xy'z$
3. a. $(xy' + x'y')' = (x' + y)(x + y)$ **b.** $(xyz + x'yz')' = (x' + y' + z')(x + y' + z)$
   c. $(xyz + xyz' + x'yz)' = (x' + y' + z')(x' + y' + z)(x + y' + z')$
   d. $(xyzu + xy'zu + xy'zu')' = (x' + y' + z' + u')(x' + y + z' + u')(x' + y + z' + u)$
5. a. $x' + y$ **b.** $x' + y + z$ **7. a.** $xy + x'z$ **b.** 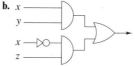   **9. a.** y' **b.** $x'y' + zu$

11. **a.** $x' + x(xy')'$ **b.** $x' + x(xy')' = x' + x(x' + y) = x' + xx' + xy = x' + xy = x' + y$ **c.** [logic gate diagram: x, y into NOR-like gate]

13. **a.** $(xy)' + xyz'$ **b.** $x'y'z' + x'yz' + xy'z' + xyz' + x'y'z + x'yz + xy'z$ **c.** $x' + y' + z'$ **d.** [logic gate diagram with x, y, z]

15. **a.**

x	y	z	s	c
0	0	0	0	0
0	0	1	1	0
0	1	0	1	0
0	1	1	0	1
1	0	0	1	0
1	0	1	0	1
1	1	0	0	1
1	1	1	1	1

b. $s = xyz + x'yz' + x'y'z + xy'z'$, $c = x'yz + xy'z + xyz' + xyz = xy + yz + xz$

c.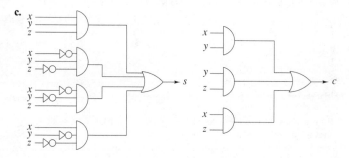

17. **a.** Use $x \downarrow x = x'$. **b.** Use $(x \downarrow x) \downarrow (y \downarrow y) = xy$. **c.** Use $x + y = (x \downarrow y)' = (x \downarrow y) \downarrow (x \downarrow y)$.
19. **a.** $x \downarrow y = (x + y)' = (y + x)' = y \downarrow x$ **b.** $x = 1, y = 1, z = 0$

Section 7.4

1. Assume $z \leq y$ and $z \leq y'$. Then, $z = zy = zy'$. Therefore, $z = zz = (zy)(zy') = z(yy') = z \cdot 0 = 0$.
3. Assume $x \leq y$. Then, $xy = x$. Hence, $x' + y' = x'$. Therefore, $y' \leq x'$.
5. $x \leq y$ if and only if $xy' = 0$ if and only if $(xy')' = 0'$ if and only if $x' + y = 1$.
7. **a.** No, since Card$(D_{32}) = 6$ and $6 \neq 2^m$. **b.** Yes, since $(D_{31}, +, \cdot, ', 1, 31)$ satisfies the Boolean algebra axioms.
9. **a.** $P(W) = \{\emptyset, \{3\}, \{5\}, \{7\}, \{3, 5\}, \{5, 7\}, \{3, 7\}, \{3, 5, 7\}\}$
   b. $f(1) = \emptyset, f(3) = \{3\}, f(5) = \{5\}, f(7) = \{7\}, f(15) = \{3, 5\}, f(35) = \{5, 7\}, f(21) = \{3, 7\}, f(105) = \{3, 5, 7\}$
11. Proof: ($\rightarrow$) Assume $(D_n, +, \cdot, ', 1, n)$ is a Boolean algebra. Suppose there is an integer $m > 1$ such that m^2 divides n. Then $m|m'$ and $\gcd(m, m') = m \neq 1$. Contradiction. ($\leftarrow$) Assume there is no integer $m > 1$ such that m^2 divides n. It is easy to verify that $\gcd(x, x') = 1$ and $\text{lcm}(x, x') = n$ for all $x \in D_n$. Furthermore, all the other axioms of a Boolean algebra are satisfied.
13. By Theorem 7.4.7, $a_1 a_2 = a_1$ or $a_1 a_2 = a_2$ or $a_1 a_2 = 0$. Therefore, $a_1 \neq a_2$ if and only if $a_1 a_2 = 0$.
15. Let $Y = \{a_1, a_2, \ldots, a_k\}$ and $y = a_1 + a_2 + \cdots + a_k$. Assume b is an atom. Then $b \leq y$ if and only if $b = a_i$ for some $i = 1, 2, \ldots, k$, by Theorem 7.4.8a and mathematical induction. Therefore, Atom$[y] = \{a_1, a_2, \ldots, a_k\}$.
17. Atom$[x \, \theta \, y]$ = Atom$[x'y + xy']$ = Atom$[x'y] \cup$ Atom$[xy']$ = (Atom$[x'] \cap$ Atom$[y]) \cup$ (Atom$[x] \cap$ Atom$[y']$) = (Atom$[y] -$ Atom$[x]) \cup$ (Atom$[x] -$ Atom$[y]$) = Atom$[x] \, \Delta \,$ Atom$[y]$

19. a. Prove $\cup$ is a binary operation.
Proof: If A and B are in S, then A is finite or A' is finite, and B is finite or B' is finite. If both A and B are finite, then $A \cup B$ is finite and is in S. If one of A and B is not finite—say, A is not finite—then A' is finite. Therefore, $(A \cup B)' = A' \cap B' \subseteq A'$, and $(A \cup B)'$ is finite. Hence, $A \cup B \in S$.
Prove $\cap$ is a binary operation.
Proof: If A and B are in S, then $A \cap B$ is finite when A or B is finite. If both A and B are not finite, then both A' and B' are finite. Thus, $(A \cap B)' = A' \cup B'$ is finite and $A \cap B \in S$.
Finally, if $A \in S$, then $A' \in S$ by definition. Hence, $'$ is a unary operation.
b. The atoms are the singleton sets. Every nonempty set has an atom as a subset.

21. a. If $0 = 1$, then, for any $x \in B$, we have $x = x \cdot 1 = x \cdot 0 = 0$. Therefore, $B = \{0\}$ and $\text{Card}(B) = 1$.
b. If, for some $x \in B$, $x = x'$, then we have $x = xx = xx' = 0$. From $x = 0$ and $x = x'$, we have $0 = 0' = 1$. Therefore, $\text{Card}(B) = 1$ by part a.
c. Suppose B has an odd number of elements. Arrange these distinct elements as follows:
$0, 1, x_1, x'_1, x_2, x'_2, \ldots, x_{n-1}, x'_{n-1}, x_n$. Then x'_n exists in B and must be an element of this list. Since $\text{Card}(B) > 1$, $x'_n \neq x_n$ by part b. If $x'_n = x_i$ for $i < n$, then $x_n = (x'_n)' = x'_i$, a contradiction to the fact that the list of elements is distinct. If $x'_n = x'_i$ for $i < n$, then $x_n = (x'_n)' = (x'_i)' = x_i$. Contradiction again. Hence, $\text{Card}(B)$ is even.

Section 7.5

1. Draw the logic network for $x + y + z' + u$.   **3.** Draw the logic network for $zu' + y'u'$.

Chapter 7 Review

1. The binary and unary operations defined here are identical to the Boolean operations for the set $\{0, 1\}$.
3. No, by Corollary 7.4.16
5. $(x' + y + z' + u')(x' + y + z + u')(x + y + z' + u)(x + y + z' + u)(x + y + z + u')(x + y + z + u)$
7. One example: $B = P(\{a, b, c\})$, $+$ is $\cup$, $\cdot$ is $\cap$, $'$ is set complement, 0 is $\emptyset$, and 1 is $\{a, b, c\}$.
   a. Let $x = \{a, b\}$, $y = \{a\}$, and $z = \{b\}$.   **b.** Let $u = \{a, b\}$, $v = \{a\}$, and $w = \{a, c\}$.
9. a. $(xy + zu)(xy + (zu)')z$ **b.** $(xy + zu)(xy + (zu)')z = (xy + zu(zu)')z = (xy + 0)z = xyz$
  c. Draw the logic network for xyz.
11. a. $xyzu' + x'yzu + x'yzu + x'y'zu + x'y'zu' + x'y'z'u + x'y'z'u' + xy'zu'$
  b. $xyz'u' + xyz'u + xyzu + xy'zu + x'yzu + xy'z'u + xy'z'u' + x'yz'u'$
  c. $(x' + y' + z + u) \cdot (x' + y' + z + u') \cdot (x' + y' + z' + u') \cdot (x' + y + z' + u') \cdot (x + y' + z' + u) \cdot$
    $(x' + y + z + u') \cdot (x' + y + z + u) \cdot (x + y' + z + u)$
13. 2^5 **15.** 2^{2^4}
17. a. It is routine to prove that R is reflexive, symmetric, and transitive.
  b. The equivalence classes are $\{1\}$, $\{2, 3, 5\}$, $\{6, 10, 15\}$, and $\{30\}$.

CHAPTER 8

Section 8.1

1. $\begin{bmatrix} 0 & 1 & 1 & 0 \\ 1 & 0 & 0 & 1 \\ 1 & 0 & 0 & 1 \\ 0 & 1 & 1 & 0 \end{bmatrix}$ **3. a.** 17   **b.** 5

5. a. 29 (There are seven 3-edge connected subgraphs, five 4-edge connected subgraphs of G, etc.)   **b.** 5

7. $\begin{bmatrix} 0 & 1 & 0 & 1 & 0 \\ 1 & 0 & 1 & 0 & 0 \\ 0 & 1 & 0 & 1 & 0 \\ 1 & 0 & 1 & 0 & 1 \\ 0 & 0 & 0 & 1 & 0 \end{bmatrix}$

9. a.

Program
Outreach
Finance
Research Acquisition

b. Friday: Program, Research
Saturday: Outreach, Finance
Sunday: Acquisition

Friday: Outreach
Saturday: Acquisition, Finance
Sunday: Program, Research

11. Three meeting times 13. Two 15. Ten edges

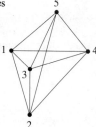

17. Let H be a connected subgraph of G. Assume that v is a vertex in H. Let K be a component of G containing v. Since H is connected and v is a vertex in H, every vertex in H is connected to v. Hence, $H \subseteq K$. If H is also maximal, then we have $H = K$. Hence, a maximal connected subgraph of G is a component of G.

19.

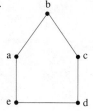

21. It is not possible by Corollary 8.1.7.

23. a. A B   b. A B   25. mn

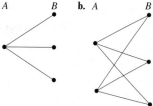

Section 8.2

1. a b

3. a. There is no Eulerian tour by Theorem 8.2.3.
b. (f, e_5, g, e_8, k, e_9, t, e_7, g, e_4, h, e_1, k, e_2, h, e_3, f, e_6, t) is an Eulerian path.
c. (h, e_3, f, e_5, g, e_7, t, e_9, k, e_1, h) is a Hamiltonian cycle.

5. $(a, e_4, b, e_9, c, e_{16}, g, e_{17}, i, e_{15}, f, e_{14}, c, e_{11}, k, e_{12}, f, e_{10}, b, e_8, k, e_{13}, i, e_7, h, e_6, k, e_5, a, e_3, h, e_2, d, e_1, a)$

7.  **9.** The complete graph K_n has an Eulerian tour when n is odd

11. Since each edge of G must be included exactly twice, we can insert an auxiliary edge to each edge of G to obtain a new multigraph G'. Thus, each vertex of G' has an even degree. By Theorem 8.2.3 and its proof, we can construct an Eulerian tour that includes each edge of G' exactly once. This Eulerian tour is the desired closed path that includes each edge of G exactly twice.

13. Let G be a connected graph containing a circuit from vertex v to v. Suppose an edge e in the circuit is deleted from G. The vertices incident to e are still connected to v by a path in the circuit. Hence, $G - e$ is connected.

15.

17. If G has an Eulerian path from v to w, then clearly G is connected. Insert the edge $e = \{v, w\}$ to the graph G. The resulting graph G' has an Eulerian tour formed by the Eulerian path from v to w in G and the edge e. By Theorem 8.2.3, every vertex of G' has even degree. Therefore, v and w are the only vertices of $G = G' - e$ with odd degree. Conversely, assume that G is connected and v and w are the only vertices of odd degree. Insert an edge $e = \{v, w\}$ to the graph G. The resulting graph G' has an Eulerian tour by Theorem 8.2.3. Removing the edge e yields an Eulerian path of G from v to w.

19. Let $e_1 = \{v_1, v_4\}$, $e_2 = \{v_4, v_3\}$, $e_3 = \{v_3, v_6\}$, $e_4 = \{v_5, v_6\}$, $e_5 = \{v_5, v_2\}$, and $e_6 = \{v_2, v_1\}$. A closest-neighbor cycle starting at v_1 is $(v_1, e_1, v_4, e_2, v_3, e_3, v_6, e_4, v_5, e_5, v_2, e_6, v_1)$. Yes.

Section 8.3

1. a. There is a unique path with no repeated edges between each pair of vertices.   **b.** $\mathrm{Card}(E) = 6 = 7 - 1 = \mathrm{Card}(V) - 1$

3. a. a, b, h, i   **b.** d, e, g, f   **c.** d, g   **d.** j, k   **5.** 76   **7. a.** 2^q **b.** $\dfrac{2 \cdot 2^q - 1}{2 - 1} = 2^{q+1} - 1$

9. a. 4   **b.** 28   **11. a.**   **b.**   **13. a.** $a\,b\,c\cdot-$ **b.** $3b\cdot 5c\cdot d\,-\,+$

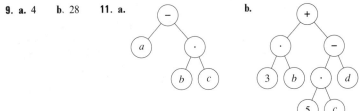

15. a. $(6 + a) \div [(b - 2 \cdot c) - (6 \cdot d + e)]$ **b.** $6\,a + b\,2\,c \cdot\, - 6\,d \cdot e\, + \, - \, \div$

17. a.   **b.** $b\,a + b\,3\,c + \cdot \div c\,d \cdot -$ **19. a.** $\div + 6\,a - - b \cdot 2\,c + \cdot 6\,d\,e$

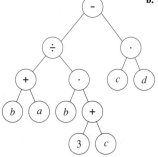

21. a. FAC **b.** ACID **c.** FASCAL

23. By Theorem 8.3.15, there are $\text{Card}(V) - 1$ edges in a tree with vertex set V. Each edge gives rise to two degrees. Hence, the sum of degrees is $2[\text{Card}(V) - 1] = 2 \cdot \text{Card}(V) - 2$.

25. *Hints:* ($\rightarrow$) Use Theorem 8.3.15 and the definition of a tree. ($\leftarrow$) Show T has no cycles and use Theorem 8.3.15.

Section 8.4

1. Let $e_1 = \{d, c\}$, $e_2 = \{c, k\}$, and $e_3 = \{d, k\}$. Spanning trees of G are $G - e_1$, $G - e_2$, and $G - e_3$.

3. If $G = (V, E)$ is a connected graph of n vertices, then G has a spanning tree T with n vertices by Algorithm 8.4.3. By Theorem 8.3.14, T has $n - 1$ edges. Hence, $\text{Card}(E) \geq n - 1$.

5. If G has n vertices and $n - 1$ edges, then G cannot have any cycles and is therefore a tree.

7. a. 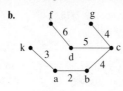 **b.**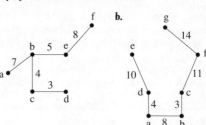

9. a. **b.**

11.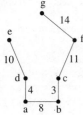

13. In Algorithm 8.4.3, any edge in G may be selected initially.

15. Let $e = \{v, w\}$ have a weight that is less than the weight of every other edge in G. Since a spanning tree must include vertex v, edge e must be chosen among all possible edges in G incident with v.

17. Since all edges in G have distinct weights, each choice of an edge to be added to the graph while building the spanning tree is unique.

Section 8.5

1.

 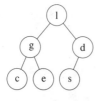 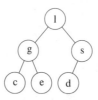 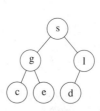

SOLUTIONS FOR THE EXERCISES   A-41

3. Remove s. Temp = d   Promote 1. 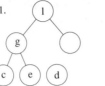   Reinsert d.

Remove 1. Temp = e 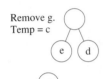   Promote g, reinsert e.

Remove g. Temp = c   Promote e, reinsert c.   Remove e. Temp = d

Reinsert d.   Remove d. Temp = c   Reinsert c.   Remove c.

Chapter 8 Review

1. 11 edges and 7 vertices

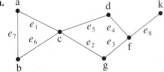

3. a.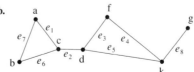

(a, e_1, c, e_5, d, e_4, f, e_3, g, e_2, c, e_6, b, e_7, a) is a circuit but is not a cycle.

b.

(a, e_1, c, e_2, d, e_3, f, e_4, k, e_5, d, e_2, c, e_6, b, e_7, a) is a closed path that is neither a circuit nor a cycle.

5.

7. Chromatic number three

9.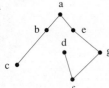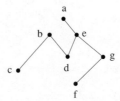

11. n **13. a.** Yes   **b.** 5   **c.** d, g   **d.** j, k, l, p, q   **15. a.** 0001101110000   **b.** 10111001   **c.** 01100000
17. a. $3 f \cdot 7 d g - - 5 h \cdot - +$ **b.** $7 a \cdot b 4 + 8 a + \cdot -$ **19.** $C(m, k)$, where $m = C(n, 2)$
21. If a vertex v of a graph G is connected to every other vertex of G, then the component containing v includes all vertices of G. Hence, G has only one component and G is connected.
23. a. [figure with vertices a, g, b, k, f, c, d]   **b.** G has no Eulerian tour since not every vertex of G has even degree.

CHAPTER 9
Section 9.1

1. a. $A_0 = 2000$, $A_n = (1 + 0.08/12)^{12}(A_{n-1})$ **b.** $A_n = 2000(1 + 0.08/12)^{12n}$
3. a. $A_1 = 3000$, $A_n = 3000 + 1.08 A_{n-1}$ **b.** $A_n = (3000/0.08)(1.08^n - 1)$ **5.** $F_{15} = 610$
7. $C(8, 0) = 1$, $C(8, 1) = C(7, 0) + C(7, 1) = 8$, $C(8, 2) = 7 + 21 = 28$, $C(8, 3) = 21 + 35 = 56$, $C(8, 4) = 70$, $C(8, 5) = 56$, $C(8, 6) = 28$, $C(8, 7) = 8$, $C(8, 8) = 1$
9. $a_1 = 1$, $a_2 = 2$, and $a_n = a_{n-1} + a_{n-2}$ **11.** $t_1 = 3$, $t_n = 2t_{n-1} + 2t_{n-2}$; set $t_0 = 1$
13. $t_1 = 10$, $t_2 = 100$, $t_n = 10 t_{n-1} + 39 t_{n-2}$; set $t_0 = 0$
15. Conjecture: $F_{n+2}F_n - F_{n+1}^2 = (-1)^{n+1}$. Proof: Basis step ($n = 0$): $F_2 F_0 - F_1^2 = 1 \cdot 0 - 1^2 = -1 = (-1)^1$. Induction hypothesis: Assume $F_{k+2}F_k - F_{k+1}^2 = (-1)^{k+1}$. Then,
$F_{k+3}F_{k+1} - F_{k+2}^2 = (F_{k+2} + F_{k+1})F_{k+1} - (F_{k+1} + F_k)F_{k+2} = F_{k+1}^2 - F_{k+2}F_k = -(-1)^{k+1} = (-1)^{k+2}$.
17. Conjecture: $F_0^2 + F_1^2 + \cdots + F_n^2 = F_n F_{n+1}$. Proof: Basis step ($n = 0$): $F_0^2 = 1 \cdot 0 = F_1 F_0$. Assume
$F_0^2 + F_1^2 + \cdots + F_k^2 = F_k F_{k+1}$. Then, $F_0^2 + F_1^2 + \cdots + F_k^2 + F_{k+1}^2 = F_k F_{k+1} + F_{k+1}^2 = F_{k+1}(F_k + F_{k+1}) = F_{k+1}F_{k+2}$.
19. a. $L_4 = 4$, $L_5 = 7$, $L_6 = 11$, $L_7 = 18$
b. Prove: $L_n = F_n + F_{n-2}$. Proof is by mathematical induction. Basis step ($n = 2$): $L_2 = 1 = 1 + 0 = F_2 + F_0$. Assume $L_j = F_j + F_{j-2}$ for $j \leq k$. Then, $L_{k+1} = L_k + L_{k-1} = (F_k + F_{k-2}) + (F_{k-1} + F_{k-3}) = F_{k+1} + F_{k-1}$.
21. Step 2. $F_1 = 1 = C(0, 0) = t_1$ and $F_2 = 1 = C(1, 0) = t_2$.
Step 3. $t_{n-1} + t_{n-2} = C(n-2, 0) + C(n-3, 1) + \cdots + C(n-k-2, k) + C(n-3, 0) + C(n-4, 1) + \cdots + C(n-j-3, j)$, where $k = j$ when n is odd and $k = j + 1$ when n is even. For n odd,
$t_{n-1} + t_{n-2} = C(n-2, 0) + [C(n-3, 1) + C(n-3, 0)] + \cdots + [C(n-k-2, k) + C(n-k-2, k-1)] +$
$C(n-k-3, k) = C(n-1, 0) + [C(n-3, 1) + C(n-3, 0)] + \cdots + [C(n-k-2, k) + C(n-k-2, k-1)] +$
$C(n-k-2, k+1) = C(n-1, 0) + C(n-2, 1) + \cdots + C(n-k-1, k) + C(n-k-2, k+1) = t_n$. The case for n even is similar.
Step 4. By steps 2 and 3, t_n and F_n satisfy the same initial conditions and recurrence relation.

23. $A(2, 5) = 13$

25. Prove $A(2, m) = 2m + 3$ for $m = 0, 1, \ldots$. We will use the result of Exercise 24.
Proof: Basis step: $A(2, 0) = 3 = 2 \cdot 0 + 3$. Assume $A(2, k) = 2k + 3$. Then,
$A(2, k + 1) = A(1, A(2, k)) = A(2, k) + 2 = 2k + 3 + 2 = 2(k + 1) + 3$.

27. As in Example 9.1.7, we see that $T(n) = 1 + T(\lfloor n/2 \rfloor + 1)$. For $2^{r-1} \leq n < 2^r$, we have $2^{r-2} < \lfloor n/2 \rfloor + 1 \leq 2^{r-1}$. Now, by iteration we obtain $T(n) = r$.

Section 9.2

1. $a_n = 3 \cdot 2^n + (-1)^n$ **3.** $a_n = (2 + \sqrt{3})^n - (2 - \sqrt{3})^n$ **5.** $a_n = 8n(-5)^{n-1} + (-5)^n$ **7.** $a_n = 3^n + 3(-2)^n$
9. $a_n = (-1)^n + 2n(-1)^n + 3^n$ **11.** $a_n = (17/8)(-1)^n - (1/8)3^n + (n/2)3^n$ **13.** $a_n = (-1 + i)^n + (-1 - i)^n$
15. $t_n = (\frac{5}{8})13^n - (\frac{5}{8})(-3)^n$ **17. a.** $G_n = a + (b - a)n$ **b.** When $a = b$, G_n is constant.
19. Let $a_n = nr^n$. Then, $a_n - 2ra_{n-1} + r^2a_{n-2} = nr^n - 2r(n-1)r^{n-1} + r^2(n-2)r^{n-2} = 0$.
21. Let $r = (1 + \sqrt{5})/2$ and $s = (1 - \sqrt{5})/2$. Then, $F_n = (\sqrt{5}/5)(r^n - s^n)$. Hence, $F_{n+1}/F_n = (r^{n+1} - s^{n+1})/(r^n - s^n)$. Divide each term by r^n to obtain $F_{n+1}/F_n = [r - (s/r)^n s]/[1 - (s/r)^n]$. The result follows from the fact that $(s/r)^n$ approaches zero as n goes to infinity because $|s/r| < 1$.
23. $q(x) = x[x(3 \cdot x \cdot x + 1) - 2] + 4$ **25.** $q(\pi) = \pi[\pi(3 \cdot \pi \cdot \pi + 1) - 2] + 4$

Section 9.3

1. $\dfrac{2 - 3x}{(1 + 5x)(1 - x)} = \dfrac{\frac{13}{6}}{1 + 5x} - \dfrac{\frac{1}{6}}{1 - x}$ **3.** $\dfrac{4x + 1}{(x - 2)(x + 1)} = \dfrac{1}{1 + x} + \dfrac{3}{x - 2}$ **5.** $a_n = 3 \cdot 2^n + (-1)^n$
7. $a_n = (2 + \sqrt{3})^n - (2 - \sqrt{3})^n$ **9.** $a_n = 8n(-5)^{n-1} + (-5)^n$ **11.** $a_n = 3 \cdot 4^n + 5n4^n$ **13.** $a_n = 3^n + 3(-2)^n$
15. Let $A(x) = a_0 + a_1x + a_2x^2 + \cdots$. First, show that $A(x) = \dfrac{a_0 + kx}{(1 - rx)^2} = \dfrac{c}{1 - rx} + \dfrac{b}{(1 - rx)^2}$ for some constants c and b.
But, $A(x)$ is the generating function for the sequence with nth term $cr^n + b(n + 1)r^n = ar^n + bnr^n$, where $a = c + b$.
17. a. The generating function is $\dfrac{1}{1 - 3x} \cdot \dfrac{1}{1 - 5x}$.   **b.** $a_n = (\frac{1}{2})(5^{n+1} - 3^{n+1})$
19. $(1, -1, 1, -1, \ldots, (-1)^n, \ldots)$ **21.** $(0, C(2, 2), C(3, 2), \ldots, C(n + 1, 2), \ldots)$ **23.** $\dfrac{1}{(1 + x)^2}$
25. Differentiate both sides of $(1 - x)^{-2} = 1 + 2x + 3x^2 + 4x^3 + \cdots$ to obtain $2(1 - x)^{-3} = 2 + 3 \cdot 2x + 4 \cdot 3x^2 + \cdots$, and then divide both sides of the resulting equation by 2.
27. $2n \cdot 3^{n-1}$

Section 9.4

1. We have $b_r = b_{r-1} + b$ and $b_0 = b$. Let $g(x)$ be the generating function for b_r and obtain $g(x) - x \cdot g(x) = b/(1 - x)$. Hence, $g(x) = b/(1 - x)^2$ and, therefore, $b_r = b(r + 1)$.
3. We have $b_r = db_{r-1} + b$. The induction hypothesis is $b_{r-1} = cd^{r-1} + k$, where $k = b/(1 - d)$. Use the recurrence relation to prove $b_r = cd^r + k$.
5. We have $b_r = db_{r-1} + bd^r$. The induction hypothesis is $b_{r-1} = b(r - 1)d^{r-1}$. Use the recurrence relation to prove $b_r = brd^r$.
7. For step c: The recurrence relation in step b yields $b_{r-j} - 2b_{r-j-1} = 2^{r-j} - 1$. Multiply both sides by 2^j.

Chapter 9 Review

1. $\dfrac{7 - 3x}{1 - x^2} = \dfrac{2}{1 - x} + \dfrac{5}{1 + x}$ **3.** $a_n = 3 \cdot 5^n$ **5.** $a_n = (-3)^n - n(-3)^{n-1}$ **7.** $a_n = (5\sqrt{6}/18)[(5 + 3\sqrt{6})^n - (5 - 3\sqrt{6})^n]$
9. $a_n = 5^n + 3(-1)^n + 2n(-1)^n$ **11.** $a_n = 5 - n + 2^n + 3n \cdot 2^n$
13. a. $t_1 = 2, t_2 = 4, t_3 = 7, t_4 = 11, t_5 = 16$ **b.** $t_n = t_{n-1} + n$ **c.** $t_n = 1 + C(n + 1, 2)$
15. a. $h_1 = 1, h_2 = 7$ **b.** $h_n = h_{n-1} + 6h_{n-2}$ **c.** $h_n = (1/5)[3^{n+1} + 2(-2)^n]$
17. Similar to the solution to Exercise 17 of Exercise Set 9.1.

CHAPTER 10

Section 10.1

1. $\lambda \vee 11(0 \vee 1)^*$ 3. $01^* \vee (110)^*$ 5. **a.** $L(L \vee D)^*$ **b.** 73
7. **a.** *abc* (No), *abacb* (Yes)   **b.** *ab* (Yes), *abc* (No), *acb* (Yes), *acbb* (Yes)   9. **a.** $1(000)^*1$ **b.** $(\lambda \vee 1111^*)00^*$
11. $0^*10^*(10^*1)^*0^*$ 13. **a.**   **b.**   15.

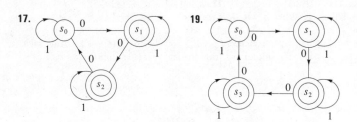

17.   19.

21. We prove $A(B \vee C) \subseteq AB \vee AC$. The proof of the other set inclusion is similar. Assume $w \in A(B \vee C)$. Then, $w = ax$, where $a \in A$ and $x \in B \cup C$. Hence, $w \in AB$ or $w \in AC$. Therefore, $w \in AB \vee AC$.
23. **a.** This follows from the theorem of set theory $A \cup \emptyset = \emptyset \cup A = A$.
    b. This follows from the theorem $A \cup B = B \cup A$, where $B = \lambda$.   **c.** One counterexample is $A = \{a\}$.
25. **a.** This follows from the theorem of set theory $A \cup A = A$.   **b.** One counterexample is $A = \{a\}$.

Section 10.2

1. **a.** @ $\Rightarrow$ QR $\Rightarrow$ ANR $\Rightarrow$ TheNR $\Rightarrow$ TheChildR $\Rightarrow$ TheChildPlays   **b.**

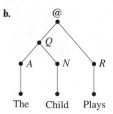

3. **a.** @ $\Rightarrow$ @ * @ $\Rightarrow$ (@) * @ $\Rightarrow$ (@ + @) * @ $\Rightarrow$ (I + @) * @ $\Rightarrow$ (I + I) * @ $\Rightarrow$ (I + I) * I $\Rightarrow$ (x + I) * I $\Rightarrow$ (x + x) * I $\Rightarrow$ (x + x) * x
   b. @ $\Rightarrow$ @ + @ $\Rightarrow$ @ * @ + @ $\Rightarrow$ I * @ + @ $\Rightarrow$ I * I + @ $\Rightarrow$ I * I + I $\Rightarrow$ x * I + I $\Rightarrow$ x * x + I $\Rightarrow$ x * x + x
   c.

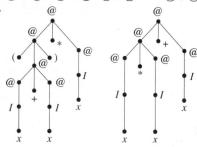

5. a. $@ \Rightarrow a@ \Rightarrow a^2@ \Rightarrow a^2bA \Rightarrow a^2b^2A \Rightarrow a^2b^3A \Rightarrow a^2b^3c$ **b.**

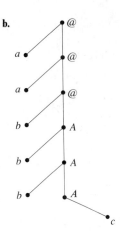

c. $@ \Rightarrow bA \Rightarrow b^2A \Rightarrow b^3A \Rightarrow b^4A \Rightarrow b^5A \Rightarrow b^5c$ **d.** $n \geq 0, m \geq 1, j = 1$

7. $G = (V, T, @, P)$; with $V = \{@, 0, 1\}$, $T = \{0, 1\}$, and $P: @ \to 01, @ \to 0@1$
9. $G = (V, T, @, P)$; with $V = \{@, A, 0, 1\}$, $T = \{0, 1\}$, and $P: @ \to \lambda, @ \to 0@, @ \to A, A \to 1, A \to A1$
11. $G = (V, T, @, P)$; with $V = \{@, B, a, b, c\}$, $T = \{a, b, c\}$, and $P: @ \to Bb, B \to aBb, B \to c$
13. $G = (V, T, @, P)$; with $V = \{@, A, B, a, b\}$, $T = \{a, b\}$, and $P: @ \to aA, @ \to a, A \to a, A \to B, A \to aB, B \to bB, B \to b$
15. $G = (V, T, @, P)$; with $V = \{@, A, B, a, x, z\}$, $T = \{a, x, z\}$, and $P: @ \to x^3A, A \to \lambda, A \to xAa, @ \to zB, B \to za, B \to zBa$

Section 10.3

1. a. λ, a^2b^2 **b.** $(m = 0 \text{ and } n = 0) \text{ or } (m \geq 1 \text{ and } n = 2)$ **c.** $\lambda \vee a^2bb^*$
3. $G = (V, T, @, P)$; with $V = \{@, A, B, a, b\}$, $T = \{a, b\}$, and $P: @ \to a, @ \to aA, A \to aA, A \to a, A \to b, A \to bB,$
$B \to bB, B \to b$
5. $G = (V, T, @, P)$; with $V = \{@, A, B, C, a, b, c\}$, $T = \{a, b, c\}$, and $P: @ \to aA, A \to aA, A \to bB, B \to b, B \to bC,$
$C \to c, C \to cC$
7. $(0 \vee 1)^*$ **9.** $(0 \vee 1)^*$
11. $G = (V, T, @, P)$; with $V = \{s_0, s_1, s_2, s_3, 0, 1\}$, $T = \{0, 1\}$, and $P: s_0 \to 0s_3, s_0 \to 1s_1, s_1 \to 0s_3, s_1 \to 1s_2, s_2 \to 0s_2,$
$s_2 \to 1s_2, s_3 \to 0s_3, s_3 \to 1s_3, s_0 \to \lambda, s_1 \to 1, s_2 \to 0, s_2 \to 1$
13. $G = (V, T, @, P)$; with $V = \{s_0, s_1, s_2, s_3, s_4, s_5, 0, 1\}$, $T = \{0, 1\}$, and $P: s_0 \to 0s_1, s_0 \to 1s_2, s_1 \to 0s_3, s_1 \to 1s_1,$
$s_2 \to 0s_3, s_2 \to 1s_4, s_3 \to 0s_3, s_3 \to 1s_3, s_4 \to 0s_5, s_4 \to 1s_3, s_5 \to 0s_3, s_5 \to 1s_2, s_0 \to \lambda, s_0 \to 0, s_1 \to 1, s_4 \to 0$
15. $G = (V, T, @, P)$; with $V = \{s_0, s_1, 0, 1\}$, $T = \{0, 1\}$, and $P: s_0 \to 1, s_1 \to 0, s_0 \to 0s_0, s_1 \to 0s_1, s_1 \to 1s_0, s_0 \to 1s_1$
17. Proof: Suppose $L = \{a^n b^n : n \in \mathbb{N}\}$ is regular. Then there is an FSR that recognizes L. The FSR must have a finite number (say m) states. Consider the string $a^{m+1}b^{m+1} \in L$. Now proceed as in the proof of Theorem 10.3.6 to obtain a contradiction.
19. We prove $AA^* \subseteq A^*A$. The proof of the other set inclusion is similar. Assume $w \in AA^*$. Then $w = av$, where $a \in A$ and v is a string of symbols from A. Either $v = \lambda$ or the rightmost symbol in v is an element of A. In either case, $w \in A^*A$.
21. We prove $A(BA)^* \subseteq (AB)^*A$. The proof of the other set inclusion is similar. Assume $w \in A(BA)^*$. Then, $w = au$, where $a \in A$ and u is a string of symbol pairs $b_i a_i$ from BA. If $u = \lambda$, then $w \in (AB)^*A$. Otherwise,
$w = ab_1 a_1 b_2 \ldots a_{n-1} b_n a_n$. Hence, $w \in (AB)^*A$.
23. We prove $(A \vee B)^* \subseteq A^*(BA^*)^*$. The proof of the other set inclusion is similar. Assume $w \in (A \vee B)^*$. If $w \in A^*$, we are done. If not, $w = vu$, where $v \in A^*$ and the leftmost symbol of u is an element of B. Then, $u = b_1 v_1 b_2 v_2 \ldots b_n v_n$, where each $b_i \in B$ and each $v_i \in A^*$. Therefore, $w \in A^*(BA^*)^*$.

Section 10.4

1. a. (i) $@ \Rightarrow A \Rightarrow AA \Rightarrow (A)A \Rightarrow (())A \Rightarrow (())()$ **b.** $\lambda, (), (())$ **c.** $\lambda, ()$ **d.** $\lambda, (), A, AA, (A)$
(ii) $@ \Rightarrow A \Rightarrow AA \Rightarrow A() \Rightarrow (A)() \Rightarrow (())()$
The parse tree for (i) and (ii):

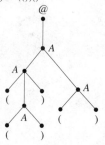

3. a. $@ \Rightarrow A \Rightarrow 1A1 \Rightarrow 11A11 \Rightarrow 111A111 \Rightarrow 1110111$ **b.** $m = n$ and $m \geq 0$ **c.** 0, 101, 11011, $1^3 01^3$, $1^4 01^4$

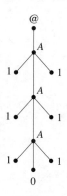

5. a. $@ \Rightarrow AA \Rightarrow 1AA \Rightarrow 11AA \Rightarrow 1^3 AA \Rightarrow 1^3 0A \Rightarrow 1^3 01A \Rightarrow 1^3 01^2 A \Rightarrow 1^3 01^2 A1 \Rightarrow 1^3 01^2 A1^2 \Rightarrow 1^3 01^2 A1^3 \Rightarrow 1^3 01^2 A1^4 \Rightarrow 1^3 01^2 01^4$
b. $m \geq 0, n \geq 0, j \geq 0$
c. 00, 100, $1^2 00$, $1^3 00$, 010, $01^2 0$, $01^3 0$, 001, 001^2, 001^3, 1010, $101^2 0$, 10101, 1001, 1001^2, 0101, $1^2 010$, $1^2 001$, $01^2 01$, 0101^2

7. a. $G = (V, T, @, P)$; with $V = \{@, A, B, a, b\}$, $T = \{a, b\}$, and P: $@ \to \lambda$, $@ \to a@$, $@ \to B$, $B \to bb$, $B \to bBb$
b. $G' = (V, T, @, P)$; with $V = \{@, B, a, b\}$, $T = \{a, b\}$, and P: $@ \to \lambda$, $@ \to a@$, $@ \to B$, $B \to bA$, $A \to b$, $A \to bB$

9. $S_0 = \{@\}$, $S_1 = \{@, A\}$, $S_2 = \{@, A, 1A1, 0\}$, $S_3 = \{@, A, 1A1, 0, 11A11\} = S_4 = S_5 = \cdots$

11. $S_0 = \{@\}$, $S_1 = \{@, AA\}$, $S_2 = \{@, AA, A1A, 1AA, AA1, A0, 0A\}$,
$S_3 = \{@, AA, A1A, 1AA, AA1, A0, 0A, 1A1A, 11AA, 1AA1, A11A, A1A1, 1A0, 01A, A10, 0A1, 00\}$,
$S_4 = S_3 \cup \{101A, 1A10, 110A, 11A0, 10A1, 1A01, 011A, A110, 01A1, A101, 100, 010, 010, 001, 0A11\}$,
$S_5 = S_4 \cup \{1010, 1100, 1001, 0110, 0101, 0011\} = S_6 = \cdots$

13. $G = (V, T, @, P)$; with $V = \{@, A, 0, 1\}$, $T = \{0, 1\}$, and P: $@ \to 0$, $@ \to 0A$, $A \to 1A$, $A \to 1$

15. $G = (V, T, @, P)$; with $V = \{@, B, a, b, c\}$, $T = \{a, b, c\}$, and P: $@ \to a^2 c^2$, $@ \to a^2 B c^2$, $B \to bB$, $B \to b$

17. a. $@ \Rightarrow @ + E \Rightarrow E + E \Rightarrow E * F + E \Rightarrow F * F + E \Rightarrow F * F + F \Rightarrow a * F + F \Rightarrow a * b + F \Rightarrow a * b + b$
b. $@ \Rightarrow E * F \Rightarrow F * F \Rightarrow a * F \Rightarrow a * (@) \Rightarrow a * (@ + E) \Rightarrow a * (E + E) \Rightarrow a * (F + E) \Rightarrow a * (b + E) \Rightarrow a * (b + F) \Rightarrow a * (b + a)$

19. a. $G = (V, T, @, P)$; with $V = \{@, d, 0, 1\}$, $T = \{d, 0, 1\}$, and P: $@ \to d$, $@ \to 0@0$, $@ \to 1@1$
b. $@ \Rightarrow 1BJ \Rightarrow 1B1BJ \Rightarrow 1B1B0AJ \Rightarrow 11BB0AJ \Rightarrow 11B0BAJ \Rightarrow 110BBAJ \Rightarrow 110BBJ0 \Rightarrow 110BJ10 \Rightarrow 110J110 \Rightarrow 110d110$

21. Proof: We will prove $(ab)^n \in L(G)$ and $@ \overset{*}{\Rightarrow} (ab)^n @$ for all $n \in \mathbb{N}$. Basis step $(n = 0)$: $\lambda = (ab)^0 \in L(G)$ and $@ \overset{*}{\Rightarrow} (ab)^0 @$ since $@ = @$. Assume $(ab)^k \in L(G)$ and $@ \overset{*}{\Rightarrow} (ab)^k @$. Since $@ \to ab@$ is a production, $@ \overset{*}{\Rightarrow} (ab)^{k+1} @$. Since $@ \to \lambda$ is a production, $(ab)^{k+1} \in L(G)$.

23. $@ ::= \lambda | A$, $A ::= (A) | AA | ()$

Section 10.5

1. $S = \{s_0, s_1\}$, $F = \{s_1\}$, $V = \{0, 1, d\}$, $W = \{0, 1, \beta\}$. The set of transitions is: (1) $(s_0, 0, \beta, s_0, 0\beta)$ (2) $(s_0, 1, \beta, s_0, 1\beta)$ (3) $(s_0, 0, 0, s_0, 00)$ (4) $(s_0, 1, 1, s_0, 11)$ (5) $(s_0, 0, 1, s_0, 01)$ (6) $(s_0, 1, 0, s_0, 10)$ (7) $(s_0, d, \beta, s_1, \lambda)$ (8) $(s_0, d, 0, s_1, 0)$ (9) $(s_0, d, 1, s_1, 1)$ (10) $(s_1, 0, 0, s_1, \lambda)$ (11) $(s_1, 1, 1, s_1, \lambda)$ (12) $(s_1, \lambda, \beta, s_1, \lambda)$

3. $S = \{s_0, s_1\}$, $F = \{s_1\}$, $V = \{a, b\}$, $W = \{1, \beta\}$. The set of transitions is: (1) $(s_0, a, \beta, s_0, 1\beta)$ (2) $(s_0, a, 1, s_0, 11)$ (3) $(s_0, b, 1, s_1, \lambda)$ (4) $(s_1, b, 1, s_1, \lambda)$ (5) $(s_1, \lambda, \beta, s_1, \lambda)$

5. Assume $M = (S, V, g, F, s_0)$ is an FSR. The PDA to emulate M consists of: a set of states S, a set of final states F, an initial state s_0, a set of input symbols V, and $W = \{\beta\}$. Transitions are defined by $(s, x, \beta, g(s, x), \beta)$, where $x \in V$ and $s \in S$.

Chapter 10 Review

1. a.

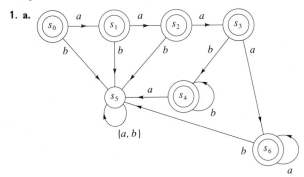

b. $G = (V, T, @, P)$; with $V = \{@, A, B, C, D, a, b\}$, $T = \{a, b\}$, and P: $@ \to \lambda$, $@ \to a$, $@ \to aA$, $A \to a$, $A \to aC$, $C \to a$, $C \to aD$, $D \to b$, $D \to a$, $D \to aF$, $D \to bB$, $F \to a$, $F \to aF$, $B \to bB$, $B \to b$

c. $G = (V, T, @, P)$; with $V = \{@, A, a, b\}$, $T = \{a, b\}$, and P: $@ \to A$, $A \to \lambda$, $A \to aA$, $@ \to a^3B$, $B \to \lambda$, $B \to bB$

3. 010*

5. Find a context-free grammar G such that $L(G) = L$: $G = (V, T, @, P)$; with $V = \{@, a, b\}$, $T = \{a, b\}$, and P: $@ \to ab$, $@ \to a^3@b^3$

7. a. $@ \Rightarrow 1@1 \Rightarrow 1^2@1^2 \Rightarrow 1^3@1^3 \Rightarrow 1^3A1^3 \Rightarrow 1^30A1^3 \Rightarrow 1^30^21^3$ **b.** $k = n$, $n \geq 0$, $m \geq 1$

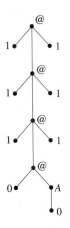

c. 0, 00, 000, 0000, 00000, 101, 1001, 10001, 11011

9. Let $D = \{0, 1, 2, \ldots, 9\}$ and $V = D \cup \{+, -\}$.
   a. 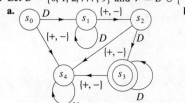   **b.** $DD^*(+ \vee -)DD^*$

11. One method of proving this is to prove that a regular expression written in reverse order is still a regular expression. Reverse order for a regular expression must be carefully defined.

13. Let V be the character set available.
   a.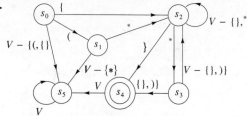

   b. Let x stand for an arbitrary symbol from $V - \{(, \{, \},), *\}$. The productions are P: $@ \to (A$, $@ \to \{B$, $A \to *B$, $B \to *C$, $B \to \}D$, $B \to \}$, $B \to xB$, $B \to (B$, $B \to)B$, $C \to xB$, $C \to *B$, $C \to \{B$, $C \to (B$, $C \to \}D$, $C \to)D$, $C \to \}$, $C \to)$.

INDEX

Abs, 113
Accept, 392, 428
Acceptance of a string, 392, 428
Ackermann's function, 360
Acquaintance relation, 55
Adder network
 full, 288
 half, 278
Addition
 of binary numbers, 247
 of matrices, 68
Addition mod d ($+_d$), 245
Additive law of probability, 166, 167
Additive rule for inequalities, 197
Adjacency matrix, 66, 71, 74, 310
 for a digraph, 66
 for a graph, 310
 for a relation, 67, 71
Admissible sequence, 174
Admissible string, 175
Algorithm, 17, 19
 analysis, 126
 closest neighbor, 321
 division, 233
 Euclidean, 238, 243
 Floyd's shortest path, 85
 greedy, 321
 heapsort, 343–347
 High–low, 382
 inorder traversal, 329
 Karnaugh map, 282
 Kruskal's, 339
 minimal spanning tree, 339, 340
 postorder traversal, 329
 Prim's, 340
 spanning tree, 338
 tree traversal, 329
 Warshall's, 80, 82, 84
Alphabet, 389, 399
Ambiguity, 418
 parsing and, 418

Ambiguous grammar, 420
Ancestor, 325
AND, 31, 122, 182, 183
AND gate, 182, 191, 275
Annuity, 353, 361
Antisymmetric relation, 53, 54
Arc, 41, 303
Arithmetic progression, 355
ARPANET, 40, 89
Articulation point, 312
Artificial intelligence (AI), 180, 223
Associative laws
 for Boolean algebra, 258, 264
 for logic, 189
 for regular expressions, 396
 for sets, 203
Atom, 291
Atom[x], 292
Atomistic Boolean algebra, 293
Automata method, 388
Automated teller machine, 393
Automated theorem proving, 180
Axioms for Boolean algebra, 258

Backus–Naur form (BNF), 426
Backwards/forwards method, 29
Base, 246
Base 2, 246
Base 10, 246
Basis step (for mathematical
 induction), 204, 207
Biconditional, 34
Big-oh notation, 219, 220, 221, 346
Bijective function (bijection), 118
Binary coded decimal (BCD), 296
Binary notation, 246
Binary numbers, 122, 246
Binary operation, 121, 122
Binary relation, 49, 52
Binary search, 357

Binary tree, 325
 full, 325
Binomial coefficients, 152, 168, 360
Binomial theorem, 153
Bipartite graph, 313
 complete, 313
Birthday problem, 170
Bit string, 110, 114, 122, 154
Boolean (Pascal type), 192
Boolean addition for {0, 1}, 68, 111
Boolean algebra, 258
 axioms, 258
 finite, 293
 laws, 260
 two-element, 258
Boolean data type, 192
Boolean expression, 22, 266
Boolean function, 265
Boolean laws of logic, 189, 200
Boolean laws for sets, 13, 200, 257
Boolean multiplication for {0, 1}, 68, 111
Boolean operations for {0, 1}, 68, 111, 258
Boolean product for matrices, 69, 72
Boolean sum for matrices, 68
Boolean-valued function, 110
Bottom of a stack, 427
Bubble sort, 356
Buffer, 89
Buffer deadlock, 89

$C(n, r)$, 149, 150
Card(S), 3
Cardinality, 3
Cardinality principle, 294
Cartesian product, 11, 50
Cases, proof by, 199
Catalan number, 176
Ceiling function, 115
Characteristic equation, 365

I-1

Characteristic equation method, 365
Characteristic function, 111, 112, 114
Child, 324
　left, 329
　right, 329
Chomsky hierarchy, 430
Chr, 122
Chromatic number, 305
Circuit, 43, 314
　in a digraph, 43
　in a graph, 314
Clause, 224
Closed-form expression, 210, 354
Closed path
　in a digraph, 43
　in a graph, 314
Closest-neighbor algorithm, 320, 321
CNF, 224
Codes, 332
　Huffman, 332
Codomain of a function, 107
Column index, 65
Combination, 149
Comment (in Pascal), 433
Commutative laws
　for Boolean algebra, 258
　for logic, 189
　for sets, 13
Comparable elements, 97
Complement, 5, 258
Complement of a graph, 306
Complement of a set, 5
Complete bipartite graph, 313
Complete graph, 306
Component, 308
Composition of functions, 119
Composition of relations, 60, 71
Computationally intractable, 253
Computer program, 23
Concatenation of strings, 394
Conclusion, 25, 32
Conditional, 32
Congruence (modulo d), 100, 202, 244
Congruent triangles, 100
Conjecture, 10
Conjunction of statements, 223
Conjunctive normal form (CNF), 224, 274
Connected, 307
Connected graph, 308
Connected multigraph, 308
Connectives, 31
　AND, 31, 182, 183

Connectives *(continued)*
　exclusive or, 185, 194
　IF AND ONLY IF, 34, 182, 183
　IMPLIES, 32, 182, 183
　NAND, 191
　NOR, 191
　NOT, 31, 182, 183
　OR, 31, 182, 183
Connectivity matrix, 74, 78
Connectivity relation, 58, 63
Consensus theorem, 300
Constant of proportionality, 125
Context-free grammar, 413
Context-free language, 413
　not regular, 414
Contradiction, 187, 188, 189
　proof by, 197
Contrapositive, 187
　proof by, 198
Control structures, 22
Converse, 33
Conversion from binary to decimal, 246
Conversion from decimal to binary, 247, 248
Convex *n*-gon, 207
Coordinates, 50
Corresponds, Boolean expression to Boolean function, 267
Counterexample, 10, 190, 217
Counting rules
　fundamental counting principle, 141, 142
　product rule, 134
　sum rule, 132
Cryptography, 230
Cryptosystem, 250
Cycle, 43, 314
　in a digraph, 43
　in a graph, 314

Deadlock, 40, 89
　detection, 89
Decimal notation, 246
Decision problem, 17
Decryption keys, 250
Degree of a vertex, 309
Degree of a relation, 50
De Morgan's laws
　for Boolean algebra, 261, 263, 264
　for logic, 187, 191, 216
　for regular expressions, 396
　for sets, 203, 209

Derivation, 404
　direct, 403
　leftmost, 420
Derived, 404
Descendant, 325
Diagonal relation, 83
Differ in one literal, 280
Difference set, 5
Digraph (directed graph), 41, 52
Direct derivation, 403
Direct proof, rule for, 187, 195
Direct proof of implication, 195
Directed graph, 41, 52
Directly derived, 403
Disjoint sets, 4
Disjunction of statements, 224
Disjunctive normal form, 270
Distributive laws
　for Boolean algebra, 258, 264
　for logic, 189
　for sets, 13
Div, 235, 237
Divide(s), 24, 56, 231
Divide and conquer relation, 382
Division algorithm, 233
Divisor, 231
Domain
　of definition, 112
　of a function, 107
Don't care conditions, 284
Double negation rule, 187
Double recurrence, 360
Dual, 258
Duality principle, 258
Dump (error) state, 393

Edge, 41, 303
　multiple, 303
Effective procedure, 17
Egrep, 411
Eight-queens configuration, 158
Eight-queens problem, 157
Elements of a set, 2
Ellipsis, 2
Empty set, 3
Empty string, 394
Encryption keys, 250
Equal sets, 2
Equal strings, 400
Equality of two sets, 2
Equiprobable space, 164
Equivalence class, 95
Equivalence relation, 94

Equivalent Boolean expressions, 268
Equivalent logic networks, 276
Error (dump) state, 393
Euclidean algorithm, 238, 242, 243
Euler's theorem, 316
Eulerian path, 317
Eulerian tour, 314
Event, 163
   simple, 163
   occurrence of, 163
Exclusive OR
   connective, 185, 194
   gate, 274
Existential quantifier, 213
Experiment, 162
Expression tree, 328
Extremal elements, 101

Factor, 56, 231
Factorial notation, $n!$, 142
False, 36
Favorable outcome, 163
Fermat's theorem, 251
Fibonacci sequence, 358, 366
Finite Boolean algebra, 293
Finite sequence, 110
Finite state recognizer (FSR), 389
First coordinate, 11
Floor function (greatest integer
   function), 115
Floyd's shortest path algorithm, 85
Follow logically, 195
Formal language, 400
FSR (finite state recognizer), 389
Full-adder network, 288
Full binary tree, 325
Full m-ary tree, 325
Function, 107
Functionally complete, 191, 271
   Boolean operations, 271, 272
   connective, 191
   set of connectives, 191
Fundamental counting principle,
   141, 142
Fundamental theorem of arithmetic,
   232, 234

Gates, 191, 275
Gcd (greatest common divisor), 237,
   238, 242, 243
Generalized De Morgan's laws, 209
Generalized distributive laws, 206,
   207
Generalized intersection, 206, 218

Generalized union, 206, 218
Generating function, 373, 379–381
Generating function method, 375
Generative method, 388
Geometric progression, 355
Geometric series, 373
Glb (greatest lower bound), 105
Golden ratio, 367
Golden section, 366
Grammar, 402
   ambiguous, 420
   context-free, 413
   noncontracting, 416
   phrase structure, 402
   regular, 407
Graph, 303
   complete, 306
   connected, 308
   directed graph (digraph), 41
   loopgraph, 303
   multigraph, 303
   state, 389
Greatest common divisor (gcd), 237,
   238, 242, 243
Greatest element, 101, 214
Greatest integer function (floor
   function), 115
Greatest lower bound (glb), 105
Greedy algorithm, 321
Grep, 411

Half-adder network, 278
Hamiltonian cycle, 314
Hasse diagram, 98
Heap, 343
   order property, 343
   shape property, 343
Heapsort, 343–347
   algorithm, 346
Height of a rooted tree, 324
High–low, algorithm, 382
Homogeneous recurrence relation,
   364
Horner's method, 371
Huffman codes, 332
Hypothesis, 25, 31

Idempotent laws, 263
Identifier (Pascal), 392
Identity function, 118
Identity laws
   for Boolean algebra, 258, 263
Identity matrix, 70
If and only if, 34, 182, 183

Image, 107
Implication, 32, 190
Implies, 32, 182, 183
Impossible event, 163
Incident, 303
Inclusion and exclusion, law of, 132
Inclusive OR, 185
Indirect proof, 197
Induction hypothesis, 204
Induction principle, 204
Induction step, 204
Infix notation, 328
Initial conditions, 353
Initial state, 389, 427
Initial vertex, 41, 43, 307
   of an edge, 41
   of a path, 43, 307
Injective function, 117
Inorder traversal, 328
   algorithm, 328
Input symbol, 389, 427
Integers, 3
Interior vertex, 78
Internal vertex, 325
Intersection of sets, 4, 5, 206, 218
Invalid argument, 195
Inverse function, 120
Inverter, 275
Involution laws, 260
Irrational number
   $\sqrt{2}$ is irrational, 236
   $\sqrt{3}$ is irrational, 235
Iteration, solution by, 355

Join, 303
Judicious substitution, 374

Karnaugh map, 280
   algorithm, 282
Kleene's theorem, 393
Knight's tour, 318
Königsberg bridge problem, 315
Kruskal's algorithm, 339

Lambda production, 415, 417
   dispensability of, 417
Language, 401, 404
   context-free, 413
   context-free that is not regular, 414
   formal, 400
   generated by a grammar, 404
   regular, 407
Language generated by a grammar,
   404

Language specification, 421
Last-in-first-out (lifo) structure, 174, 427
Law(s)
  of Boolean algebra, 260
  of inclusion and exclusion, 132
  of logic, 189, 200
  of set theory, 257
Lcm (least common multiple), 240, 243
Leaf (leaves), 325, 420
Least common multiple (lcm), 240, 243
Least element, 101, 222
Least upper bound (lub), 105
Leaves of a tree, 325, 420
Left child, 329
Left subtree, 329
Leftmost derivation, 420
Length of path, 43, 307
Length of a string, 400
Level of a vertex, 324
Lexical analyzer, 387, 396
Lexicographic order, 101
Lifo (last-in-first-out) structure, 174, 427
Linear ordering (linear order), 97
Linear recurrence relation with constant coefficients, 364
Literal, 223, 269
Logic gate, 275
Logic network, 256, 275
Logical connectives, 31, 32, 34, 182, 183
Logical equivalence, 187, 189
Loop, 303
Loop invariant, 208, 209
Loopgraph, 303
Lower bound, 104, 222
Lub (least upper bound), 105
Lucas sequence, 363

m-ary tree, 325
  full, 325
Mail carrier problem, 317
Main diagonal of a matrix, 66
Mathematical induction, principle of, 204
Matrix, 65
  identity matrix, 70
  square matrix, 66
  zero matrix, 70
Maximal connected
  subgraph, 308

Maximal connected *(continued)*
  submultigraph, 308
Maximal element in a poset, 102
Maxterm, 273
Members of a set, 2
Membership problem, 415
Method of judicious substitution, 374
Minimal element in a poset, 102
Minimal spanning tree, 339
  algorithm, 339, 340
Minimal sum of products, 279
Minimization rule, 280
Minterm, 269
Mod, 236, 237
Mod d, modulo d, 202, 243
Modular arithmetic, 244
Modus ponens, 188
Multigraph, 303
Multiple, 56, 231
Multiple edges, 303
Multiplication mod d ($*_d$), 245
Multiplicative inverse mod n, 250
Mutually exclusive events, 163

$n!$ (n factorial), 142, 353, 361
n-ary relation, 50
n-colorable, 305
n-tuple, 50
NAND, 191
  connective, 191
  gate, 192, 194, 272, 274
  operation, 272, 274
Natural numbers, 3
Necessary and sufficient condition, 190
Necessary condition, 190
Negation, 31, 187, 216
Network, 256
  project, 44
Next-state function, 389
Node, 41, 303
Noncontracting grammar, 416
Nonnegative integers, 3
Nonterminal, 402
NOR, 191
  connective, 191
  gate, 191, 272
  operation, 272
NOT, 31, 122, 182, 183
NOT gate (inverter), 182, 191, 275
Notation
  infix, 328
  Polish, 336
  postfix, 329

Notation *(continued)*
  prefix, 336
  reverse Polish, 329
Null set, 3

O-notation (big-oh notation), 219, 346
Octal numbers, 359
Odds, 171
One element, 258
One-to-one (1-1) function, 117
One-to-one onto function, 118
Onto (surjective) function, 117
Operand, 328
Operation, binary, 121, 122
Operator, 328
OR, 31, 122, 182, 183
Or form of an implication, 187
OR gate, 182, 191, 275
Ord, 122
Order property of the heap, 343
Ordered pair, 11

$P(n, r)$, 144, 147
Pairwise disjoint, 95
Parent, 324, 325
Parity
  bit, 388
  checker, 388
  even, 388
  odd, 388
Parse, 418
Parse tree, 405
Parsing and ambiguity, 418
Partial fraction decomposition, 373
Partial function, 112
Partial ordering (partial order), 97, 289
  for a Boolean algebra, 289
Partially ordered set, 97
Partition, 95
Pascal identifier, 392
Pascal's triangle, 155, 158
Path, 43
  closed, 314
  in a digraph, 43
  in a graph, 307
  simple, 314
Path length, 43, 307
PDA (push-down automaton), 427
Permutation, 118, 140, 141, 143
Perpendicular, 100
Phrase structure grammar, 402
Pick-a-point method, 26, 199
Pigeonhole principle, 62, 64, 117

Polish notation, 336
Pop, 174, 427
Poset (partially ordered set), 97
Positive real numbers, 3
Postfix notation, 329
Postorder traversal, 329
Power set, 12
Precedence relationship, 103
Pred, 122
Predicate, 211
Predicate logic, 212
Prefix notation, 336
Preimage, 107
Preorder traversal, 336
Prim's algorithm, 340
Prime (prime number), 20, 190, 231, 237
   factorization, 232, 234
   infinitely many, 232
Prime statement, 223
Principle of duality, 258
Principle of mathematical induction, 204
Principle of strong induction, 228
Probability function, 163, 165
Product (Cartesian product), 11, 50
Product of two sets, 11, 50, 134
Product rule (for counting), 134
Production, 402
Program, 23
Progression
   arithmetic, 355
   geometric, 355
Project network, 44
Projection, 52
Proof, 195
   by cases, 199
   by contradiction, 197
   by contrapositive, 198
   direct, 195
   by mathematical induction, 204
   by pick-a-point method, 199
Proper subgraph, 306
Proper submultigraph, 306
Proper subset, 3
Pseudocode, 18
Public-key cryptosystem, 250
Push, 174, 427
Push-down automaton (PDA), 427

Quantifier
   existential, 213
   restricted, 218
   universal, 213

Questions to answer while reading a proof, 26, 28
Quotient, 233

r-combination, 149, 157
   with replacement (or repetition), 156, 157
r-permutation, 143, 144, 146
   with replacement (or repetition), 145, 146
Reachability, 78
   matrix, 83
   problem, 43, 78
   relation, 83
Reachable, 44
Real numbers, 3
Recognize, 392, 428
Recognizing states, 389, 427
Recurrence relation, 353, 364
   double, 360
   homogeneous, 364
   linear, 364
   of order k, 364
   to solve, 355
Recursive definition of a polynomial, 371
Reflexive closure, 55
Reflexive relation, 53, 54
Reflexive transitive closure, 83
Regular expression, 390, 393, 394
   identities, 396, 411
Regular grammar, 407
   construction from a given FSR, 409
Regular language, 407, 408
Regular set, 393, 394
Relation, 49, 51
Remainder, 233
Resolution principle (RP), 225
Resolvent, 225
Restricted quantifier, 218
Reverse Polish notation (RPN), 329, 429
Right child, 329
Right subtree, 329
Root of a tree, 324
Rooted tree, 324
Row index, 65
RPN (reverse Polish notation), 329, 429
RSA scheme (cryptosystem), 250, 253

Sample point, 162
Sample space, 162
Second coordinate, 11

Self-loop, 43
Semantics, 402
Sequence, 109
Set, 2
Set-builder notation, 3
Set recognized, 392, 427
   by the automaton, 427
   by an FSR, 392
Seven-segment display, 296
Shape property of the heap, 343
Sheffer stroke, 194, 274
Shortest path algorithm, 85
Shortest path matrix, 86
Sibling, 325
Similar triangles, 94
Simple event, 163
Simple path, 43, 314
Singleton set, 3
Solution by iteration, 355
Solution set, 212
Solve a recurrence relation, 355
Sort
   bubble, 356
   heap, 343–347
   merge, 385
Spanning subgraph, 308
Spanning tree, 337
   algorithm, 338
   minimal, 339
Sqr, 113
Square matrix of order n, 66
Stack (push-down store), 174, 427
   bottom of, 427
   top of, 174, 427
Stack permutation, 174
Stack symbol, 427
Standard product of sums, 274
Standard sum of products, 270
Start symbol, 402
State, 389, 427
   dump, 393
   error, 393
   initial, 389, 427
   recognizing, 389, 427
State graph, 389
Statement, 30
Statement block, 22
Statement connectives, 31
   AND, 31
   exclusive or, 185, 194
   IF AND ONLY IF, 34
   IMPLIES, 32
   NAND, 191
   NOR, 191

Statement connectives *(continued)*
  NOT, 31
  OR, 31
Statement variable, 223
Strictly greater than, 102
Strictly less than, 102, 291
String, 111, 390, 400
  bit, 110, 114, 122, 154
  empty, 394
  equal, 400
  length of, 400
  sub-, 403
Subgraph, 306
  maximal connected, 308
  proper, 306
  spanning, 308
Submultigraph, 306
  maximal connected, 308
  proper, 306
Subset, 3
  proper, 3
Substitution principle, 193, 196
Substring, 403
Succ, 122
Sufficient condition, 190
Sum rule for sets, 132
Surjective (onto) function, 117
Symbol
  input, 389, 427
  stack, 427
Symmetric difference of two sets, 16
Symmetric relation, 53, 54
Syntax, 402
Syntax rules, 421

Table, 50
Tautology, 188, 189

Telephone tree, 327
Terminal, 402, 404
Terminal vertex, 41, 43, 307
Ternary relation, 50
Ternary tree, 325
Time-complexity function, 125, 219
Top of a stack, 174, 427
Top-down design, 21
Topological sorting, 103, 105
Transition, 427
Transitive closure, 55, 59, 63
Transitive relation, 53, 54
Traveling salesperson problem, 318, 319
Traversal
  in order, 328, 329
  postorder, 329
  preorder, 336
Traverse, 328
Tree, 324
  binary, 325
  expression, 328
  full binary, 325
  full m-ary, 325
  m-ary, 325
  spanning, 337
  ternary, 325
Tree traversal, 328
True, 36
Trunc, 113
Truth table, 31, 182, 184, 265
Truth table method, 270

Unary operation, 121, 122
Unary relation, 50
Undefined, 112
Union, 4, 5, 206, 218
Unique complement law, 260

Universal quantifier, 213
Universal set (or universe), 5
Upper bound, 104, 214

Vacuously true case, 32
Vacuously true statement, 32
Valid argument, 194
Variable, 211
Venn diagram, 8
Vertex (vertices), 41, 303
  of a digraph, 41
  of a graph, 330
  initial, 41, 43, 307
  interior, 78
  internal, 325
  terminal, 307
Vertex degree, 309

Warshall's algorithm, 78, 80, 82, 84
Weight
  of an edge, 318
  function, 318
Weighted graph, 318
Well-balanced parentheses, 176
  checker, 428
Well-ordering principle, 228, 233
With replacement, 145
Without replacement, 146
Word, 388

XOR gate, 274

Yield of a parse tree, 420

Z_d, 244, 245
Zero element, 258
Zero matrix, 70

TABLE OF SYMBOLS

Sets
		Page
$\mathbb{R}$	Real numbers, real line	3
$\mathbb{R}^+$	Positive real numbers $(0, \infty)$	3
$\mathbb{N}$	Natural numbers, nonnegative integers	3
$\mathbb{Z}$	Integers	3
$x \in B$	Set membership, an element of	2
$x \notin B$	Not an element of	2
$\emptyset$	Empty set	3
U	Universal set	5
$A = B$	Set equality	2
$A \subseteq B$	Set inclusion (subset)	3
$A \nsubseteq B$	Not a subset	3
$A \subset B$	Proper subset	3
$A \cap B$	Set intersection	4
$A \cup B$	Set union	4
A'	Set complement	5
$A - B$	Set difference	5
$\text{Card}(A)$	Number of elements in A, the cardinality of A	3
$P(A)$	Power set, set of all subsets of A	12
$A \times B$	Cartesian product	11
A^n	$A \times A \times \cdots \times A$ (n times)	50
$A \triangle B$	Symmetric difference	16

Logic
$p \to q$	Implication, p implies q, if p then q	32
$p \wedge q$	Conjunction, p and q	31
$p \vee q$	Disjunction, p or q	31
$p \veebar q$	Exclusive OR	185
$\neg p$	Negation, not p	31
$p \leftrightarrow q$	Biconditional, p if and only if q	34
$p \equiv q$	p is logically equivalent to q	187
$\exists x$	Existential quantifier; there exists an x	213
$\forall x$	Universal quantifier; for all x, for any x	213
$\downarrow$	NOR	191
$\uparrow$	NAND	194

Relations
R^+	Connectivity relation, transitive closure of R	58, 59
R^*	Reachability relation, reflexive transitive closure of R	83
$R \circ S$	Composition of the relations R and S	60
R^k	$R \circ R \circ \cdots \circ R$ (k times)	61
$[a]$	Equivalence class of a	95
$\preceq$	Partial ordering	97, 98